Sustainability Sciences in Asia and Africa

Agroecosystem Dynamics and Sustainable Practices

Series Editors

Kamel A. Abd-Elsalam, Plant Pathology Research Institute
Agricultural Research Center
Giza, Egypt

Olubukola O. Babalola, Faculty of Natural & Agricultural Sciences
North-West University
Mmabatho, South Africa

Sustainability as a scientific endeavor requires broad understanding of the interconnections in our global environment. It is therefore positioned at the crossroads of research in natural sciences, technology and human sciences. This book series supports the global efforts by providing timely coverage of the progress, opportunities, and challenges towards achieving sustainable development goals (SDGs) in Asia and Africa. It fosters the research in transdisciplinary academic fields spanning across agriculture and global food security, health and diseases, access to clean water and affordable energy, climate resilience and effects of global climate change on biodiversity, water and land. This series is planned to narrate the stories of scientific development and progression towards SDGs from the regions of Asia and Africa. The overall aim of the series is to provide a comprehensive publication platform and act as a knowledge engine in the growth of sustainability sciences with a special focus on developing nations. The series constitute of four sub-series each focusing on a specific theme closely linked to a set of SDGs. Each sub-series is leaders and experts in agricultural sciences, environmental sciences, biomedicine and medical sciences respectively. Books in this series are useful for environmentalists, biotechnologists, plant scientists, medical practitioners, conservation biologists, policy makers, NGOs and scientific bodies working in the areas of sustainability sciences in the developing regions.

This book series publishes research associated with the prevailing opportunities and challenges faced by agroecosystems in Asia and Africa. It also explores the future scope of agroecological techniques and offers guidance on sustainable management of these systems. The series discusses important areas such as sustainable agricultural production, soil health, water resource management, biodiversity conservation, animal health and productivity, insect management strategies, genome editing and climate change mitigation. Further, it includes the challenges faced by African agriculture in arid and semi-arid regions, particularly in response to climate change. It highlights issues such as water shortages, soil degradation, desertification, and the need for sustainable farming techniques.

Sustainable agroecosystem maintenance entails strategies that improve overall environmental health and economic viability while limiting negative environmental impacts. The series takes a multidisciplinary approach to investigate creative ways for boosting sustainable agriculture, increasing food security, and improving lives for people in global south. This book series acts as a global publishing platform to collate and disseminate knowledge on fostering resilient and productive agroecosystems, thus enhancing accessibility, and fostering informed decision-making.

Kamel A. Abd-Elsalam
Editor

Plant-Microbiome Nanotechnology

Gateway to Sustainable Crop Health

Springer

Editor
Kamel A. Abd-Elsalam
Agricultural Research Center
Plant Pathology Research Institute
Giza, Egypt

ISSN 2730-6771 ISSN 2730-678X (electronic)
Sustainability Sciences in Asia and Africa
ISSN 3059-247X ISSN 3059-2488 (electronic)
Agroecosystem Dynamics and Sustainable Practices
ISBN 978-981-95-5055-5 ISBN 978-981-95-5056-2 (eBook)
https://doi.org/10.1007/978-981-95-5056-2

This Springer imprint is published by the registered company Springer Nature Singapore Pte Ltd.
The registered company address is: 152 Beach Road, #21-01/04 Gateway East, Singapore 189721, Singapore

Preface

The rapidly emerging plant–microbiome nanotechnology is a new frontier in sustainable agriculture and crop protection. The recent decades have witnessed nanoscience progress offer unprecedented opportunities for research and manipulation of highly complex interactions between plants, their in-dwelling microbiomes, and the environment. This book, *Plant-Microbiome Nanotechnology: Gateway to Sustainable Crop Health*, provides a comprehensive coverage of how nanoscale technologies can be applied to improve plant resistance, soil fertility, and abiotic and biotic stress mitigation.

The chapters as a whole address both the possibilities and challenges of applying nanotechnology in the context of plant–microbe associations. The book begins by addressing plant health in the context of nanotechnology and microbiome equilibrium, before going on to address both the desired and the unwanted impacts of metal nanoparticles on valuable soil microbes, fungi, and biocontrol agents. Special emphasis is placed on the toxicological properties of metal-based nanoparticles, stressing the challenge of how to reconcile innovation and environmental protection.

Novel strategies are deliberated, including application of nanoparticles to strengthen plant–microbe symbiosis, their use in heavy-metal bioremediation, and crop biofertilization potential through synergistic interaction with beneficial microorganisms. A few of the chapters discuss twin roles of nanoscale solutions in disease control and stress management, ranging from nanofungicides and nanoparticle-supported epiphytic microbes to synergistic action of plant growth-promoting rhizobacteria (PGPR) with nanoparticles under salinity stress, drought stress, and heavy metal stress conditions.

The book also presents possibilities and challenges of nano-agrochemicals, emphasizing an equitable review to avoid unwanted impacts on plant microbiomes and ecosystems. With applications ranging from stress management with nanoparticles to microbial nanoscale innovations for crop health management, the book provides a comprehensive handbook for researchers, practitioners, and policymakers who are interested in integrating nanotechnology with values of sustainability, food security, and environmental stewardship.

We would also like to express our gratitude to Springer's team, and specifically thank Sanchi Bhimrajka, our committed Publishing Editor. We are eternally thankful to the Series Editor, Prof. Olubukola O. Babalola, Vice-President of The World Academy of Sciences (TWAS) and an esteemed Research Director at North-West University, South Africa. Your unmatched expertise and insight contribution has immensely enriched the Agroecosystem Dynamics and Sustainable Practices series, such that it is an excellent reference in the academe. We also acknowledge Prof. Adel Abdel Azeem, President of Agricultural Research Center, Giza, Egypt, and editorial board members of the series for their persistent effort toward knowledge and development progress in agroecosystem dynamics and sustainable management. Your persistent effort has been critical in facilitating the possibility of the achievement of this monumental task.

Giza, Egypt Kamel A. Abd-Elsalam

Contents

Editor and Contributors

About the Editor

Kamel A. Abd-Elsalam, Ph.D., is currently a Research Professor and Deputy Director for Research at the Plant Pathology Research Institute, Agricultural Research Center, Giza, Egypt. His research interests include developing and improving plant biosecurity diagnostic tools, understanding fungal pathogen genomes, and creating eco-friendly hybrid nanomaterials for controlling toxigenic fungi and plant diseases as well as applications in agroecosystems. He earned his Ph.D. in Molecular Plant Pathology from Christian Albrechts University of Kiel (Germany) and Suez Canal University (Egypt). In 2008, he was awarded a postdoctoral fellowship by the same institution. He served as a Visiting Associate Professor at Mae Fah Luang University in Thailand, the Institute of Microbiology at TUM in Germany, the Laboratory of Phytopathology at Wageningen University in the Netherlands, and the Plant Protection Department at Sassari University in Italy.

He has published 40 books related to nanobiotechnology applications in agriculture and plant protection with major publishing houses such as RSC, Springer, Taylor & Francis, and Elsevier. Since 2019, he has served as the Editor-in-Chief of the Elsevier book series Nanobiotechnology for Plant Protection and as the Series Editor for Applications of Genome Modified Plants and Microbes in Food and Agriculture.

Additionally, he has been an active member of the Elsevier Advisory Panel since 2020, providing feedback and suggestions to improve Elsevier's products and services. He has published more than 327 scientific research articles in international and regional specialized journals with high impact factors, and has an H-index of 55, an i-10 index of 170, and over 11,346 citations. He served as a Guest Editor for the *Journal of Fungi, Plants, and Microorganisms*, and as a Review Editor for *Frontiers in Genomic Assay Technology*, *Agrochemicals*, and *Plant Nano Biology*. He was ranked in the top 2% of the most influential scientists in the world in nanobiotechnology for 2021, 2022, 2023, and 2024 by Elsevier and Stanford University. In 2014, he was awarded the Federation of Arab Scientific Research Councils Prize for excellent scientific research in biotechnology (fungal genomics), ranking first. Additionally, he received the Egyptian State Award for Agricultural Scientific Excellence from ASRT, Egypt, in 2022 with a silver medal.

Contributors

Reham I. AbdelHamid Sugar Crops Research Institute, Agriculture Research Center (ARC), Giza, Egypt

Toka E. Abdelkhalek Biotechnology English Program, Faculty of Agriculture, Cairo University, Giza, Egypt

Kamel A. Abd-Elsalam Plant Pathology Research Institute, Agricultural Research Center (ARC), Giza, Egypt

Hakima Achetoui Phytopathology Unit, Department of Plant Protection, Ecole Nationale d'Agriculture de Meknès, Meknès, Morocco

Farah K. Ahmed Nanobiotechnology Department, Faculty of Postgraduate Studies for Nanotechnology, Cairo University, El-Sheikh Zayed, Egypt

Muhammad Mahmood Ahmed Department of Bioinformatics, Institute of Biochemistry, Biotechnology and Bioinformatics, The Islamia University of Bahawalpur, Bahawalpur, Punjab, Pakistan

Muhammad Saleem Arif Department of Environmental Sciences, Government College University Faisalabad, Faisalabad, Pakistan

Muhammad Ashfaq Plant Pathology, Institute of Plant Protection, Muhammad Nawaz Shareef University of Agriculture, Multan, Punjab, Pakistan

Malaika Ashiq Department of Environmental Sciences, Government College University Faisalabad, Faisalabad, Pakistan

Aspak School of Agriculture, Lovely Professional University, Phagwara, Punjab, India

Elif Özlem Arslan Aydoğdu Istanbul University, Faculty of Science, Department of Biology, Istanbul, Turkey

Kavya Bakka Department of Microbiology, School of Life Sciences, Central University of Tamil Nadu, Thiruvarur, Tamil Nadu, India

Tara Budimac University of Novi Sad, Faculty of Technology Novi Sad, Novi Sad, Serbia

Dinakar Challabathula Department of Biotechnology, School of Integrative Biology, Central University of Tamil Nadu, Thiruvarur, Tamil Nadu, India

B. Chithradevi Department of Microbiology, School of Life Sciences, Central University of Tamil Nadu, Thiruvarur, Tamil Nadu, India

Teodora Marić University of Novi Sad, Faculty of Technology Novi Sad, Novi Sad, Serbia

Ivana Danilov Faculty of Technology Novi Sad, University of Novi Sad, Bulevar cara Lazara 1, Novi Sad, Serbia

Barnik Debnath ICAR-Central Institute for Cotton Research, Nagpur, Maharashtra, India

Sozan E. El-Abeid Plant Pathology Research Institute, Agricultural Research Center (ARC), Giza, Egypt

Fahad Faheem Plant Pathology, Institute of Plant Protection, Muhammad Nawaz Shareef University of Agriculture, Multan, Punjab, Pakistan

Abdelaaziz Farhaoui Phytopathology Unit, Department of Plant Protection, Ecole Nationale d'Agriculture de Meknès, Meknès, Morocco

Equipe de Recherche en Didactique des Sciences et Sciences de l'Education et TICE. (DSSE_TICE), Centre Régional des Métiers de l'Education et de la Formation de Fès Meknès, Annexe Meknès, Maroc, Fes, Morocco

Fatima Department of Environmental Sciences, Government College University Faisalabad, Faisalabad, Pakistan

Elizabeth Gonzalez-Terreros Universidad de la Sierra Juárez, Oaxaca, Mexico

Jovana Grahovac Faculty of Technology Novi Sad, University of Novi Sad, Bulevar cara Lazara 1, Novi Sad, Serbia

Paula Cecilia Guadarrama-Mendoza Instituto de Agroindustrias, Universidad Tecnológica de la Mixteca, Oaxaca, México

Waqas Haider Department of Environmental Sciences, Government College University Faisalabad, Faisalabad, Pakistan

Megha Handa Department of Microbiology, College of Basic Sciences and Humanities, Punjab Agricultural University, Ludhiana, Punjab, India

Rawan K. Hassan Biotechnology English Program, Faculty of Agriculture, Cairo University, Giza, Egypt

Noman Ijaz Plant Pathology, Institute of Plant Protection, Muhammad Nawaz Shareef University of Agriculture, Multan, Punjab, Pakistan

Faisal Islam International Genome Center, Faculty of Life Sciences, Jiangsu University, Zhenjiang, China

Jyoti Jain Department of Plant Pathology, Punjab Agricultural University, Ludhiana, India

Sandeep Jain Department of Plant Pathology, Punjab Agricultural University, Ludhiana, India

Anu Kalia Electron Microscopy and Nanoscience Laboratory, Punjab Agricultural University, Ludhiana, Punjab, India

Simarjot Kaur Department of Plant Pathology, Punjab Agricultural University, Ludhiana, India

Samah Bashir Kayani Govt. Graduate College (W) Mumtazabad, Multan, Punjab, Pakistan

Hysen Kokiçi Institute of Plant Genetic Resources, Agricultural University of Tirana, Tirana, Albania

Rachid Lahlali Phytopathology Unit, Department of Plant Protection, Ecole Nationale d'Agriculture de Meknès, Meknès, Morocco

A. K. Malavika Ram Hawkesbury Institute for the Environment, Western Sydney University, Sydney, NSW, Australia

Mirza Abid Mehmood Plant Pathology, Institute of Plant Protection, Muhammad Nawaz Shareef University of Agriculture, Multan, Punjab, Pakistan

State Key Laboratory of Vegetation Structure, Function and Construction; School of Ecology and Environmental Science, Yunnan University, Kunming, People's Republic of China

Ministry of Education Key Laboratory for Transboundary Ecosecurity of Southwest China, Crop Disease & Pest Biocontrol Engineering Research Center of Yunnan Province, Kunming, Yunnan, China

Swetha Muduku Department of Microbiology, College of Basic Sciences and Humanities, Punjab Agricultural University, Ludhiana, Punjab, India

Mary Carmen Pacheco-Esteva Instituto Politécnico Nacional, CIIDIR Unidad Oaxaca, Oaxaca, Mexico

M. Petheeswaran Department of Biotechnology, School of Integrative Biology, Central University of Tamil Nadu, Thiruvarur, Tamil Nadu, India

Mohammed Radi Phytopathology Unit, Department of Plant Protection, Ecole Nationale d'Agriculture de Meknès, Meknès, Morocco

Environment and Valorization of Microbial and Plant Resources Unit, Faculty of Sciences, Moulay Ismail University, Meknes, Morocco

Heriberto Fortino Ramírez-Cariño Universidad de la Sierra Juárez, Avenida Universidad S/N, Ixtlán de Juárez, Oaxaca, Mexico

Daniel Alejandro Ramírez-Villanueva Universidad de la Sierra Juárez, Oaxaca, Mexico

Meenakshi Rana School of Agriculture, Lovely Professional University, Phagwara, Punjab, India

Ritu Rani Department of Plant Pathology, Punjab Agricultural University, Ludhiana, India

Hamza Rhazal Phytopathology Unit, Department of Plant Protection, Ecole Nationale d'Agriculture de Meknès, Meknès, Morocco

Laboratory of Functional Ecology and Environmental Engineering, Sidi Mohamed Ben Abdellah University, Fez, Morocco

Muhammad Riaz Department of Environmental Sciences, Government College University Faisalabad, Faisalabad, Pakistan

Hadjer Saifi Biogeochemistry of Desert Environments Laboratory, Scientific and Technical Research Center in Physicochemical Analysis, Department of Biological Sciences, Kasdi Merbah University, Ouargla, Algeria

Rayane Saifi Research Laboratory of Sciences and Environment: Bioresources, Geochemical-Physics, Legislation and Socio-Economic Development, University of Tamanrasset, Tamanrasset, Algeria

Pooja Salaria Department of Plant Pathology, Punjab Agricultural University, Ludhiana, India

Gülnihan Selim Istanbul University, Institute of Graduate Studies in Science, Istanbul, Turkey

Wafae Sfa Phytopathology Unit, Department of Plant Protection, Ecole Nationale d'Agriculture de Meknès, Meknès, Morocco

Laboratory of Functional Ecology and Environmental Engineering, Sidi Mohamed Ben Abdellah University, Fez, Morocco

Muhammad Shaaban College of Agriculture, Henan University of Science and Technology, Luoyang, China

Jayesh Singh Department of Plant Breeding and Genetics, College of Agriculture, Punjab Agricultural University, Ludhiana, Punjab, India

Olja Šovljanski University of Novi Sad, Faculty of Technology Novi Sad, Novi Sad, Serbia

Seweta Srivastava School of Agriculture, Lovely Professional University, Phagwara, Punjab, India

Mohammed Taoussi Phytopathology Unit, Department of Plant Protection, Ecole Nationale d'Agriculture de Meknès, Meknès, Morocco

Environment and Valorization of Microbial and Plant Resources Unit, Faculty of Sciences, Moulay Ismail University, Meknes, Morocco

Mohsin Tariq Department of Bioinformatics and Biotechnology, Government College University Faisalabad, Faisalabad, Pakistan

Ana Tomić University of Novi Sad, Faculty of Technology Novi Sad, Novi Sad, Serbia

Rogelio Valadez-Blanco Instituto de Agroindustrias, Universidad Tecnológica de la Mixteca, Oaxaca, México

Vanja Vlajkov Faculty of Technology Novi Sad, University of Novi Sad, Bulevar cara Lazara 1, Novi Sad, Serbia

Anja Vučetić University of Novi Sad, Faculty of Technology Novi Sad, Novi Sad, Serbia

Bekri Xhemali Laboratory of Plant Protection, Kosovo Institute of Agriculture, Pejë, Kosovo

Tahira Yasmeen Department of Environmental Sciences, Government College University Faisalabad, Faisalabad, Pakistan

Heba Yousef Plant Pathology Research Institute, Agricultural Research Center, (ARC), Giza, Egypt

Chapter 1
Microbial Communities and Plant Health: Insights into the Plant Microbiome

Mohammed Radi, Mohammed Taoussi, Abdelaaziz Farhaoui, Hakima Achetoui, and Rachid Lahlali

Abstract The plant microbiome, which is the diverse community of microorganisms that exists in and around plant tissue, influences the health, productivity, and stress resistance of plants. These microorganisms, especially in the rhizosphere, phyllosphere, and endosphere, contribute to nutrient uptake, hormone regulation, immune function modulation, and resistance against pathogens. The structure, origin, and functional diversity of the microbiota inhabiting plants, and particularly their roles in growth promotion and in the enhancement of plant tolerance to biotic and/or abiotic stresses, will be extensively discussed in this chapter. Special attention is given to the mechanisms by which beneficial microbes suppress plant pathogens, including parasitism, competition, secretion of lytic enzymes and antibiotics, and induction of host defenses. The main challenges surrounding the microbiome-plant host-pathogen interactions linked to abiotic stress are also considered. The chapter finally concludes with the opportunities of using the plant microbiome for sustainable agriculture and focuses on the future perspectives of microbiome engineering and bio-control in the integrated crop management.

M. Radi · M. Taoussi
Phytopathology Unit, Department of Plant Protection, Ecole Nationale d'Agriculture de Meknès, Meknès, Morocco

Environment and Valorization of Microbial and Plant Resources Unit, Faculty of Sciences, Moulay Ismail University, Meknes, Morocco

A. Farhaoui
Phytopathology Unit, Department of Plant Protection, Ecole Nationale d'Agriculture de Meknès, Meknès, Morocco

Equipe de Recherche en Didactique des Sciences et Sciences de l'Education et TICE. (DSSE_TICE), Centre Régional des Métiers de l'Education et de la Formation de Fès Meknès, Annexe Meknès, Maroc, Fes, Morocco

H. Achetoui · R. Lahlali (✉)
Phytopathology Unit, Department of Plant Protection, Ecole Nationale d'Agriculture de Meknès, Meknès, Morocco
e-mail: rlahlali@enameknes.ac.ma

K. A. Abd-Elsalam (ed.), *Plant-Microbiome Nanotechnology*, Sustainability Sciences in Asia and Africa, https://doi.org/10.1007/978-981-95-5056-2_1

Keywords Plant microbiome · Rhizosphere · Phyllosphere · Endosphere · Biocontrol · Plant growth promotion · Microbial diversity · Abiotic stress tolerance · Pathogen suppression · Sustainable agriculture · Microbiome engineering

1.1 Introduction

The microorganism complex, commonly referred to as the second genome of the plant, includes bacteria, fungi, archaea, and viruses, all of which inhabit the various parts of the plant (the rhizosphere (root-associated soil), the phyllosphere (aboveground plant surfaces), and endosphere (internal tissues)) (Trivedi et al. 2020). This principle highlights the idea that the microorganisms that accompany plants are not mere external inhabitants but an inseparable part of the plant and its genome that will considerably affect its physiology, metabolism, and well-being (Compant et al. 2019; Glick and Gamalero 2021).

In a process of co-evolution lasting millions of years, plants have evolved highly specific interactions with their microbiomes, creating an interdependent ecological network, which increases their survival and the ability to adapt to variety of environments. Microbial communities also serve to obtain the necessary nutrients of nitrogen by the process of biological nitrogen fixation, and phosphorus by solubilizing insoluble phosphate compounds. Moreover, they synthesize phytohormones auxins, cytokinins, and gibberellins, whose direct regulating effect on root architectures, cell division, and elongation facilitates plant growth and development (Liu et al. 2020; Noman et al. 2021).

In addition, the plant microbiome is important in immune regulation. Some taxa of microbes stimulate systemic resistance potential through complex signaling processes and synthesis of defensive metabolites, antimicrobials, and cell wall-strengthening agents that overall enhance plant defense against pathogens (Afridi et al. 2022). As an example, useful rhizobacteria, such as *Pseudomonas* or *Bacillus* species, release antibiotics, siderophores, lytic enzymes, and volatile organic compounds that suppress the pathogens in plants and promote the immunity of the plants (Glick and Gamalero 2021).

Along with the mentioned advantages, the structure and activity of microbial communities remain highly vulnerable to external disturbances. Studies report that intensive agricultural practices, including the excessive use of chemical fertilizers, herbicides, and pesticides, can reduce microbial diversity, alter community composition, disrupt beneficial plant-microbe interactions, and ultimately compromise soil health and plant productivity (Munir et al. 2022). Furthermore, environmental stressors such as drought, salinity, heat waves, and soil contamination, all of which are intensifying under climate change, can alter microbiome assemblages, reduce their resilience, and diminish their capacity to provide essential ecosystem services (Liu et al. 2020; Noman et al. 2021).

Since these challenges, the plant microbiome has become a strategic target of present research in the agricultural world, and the interest in its capability to contribute to the development of sustainable systems of crop management is increasingly clarified. They have in-development new approaches—including the engineering of microbiomes, application of synthetic microbial consortia, and bioinoculant goods in the exploitation of beneficial microbes to improve nutrient uptake efficiency, plant well-being, and resistance to both biotic and abiotic stress—and mitigate the use of agrochemicals and related environmental impoverishment (Arif et al. 2020; Glick and Gamalero 2021).

The objectives of the chapter are to present a strong synthesis of what is currently known about the composition and origin, functional diversity, and mode of action of plant-associated microbial communities, and explain their crucial role in sustaining health, productivity, and the possibility of their use in creating sustainable and resilient agricultural systems globally.

1.2 The Plant Microbiome: Composition, Origins, and Diversity

1.2.1 The Plant Microbiome and Its Role in Plant Health

Plant microbiome, often referred to as plant's second genome, is a dynamic and diverse community of bacteria, fungi, archaea, and viruses of the microorganism that live in the various environments of the plants, such as rhizosphere, phyllosphere, and endosphere. These microbes are critical to plant life because they influence nutrient cycling, hormonal signaling, resistance, and tolerance against environmental adversaries (Trivedi et al. 2020).

In the rhizosphere, microorganisms are important for nutrient mobilization and acquisition at the same time. Nitrogen-fixing bacteria such as *Rhizobium* and *Azospirillum* facilitate the assimilation of atmospheric nitrogen; phosphate-solubilizing bacteria like *Pseudomonas* and *Bacillus* improve phosphorus availability by excreting organic acids and enzymes (Compant et al. 2019). Arbuscular mycorrhizal fungi (AMF) further extend plants' nutrient acquisition using the large hyphal networks of which the plants are otherwise deprived by hyphal size characteristics, which enhance water and phosphorus and micronutrient capture (Smith and Read 2010).

Besides the enhancement of nutrient acquisition, some microorganisms that produce phytohormones may also induce plant growth. For example, in some reports, indole-3-acetic acid (IAA), gibberellins, and cytokinins excreted by plant growth-promoting rhizobacteria (PGPR) promote root elongation, shoot growth, and cell division (Glick 2012). Furthermore, bacteria expressing ACC deaminase can reduce plant ethylene levels under stress conditions, avoiding precocious growth arrest and senescence (Olanrewaju et al. 2017).

Microbial communities also improve plant immunity through diverse defense pathways. While the systemic resistance (ISR) against pathogens, mediated by certain microorganisms, activates jasmonic acid (JA) and ethylene (ET) signaling pathways and primes plants to respond more effectively to future pathogen attacks (Pieterse et al. 2014). In addition, perception of MAMPs by plant receptors triggers MAMP-inducing immunity (MTI), which is a general non-specific protective immunity. Some beneficial microorganisms and macroorganisms can also antagonize the pathogens directly by producing antibiotics, siderophores, and lytic enzymes (Berendsen et al. 2012; Raaijmakers et al. 2009).

The plant microbiome structure and function are not static but rather are modulated by plant genotype, developmental stage, and also environmental conditions. Sugars, amino acids, organic acids, and flavonoids that comprise root exudates serve as chemical signals that drive microbial recruitment (Sasse et al. 2018). In addition, seed transmittance of microbes indicates that a portion of the microbiome is heritable, giving rise to a "core microbiome" that helps in early plant development (Shade et al. 2017).

With regard to abiotic stresses, microbial symbionts contribute to plant stress tolerance by increasing antioxidant levels, altering stress hormones' signal pathway expressions, and increasing water use efficiency. Recent reviews also present how PGPR can induce the antioxidant enzyme systems, maintain osmotic balance, and regulate ABA signaling in the plant under drought stress (Liu et al. 2025). Moreover, the soil microbes can help to ameliorate the stress resistance by controlling nutrient uptake and osmolyte production and maintaining membrane stability (Grover et al. 2011).

Nevertheless, environmental perturbations, including monoculture, pesticide abuse, and nutrient imbalance, can disturb this homeostasis of microbes (dysbiosis), leading to the impairment of the beneficial interspecies interactions and also to the susceptibility of plant diseases (Berg et al. 2020). This way, being able to comprehend these interactions, is of key to several importance in sustainable agriculture and plant health management.

1.2.2 Composition and Functions of the Plant Microbiome (Rhizosphere, Phyllosphere, Endosphere)

The diversity of microorganisms living on plants' roots is immense: we know of tens of thousands of species. This complex microbial community (called the plant's "second genome") is necessary to maintain a plant's health. Recent research on plant and microbe interactions demonstrated that plants can also shape their rhizosphere microbiome. Different groups of microbes colonize plants of different species, which are grown in the same soil (Berendsen et al. 2012).

Select microorganisms in microbial sources, such as soil and rhizosphere, among other sources, are reinforced by plants. This includes the microbial surroundings or the plant outer habitat or the leaf surface as it pertains to the environment and was referred to as the phyllosphere. They are also connected with the anthosphere, space around the flowers, the spermosphere or the surface of the germinating seeds, and the carposphere, the space around fruits, as well (Hardoim et al. 2015). The soil also plays the main role in the movement of rhizosphere microbiota, which contains a high number of individuals belonging to microorganisms such as *Acidobacteria*, *Verrucomicrobia*, *Bacteroidetes*, *Proteobacteria*, *Planctomycetes*, and *Actinobacteria* (Fierer 2017). This is also interesting to note here that bacteria too can be propagated generation after generation through seeds. There is a significant microbial bank in seeds that can colonize the root of a new plant (Liu et al. 2012).

Root systems of plants establish a specific environment that suits soil microbiota that can exist in the rhizosphere, roots, and to a certain extent, the above-ground part of the plants (Hartmann et al. 2009). The rhizosphere, the zone in the ground that is directly affected by the roots of the plant, is considered to be an area of high microbial activity and the most complex ecosystem (Kawasaki et al. 2016). A new study published by Donn et al. (2015) demonstrated that a shift in bacterial community driven by roots in the wheat rhizosphere resulted in an elevated count of *actinobacteria*, *pseudomonads*, *oligotrophs*, and *copiotrophs* tenfold higher than in the neighboring soil.

Further, researchers observed that not only the rhizosphere but also rhizoplane communities altered over a period of time, but the population in bulk soils remained unaltered. Similarly, Kawasaki et al. (2016) found that rhizosphere of *Brachypodium distachyon*, a model organism of wheat, majorly contained *Burkholderiales*, *Sphingobacteriales*, and *Xanthomonadales*, whereas *Bacillales* were more abundant in the bulk soil. It is also common knowledge that root exudates, including organic acids, amino acids, fatty acids, phenols, plant growth hormones, nucleotides, sugars, putrescine, sterols, and vitamins, may influence the composition of microbes adjacent to the roots, a phenomenon referred to as the effect of the rhizosphere (Hartmann et al. 2008).

For example, certain defensive secondary metabolites known as benzoxazinoids (BXs), released by maize roots, have been shown to alter the composition of root-associated microbial communities, specifically affecting members of the *Actinobacteria* and *Proteobacteria* groups (Hu et al. 2018).

Moreover, Zhalnina et al. (2018) determined the development of bacterial communities in the rhizosphere of *Avena barbata*. They found that the combination of root exudate chemistry and the bacterial preferences of specific substrates affects the formation of these communities. In another experiment, Fitzpatrick et al. (2018) shone light on some bacterial taxa detected in the rhizosphere and specifically those of *Pseudoxanthomonas* that were highly different in abundance on 30 angiosperms. To conclude, it should be mentioned that different plant populations and genotypes influence the composition of rhizosphere microbiome, which alters depending on the nature and characteristics of exudates produced by their roots.

Studies have revealed that a vast range of bacterial endophytes reside in the internal part of the roots (root endosphere). The mechanism through which these endophytes are actually entering these root cells is usually passive, e.g., through cracks in the roots, at the side of lateral root emergence, as well as through active mechanisms (Compant et al. 2005). The colonization and movement of these endophytes in plants are influenced by many factors such as the manner in which the plants distribute their resources and the capacity of the endophytes to colonize. Many bacterial taxa have been identified to enter the root tissues. The most common phyla present in grapevine root are the *Proteobacteria*, *Acidobacteria*, *Actinobacteria*, *Bacteroidetes*, *Verrucomicrobia*, *Planctomycetes*, *Chloroflexi*, *Firmicutes*, and *gemmatimonadetes* (Burns et al. 2015). The most occurring family in the rice root was the *Rhizobiaceae*, *Comamonadaceae*, *Streptomycetaceae*, and *Bradyrhizobiaceae* (Edwards et al. 2015). In a similar study, Correa-Galeote et al. (2018) reported that the dominant phyla in maize roots were *Proteobacteria*, *Firmicutes*, and *Bacteroidetes*, and that the prevalence of these groups was influenced by the soil's cultivation history.

The plant microbiome can be categorized into three organisms, namely, helpful, neutral, or harmful microorganisms. Plant growth-promoting bacteria (PGPB) are the kind of bacteria that cause the plant to grow and also have the capability of causing the plant to grow in a number of direct and indirect ways. Some PGPB also have the capacity to synthesize plant hormones such as auxin, gibberellin, and indole-3-acetic acid (IAA), which have the potential of influencing plant growth by changing the internal hormone levels of the plants. Also, a few PGPB produce an enzyme known as 1-aminocyclopropane-1-carboxylate (ACC) deaminase that has been found to decrease the levels of the stress hormone ethylene in plants. Plant growth-promoting strains, including *Pseudomonas* spp., *Arthrobacter* spp., and *Bacillus* spp., have been reported to produce ACC deaminase. Rascovan et al. (2016) revealed a diverse content of bacteria, including *Pseudomonas* spp., *Paraburkholderia* spp., and *Pantoea* spp., in the roots of wheat and soybean, which had significant capacities to foster the growth of plants by means of numerous activities, such as phosphate solubilization, nitrogen fixation, production of indole acetic acid, and ACC deaminase release. These techniques lead to better nutrient retention, growth, and resistance to stress. Another important thing to know is that certain kinds of bacterial strains are also able to provoke the disease symptoms due to the production of dangerous chemicals, proteins, and phytohormones. Likewise, the plant pathogen *Pseudomonas syringae*, affecting a wide variety of hosts such as tomato, tobacco, olive, and green bean, is widespread. Another significant harmful bacterium is *Erwinia amylovora*, which causes fire blight disease in fruit trees and ornamental plants. *Xanthomonas species*, *Ralstonia solanacearum*, and *Xylella fastidiosa* are linked to several significant crop diseases affecting plants such as potatoes and bananas (Mansfield et al. 2012).

The seriousness of a plant disease is influenced by several elements. These include the number of pathogens present, the vulnerability of the host plant, the presence of favorable environmental conditions, and biotic factors such as the plant's microbiota. Together, these aspects shape how the interaction between the plant and the pathogen turns out (Brader et al. 2017). Many research projects have shown that bacteria associated with plants, both in the soil and on the surface, can improve a plant's ability to resist infectious pathogens. This occurs by direct encounters with non-pathogenic microorganisms or by making alterations to the host plant defense system (Rudrappa et al. 2008). Several cases of biocontrol to overcome pathogen attacks and diseases have been reported. These involve the production of antibiotics, lytic enzymes, volatile compounds that inhibit pathogens, and siderophores (Berg and Koskella 2018). It has been revealed that some bacteria aid plants in guarding against harmful pathogens by regulating the plant hormones and triggering general plant resistance (Mansfield et al. 2012). It is vital to realize that the repetitive utilization of agricultural soils may lead to the development of pathogen pressure. This can lead to the establishment of disease-suppressive soils, which are soils containing microorganisms with the possibility of being used to suppress diseases (Durán et al. 2018). Genera such as *Pseudomonas, Streptomyces, Bacillus, Paenibacillus, Enterobacter, Pantoea, Burkholderia*, and *Paraburkholderia* have been reported to suppress plant pathogens (Schlatter et al. 2017). According to Trivedi et al. (2017), *Acidobacteria*, *Actinobacteria*, and *Firmicutes* all had three major bacterial groups that controlled the transmission of *Fusarium* wilt over a broad landscape. The ability of *Paraburkholderia graminis PHS1* to combat the fungal root pathogen is already described, and it has been noted that soil suppressiveness related to the production of sulfurous volatile compounds, such as dimethyl sulfoxide reductase and cysteine desulfurase (Carrión et al. 2018). Durán et al. (2018) highlighted the importance of the endosphere bacterial community in suppressing take-all disease caused by *Gaeumannomyces graminis*, and identified endophytes belonging to the genera Serratia and *Enterobacter* as promising candidates for countering this pathogen.

1.2.3 Origin and Drivers of Microbial Diversity

1.2.3.1 Microbiome Origin: Seed Transmission, Soil Recruitment

1.2.3.1.1 Vertical Transmission via Seeds

Endophytic and epiphytic microbial communities are transmitted to seeds from the mother plant by flower organs or vascular tissues. Such seed-borne microbes serve as pioneers for the early development of the plants and contribute to the initial assembly of the microbiome (Rashmi et al. 2025). Recent research has demonstrated that the seed microbiome is transmissible over generations, thus promoting plant survival under adverse conditions (Romão et al. 2025). For instance, the seed microbiota of beans were reported to be conserved and vertically transferred across generations, molding the early root and shoot microbiome (Garrido-Sanz and Keel 2025).

1.2.3.1.2 Horizontal Recruitment from the Soil

Though seed transmission provides a foundational inoculum, the bulk of the plant root–associated microbial diversity is sourced from the soil microbiome, especially with respect to roots in the rhizosphere. Plants attract certain microbes from the rhizospheric community using chemical signals and nutrient exchanges. The potential microorganisms available for recruitment are primarily influenced by soil type, texture, and chemical properties (Sarsaiya et al. 2025).

1.2.3.2 Key Factors Shaping Microbiome Assembly

1.2.3.2.1 Plant Genotype

The genotype of the host is a major factor shaping the microbiome. Rebar et al. proposed that plants recruit specific microorganisms that promote growth or defense of the host. The strong phylosymbiosis is also supported by the evidence that the evolution of plants could be linked to the structure of the microbial community (Romão et al. 2025). Genetic variations in root architecture, metabolic pathways, and immune responses significantly influence microbial selection.

1.2.3.2.2 Root Exudates

Root exudates, which are composed of sugars, amino acids, organic acids, secondary metabolites, and signal molecules (Robert et al. 2025), serve as both nutrient supplies and biochemical inducers for microbial colonization. For example, flavonoids and phenolics are able to specifically promote symbiotic rhizobia and inhibit pathogenic one(s). Under stress (e.g., drought or salinity), the profile of root exudates changes, resulting in a selective recruitment of stress-adapted microbial taxa (Wankhade et al. 2025).

1.2.3.2.3 Environment (Soil, Climate, and Agronomic Practices)

Environmental factors—the type of soil, its pH, moisture, and temperature conditions—and agricultural practices greatly determine microbial diversity. The environment overrides plant genotype in many cases (Wankhade et al. 2025) (Khasanova et al. 2023). Actors of climate, such as elevated CO_2 or heat stress, reorganize the microbial community via changing plant physiology and patterns of root exudation (Zeng et al. 2025). Seed geographic origin has been found to affect the initial microbiome, irrespective of the species of plants (Zeng et al. 2025).

1.2.3.2.4 Integration of Origins and Drivers

The plant microbiome is initiated by an active relationship between vertical and horizontal transmission paths. In this way, seed-borne microbes offer an initial community that has great potential benefit, while soil microbes contribute both diversity and local adaptation (Hanif et al. 2024). The plant genotype filters these microbes through root exudation and immune recognition; environmental conditions further modulate both the availability and activity of microbial taxa. All of these interactions, networked together, finally define both the structure and the functional potential of the plant microbiome (Chen and Liu 2024).

1.3 Microbial Functions in Plant Growth and Health

1.3.1 Microbial Roles in Plant Growth Promotion and Disease Resistance (Direct and Indirect Mechanisms)

- *Direct Mechanisms*

Many species of fungi, plant growth-promoting rhizobacteria (PGPR), and other beneficial microorganisms play a major role in optimizing plant growth and defense against pathogens through direct or indirect mechanisms (Fig. 1.1) (Farhaoui et al. 2023a, b; Kumar and Choudhury 2025).

Biofertilization By solubilizing phosphates, fixing atmospheric nitrogen, and mobilizing other mineral nutrients, PGPR are well known for their ability to improve plant nutrition (Farhaoui et al. 2023a, b; Mitra et al. 2024). Atmospheric N_2 is converted to ammonia by several bacterial species belonging to the genera *Rhizobium*, *Azotobacter*, and *Azospirillum* (Ibrahim and El-Sawah 2022). Phosphate-solubilizing bacteria, which produce organic acids and efficiently release inorganic phosphorus from insoluble soil compounds, include *Bacillus megaterium* and *Pseudomonas fluorescens*. This promotes phosphorus uptake by plants (Saeid et al. 2018).

Phytohormone Production Many PGPR produce phytohormones, including indole-3-acetic acid (IAA), which promote lateral root formation, cell division, and root extension. This leads to a more developed root system and improved nutrient and water uptake (Maheshwari et al. 2015). *Azospirillum brasilense*, for example, produces significant amounts of IAA, which promotes the growth of root hairs and thus improves plant growth (Santos et al. 2021). In addition, PGPR can produce gibberellins, which promote stem elongation, and cytokinins, which stimulate cell division and chloroplast formation (Orozco-Mosqueda et al. 2023).

Reduction of Ethylene-Induced Stress Plants often produce large amounts of ethylene under stress conditions, which inhibits root growth. By breaking down the

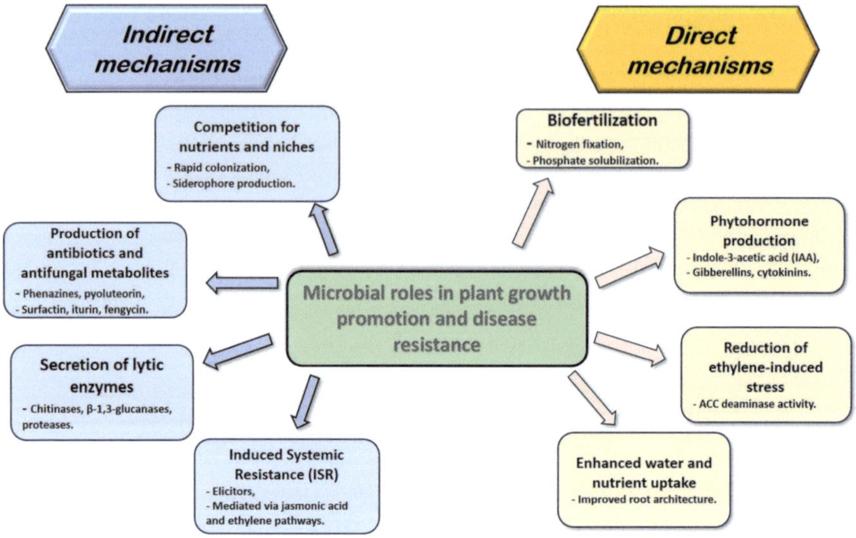

Fig. 1.1 Summary of the main direct and indirect mechanisms by which beneficial microorganisms, including plant growth-promoting rhizobacteria (PGPR) and beneficial fungi, confer resistance to phytopathogens and stimulate plant growth

ethylene precursor, ACC (1-aminocyclopropane-1-carboxylate) and PGPR that produce the enzyme ACC deaminase reduce ethylene concentration and mitigate its growth-inhibiting effects (Glick 2012). De Andrade et al. (2023) report that under salt or metal stress, *Pseudomonas putida* and *Enterobacter cloacae*, which exhibit high ACC deaminase activity, ensure root elongation, and increase biomass.

Enhanced Water and Nutrient Uptake PGPR optimize water and nutrient uptake by optimizing the structure of the root system, particularly with regard to the length, surface area, and density of root hairs (Grover et al. 2021). A well-developed root system improves exploration of deeper soil layers, which promote water and nutrient absorption and, consequently, plant growth. This effect is particularly significant in dry or nutrient-deficient areas (Comas et al. 2013).

- *Indirect Mechanisms*

Competition for Niches and Nutrients In the rhizosphere, phytopathogens compete with various beneficial fungi and PGPR for colonization sites and nutrients. The establishment of many pathogens is hampered by the rapid growth of these beneficial microorganisms and their high affinity for root secretions (Farhaoui et al. 2024). For example, *Bacillus subtilis* and other siderophore-producing PGPR absorb iron from the environment, preventing phytopathogenic bacteria and fungi from accessing this essential micronutrient (Benaissa 2024).

Secretion of Antibiotics and Antifungal Molecules Several beneficial fungi and bacteria are known to produce various secondary metabolites with antimicrobial effects. The multiplication of many soil-borne fungal pathogens, such as *Macrophomina phaseolina*, *Sclerotinia sclerotiorum*, *Rhizoctonia solani*, and *Fusarium* spp., is inhibited by phenazines, pyrrolnitrin, and pyoluteorin, released by species belonging to the genus *Pseudomonas* (Biessy and Filion 2018). The production of lipopeptide antibiotics such as bacillomycin, surfactin, iturin, and fengycin by several *Bacillus* species, including *B. subtilis* and *B. amyloliquefaciens*, also results in a loss of membrane integrity, leading to cell lysis and, ultimately, pathogen elimination (Farhaoui et al. 2022) (Fig. 1.1).

Synthesis of Hydrolytic Enzymes Several studies have shown the ability of PGPR and other microorganisms to release lytic enzymes such as β-1,3-glucanase, protease, and chitinase. (Fig. 1.1). These active molecules can function independently or synergistically and cause a loss of cell wall integrity of pathogenic fungi, thus slowing their multiplication and colonization (Panicker and Sayyed 2022). Several species of the genus *Trichoderma*, such as *T. harzianum*, release chitinase and glucanase, both of which are known for their effectiveness against a multitude of fungi responsible for root rot and damping-off of several cultivated plants (Farhaoui et al. 2023a, b).

Induced Systemic Resistance (ISR) Several PGPR activate plant defense mechanisms through ISR reactions by generating active molecules such as flagellins, siderophores, lipopolysaccharides, and volatile organic compounds (VOCs). By enhancing plant innate immunity, this "priming" elicits faster and stronger responses to pathogen invasion. Unlike systemic acquired resistance (SAR), which depends on salicylic acid, ISR is generally mediated by ethylene and jasmonic acid (Lahlali et al. 2025).

According to de Andrade et al. (2023), these direct and indirect mechanisms (Fig. 1.1) will demonstrate the versatile role of PGPR and several beneficial fungi as biofertilizers and biological control agents, promoting organic and sustainable agriculture by stimulating plant growth and reducing dependence on synthetic pesticides and chemical fertilizers.

1.3.2 Functional Types of Beneficial Microbes

Plant-beneficial microorganisms encompass many functional categories that can improve plant growth, nutrition, and stress tolerance. *Bacillus*, *Pseudomonas*, *Azotobacter*, and *Azospirillum* are examples of PGPR known for their ability to invade the rhizosphere. By fixing nitrogen, solubilizing phosphates, producing growth hormones (auxin and cytokinins), or releasing ACC-deaminase to reduce ethylene levels, these microorganisms stimulate plant growth. By producing antibiotics and siderophores capable of inhibiting pathogen growth, they also provide root protection against several diseases (Farhaoui et al. 2025a; Lahlali et al. 2025).

Arbuscular mycorrhizal fungi (AMF) are symbiotic organisms (*Glomeromycotina*) associated with approximately 80% of terrestrial plant species. AMF form arbuscules in root cortical cells and enhance the uptake of water, phosphorus, and a multitude of micronutrients (Farhaoui et al. 2025a). These beneficial organisms increase plant yield by improving growth and resistance to salinity and drought (Begum et al. 2019). Co-inoculation of AMF and PGPR is generally associated with synergistic effects, increasing crop yields and reducing the need for chemical inputs (Farhaoui et al. 2025a). *Glomus*, *Funneliformis*, and *Rhizophagus* are among the AMF genera widely used to improve phosphorus nutrition and stress resilience in crops such as maize, wheat, and vegetables (Begum et al. 2019).

Endophytic fungi colonize internal plant tissues, providing numerous benefits without harming plant health. *Trichoderma*, *Beauveria*, and *Piriformospora* are among the genera that produce phytohormones (such as gibberellins and IAA), solubilize nutrients, and induce systemic resistance against several phytopathogens (Majhi et al. 2025). Moreover, endophytic fungi are known for their aptitude to improve resistance to abiotic stressors such as heavy metals, drought, and salinity (Bilal et al. 2020; Khadiri et al. 2024). For example, many *Trichoderma* species colonize the roots of numerous crops, such as beans, wheat, and corn, promoting nutrient solubilization, modulating plant hormone signaling, and inducing systemic defense (Majhi et al. 2025).

Actinobacteria are Gram-positive, high-G-C bacteria belonging to the genera *Frankia*, *Streptomyces*, and *Arthrobacter*. *Streptomyces* species produce phytohormones and antibiotics, decompose complex soil organic matter, and enhance plant defenses against several pathogens. *Frankia* species form nitrogen-fixing nodules on the roots of several woody plants, such as alder (Abdellatif et al. 2024). *Actinobacteria* play a key role in soil fertility, health, and sustainability by contributing to nutrient recycling and increasing soil organic matter. Consequently, these microorganisms stimulate plant growth and yield (Abdellatif et al. 2024).

In general, each group exhibits distinct functional characteristics that are beneficial for sustainable agriculture. Table 1.1 presents a summary of the most important examples, mechanisms, and target crops.

1.3.3 Microbial Contribution to Abiotic Stress Tolerance

Beneficial microorganisms play a vital role in enhancing plant resistance to various abiotic stresses such as salinity, water deficit, and toxic amounts of heavy metals. They have the ability to modify phytohormone pathways. PGPRs such as *Bacillus and Pseudomonas* can produce cytokinins, IAA, and gibberellins. These active metabolites alter stomatal conductance and root system structure to maximize

Table 1.1 Functional types of beneficial microbes, representative genera, modes of action, and target crops

Type of microbe	Example genera	Target crops (examples)	Modes of action
Endophytic fungi	*Trichoderma*, *Piriformospora*, *Beauveria*	Cereals, pulses, fruit trees	Phytohormone production; nutrient solubilization; pathogen and pest antagonism
AMF (symbiotic fungi)	*Rhizophagus*, *Funneliformis*, *Glomus*	Maize, wheat, legumes, and vegetables	Enhanced P and water uptake; improved stress tolerance; induced systemic resistance
Actinobacteria	*Streptomyces*, *Frankia*, and *Arthrobacter*	Legumes, woody plants, and various crops	Antibiotic and enzyme production; nitrogen fixation; organic matter decomposition
PGPR (rhizobacteria)	*Pseudomonas*, *Bacillus*, *Azospirillum*	Cereals (wheat, rice), legumes (soybean), and vegetables	Nitrogen fixation, phosphate solubilization, hormone production, and pathogen inhibition

nutrient and water uptake under salinity or drought stress (Ansabayeva et al. 2025). Additionally, recent research has shown the ability of some PGPR strains to generate ACC deaminase, which breaks down 1-aminocyclopropane-1-carboxylic acid (ACC), an ethylene precursor. This reduces stress-induced ethylene concentrations and decreases plant growth disorders under abiotic stress (Yuan et al. 2022).

AMF also play a crucial role in stress resistance by promoting the uptake of several nutrients and water through their extensive hyphal networks, regulating phytohormone levels, and increasing root hydraulic conductivity (Farhaoui et al. 2025a). According to Begum et al. (2019), AMF colonization frequently results in significant increases in proline and abscisic acid (ABA) levels in host plants, thus optimizing their resistance to water deficit. Additionally, beneficial microorganisms participate in defense against reactive oxygen species (ROS) by stimulating plant antioxidant protection mechanisms. During abiotic stress, excessive accumulation of ROS damages cellular elements, thereby causing oxidative stress. Endophytic fungi and PGPR have the ability to positively activate antioxidant enzymes such as catalase (CAT), superoxide dismutase (SOD), ascorbate peroxidase (APX), and peroxidases (POD). This strengthens their defenses against ROS and shields plants from oxidative stress (Gupta and Kumar 2023). According to Ahmad et al. (2015), a number of *Trichoderma harzianum* isolates have demonstrated the capacity to increase the activity of antioxidant enzymes and the buildup of osmolytes in plants, enhancing their resistance to salinity and water deficit. According to Narsing Rao et al. (2022), *Streptomyces* species are actinobacteria that can increase resistance to abiotic stress by generating osmoprotectants, antioxidants, and volatile organic metabolites that control plant metabolism under stress.

Overall, the main mechanisms mitigating the adverse effects of abiotic stress include microbial modulation of phytohormones and activation of antioxidant defense systems (Table 1.2). These mechanisms contribute significantly to sustainable crop productivity.

Table 1.2 Microbial contributions to abiotic stress tolerance: mechanisms and examples

Microbe type	Abiotic stress mitigated	Representative genera	Example crop	Mechanisms
Endophytic fungi	Drought, salinity, heavy metals	*Trichoderma*, *Piriformospora*	Beans, tomato, cereals	Antioxidant enzyme activation, osmolyte accumulation, phytohormone modulation
PGPR	Drought, salinity	*Pseudomonas*, *Bacillus*, *Azospirillum*	Wheat, rice, maize	Phytohormone production (IAA, gibberellins), ACC deaminase activity, antioxidant induction
AMF	Drought, salinity	*Rhizophagus*, *Funneliformis*, *Glomus*	Maize, soybean, tomato	Enhanced water and nutrient uptake, ABA accumulation, and proline synthesis
Actinobacteria	Drought, salinity	*Streptomyces*, *Frankia*	Legumes, cereals	Osmoprotectant and antioxidant production, volatile compound secretion

1.4 Microbial Roles in Plant Growth Promotion and Disease Resistance

Microbial communities that surround the environment of plants are friendly allies that can give many advantages to the growth and survival of the plants. The rhizospheric region is known to be the most microbially enriched area, but microorganisms can also be found in the phyllosphere, either on the surface of aerial plant parts as epiphytes or within plant tissues as endophytes (Lemanceau et al. 2017). The counterpart is the energy supplied by plant roots from the breakdown of organic matter. This process is crucial for thriving microbial communities in the rhizosphere, as it attracts specifically biotrophs, including mycorrhizal fungi and saprotrophs, that can grow in the same area as pathogens that can potentially infect vulnerable plants (Priyadharsini et al. 2016). Plant microbiome helps in stress management and favors disease resistance, thereby it is essential for the mitigation of negative agroecological issues consequently helping to lessen economic losses caused by plant disease and abiotic stresses and improves agricultural output (Compant et al. 2019).

Recently, various studies have put the focus on the interplay between plant microbiome and plant growth, especially when it comes to the surrounding rhizosphere area that contains PGPRs, which are bacteria that enhance plant wellbeing (Berendsen et al. 2012). The frequently observed genera of PGPRs are *Acinetobacter*, *Agrobacterium*, *Bradyrhizobium*, *Rhizobium*, *Azotobacter*, *Arthrobacter*, *Burkholderia*, *Frankia*, *Serratia*, *Thiobacillus*, *Pseudomonas*, *Azospirillum*, and *Bacillus* (Glick 2014). The *Burkholderia* and *Pseudomonas* species demonstrated their ability to enhance plant growth and provide biocontrol (Wakelin et al. 2017).

For instance, *Streptomyces mutabilis* IA1, isolated from soil in the Sahara, has shown a positive impact on the growth of wheat seedlings (Toumatia et al. 2016).

In addition to bacteria, fungal endophytes form numerous relationships with their hosts, such as commensalism, mutualism, symbiosis, and parasitism; such interactions hinge on the genetic load of the plant as well as other environmental parameters. The interaction of a fungal endophyte and its host plant helps the plant to grow and reproduce due to a lessening of various kinds of biological and environmental stressors (Gowtham et al. 2024). Some varieties of *Phialophora graminicola* have also been found to assist plants in absorbing a greater proportion of mineral nutrients. *Trichoderma* is often utilized in commercial products as a biopesticide and biofertilizer due to its very fast rate of growth and production of a diversity of antimicrobial compounds. It is also a fungus that has applications in biocontrol (Verma et al. 2007).

An association between a fungal endophyte and the host plant positively affects the host plant by allowing it to grow and complete its reproduction process due to the reduction of various forms of environmental and biological stress (Gowtham et al. 2024). Individual strains of *Phialophora graminicola* are reported to assist the plants in the uptake of extra mineral nutrients. *Trichoderma* is a well-known biocontrol agent, which is frequently applied in commercial products in the form of biofertilizers and biopesticides. The successful capacity to manage the pathogens is because of the ability of this fungus to produce numerous antimicrobial metabolites as well as rapid growth. It has been found that 70% of plant types associate with arbuscular mycorrhizal fungi (AMF), and the association is considered to be crucial to the growth, development, nutrient assimilation, and defending abilities of this plant variety against environmental and organismal stressors (Zhang et al. 2016).

Through a highly branched structure known as arbuscules, AMF deliver N and P to plant directly inside the root cells using an extraradical hyphae (Parniske 2008). In addition to interacting with plants, AMF are also known to associate with numerous soil microbes inhabiting the rhizosphere. Together, they form a complex mutualistic network, and several studies have shown that AMF can interact with biocontrol agents, nitrogen-fixing bacteria, and phosphate-solubilizing bacteria (Farhaoui et al. 2025b). Studies have shown that PGPR promote plant growth through both direct and indirect mechanisms. Direct mechanisms act within the plant by modulating its metabolism (Glick 2014). It is an immediate process that involves releasing different plant growth regulators and growth factors that interact directly with plants. Additionally, other processes act as a reservoir for hormones that plants release, stimulating the plant's metabolism and enhancing its ability to adapt. This interaction includes processes such as fixing nitrogen biologically, making phosphate more available, producing siderophores, and creating phytohormones (Gutiérrez-Mañero et al. 2001).

On the other hand, indirect methods are linked to plant's defensive mechanisms that interact which signaling molecules released by bacteria (Glick 2014). This, in turn, affects metabolic functions and growth indirectly. There are two main mechanisms in this area: one involves plants' defense against abiotic stress, while the other is specific to biotic stress caused by harmful pathogens (Glick 2014). The shift from soil to root or the surrounding areas is based on many suggestions that have been made by researchers lately (Bais et al. 2006; Doornbos et al. 2012), one of which is the idea that carbohydrates and amino acids act as chemical signals in the rhizosphere (Moe 2013). Additionally, the recognition of flavonoids, which are plant-specific secondary metabolites and serve as key signaling molecules guiding the assembly of microbial communities in the rhizosphere, further supports this assertion (Weston and Mathesius 2013). Emergent studies in genomics, transcriptomics, metagenomics, and proteomics can put spotlight on new genes and proteins that play a role in managing stress responses (Manzoni et al. 2018). For instance, Alavi et al. (2013) found out about spermidine, a new regulator of plant abiotic stress as a PGPM through transcriptome analysis of oilseed rape and its partner *Stenotrophomonas rhizophila* (Alavi et al. 2013).

Many plant diseases are caused by plant pathogens, like fungi, bacteria, nematodes, and viruses, but at the same time they can be managed using biocontrol agents. These agents suppress pathogens through multiple mechanisms, including the production of enzymes and antibiotics that degrade cell walls, the enhancement of plant defense responses, and competition for essential nutrients such as iron, carbon, and nitrogen (Valencia et al. 2011). In biocontrol, PGPMs can use these methods at the same time, and there are studies that encourage genetically modified agents that have improved biocontrol capabilities (Glick 2014). It has also been shown that using various biocontrol agents together can help plants grow and reduce fungal diseases in them (Akhtar et al. 2011).

1.4.1 Parasitism

Parasitism is an interaction between two different organisms that are phylogenetically unrelated, in which one is the parasite that benefits and the other is the host that is harmed or from which the parasite benefits (Verma et al. 2019). Many studies have pointed to *Trichoderma fungus*, a type that acts as an opportunistic, non-pathogenic plant friend and pathogen microbe's enemy, as it is now perceived as a biocontrol method against many pathogenic microbes and also nematodes (Reddy et al. 1996). Numerous studies have shown that different species of *Trichoderma* have strong abilities to control *M. javanica* effectively (Spiegel et al. 2007). Additionally, it has been noted that *T. asperellum-203* and *T. atroviride* can attack nematode eggs (Sharon et al. 2007). This phenomenon is specifically called mycoparasitism, where *Trichoderma* wraps around the pathogen hyphae, which produces structures similar to appressoria (Lu et al. 2004). This prevents the growth and spread of numerous fungal pathogens. The biocontrol effectiveness of *Trichoderma* has been thoroughly recorded regarding different plant harmful fungi, which include *Rhizoctonia solani*, *Botrytis cinerea*, *Sclerotium rolfsii*, *Sclerotinia sclerotiorum*, *Macrophomina phaseolina*, *Pythium* species, and *Fusarium* species (Gajera 2012).

Table 1.3 Case studies of parasitism induced by *Trichoderma* species

Trichoderma species	Target organism	Observation	Source
T. harzianum	*Globodera rostochiensis* (potato cyst nematode)	Direct in vitro interaction documented	Saifullah and Thomas (1996)
T. asperellum-203, *T. atroviride IMI 206040*	*Meloidogyne javanica* (root-knot nematode)	Show biocontrol effects in soil	Sharon et al. (2007)
T. asperellum-203, *T. atroviride*	Nematode eggs	Capable of attacking and parasitizing nematode eggs	Sharon et al. (2007)
Trichoderma spp.	*Meloidogyne javanica*	Parasitism facilitated by jelly-like matrix	Sharon et al. (2007)
Trichoderma spp.	*Rhizoctonia solani*	Secretes factor triggering ech42 gene transcription (endochitinase 42) before contact	Zeilinger et al. (1999)

According to reports, *Trichoderma*'s biocontrol function makes use of a number of well-known strategies, including releasing antifungal compounds and competing for nutrients (Table 1.3) (Lorito et al. 1996).

1.4.2 Competition

Soil-dwelling organisms often compete with one another for resources like oxygen and nutrients. The efficiency of biocontrol is impacted by the direct competition between pathogens and hostile organisms for these resources. Microorganisms that live in roots also compete with one another for desirable spots on the root surface. Studies have indicated that the primary cause of fungistasis, which results in reduced spore germination in the soil, is the competition for mineral resources, particularly carbon (Alabouvette et al. 2009). Studies reveal that different kinds of *Pseudomonas* make excellent use of glucose. They have an advantage over other microbes that are unable to consume glucose because they convert it into gluconic acid and 2-ketogluconic acid (Gerritse and Gottschal 1993). In soil environments, competition has been observed for trace elements like iron, copper, zinc, and manganese. According to research, there is fierce competition as the amount of soluble and bioavailable iron in the soil decreases (Loper and Henkels 1999). Microorganisms, including biocontrol microbes, have been shown to limit the growth of other microbes. They do this by replenishing iron through the creation of siderophores, which are small compounds that attract iron when there is a shortage of it (Loper and Henkels 1999). There have been cases where Pseudomonads that produce siderophores help to control plant pathogens found in the soil (Loper et al. 1997).

1.4.3 Production of Lytic Enzymes

Biocontrol agents use enzymes like β-1,3-glucanase, chitinase, cellulase, and protease to manage soilborne pathogens. These enzymes, produced by beneficial plant growth-promoting microbes, directly hinder fungal growth by breaking down their cell walls. Chitinase breaks down chitin, a tough polymer in fungi's cell wall, while β-1,3-glucanase, produced by *Paenibacillus* and *Streptomyces* species, effectively breaks down the cell walls of harmful *F. oxysporum* fungi (Singh and Jha 2016). The potential to produce natamycin and chitinase has been demonstrated by *Streptomyces lydicus* strain A01, which was isolated from the soil of a vegetable farm in a Beijing, China, suburb. It means that it is very effective in suppressing the growth of *Botrytis cinerea* (Lu et al. 2004). Li et al. (2015) expanded the expression of the 1,4-glucanase and 1,3-glucanase of *Paenibacillus polymyxa* in *Streptomyces lydicus* A01, with an increased capacity to curb *B. cinereal* being observed. It had previously been demonstrated that *Bacillus cepacia* is capable of synthesizing β-1,3-glucanase (Fig. 1.2), which has the ability to degrade the cell walls of some soil-borne pathogens, such as *R. solani*, *P. ultimum*, and *S. rolfsii* (Fridlender et al. 1993).

Besides, other *Bacillus* species were also found as potential biocontrol agents because of their properties to degrade chitin. This group of bacteria includes the *B. subtilis*, *B. cereus*, *B. circulans*, *B. licheniformis*, and *B. thuringiensis.* Moreover, a variety of Gram-negative bacteria have been reported to display chitinolytic properties, such as *Pseudomonas fluorescens*, *Serratia marcescens*, *Enterobacter agglomerans*, and *P. aeruginosa* (Frankowski et al. 2001). The creation of chitinase by *Serratia plymuthica C48* was seen, which prevented the spores from germinating and the germ tubes from growing longer in *Botrytis cinerea* (Muhammad et al. 2024). In addition, mutant versions of *Stenotrophomonas maltophilia W81* have been noted to produce excessive amounts of extracellular protease, thus improving their effectiveness in controlling *Pythium* (Dunne et al. 2000).

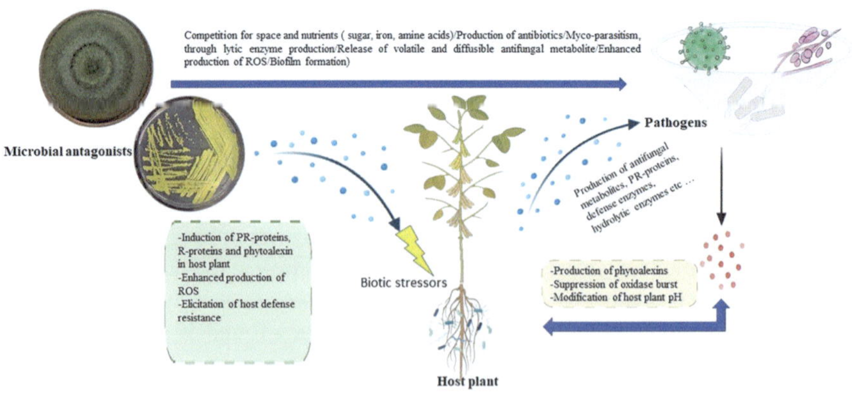

Fig. 1.2 Mechanism used by biocontrol agents in host plant protection

1.4.4 Antibiotic Production

Antibiotics are active compounds produced by microbes that affect plant interactions with pathogens. Bacillus produces various antibacterial and antifungal antibiotics, including subtilin, subtilosin A, TasA, and sublancin, which are ribosomally recognized. Other groups of antimicrobial compounds include rhizocticins, mycobacillin, chlorotetain, and bacilysin, as well as the iturin, fengycin, and surfactin families. These lipopeptides are synthesized by non-ribosomal polyketide synthases and peptide synthetases (Leclère et al. 2005). Kanosamine and zwittermicin A are reported to be produced by strains of *Bacillus subtilis* (Ongena et al. 2005). A study on *B. amyloliquefaciens FZB42*, where nine gene clusters (dhb, bmy, fen, srf, nrs, mln, bae, bac, and dfn) have been identified as the primary sources for the production of polyketides and peptides synthesized by the enzymes PKS and NRPSs (Chang et al. 2007).

Pseudomonas species, including *Pseudomonas fluorescens* and *Pseudomonas aeruginosa*, produce antibiotics like PCA, PCN, Plt, Prn, DAPG, oomycin A, viscosinamide, butyrolactones, kanosamine, zwittermicin A, aerugine, rhamnolipids, cepaciamide A, ecomycins, pseudomonic acid, azomycin, FR901463, cepafungins, and karalicin, which are responsible for their production.

Trichoderma pseudokoningii SMF2 generates three key peptaibols: trichokonin VI (TK VI), trichokonin VII, and trichokonin VIII (Shi et al. 2012). Research has shown that certain fungi used in biocontrol can create various antibiotics. *Trichoderma* species generate peptaibols, gliovirin, gliotoxin, pyrones, and viridin, and combat a range of fungal pathogens (Harman et al. 2004). *Trichoderma* and other similar genera, like *Gliocladium* and *Emericellopsis*, are recognized for generating peptaibols, which are a broad group of antibiotic peptides (Song et al. 2011).

1.4.5 Induction of Plant Defenses

Biocontrol types of PGPM use induced systemic resistance (ISR) to protect plants from disease. This resistance, far from pathogens, reduces the spread of disease-causing agents and infection severity. Plants have physical and chemical barriers, and advanced immune reactions that identify pathogen- or microbe-associated molecular patterns (PAMPs or MAMPs). These defenses occur quickly and last a short time, with plants producing molecules like reactive oxygen species (ROS), nitric oxide, and ethylene signals. This is followed by the buildup of callose and the creation of antimicrobial substances related to defense (Hermosa et al. 2012). For PGPR, several MAMPs have been identified, including lipopolysaccharides and flagellin. However, it has been demonstrated that substances that are released, such as volatile organic compounds, biosurfactants, and antibiotics, can cause systemic resistance. Additionally, microorganisms in the rhizosphere have been found to increase the formation of secondary metabolites in plants (Sekar and Kandavel 2010).

Ajmalicine, serpentine, picrocrocin, crocetin, hyoscyamine, scopolamine, safranal compounds, and tanshinone are significant metabolites produced by PGPR species that cause physiological and morphological reactions in crop plants. A variety of MAMPs are produced by distinct strains of *Trichoderma*. As an illustration, consider ET-inducing xylanase (Xyn2/Eix), a strong defense response elicitor found in particular tomato and tobacco cultivars that was first identified in Trichoderma (Rotblat et al. 2002). Because of enhanced defense mechanisms brought on by their presence, certain AM fungi have an effect on promoting plant development and triggering defense responses. It has been shown that the interaction between plants and AM fungus activates defense mechanisms, making plants more susceptible to root pathogen infections (Gianinazzi-Pearson et al. 1994).

1.5 Microbiome–Pathobiome Interactions Under Environmental Stress

Plants interact with microbiome at many different levels, and this highly dynamic relationship is strongly influenced by environmental factors. Ideally, in its preferred state, the plant–microbiome mutualistic and commensal species perform a balanced function, favoring nutrient uptake or growth promotion and immunological priming, among others (Vannier et al. 2018). Stress from toxins and environment can break this equilibrium, leading to more pathogenic organisms and weakened microbiome defense systems. The change, or shift, from a healthy microbiome to a pathobiome is key in explaining increased plant susceptibility in changing environments (Trivedi et al. 2013).

1.5.1 Disruptions Caused by Pathogens and Environmental Stress

Under environmental stresses such as salinity, drought, and soil pollution, the assembly and stability of microbial communities can be significantly disrupted, leading to reduced microbial diversity and shifts in community composition (Xu et al. 2018). For instance, the richness of beneficial taxa (e.g., P*seudomonas, Bacillus*, and mycorrhizal fungi) can decrease by drought stress, while promoting the proliferation of latent or opportunistic pathogens like *Fusarium*, *Alternaria*, and *Rhizoctonia* (Santos-Medellín et al. 2017).

The functionality and structural system can be also affected by these disruptions. Oxidative stress can unintentionally create a favorable niche for pathogens by damaging root tissues and altering root exudation profiles (De Vries et al. 2020).

Conversely, the pathogens can upregulate virulence genes, boosting colonization efficiency and aggressiveness. Also, plants show disrupted immunity pathways, including altered formation of phytohormones such as salicylic acid, jasmonic acid, and ethylene, which further impairs their ability to modulate microbiome composition and mount effective defense responses to stress signals (Microbiota et al. 2019) (Berendsen et al. 2012). Environmental stress acts both as an inhibitor of beneficial interactions and also as a catalyst for pathogen activation (Berendsen et al. 2012). The result is a feedback loop, where environmental stress, an imbalance of the microbiome, and pathogen pressure reinforce each other to increase the incidence of disease outbreaks.

1.5.2 Competitive Dynamics Between Beneficial and Pathogenic Microbes

Beneficial and pathogenic microorganisms in the rhizosphere and endosphere can develop a complex ecosystem. It is defined by competition for resources, chemical antagonism, and host influence (Espinosa-Palomeque et al. 2025). Symbiotic microbes, especially plant growth-promoting rhizobacteria (PGPR) and arbuscular mycorrhizal fungi (AMF), can suppress pathogens through multiple mechanisms: niche exclusion, competition for iron via siderophore production, secretion of antibiotics and lytic enzymes, and activation of host immune responses via induced systemic resistance (ISR) (Pieterse et al. 2014).

Plant protection mechanisms can be significantly influenced by environmental stress. For example, drought can suppress gene expression related to siderophore biosynthesis or the production of antimicrobial metabolites by PGPR, while reducing the recruitment of beneficial taxa by plants (Trivedi et al. 2022). Likewise, salinity stress may disrupt mycorrhizal symbiosis and decrease the functional integrity of beneficial fungal networks.

In contrast, pathogens may be more robust to such changes or even utilize them. Some pathogens may also be able to mimic host signals or produce phytotoxins that suppress microbial competitors or manipulate host responses for their own benefits or prevent the host from perceiving it appropriately (Microbiota et al. 2019). This result leads to an ecological competitive point, where pathogenic taxa increasingly dominate beneficial microbes, particularly microbiomes that do not contain keystone regulatory species.

Increasingly, there is evidence to understand the importance of keystone taxa microorganism species with disproportionate effect on community structure and stability in promotion of plant health. Under stress, the decline or loss of these taxa can lead to cascading effects, destabilization of the community and allowing for the emergence of a pathobiome-dominated state (Banerjee et al. 2018) (Fig. 1.3).

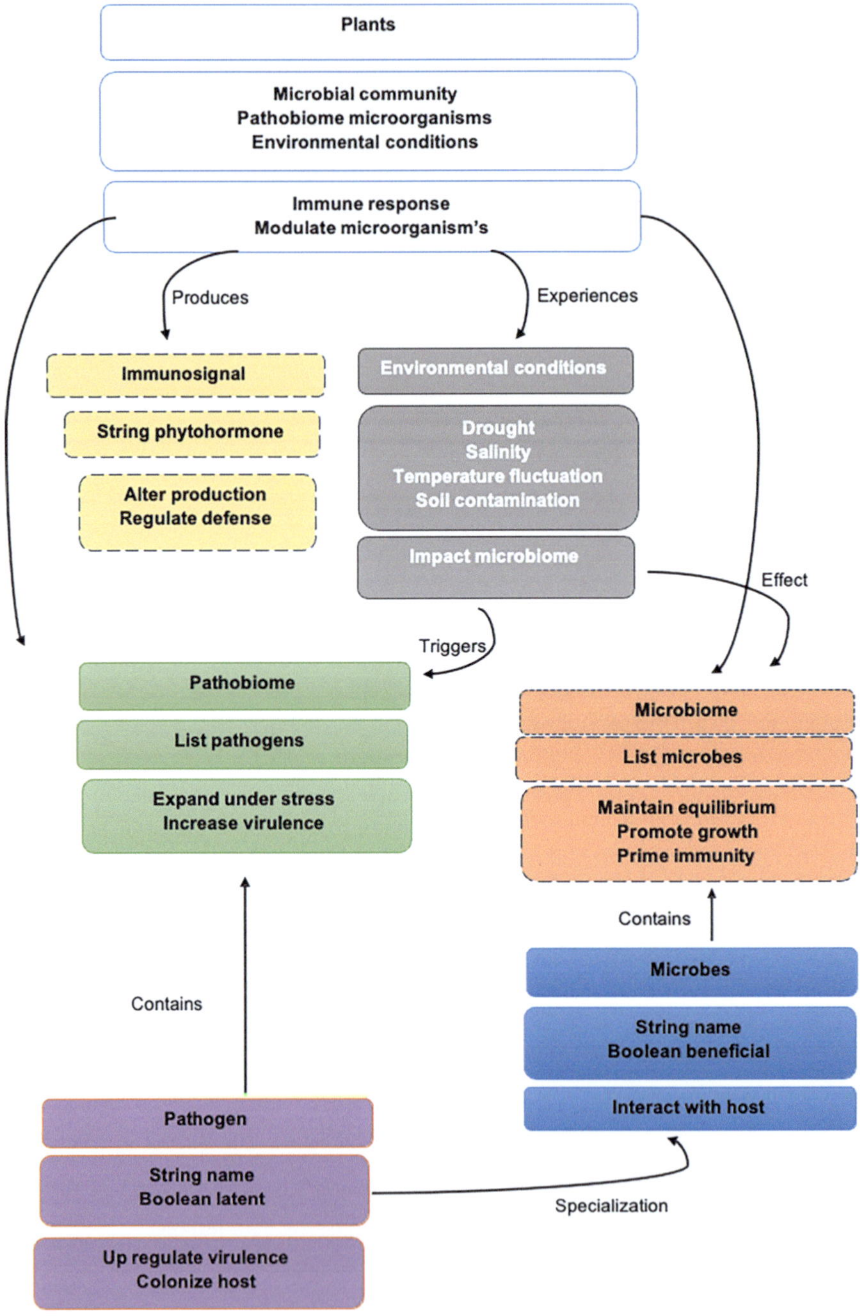

Fig. 1.3 UML diagram of plant–microbiome–pathobiome interaction conditions

1.6 Immune Activation and Microbial Recognition Mechanisms

1.6.1 PRRs-Mediated MAMP-Triggered Immunity (MTI)

The recognition of microorganism-associated molecular patterns (MAMPs) by membrane receptors called pattern recognition receptors (PRRs) forms the basis of the first line of defense in plants. By identifying patterns conserved in bacteria, fungi, and other microorganisms, these PRRs trigger MAMP-mediated immunity (MTI), which inhibits pathogen invasion (Du et al. 2024). For example, bacterial flagellin is identified by FLS2, a PRR functioning as a receptor-like kinase (RLK), via its epitope flg22. On the other hand, EFR is able to recognize EF-Tu, a bacterial elongation factor (Jones et al. 2024). In fungi, chitin, the main component of their cell wall, is detected by the receptor OsCERK1 in rice and CERK1 in *Arabidopsis*. This triggers a comparable defense sequence (Chang et al. 2025). When MAMP is detected by PRRs, receptor-coreceptor complexes (such as FLS2-BAK1) are formed, which in turn activate cytoplasmic kinases such as BIK1, promoting calcium entry, reactive oxygen species (ROS) generation, and MAP kinase activation, which in turn triggers the transcription of defense genes (Nabi et al. 2024). These reactions limit microbial growth, generate phytoalexins, and strengthen the cell wall (Du et al. 2024). To prevent excessive energy costs from affecting plant growth, the intensity of MTI should be controlled (Dodds et al. 2024). According to recent research, even commensal or beneficial microorganisms producing comparable MAMPs cause more attenuated IMT, suggesting selective immune tolerance (Yang et al. 2025). For example, some strains of *Pseudomonas fluorescens* promote plant growth while eliciting moderate PRR responses (Mitra et al. 2024). This coexistence, which allows the plant to protect itself against pathogens while preserving its advantageous microbiome, depends on a delicate balance between tolerance and recognition (Yu et al. 2021).

1.6.2 NLR-Mediated Effector-Triggered Immunity (ETI)

To counter the immunity triggered by MAMP (MTI), a variety of pathogens generate effectors that are injected into plant cells by specialized secretion systems (such as type III in bacteria). According to Duxbury et al. (2021), through the suppression of PRR signaling, these effectors promote infection. These effectors are detected either directly or indirectly by intracellular receptors known as nucleotide-binding leucine-rich repeat receptors (NLRs). Plants have thus developed a second line of defense (Dodds et al. 2024). NLRs have an N-terminal domain (TIR or CC) involved in immune activation, an LRR domain for recognition, and a central NB-ARC domain for signaling (Wang and Luan 2024). Effector-triggered immunity (ETI), which is characterized by massive ROS accumulation, defense gene expression, and frequently a hypersensitive response (HR) that includes localized programmed cell death to limit pathogen spread, is triggered when an NLR identifies an effector. This response is rapid and intense (Nabi et al. 2024). According to recent research, ETI responses are more robust and quicker than MTI responses, but they share many components, such as MAP kinase activation and phytoalexin production (Yu et al. 2024). The *Pseudomonas syringae* effector AvrRpt2 is detected by the *Arabidopsis* NLR RPS2, for instance, and HR is induced within hours (Kim et al. 2009). Furthermore, it has been demonstrated that certain NLRs, like "helper NLRs," work in networks or pairs to enhance immune signaling (Gong et al. 2023). ETI activation, however, must be carefully controlled, as excessive reactions can collaterally harm plant tissues, impairing growth and yield (Kourelis and Adachi 2022). Therefore, exploiting ETI in plant breeding still requires spatiotemporal control of NLR activation (Yu et al. 2024).

1.6.3 Tolerance and Defense Activation in Balance

The plant immune system must strike a delicate balance between maintaining tolerance to beneficial microorganisms and triggering defense responses. Plant growth and development can be affected by overactivation of defense mechanisms, even though immune responses are essential for protection against infections. Under environmental stress, where the cost of a robust immune response can hamper other essential functions, such as nutrient acquisition and reproduction, this balance between defense and growth is particularly crucial (Yu et al. 2022). Recent research on how plants control this balance suggests that hormonal interaction is critical for adjusting immune activation. For example, two hormones that modulate defense responses are salicylic acid (SA) and jasmonic acid (JA). JA generally promotes resistance to necrotrophic pathogens, whereas SA is more related to resistance to biotrophic pathogens (Myers Jr et al. 2023). Furthermore, the idea of "immune priming" has gained popularity, which suggests that plants can strengthen their immune systems in response to prior pathogen exposure, resulting in faster and more effective defense responses to comparable threats (Cooper and Ton 2022).

1.7 Future Perspectives and Agroecological Applications

Plant microbiome engineering is becoming a growing research and development area with a view to exploiting and maximizing the favorable aspects of microbe populations in order to increase the well-being and productivity of crops, and farm system sustainability. The rationale of this strategy is that a targeted modulation of plant-related microbiomes may become the tool that will grant a radical change toward better agronomic performance and environmental sustainability (Afridi et al. 2022; Kaul et al. 2021).

The first possible measure, as far as the future is concerned, is the creation of more sophisticated microbial engineering approaches. These are a combination of genome editing technologies like CRISPR/Cas9, which precise implantation or alteration of the gene of interest in a strain of microbes would allow the production of the characteristics of nitrogen-fixation, phosphorus solubilization, phytohormones synthesis, or antibiotic synthesis of secondary metabolites to protect crops (Kaul et al. 2021). These alterations can also reduce the ability of microbes to withstand adverse environmental conditions, affecting not only their survival but also their performance in soils exposed to climatic or human-induced stresses (Foo et al. 2017).

A different large direction is the creation of synthetic microbial consortia, where microbial species are selected with complementary functions to create synergistic interactions. These consortia can be designed to serve as bioinoculants that can be used to drive plant growth, enhance nutrient utilization, and enhance plant immune response to a wide range of pathogens (Afridi et al. 2022; Barea 2015) Nonetheless, before applying such formulations in large scales, detailed investigations of the dynamics of microbial communities after the inoculation, their adaptation to the existing microbiomes, and long-term survival in the fields are necessary (Foo et al. 2017).

Microbiome engineering has an agroecological advantage of fitting perfectly well with the concepts of sustainable and regenerative agriculture because it takes advantage of soil functional biodiversity as a driving force to achieve agricultural robustness. Microbiological improvement of the fertility of degraded soils using local microbial inoculants, microbial population inoculation (indigenous microbes) using specific organic amendments (prebiotics), and optimization of farming systems (e.g., rotations, intercrops, reduced tillage) are the strategies that should facilitate the development of healthy and resilient microbial communities (Chave et al. 2014; Munir et al. 2022).

In this context, the concept of microbiome-assisted breeding, in which plant varieties are selected based on their ability to recruit beneficial and efficient microbiomes, represents a promising new direction. This approach uses plant genetics and microbial ecology in a synergetic fashion to maximize the crop performance with respect to a changing global climate and reduced natural resources (Kaul et al. 2021).

However, several major challenges remain for the operational implementation of these innovative technologies. To begin with, little is known about how microbes engage with both plants and environment in complex ways, especially about how their traits can be stable and active in heterogeneous field situations (Foo et al. 2017). Secondly, the problem of environmental and ethical implications of releasing genetically modified microorganisms entails the necessity to develop efficient regulatory mechanisms and social acceptance of the application of such biotechnologies (Afridi et al. 2022).

Thus, further insights include the development of cross-disciplinary methods of integrating microbiology, genomics, ecology, agronomy, predictive modelling, and social sciences in ways that guarantee to address safe, effective, and socially acceptable solutions. An effective embedding of such innovations in the agricultural systems will also involve the development of technical capacities of the farmer and innovating together with all interested stakeholders (Verma et al. 2020).

Finally, plant microbiome engineering has significant potential as a way to provide food security in world, remediate soil, and lead to agroecological transformation. It forms a strategic pole to resolve modern agricultural issues brought about by global environmental change, as long as it is carried out in a systemic and ethical framework, balancing innovations and principles of sustainability and preservation of biodiversity.

1.8 Conclusion

Recent decades have seen the plant microbiome come to the fore as a marker of plant health, growth, and resilience, which may be termed the second genome of the plant. Different communities of microbes related to plant compartments are important in nutrient acquisition, synthesis of phytohormones, toleration of stress, and defense against pathogens. These depend on plant genotype, growth conditions, farming management, and interactions in the soil.

This chapter also indicated the importance of beneficial microorganisms such as plant growth-promoting rhizobacteria (PGPR), mycorrhizal fungi, endophytes, and biocontrol agents in enhancing plant nutrition and defense systems via mechanisms, including biological fixation of nitrogen, phosphate solubilization, siderophore synthesis, enzymatic destruction of the pathogen cell wall, biosynthesis of antibiotics, and induced systemic response. The functions also increase the fitness of plants besides minimizing dependency on agrochemicals, which promote ecofriendly farming.

However, agricultural activities such as intensive farming, stresses imposed by climate change, and soil degradation endanger the stability, diversity, and functionality of plant-associated microbiomes, which comprise an impact on plant health and productivity. In turn, future projections are focusing more and more on microbiome engineering, synthetic microbial consortia, microbiome-assisted breeding, and the design of customized bioinoculants to rebuild and exploit microbial functions to sustainably produce crops.

Ultimately, a higher level of comprehension of plant-microbiome interactions and their ecological controls will be necessary in the development of novel approaches that incorporate microbiology, genomics, agronomy, and agroecology. This type of integration will facilitate the creation of robust, productive, and sustainable agricultural systems. It can also solve global challenges, including food security, climate change adaptation, and environmental conservation.

Acknowledgments This study was financially supported by the Phytopathology Unit of the Department of Plant Protection at the Ecole Nationale d'Agriculture de Meknès, Morocco.

References

Abdellatif AAM, Gebily DAS, Elmaghraby MMK, Sahu PK, Thakur B, Kaur S (2024) Transforming roles of Actinobacteria in sustainable agriculture: from soil health and plant productivity perspective. In: Metabolomics, proteomics and gene editing approaches in biofertilizer industry: volume II, pp 299–338

Afridi MS, Javed MA, Ali S, De Medeiros FHV, Ali B, Salam A, Sumaira, Marc RA, Alkhalifah DHM, Selim S (2022) New opportunities in plant microbiome engineering for increasing agricultural sustainability under stressful conditions. Front Plant Sci 13:899464

Ahmad P, Hashem A, Abd-Allah EF, Alqarawi AA, John R, Egamberdieva D, Gucel S (2015) Role of Trichoderma harzianum in mitigating NaCl stress in Indian mustard (Brassica juncea L) through antioxidative defense system. Front Plant Sci 6:868

Akhtar MS, Siddiqui ZA, Wiemken A (2011) Arbuscular mycorrhizal fungi and rhizobium to control plant fungal diseases. In: Alternative farming systems, biotechnology, drought stress and ecological fertilisation. Springer, pp 263–292. https://doi.org/10.1007/978-94-007-0186-1_9

Alabouvette C, Olivain C, Migheli Q, Steinberg C (2009) Microbiological control of soil-borne phytopathogenic fungi with special emphasis on wilt-inducing *Fusarium oxysporum*. New Phytol 184(3):529–544. https://doi.org/10.1111/j.1469-8137.2009.03014.x

Alavi P, Starcher MR, Zachow C, Müller H, Berg G (2013) Root-microbe systems: the effect and mode of interaction of Stress Protecting Agent (SPA) Stenotrophomonas rhizophila DSM14405T. Front Plant Sci 4. https://doi.org/10.3389/fpls.2013.00141

Ansabayeva A, Makhambetov M, Rebouh NY, Abdelkader M, Saudy HS, Hassan KM, Nasser MA, Ali MAA, Ebrahim M (2025) Plant growth-promoting microbes for resilient farming systems: mitigating environmental stressors and boosting crops productivity—a review. Horticulturae 11(3):260

Arif I, Batool M, Schenk PM (2020) Plant microbiome engineering: expected benefits for improved crop growth and resilience. Trends Biotechnol 38(12):1385–1396

Bais HP, Weir TL, Perry LG, Gilroy S, Vivanco JM (2006) The role of root exudates in rhizosphere interactions with plants and other organisms. Annu Rev Plant Biol 57(1):233–266. https://doi.org/10.1146/annurev.arplant.57.032905.105159

Banerjee S, Schlaeppi K, van der Heijden MGA (2018) Keystone taxa as drivers of microbiome structure and functioning. Nat Rev Microbiol 16(9):567–576. https://doi.org/10.1038/s41579-018-0024-1

Barea JM (2015) Future challenges and perspectives for applying microbial biotechnology in sustainable agriculture based on a better understanding of plant-microbiome interactions. J Soil Sci Plant Nutr 15(2):261–282

Begum N, Qin C, Ahanger MA, Raza S, Khan MI, Ashraf M, Ahmed N, Zhang L (2019) Role of arbuscular mycorrhizal fungi in plant growth regulation: implications in abiotic stress tolerance. Front Plant Sci 10:1068

Benaissa A (2024) Rhizosphere: role of bacteria to manage plant diseases and sustainable agriculture—a review. J Basic Microbiol 64(3):2300361

Berendsen RL, Pieterse CMJ, Bakker PAHM (2012) The rhizosphere microbiome and plant health. Trends Plant Sci 17(8):478–486

Berg M, Koskella B (2018) Nutrient- and dose-dependent microbiome-mediated protection against a plant pathogen. Curr Biol 28(15):2487–2492.e3. https://doi.org/10.1016/j.cub.2018.05.085

Berg G, Rybakova D, Fischer D, Cernava T, Vergès M-CC, Charles T, Chen X, Cocolin L, Eversole K, Corral GH (2020) Microbiome definition re-visited: old concepts and new challenges. Microbiome 8(1):103

Biessy A, Filion M (2018) Phenazines in plant-beneficial pseudomonas spp.: biosynthesis, regulation, function and genomics. Environ Microbiol 20(11):3905–3917

Bilal S, Shahzad R, Imran M, Jan R, Kim KM, Lee I-J (2020) Synergistic association of endophytic fungi enhances Glycine max L. resilience to combined abiotic stresses: heavy metals, high temperature and drought stress. Ind Crop Prod 143:111931

Brader G, Compant S, Vescio K, Mitter B, Trognitz F, Ma L-J, Sessitsch A (2017) Ecology and genomic insights into plant-pathogenic and plant-nonpathogenic endophytes. Annu Rev Phytopathol 55(1):61–83. https://doi.org/10.1146/annurev-phyto-080516-035641

Burns KN, Kluepfel DA, Strauss SL, Bokulich NA, Cantu D, Steenwerth KL (2015) Vineyard soil bacterial diversity and composition revealed by 16S rRNA genes: differentiation by geographic features. Soil Biol Biochem 91:232–247. https://doi.org/10.1016/j.soilbio.2015.09.002

Carrión VJ, Cordovez V, Tyc O, Etalo DW, de Bruijn I, de Jager VCL, Medema MH, Eberl L, Raaijmakers JM (2018) Involvement of Burkholderiaceae and sulfurous volatiles in disease-suppressive soils. ISME J 12(9):2307–2321. https://doi.org/10.1038/s41396-018-0186-x

Chang W-T, Chen Y-C, Jao C-L (2007) Antifungal activity and enhancement of plant growth by Bacillus cereus grown on shellfish chitin wastes. Bioresour Technol 98(6):1224–1230. https://doi.org/10.1016/j.biortech.2006.05.005

Chang T-H, Cardona Gloria Y, Hellmann MJ, Richardo T, Greve CL, Le Roy D, Roger T, Bork F, Bugl S, Jakob J (2025) Transkingdom mechanism of MAMP generation by chitotriosidase feeds oligomeric chitin from fungal pathogens and allergens into TLR2-mediated innate immune sensing. Front Immunol 16:1497174

Chave M, Tchamitchian M, Ozier-Lafontaine H (2014) Agroecological engineering to biocontrol soil pests for crop health. In: Sustainable agriculture reviews 14: agroecology and global change. Springer, pp 269–297

Chen L, Liu Y (2024) The function of root exudates in the root colonization by beneficial soil Rhizobacteria. Biology 13(2):1–18. https://doi.org/10.3390/biology13020095

Comas LH, Becker SR, Cruz VMV, Byrne PF, Dierig DA (2013) Root traits contributing to plant productivity under drought. Front Plant Sci 4:442

Compant S, Reiter B, Sessitsch A, Nowak J, Clément C, Ait Barka E (2005) Endophytic colonization of *Vitis vinifera* L. by plant growth-promoting bacterium *Burkholderia* sp. strain PsJN. Appl Environ Microbiol 71(4):1685–1693. https://doi.org/10.1128/AEM.71.4.1685-1693.2005

Compant S, Samad A, Faist H, Sessitsch A (2019) A review on the plant microbiome: ecology, functions, and emerging trends in microbial application. J Adv Res 19:29–37

Cooper A, Ton J (2022) Immune priming in plants: from the onset to transgenerational maintenance. Essays Biochem 66(5):635–646

Correa-Galeote D, Bedmar EJ, Arone GJ (2018) Maize endophytic bacterial diversity as affected by soil cultivation history. Front Microbiol 9. https://doi.org/10.3389/fmicb.2018.00484

de Andrade LA, Santos CHB, Frezarin ET, Sales LR, Rigobelo EC (2023) Plant growth-promoting rhizobacteria for sustainable agricultural production. Microorganisms 11(4):1088

De Vries FT, Griffiths RI, Knight CG, Nicolitch O, Williams A (2020) 根际微生物群和抗旱作用. Science 368(6488):270–274

Dodds PN, Chen J, Outram MA (2024) Pathogen perception and signaling in plant immunity. Plant Cell 36(5):1465–1481

Donn S, Kirkegaard JA, Perera G, Richardson AE, Watt M (2015) Evolution of bacterial communities in the wheat crop rhizosphere. Environ Microbiol 17(3):610–621. https://doi.org/10.1111/1462-2920.12452

Doornbos RF, van Loon LC, Bakker PAHM (2012) Impact of root exudates and plant defense signaling on bacterial communities in the rhizosphere. A review. Agron Sustain Dev 32(1):227–243. https://doi.org/10.1007/s13593-011-0028-y

Du B, Haensch R, Alfarraj S, Rennenberg H (2024) Strategies of plants to overcome abiotic and biotic stresses. Biol Rev 99(4):1524–1536

Dunne C, Moënne-Loccoz Y, de Bruijn FJ, O'Gara F (2000) Overproduction of an inducible extracellular serine protease improves biological control of Pythium ultimum by Stenotrophomonas maltophilia strain W81. Microbiology 146(8):2069–2078

Durán P, Tortella G, Viscardi S, Barra PJ, Carrión VJ, de la Luz Mora M, Pozo MJ (2018) Microbial community composition in take-all suppressive soils. Front Microbiol 9. https://doi.org/10.3389/fmicb.2018.02198

Duxbury Z, Wu C, Ding P (2021) A comparative overview of the intracellular guardians of plants and animals: NLRs in innate immunity and beyond. Annu Rev Plant Biol 72(1):155–184

Edwards J, Johnson C, Santos-Medellín C, Lurie E, Podishetty NK, Bhatnagar S, Eisen JA, Sundaresan V (2015) Structure, variation, and assembly of the root-associated microbiomes of rice. Proc Natl Acad Sci 112(8). https://doi.org/10.1073/pnas.1414592112

Espinosa-Palomeque B, Jiménez-Pérez O, Ramírez-Gottfried RI, Preciado-Rangel P, Buendía-García A, Sifuentes GZ, Sariñana-Navarrete MA, Rivas-García T (2025) Biocontrol of Phytopathogens using plant growth promoting Rhizobacteria: bibliometric analysis and systematic review. Horticulturae 11(3):1–27. https://doi.org/10.3390/horticulturae11030271

Farhaoui A, Adadi A, Tahiri A, El Alami N, Khayi S, Mentag R, Ezrari S, Radouane N, Mokrini F, Belabess Z (2022) Biocontrol potential of plant growth-promoting rhizobacteria (PGPR) against Sclerotiorum rolfsii diseases on sugar beet (Beta vulgaris L.). Physiol Mol Plant Pathol 119:101829

Farhaoui A, El Alami N, Khadiri M, Ezrari S, Radouane N, Baala M, Tahiri A, Lahlali R (2023a) Biological control of diseases caused by Rhizoctonia solani AG-2-2 in sugar beet (Beta vulgaris L.) using plant growth-promoting rhizobacteria (PGPR). Physiol Mol Plant Pathol 124:101966

Farhaoui A, Tahiri A, Khadiri M, El Alami N, Lahlali R (2023b) Fungal root rots of sugar beets: a review of common causal agents and management strategies. Gesunde Pflanzen 75(5):1411–1440

Farhaoui A, Tahiri A, Khadiri M, El Alami N, Lahlali R (2024) Assessing management strategies for mitigating Rhizoctonia damping-off in sugar beet cultivation. Microbe 5:100164

Farhaoui A, Khadiri M, Taoussi M, Ezzouggari R, Benaissa O, Radi M, Belabess Z, Lahlali R (2025a) Biocontrol strategies: microbial defenders against plant pathogens. In: Microbial Allies, pp 205–236

Farhaoui A, Taoussi M, Laasli SE, Legrifi I, El Mazouni N, Meddich A, Hijri M, Lahlali R (2025b) Arbuscular mycorrhizal fungi and their role in plant disease control: a state-of-the-art. Microbe (Netherlands) 8. https://doi.org/10.1016/j.microb.2025.100438

Fierer N (2017) Embracing the unknown: disentangling the complexities of the soil microbiome. Nat Rev Microbiol 15(10):579–590. https://doi.org/10.1038/nrmicro.2017.87

Fitzpatrick CR, Copeland J, Wang PW, Guttman DS, Kotanen PM, Johnson MTJ (2018) Assembly and ecological function of the root microbiome across angiosperm plant species. Proc Natl Acad Sci 115(6). https://doi.org/10.1073/pnas.1717617115

Foo JL, Ling H, Lee YS, Chang MW (2017) Microbiome engineering: current applications and its future. Biotechnol J 12(3):1600099

Frankowski J, Lorito M, Scala F, Schmid R, Berg G, Bahl H (2001) Purification and properties of two chitinolytic enzymes of Serratia plymuthica HRO-C48. Arch Microbiol 176(6):421–426. https://doi.org/10.1007/s002030100347

Fridlender M, Inbar J, Chet I (1993) Biological control of soilborne plant pathogens by a β-1,3 glucanase-producing pseudomonas cepacia. Soil Biol Biochem 25(9):1211–1221. https://doi.org/10.1016/0038-0717(93)90217-Y

Gajera HP (2012) Antagonism of Trichoderma spp. against Macrophomina phaseolina: evaluation of coiling and Cell Wall degrading enzymatic activities. J Plant Pathol Microbiol 03(07). https://doi.org/10.4172/2157-7471.1000149

Garrido-Sanz D, Keel C (2025) Seed-borne bacteria drive wheat rhizosphere microbiome assembly via niche partitioning and facilitation. Nat Microbiol 10(5):1130–1144. https://doi.org/10.1038/s41564-025-01973-1

Gerritse J, Gottschal JC (1993) Oxic and anoxic growth of a new Citrobacter species on amino acids. Arch Microbiol 160:51–61

Gianinazzi-Pearson V, Gollotte A, Dumas-Gaudot E, Franken P, Gianinazzi S (1994) Gene expression and molecular modifications associated with plant responses to infection by arbuscular mycorrhizal fungi. In: Advances in molecular genetics of plant-microbe interactions: Vol. 3 proceedings of the 7th international symposium on molecular plant-microbe interactions, Edinburgh, UK, June 1994, pp 179–186

Glick BR (2012) Plant growth-promoting bacteria: mechanisms and applications. Scientifica 2012(1):963401
Glick BR (2014) Bacteria with ACC deaminase can promote plant growth and help to feed the world. Microbiol Res 169(1):30–39. https://doi.org/10.1016/j.micres.2013.09.009
Glick BR, Gamalero E (2021) Recent developments in the study of plant microbiomes. Microorganisms 9(7):1533
Gong Y, Tian L, Kontos I, Li J, Li X (2023) Plant immune signaling network mediated by helper NLRs. Curr Opin Plant Biol 73:102354
Gowtham HG, Hema P, Murali M, Shilpa N, Nataraj K, Basavaraj GL, Singh SB, Aiyaz M, Udayashankar AC, Amruthesh KN (2024) Fungal endophytes as Mitigators against biotic and abiotic stresses in crop plants. J Fungi 10(2):1–34. https://doi.org/10.3390/jof10020116
Grover M, Ali SZ, Sandhya V, Rasul A, Venkateswarlu B (2011) Role of microorganisms in adaptation of agriculture crops to abiotic stresses. World J Microbiol Biotechnol 27(5):1231–1240
Grover M, Bodhankar S, Sharma A, Sharma P, Singh J, Nain L (2021) PGPR mediated alterations in root traits: way toward sustainable crop production. Front Sustain Food Syst 4:618230
Gupta A, Kumar P (2023) Microbial oxidative enzymes: biotechnological applications. Walter de Gruyter GmbH & Co KG
Gutiérrez-Mañero FJ, Ramos-Solano B, Probanza A, Mehouachi JR, Tadeo F, Talon M (2001) The plant-growth-promoting rhizobacteria *Bacillus pumilus* and *Bacillus licheniformis* produce high amounts of physiologically active gibberellins. Physiol Plant 111(2):206–211. https://doi.org/10.1034/j.1399-3054.2001.1110211.x
Hanif MS, Tayyab M, Baillo EH, Islam MM, Islam W, Li X (2024) Plant microbiome technology for sustainable agriculture. Front Microbiol 15(November):1–10. https://doi.org/10.3389/fmicb.2024.1500260
Hardoim PR, van Overbeek LS, Berg G, Pirttilä AM, Compant S, Campisano A, Döring M, Sessitsch A (2015) The hidden world within plants: ecological and evolutionary considerations for defining functioning of microbial endophytes. Microbiol Mol Biol Rev 79(3):293–320. https://doi.org/10.1128/MMBR.00050-14
Harman GE, Howell CR, Viterbo A, Chet I, Lorito M (2004) Trichoderma species – opportunistic, avirulent plant symbionts. Nat Rev Microbiol 2(1):43–56. https://doi.org/10.1038/nrmicro797
Hartmann A, Rothballer M, Schmid M (2008) Lorenz Hiltner, a pioneer in rhizosphere microbial ecology and soil bacteriology research. Plant Soil 312(1–2):7–14. https://doi.org/10.1007/s11104-007-9514-z
Hartmann A, Schmid M, van Tuinen D, Berg G (2009) Plant-driven selection of microbes. Plant Soil 321(1–2):235–257. https://doi.org/10.1007/s11104-008-9814-y
Hermosa R, Viterbo A, Chet I, Monte E (2012) Plant-beneficial effects of Trichoderma and of its genes. Microbiology 158(1):17–25. https://doi.org/10.1099/mic.0.052274-0
Hu L, Robert CAM, Cadot S, Zhang X, Ye M, Li B, Manzo D, Chervet N, Steinger T, van der Heijden MGA, Schlaeppi K, Erb M (2018) Root exudate metabolites drive plant-soil feedbacks on growth and defense by shaping the rhizosphere microbiota. Nat Commun 9(1):2738. https://doi.org/10.1038/s41467-018-05122-7
Ibrahim HM, El-Sawah AM (2022) The mode of integration between azotobacter and rhizobium affect plant growth, yield, and physiological responses of pea (Pisum sativum L.). J Soil Sci Plant Nutr 22(2):1238–1251
Jones JDG, Staskawicz BJ, Dangl JL (2024) The plant immune system: from discovery to deployment. Cell 187(9):2095–2116
Kaul S, Choudhary M, Gupta S, Dhar MK (2021) Engineering host microbiome for crop improvement and sustainable agriculture. Front Microbiol 12:635917
Kawasaki A, Donn S, Ryan PR, Mathesius U, Devilla R, Jones A, Watt M (2016) Microbiome and exudates of the root and rhizosphere of Brachypodium distachyon, a model for wheat. PLoS One 11(10):e0164533. https://doi.org/10.1371/journal.pone.0164533
Khadiri M, Boubaker H, Askarne L, Farhaoui A, Taoussi M, Ait Haddou L, Cherrate M, Ouchari W, Zerrouk MH, Sobeh M (2024) Exploration of the antifungal potential of aqueous extracts

from two agave species against major postharvest diseases in apples. Postharvest Biol Technol 214:112992

Khasanova A, Edwards J, Bonnette J, Singer E, Haque T, Juenger TE (2023) Quantitative genetic-by-soil microbiome interactions in a perennial grass affect functional traits. Proc R Soc B Biol Sci 290(1991). https://doi.org/10.1098/rspb.2022.1350

Kim MG, Geng X, Lee SY, Mackey D (2009) The Pseudomonas syringae type III effector AvrRpm1 induces significant defenses by activating the Arabidopsis nucleotide-binding leucine-rich repeat protein RPS2. Plant J 57(4):645–653

Kourelis J, Adachi H (2022) Activation and regulation of NLR immune receptor networks. Plant Cell Physiol 63(10):1366–1377

Kumar P, Choudhury D (2025) Functional role and mechanism of PGPRs, AMFs and Endophytes in improving host resistance against plant pathogens. In: Plant microbiome and biological control: emerging trends and applications. Springer, pp 239–271

Lahlali R, Kouighat M, Baala M, Ighissane Y, Khadiri M, Ezzouggari R, El Jarroudi M, Farhaoui A (2025) Battlefronts in agriculture: understanding plant-pathogen interactions. In: Microbial Allies. Springer, pp 237–273

Leclère V, Béchet M, Adam A, Guez J-S, Wathelet B, Ongena M, Thonart P, Gancel F, Chollet-Imbert M, Jacques P (2005) Mycosubtilin overproduction by *Bacillus subtilis* BBG100 enhances the organism's antagonistic and biocontrol activities. Appl Environ Microbiol 71(8):4577–4584. https://doi.org/10.1128/AEM.71.8.4577-4584.2005

Lemanceau P, Barret M, Mazurier S, Mondy S, Pivato B, Fort T, Vacher C (2017) Chapter 5: Plant communication with associated Microbiota in the Spermosphere, rhizosphere and Phyllosphere. In: Becard BR (ed) How plants communicate with their biotic environment, vol 82. Academic, pp 101–133. https://doi.org/10.1016/bs.abr.2016.10.007

Li J, Liu W, Luo L, Dong D, Liu T, Zhang T, Lu C, Liu D, Zhang D, Wu H (2015) Expression of Paenibacillus polymyxa β-1,3-1,4-glucanase in Streptomyces lydicus A01 improves its biocontrol effect against Botrytis cinerea. Biol Control 90:141–147. https://doi.org/10.1016/j.biocontrol.2015.06.008

Liu Y, Zuo S, Xu L, Zou Y, Song W (2012) Study on diversity of endophytic bacterial communities in seeds of hybrid maize and their parental lines. Arch Microbiol 194(12):1001–1012. https://doi.org/10.1007/s00203-012-0836-8

Liu H, Brettell LE, Qiu Z, Singh BK (2020) Microbiome-mediated stress resistance in plants. Trends Plant Sci 25(8):733–743

Liu K, Deng F, Zeng F, Chen Z-H, Qin Y, Chen G (2025) Plant growth-promoting rhizobacteria improve drought tolerance of crops: a review. Plant Growth Regul:1–15. https://doi.org/10.3844/ajabssp.2011.185.188

Loper JE, Henkels MD (1999) Utilization of heterologous Siderophores enhances levels of iron available to *Pseudomonas putida* in the rhizosphere. Appl Environ Microbiol 65(12):5357–5363. https://doi.org/10.1128/AEM.65.12.5357-5363.1999

Loper J, Nowak-Thompson B, Whistler C, Hagen M, Corbell N, Henkels M, Stockwell V (1997) Biological control mediated by antifungal metabolite production and resource competition: an overview. In: International plant growth promoting rhizobacteria workshop

Lorito M, Woo SL, D'ambrosio M, Harman GE, Hayes CK, Kubicek CP, Scala F (1996) Synergistic interaction between cell wall degrading enzymes and membrane affecting compounds. MPMI-Mol Plant Microbe Interact 9(3):206–213

Lu Z, Tombolini R, Woo S, Zeilinger S, Lorito M, Jansson JK (2004) In vivo study of *Trichoderma* -pathogen-plant interactions, using constitutive and inducible Green fluorescent protein reporter systems. Appl Environ Microbiol 70(5):3073–3081. https://doi.org/10.1128/AEM.70.5.3073-3081.2004

Maheshwari DK, Dheeman S, Agarwal M (2015) Phytohormone-producing PGPR for sustainable agriculture. In: Bacterial metabolites in sustainable agroecosystem. Springer, pp 159–182

Majhi P, Pradhan U, Toppo A, Shukla AK (2025) Fungal endophytes: An insight into diversity, stress tolerance, biocontrol and plant growth-promoting potentials. Curr Microbiol 82(6):283

Mansfield J, Genin S, Magori S, Citovsky V, Sriariyanum M, Ronald P, Dow M, Verdier V, Beer SV, Machado MA, Toth I, Salmond G, Foster GD (2012) Top 10 plant pathogenic bacteria in molecular plant pathology. Mol Plant Pathol 13(6):614–629. https://doi.org/10.1111/j.1364-3703.2012.00804.x

Manzoni C, Kia DA, Vandrovcova J, Hardy J, Wood NW, Lewis PA, Ferrari R (2018) Genome, transcriptome and proteome: the rise of omics data and their integration in biomedical sciences. Brief Bioinform 19(2):286–302

Microbiota P, Carrión VJ, Perez-jaramillo J, Cordovez V, Tracanna V, Gomez-exposito R, Elsayed SS, Mohanraju P, Arifah A (2019) Pathogen-induced activation of disease-suppressive functions in the endophytic root microbiome. Science 612(November):606–612

Mitra D, Pellegrini M, Guerra-Sierra BE (2024) Interaction between plants and growth-promoting rhizobacteria (PGPR) for sustainable development. In: Bacteria, vol 3, Issue 3. MDPI, pp 136–140

Moe LA (2013) Amino acids in the rhizosphere: from plants to microbes. Am J Bot 100(9):1692–1705. https://doi.org/10.3732/ajb.1300033

Muhammad M, Ahmad MW, Basit A, Ullah S, Mohamed HI, Nisar N, Khan A (2024) Plant growth-promoting rhizobacteria and their applications and role in the management of soilborne diseases. In: Bacterial secondary metabolites. Elsevier, pp 59–82

Munir N, Hanif M, Abideen Z, Sohail M, El-Keblawy A, Radicetti E, Mancinelli R, Haider G (2022) Mechanisms and strategies of plant microbiome interactions to mitigate abiotic stresses. Agronomy 12(9):2069

Myers RJ Jr, Fichman Y, Zandalinas SI, Mittler R (2023) Jasmonic acid and salicylic acid modulate systemic reactive oxygen species signaling during stress responses. Plant Physiol 191(2):862–873

Nabi Z, Manzoor S, Nabi SU, Wani TA, Gulzar H, Farooq M, Arya VM, Baloch FS, Vlădulescu C, Popescu SM (2024) Pattern-Triggered Immunity and Effector-Triggered Immunity: crosstalk and cooperation of PRR and NLR-mediated plant defense pathways during host–pathogen interactions. Physiol Mol Biol Plants 30(4):587–604

Narsing Rao MP, Lohmaneeratana K, Bunyoo C, Thamchaipenet A (2022) Actinobacteria–plant interactions in alleviating abiotic stress. Plants 11(21):2976

Noman M, Ahmed T, Ijaz U, Shahid M, Azizullah, Li D, Manzoor I, Song F (2021) Plant–microbiome crosstalk: dawning from composition and assembly of microbial community to improvement of disease resilience in plants. Int J Mol Sci 22(13):6852

Olanrewaju OS, Glick BR, Babalola OO (2017) Mechanisms of action of plant growth promoting bacteria. World J Microbiol Biotechnol 33(11):197

Ongena M, Jacques P, Touré Y, Destain J, Jabrane A, Thonart P (2005) Involvement of fengycin-type lipopeptides in the multifaceted biocontrol potential of Bacillus subtilis. Appl Microbiol Biotechnol 69(1):29–38. https://doi.org/10.1007/s00253-005-1940-3

Orozco-Mosqueda MDC, Santoyo G, Glick BR (2023) Recent advances in the bacterial phytohormone modulation of plant growth. Plants 12(3):606

Panicker S, Sayyed RZ (2022) Hydrolytic enzymes from PGPR against plant fungal pathogens. In: Antifungal metabolites of Rhizobacteria for sustainable agriculture. Springer, pp 211–238

Parniske M (2008) Arbuscular mycorrhiza: the mother of plant root endosymbioses. Nat Rev Microbiol 6(10):763–775. https://doi.org/10.1038/nrmicro1987

Pieterse CMJ, Zamioudis C, Berendsen RL, Weller DM, Van Wees SCM, Bakker PAHM (2014) Induced systemic resistance by beneficial microbes. Annu Rev Phytopathol 52(1):347–375

Priyadharsini P, Rojamala K, Ravi RK, Muthuraja R, Nagaraj K, Muthukumar T (2016) Mycorrhizosphere: the extended rhizosphere and its significance. In: Choudhary DK, Varma A, Tuteja N (eds) Plant-microbe interaction: An approach to sustainable agriculture. Springer Singapore, pp 97–124. https://doi.org/10.1007/978-981-10-2854-0_5

Raaijmakers JM, Paulitz TC, Steinberg C, Alabouvette C, Moënne-Loccoz Y (2009) The rhizosphere: a playground and battlefield for soilborne pathogens and beneficial microorganisms. Springer

Rascovan N, Carbonetto B, Perrig D, Díaz M, Canciani W, Abalo M, Alloati J, González-Anta G, Vazquez MP (2016) Integrated analysis of root microbiomes of soybean and wheat from agricultural fields. Sci Rep 6(1):28084. https://doi.org/10.1038/srep28084
Rashmi J, Manonmani V, Sundaralingam K, Vanitha S, Gnanachitra M, Kalaiselvi T, Sk Asraful Ali (2025) The seed microbiome: microbial hashes for plant wellbeing. Open J Environ Biol 10(1):007–022. https://doi.org/10.17352/ojeb.000046
Reddy PP, Rao MS, Nagesh M (1996) Management of citrus nematode, Tylenchulus semipenetrans, by integration of Trichoderma harzianum with oil cakes. Nematol Mediterr 24(2):265–267
Robert CAM, Himmighofen P, McLaughlin S, Cofer TM, Khan SA, Siffert A, Sasse J (2025) Environmental and biological drivers of root exudation. Annu Rev Plant Biol 76(1):317–339. https://doi.org/10.1146/annurev-arplant-083123-082752
Romão IR, do Carmo Gomes J, Silva D, Vilchez JI (2025) The seed microbiota from an application perspective: an underexplored frontier in plant–microbe interactions. Crop Health 3(1). https://doi.org/10.1007/s44297-025-00051-6
Rotblat B, Enshell-Seijffers D, Gershoni JM, Schuster S, Avni A (2002) Identification of an essential component of the elicitation active site of the EIX protein elicitor. Plant J 32(6):1049–1055. https://doi.org/10.1046/j.1365-313X.2002.01490.x
Rudrappa T, Czymmek KJ, Paré PW, Bais HP (2008) Root-secreted malic acid recruits beneficial soil bacteria. Plant Physiol 148(3):1547–1556. https://doi.org/10.1104/pp.108.127613
Saeid A, Prochownik E, Dobrowolska-Iwanek J (2018) Phosphorus solubilization by bacillus species. Molecules 23(11):2897
Saifullah S, Thomas BJ (1996) Studies on the parasitism of Globodera rostochiensis by Trichoderma harzianum using low temperature scanning electron microscopy. Pak J Bot 46(1):357–361
Sasse J, Martinoia E, Northen T (2018) Feed your friends: do plant exudates shape the root microbiome? Trends Plant Sci 23(1):25–41. https://doi.org/10.1016/j.tplants.2017.09.003
Santos MS, Nogueira MA, Hungria M (2021) Outstanding impact of Azospirillum brasilense strains Ab-V5 and Ab-V6 on the Brazilian agriculture: lessons that farmers are receptive to adopt new microbial inoculants. Rev Bras Ciênc Solo 45:e0200128
Santos-Medellín C, Edwards J, Liechty Z, Nguyen B, Sundaresan V (2017) Drought stress results in a compartment-specific restructuring of. MBio 8:e00764-17, 1–15
Sarsaiya S, Jain A, Singh R, Gong Q, Wu Q, Chen J, Shi J (2025) Unveiling the rhizosphere microbiome of dendrobium: mechanisms, microbial interactions, and implications for sustainable agriculture. Front Microbiol 16(January):1–18. https://doi.org/10.3389/fmicb.2025.1531900
Shade A, Jacques MA, Barret M (2017) Ecological patterns of seed microbiome diversity, transmission, and assembly. Curr Opin Microbiol 37:15–22. https://doi.org/10.1016/j.mib.2017.03.010
Schlatter D, Kinkel L, Thomashow L, Weller D, Paulitz T (2017) Disease suppressive soils: new insights from the soil microbiome. Phytopathology 107(11):1284–1297. https://doi.org/10.1094/PHYTO-03-17-0111-RVW
Sekar S, Kandavel D (2010) Interaction of plant growth promoting rhizobacteria (PGPR) and endophytes with medicinal plants–new avenues for phytochemicals. J Phytology 2(7):91–100
Sharon E, Chet I, Viterbo A, Bar-Eyal M, Nagan H, Samuels GJ, Spiegel Y (2007) Parasitism of Trichoderma on Meloidogyne javanica and role of the gelatinous matrix. Eur J Plant Pathol 118(3):247–258. https://doi.org/10.1007/s10658-007-9140-x
Shi M, Chen L, Wang X-W, Zhang T, Zhao P-B, Song X-Y, Sun C-Y, Chen X-L, Zhou B-C, Zhang Y-Z (2012) Antimicrobial peptaibols from Trichoderma pseudokoningii induce programmed cell death in plant fungal pathogens. Microbiology 158(1):166–175. https://doi.org/10.1099/mic.0.052670-0
Singh RP, Jha PN (2016) The multifarious PGPR Serratia marcescens CDP-13 augments induced systemic resistance and enhanced salinity tolerance of wheat (Triticum aestivum L.). PLoS One 11(6):e0155026. https://doi.org/10.1371/journal.pone.0155026
Smith SE, Read DJ (2010) Mycorrhizal symbiosis. Academic press
Song X, Zhang Y, Wang Y (2011) Antimicrobial peptides peptaibols from trichoderma–a review. Wei Sheng Wu Xue Bao/Acta Microbiol Sin 51(4):438–444

Spiegel Y, Sharon E, Bar-Eyal M, Maghodia A, Vanachter A, van Assche A, van Kerckhove S, Viterbo A, Chet I (2007) Evaluation and mode of action of Trichoderma isolates as biocontrol agents against plant-parasitic nematodes. Bull OILBSROP 30(2):129–133

Toumatia O, Compant S, Yekkour A, Goudjal Y, Sabaou N, Mathieu F, Sessitsch A, Zitouni A (2016) Biocontrol and plant growth promoting properties of Streptomyces mutabilis strain IA1 isolated from a Saharan soil on wheat seedlings and visualization of its niches of colonization. S Afr J Bot 105:234–239. https://doi.org/10.1016/j.sajb.2016.03.020

Trivedi P, Anderson IC, Singh BK (2013) Microbial modulators of soil carbon storage: integrating genomic and metabolic knowledge for global prediction. Trends Microbiol 21(12):641–651. https://doi.org/10.1016/j.tim.2013.09.005

Trivedi P, Delgado-Baquerizo M, Trivedi C, Hamonts K, Anderson IC, Singh BK (2017) Keystone microbial taxa regulate the invasion of a fungal pathogen in agro-ecosystems. Soil Biol Biochem 111:10–14. https://doi.org/10.1016/j.soilbio.2017.03.013

Trivedi P, Leach JE, Tringe SG, Sa T, Singh BK (2020) Plant–microbiome interactions: from community assembly to plant health. Nat Rev Microbiol 18(11):607–621

Trivedi P, Batista BD, Bazany KE, Singh BK (2022) Plant–microbiome interactions under a changing world: responses, consequences and perspectives. New Phytol 234(6):1951–1959. https://doi.org/10.1111/nph.18016

Valencia GB, Vargas VH, Soto JNU, Jimenez NN, Corral JH (2011) Trichoderma sp native from chili region of Poanas, Durango, Mexico antagonist against phytopathogen fungi. Am J Agric Biol Sci:185–188

Vannier N, Mony C, Bittebiere AK, Michon-Coudouel S, Biget M, Vandenkoornhuyse P (2018) A microorganisms' journey between plant generations. Microbiome 6(1):79. https://doi.org/10.1186/s40168-018-0459-7

Verma M, Brar SK, Tyagi RD, Surampalli RY, Valéro JR (2007) Starch industry wastewater as a substrate for antagonist, Trichoderma viride production. Bioresour Technol 98(11):2154–2162. https://doi.org/10.1016/j.biortech.2006.08.032

Verma PP, Shelake RM, Das S, Sharma P, Kim J-Y (2019) Plant growth-promoting rhizobacteria (PGPR) and fungi (PGPF): potential biological control agents of diseases and pests. In: Microbial interventions in agriculture and environment: volume 1: research trends, priorities and prospects. Springer, pp 281–311

Verma JP, Macdonald C, Gupta VK, Podile AR (2020) New and future developments in microbial biotechnology and bioengineering: phytomicrobiome for sustainable agriculture. Elsevier

Wakelin S, Young S, Gerard E, Mander C, O'Callaghan M (2017) Isolation of root-associated *Pseudomonas* and *Burkholderia* spp. with biocontrol and plant growth-promoting traits. Biocontrol Sci Tech 27(1):139–143. https://doi.org/10.1080/09583157.2016.1248899

Wang C, Luan S (2024) Calcium homeostasis and signaling in plant immunity. Curr Opin Plant Biol 77:102485

Wankhade A, Wilkinson E, Britt DW, Kaundal A (2025) A review of plant–microbe interactions in the rhizosphere and the role of root exudates in microbiome engineering. Appl Sci 15(13):7127

Weston LA, Mathesius U (2013) Flavonoids: their structure, biosynthesis and role in the rhizosphere, including allelopathy. J Chem Ecol 39(2):283–297. https://doi.org/10.1007/s10886-013-0248-5

Xu L, Naylor D, Dong Z, Simmons T, Pierroz G, Hixson KK, Kim YM, Zink EM, Engbrecht KM, Wang Y, Gao C, DeGraaf S, Madera MA, Sievert JA, Hollingsworth J, Birdseye D, Scheller HV, Hutmacher R, Dahlberg J et al (2018) Drought delays development of the sorghum root microbiome and enriches for monoderm bacteria. Proc Natl Acad Sci USA 115(18):E4284–E4293. https://doi.org/10.1073/pnas.1717308115

Yang C-X, Chen S-J, Hong X-Y, Wang L-Z, Wu H-M, Tang Y-Y, Gao Y-Y, Hao G-F (2025) Plant exudates-driven microbiome recruitment and assembly facilitates plant health management. FEMS Microbiol Rev 49:fuaf008

Yu T-Y, Sun M-K, Liang L-K (2021) Receptors in the induction of the plant innate immunity. Mol Plant-Microbe Interact 34(6):587–601

Yu Y, Gui Y, Li Z, Jiang C, Guo J, Niu D (2022) Induced systemic resistance for improving plant immunity by beneficial microbes. Plants 11(3):386

Yu X, Niu H, Liu C, Wang H, Yin W, Xia X (2024) PTI-ETI synergistic signal mechanisms in plant immunity. Plant Biotechnol J 22(8):2113–2128

Yuan Y, Zu M, Sun L, Zuo J, Tao J (2022) Isolation and screening of 1-aminocyclopropane-1-carboxylic acid (ACC) deaminase producing PGPR from Paeonia lactiflora rhizosphere and enhancement of plant growth. Sci Hortic 297:110956

Zeilinger S, Galhaup C, Payer K, Woo SL, Mach RL, Fekete C, Lorito M, Kubicek CP (1999) Chitinase gene expression during Mycoparasitic interaction of Trichoderma harzianum with its host. Fungal Genet Biol 26(2):131–140. https://doi.org/10.1006/fgbi.1998.1111

Zeng Q, Hu HW, Ge AH, Xiong C, Zhai CC, Duan GL, Han LL, Huang SY, Zhang LM (2025) Plant–microbiome interactions and their impacts on plant adaptation to climate change. J Integr Plant Biol 67(3):826–844. https://doi.org/10.1111/jipb.13863

Zhalnina K, Louie KB, Hao Z, Mansoori N, da Rocha UN, Shi S, Cho H, Karaoz U, Loqué D, Bowen BP, Firestone MK, Northen TR, Brodie EL (2018) Dynamic root exudate chemistry and microbial substrate preferences drive patterns in rhizosphere microbial community assembly. Nat Microbiol 3(4):470–480. https://doi.org/10.1038/s41564-018-0129-3

Zhang S, Gan Y, Xu B (2016) Application of plant-growth-promoting fungi Trichoderma longibrachiatum T6 enhances tolerance of wheat to salt stress through improvement of Antioxidative defense system and gene expression. Front Plant Sci 07. https://doi.org/10.3389/fpls.2016.01405

Chapter 2
Plant-Microbiome Nanotechnology and Plant Health

Kamel A. Abd-Elsalam, Rawan K. Hassan, Toka E. Abdelkhalek, Heba Yousef, and Farah K. Ahmed

Abstract The union of plant microbiome science and nanotechnology offers a revolutionary path to sustainable agriculture and enhanced plant health. This chapter addresses the diversity, structure, and role of plant-associated microbial communities' rhizosphere, phyllosphere, and endosphere microbiota and highlights their crucial roles in plant growth, disease resistance, and ecological adaptation. Functional benefits of the microbiome are also enhanced by advances in multi-omics technologies, through which microbial host interactions and symbiotic processes can be viewed in a holistic manner. At the same time, nanotechnology is revolutionizing modern agriculture with the enhancement of nutrient supply, stress tolerance, and microbial community structure within the soil. This chapter further explores the influence of engineered nanomaterials (NMs) on microbial processes, plant metabolism, and soil characteristics in favor of improved crop yield. The combination of nanomaterials and useful microbes has the potential to bring about precision agriculture with the help of artificial intelligence (AI) and Internet of Things (IoT) platforms for real-time control and monitoring. Finally, the chapter illustrates the promise of solutions based on microbiome and nano-facilitated solutions to address the world's agrifood problems food security, climate stress, and degraded soil health, making them important tools for achieving resilient and sustainable agrifood systems in the One Health framework.

Keywords Plant microbiome · Rhizosphere · Nanotechnology · Engineered nanomaterials (ENMs) · Plant · microbe interaction · Sustainable agriculture · Microbial diversity · Soil health · Crop yield · Stress tolerance · Precision agriculture · Microbiome · nanotechnology synergy · Food security

K. A. Abd-Elsalam (✉) · H. Yousef
Plant Pathology Research Institute, Agricultural Research Center, Giza, Egypt

R. K. Hassan · T. E. Abdelkhalek
Biotechnology English Program, Faculty of Agriculture, Cairo University, Giza, Egypt

F. K. Ahmed
Nanobiotechnology Department, Faculty of Postgraduate Studies for Nanotechnology, Cairo University, El-Sheikh Zayed, Egypt

K. A. Abd-Elsalam (ed.), *Plant-Microbiome Nanotechnology*, Sustainability Sciences in Asia and Africa, https://doi.org/10.1007/978-981-95-5056-2_2

2.1 Introduction

Agriculture is confronted with numerous global challenges, from climate change, environmental degradation, soil infertility, pest outbreaks, and an increasing human population demanding increased food production. These stresses threaten the sustainability and resilience of modern agricultural systems (Hussain et al. 2023). The traditional chemical fertilizer and pesticide-based systems have not only led to environmental problems such as soil erosion and biodiversity loss but also induced the development of pathogen resistance (Ahmed et al. 2023). Against this backdrop, there is a growing sense of the need for new, environmentally friendly solutions to plant productivity and health with soil and environmental sustainability (Khan 2020). One of the most promising solutions to emerge in recent years is the identification and manipulation of the plant microbiome, the diverse community of microorganisms (bacteria, fungi, archaea, and viruses) that inhabit various plant compartments like the rhizosphere (soil immediate to the roots), phyllosphere (leaf surface), and endosphere (internal tissues) (Khan et al. 2021). These microbial assemblages form a functional unit with the plant, the above-mentioned plant holobiont, and are involved in nutrient acquisition, stress tolerance, pathogen resistance, and overall growth and development (Rajput et al. 2022). By using advanced omics technologies (metagenomics, transcriptomics, proteomics, metabolomics), researchers have unearthed the central roles of microbial taxa such as Proteobacteria, Actinobacteria, Firmicutes, and Bacteroidetes in maintaining plant health by nitrogen fixation, phosphate solubilization, and phytohormone production (Bloch et al. 2021). In the meantime, nanotechnology has also emerged as a revolutionary technology in agriculture by enabling precision targeting at the molecular level to enhance plant-microbe-soil interactions (Haider et al. 2023). Nanomaterials (NMs), with their tiny size and extensive surface area, have the potential to influence soil physicochemical properties, manage microbial populations, and induce controlled release of nutrients and agrochemicals. Engineered NMs such as zinc oxide (ZnO), titanium dioxide (TiO_2), and carbon nanotubes have been shown to be potential for the promotion of crop growth, nutrient uptake, and suppression of phytopathogens (Ghosh and Webster 2021). Moreover, nanotechnology is capable of balancing and activating the soil microbiome and hence long-term soil fertility and crop tolerance can be attained (Ali et al. 2023). The intersection of plant microbiome science and nanotechnology opens up new avenues for sustainable agriculture by the potential to develop smart, synergistic systems that maximize microbial activities and plant responses (Verma et al. 2024). Their combination with artificial intelligence (AI), Internet of Things (IoT), and other digital technologies further allows real-time monitoring, precision agriculture, and data-driven decision-making (Nawaz et al. 2024). This chapter discusses the plant microbiome composition and function, how nanotechnology affects plant and microbial response, and how synergy between the technologies can be used to create a new green revolution. By presenting current developments, applications, and perspectives, it aims to highlight how the intersection of nanotechnology and microbiomes will enable the solution of the world's current agriculture-related challenges and realization of resilient, productive, and sustainable food systems.

2.2 Plant Microbiomes: Diversity and Function

Recent research studies have opened up the question of the complexity of microbes that are associated with different plant organs that are collectively termed the plant holobiont. This includes the microbiota (microorganisms) and microbiome (collective genomes) of the rhizosphere, phyllosphere, and endosphere, which are vital to plant growth and health as well as resilience (Compant et al. 2019; Chauhan et al. 2023). Due to increasing environmental pressures and agricultural needs, such as carbon neutrality and sustainable food production (Brader et al. 2017), microalgae are becoming more and more popular as biofertilizers and biopesticides, as they can provide an ecological alternative to using chemical fertilizers. Specifically, bacteria are of interest, as they hold practical possibilities and business interest (Reinhold-Hurek et al. 2015). As an example, wheat roots enhance changes in the bacterial community of rhizospheres, in which actinobacteria and pseudomonads populations are higher than in bulk soil (Lemanceau et al. 2017). Such communities transform as well (Hacquard 2016). Bacterial endophytes enter into the internal tissues of roots, by penetrating the surface either passively or actively, and their action depends on both plant properties and microbial characteristics (Dastogeer et al. 2022; Chauhan et al. 2023). In plants, roots can host a variety of phyla, and that is the case with roots of the grapevine, which harbors bacteria such as Proteobacteria, Actinobacteria, and Firmicutes, to mention a few (Vandenkoornhuyse et al. 2015), indicating that the microbiome is diverse and plays an important role in plant health.

2.2.1 Structure and Composition of the Plant Microbiome

Farming tendencies such as monoculture and phytosanitary-based farming have a tremendous effect on plant and soil microbiomes. Continuous monocropping can not only shift microbial communities and decrease beneficial biocontrol bacteria in favor of pathogenic microbes but also change the properties of the soil (Shu and Huang 2022; Schmid et al. 2020). The drying process, which is used during the postharvest treatment of *Glycyrrhiza uralensis* (licorice), modulates the endophytic bacterial and metabolite abundances, and the microbial and product quality are more preserved when fresh processing is used (Power et al. 2018). In general, the plant microbiome is determined by the complicated interplay of plant species, microbes, soil composition, climate, and cultivation methods (Fig. 2.1) requiring the combined crop management implementation.

Species of host plants play a key role in determining their related microbiome. Significant differences in microbial community exist within neighboring plant

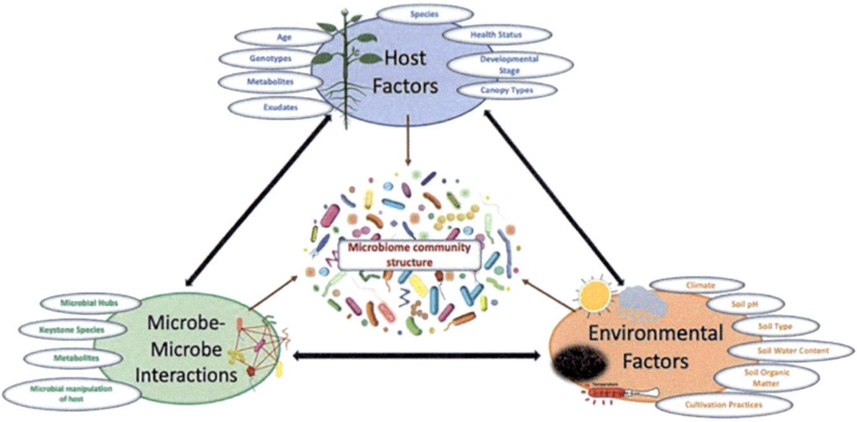

Fig. 2.1 Various factors that shape microbiome community structures (Dastogeer et al. 2020)

species under the same conditions (Berg et al. 2016). In comparative studies, different crops such as wheat, sorghum, and maize have been found to have other root microbiome compositions, together with grapevine and weeds (Samad et al. 2017). The presence of higher microbial variation is seen in phylogenetically distant plants, which suggests that there is a substantial influence of plant phylogeny on the root microbiome (Berendsen et al. 2012). Plants have undergone certain characteristic changes, which control microbial recruitment, mostly in the rhizosphere and endosphere (Bulgarelli et al. 2013). The microbiota that exist in the rhizosphere are mostly shaped by soil type, whereas endophyte composition is more defined by plant species (Xiao et al. 2017a). These differences are caused by the difference in the level of resources used by plants and also the characteristics of plants like leaf permeability, topography of the surface, root exudates, and immune responses (Rich et al. 2017). Root exudates have an effect on root metabolism and root permeability, and the microbes can take up these exudates and produce transcripts that have effects on plant signaling and metabolism (Bodenhausen et al. 2013). Moreover, the release of organic matter and changes in cellular permeability, caused by the synthesis of antibiotics, e.g., *penicillin* and *polymyxin*, by some microbes, also prove the dynamic reciprocal nature of plant-microbe interactions (Dastogeer et al. 2018).

2.2.2 Functions and Benefits of the Plant Microbiome

The last 10 years have seen increasing activity in plant microbiomes in natural as well as agricultural systems. Scientists consider a microbial community organization, its formation, evolution, and functioning (Venturi and Keel 2016). This has been aided by the innovations of high-throughput omics technologies which include metagenomics, metatranscriptomics, and metaproteomics and computational tools and bioinformatics (Yang et al. 2017). These technologies have led to improved insights about the plant and microbiome interactions, especially about the functional roles microbes play in stimulating plant growth and defense. As an example, integration of several omics methods has shown that microbial genes are transactivated in root zones, in favor of their fundamental role in the rhizosphere technology (Compant et al. 2019). The creation of synthetic microbial communities (SynComs) has also shown how certain microbes can increase plant adaptability and productivity (Sasse et al. 2018). The combination of transcriptomics and proteomics concerning genomes enables researchers to identify crucial gene clusters and metabolic pathways related to microbial symbiosis and plant health.

PGPRs, including *Bacillus subtilis*, *B. amyloliquefaciens*, and *Pseudomonas stutzer*i, were found to promote growth by effectively colonizing roots and inhibiting the pathogens *Phytophthora capsici* and *Fusarium oxysporum* (Yaish and Kumar 2015). Induced systemic resistance (ISR) can also be stimulated by PGPR, which triggers ethylene and jasmonic acid pathways that enhance the resilience of plants (Vishwakarma et al. 2020). Moreover, in a similar vein, mycorrhizal fungi are crucial because they enhance root growth, alter the shape of roots, or strengthen the resistance to soil pathogens (Chauhan et al. 2023; Muhammad et al. 2024). All these microbes make up a phytomicrobiome, which is a dynamic factor of the plant holobiont, and the relationships among them are mutualistic to antagonistic (Hakim et al. 2021). On the whole, combining microbiome science with omics and microbial engineering has potential uses in the framework of sustainable agriculture and protection of crops.

2.3 Nanotechnology: Applications in Agriculture

Nanotechnology, which deals with the handling of materials on an atomic and molecular level, has become a revolution in the agricultural field. It has massive potential to increase crop yields, reduce environmental degradation, and improve food security (Jatav et al. 2020; Lal 2016). Nanotechnology is converting the paradigm of plant health and productivity management by the creation of nutrient absorption in stressful circumstances, early disease recognition, or disease-resistant crops (Chung et al. 2015; Yasmin et al. 2021). The first strength of nanotechnology is that it is used to preserve and promote the health of the soil microbiome. Soil microbes could contribute to fertility, but plants can pick on beneficial microbes to grow (Rajput et al. 2021). These interactions can be enhanced by the use of nanomaterials (NMs) and can help nutrient cycling or soil structure. Nanotechnology in enhancing soil health includes the advantages of expediting biochemical processes and microbial actions, from which soil organic carbon is considered a critical constituent (Alamri et al. 2022; Kumari et al. 2023).

Although microbial biomass is initially affected by being temporarily inhibited by other materials, even graphene oxide, nanotechnology does not have any long-term effects that affect the biomass negatively (Farhana et al. 2022). As soil degrades further due to the use of intensive agriculture, nanotechnology presents some sustainable solutions by way of minimizing input waste and increasing resource utilization (Hossain et al. 2024). Moreover, nanotechnology could also assist in the mitigation of such soil stress conditions as salinity, acidity, and poor aeration, resulting in improved root-microbe interactions and phytostimulation (Sabra et al. 2025; Fotoukian et al. 2024). NMs can be used in combination with other measures that promote soil health, leading to improved soil biodiversity and developing cultivation systems that are resilient to climate change, thereby providing long-term yield stability. The chapter examines the important role of nanotechnology in crop improvement, rehabilitation of the soil, and creation of a greener agricultural future.

2.3.1 Types of Nanomaterials Used in Plant Microbiome Applications

Plant microbiome refers to a collection of all microorganisms present on the various plant surfaces, such as the phyllosphere (leaves), rhizoplane (root), rhizosphere (soil exposed to root exudates), and endosphere (plant tissue interior) (Upadhyay et al. 2022). This complicated microbiome is determined by a lot of parameters, including the type of plants, their health, their age, their secretions, the environment, and various physicochemical features of the soil (Compant et al. 2019). It is quite remarkable that plants are capable of actively choosing microbes, in particular, in the rhizosphere, and of changing microbial composition in various compartments even under similar conditions (Gupta et al. 2021). Microbes are very important in nutrient fixation, nutrient solubilization, supply of micronutrients, stress mitigation, and defense against diseases (Mariyam et al. 2023). In the acquisition of green agriculture practices, deciphering and modulating soil microbiomes is crucial as the climate change episodes worsen. Microbes in the soil help control biogeochemical cycles of macro- and micronutrients and can diminish the dependency on pesticides, facilitating the natural suppression of diseases (Khan et al. 2021). The most significant of them are rhizobia, mycorrhizal, and phosphate-solubilizing bacteria such as *Pseudomonas* and *Pantoea* that can transform the unavailable phosphate to a plant-available form. Most of them also make plant growth regulators such as IAA and siderophores to mobilize micronutrients (Rakgotho et al. 2022).

Engineered nanomaterials (NMs) improve the health of the soil by regulating physicochemical and biological characteristics. These parameters affect microbial associations, nutrient release, and pH of the soil. As an illustration, NMs based on Cu, TiO_2, and Fe_3O_4 induce root exudation and activate nutrients in salty and alkaline landscapes and maintain pH efficiency in acidic landscapes (Badawy et al. 2021). NMs also activate phytohormone production, gene expression, antioxidant activity, and secondary metabolite synthesis. CuO and ZnO NMs stimulate the generation of antioxidants in *Glycyrrhiza glabra*, whereas TiO_2 NMs promote nutrient uptake in rice. Additionally, to solubilize Fe_2O_3 and facilitate its uptake by plants, NMs regulate the production of microbial siderophores (Pallavi et al. 2023).

2.3.2 Advances in Nanotechnology for Plant Growth Promotion

Enhancing plant growth and productivity can address the issue and prevent such a crisis in the future. A diverse population of microorganisms present in soil, which directly or indirectly interacts with plants and its hosts (fungi, archaebacteria, viruses, nematodes, and picoeukaryotes) collectively presents itself as a network system of rhizosphere biology (Meel and Saharan 2024). The rhizosphere is an active partner in the transformation of organic and inorganic materials and make them available as nutritional elements for plant development. Thus, a study of rhizospheric science and its interface with nanotechnology, metagenomics, and allied omics methods would be beneficial to enhance the ecosystem health and plant yield in the terrestrial biosphere (Kumar et al. 2019). For gaining better crop productivity and food security, advanced nanotechnology and multi-omics approaches tools are the guarantee for ensuring sustainability. Plant development, growth, defense, and productivity have all been enriched by nanoparticles. Nanoparticles offer several benefits, including enhanced specificity of pesticide delivery, improved nutrient supply, regulation of pathogenicity, increased germination rates, and enhanced photosynthetic efficiency (Ahmed et al. 2023). Nanoparticles play an important role in enhancing plant growth, development, and yield by improving input efficiency and minimizing losses. It also provides an increased specific surface area for pesticide and fertilizers. Protection of the environment from pollution is a major objective for the industry in green agriculture, and the green synthesis of nanomaterials provides efficient management as well as preservation of resources to cultivate crops (Rajput et al. 2022). Several studies have established the beneficial impact of using nanomaterials derived from plants on plant germination, growth, and development. In recent years, more utilization of nanotechnology has been witnessed in the formulation of antimicrobial products and as micronutrients to combat plant diseases and promote growth and development of plants (Chen et al. 2024). Plant growth and development are also greatly affected by the plant root microenvironment called the rhizosphere. The rhizosphere contains several microbes that can be beneficial or non-beneficial to plant growth. It needs to be understood so that it can be managed for sustainable agriculture. Multi-omics technology during the recent past has been able to unveil different aspects of rhizosphere biology by converging different omics such as metagenomics (to reveal the diversity of rhizobium), metatranscriptomics and metaproteomics (to obtain information regarding the functional aspect of the rhizobium), and metabolomics (to obtain information regarding signaling pathways that take place in the rhizosphere), which has helped in analyzing the entire scenario of rhizobium (Argentel-Martínez et al. 2025). Here, we are going to discuss about how nanotechnology and multi-omics technology have enhanced the understanding of rhizospheric biology and through which ways they could be implemented in plant growth and development. We will also discuss about the challenges, issues, and future outlooks of these technologies (Khan et al. 2021).

2.4 Mechanisms of Interaction Between Nanoparticles and Microbiomes

ENMs may have a direct or an indirect impact on the plant growth—by impacting the soil microbial community or the plant (Fig. 2.2). Nanotechnology also enables the production of agrochemicals as nanoformulations, which release gradually to the target site over an extended period, reducing the environmental losses associated with conventional delivery methods (Srivastava and Singh 2021). Nanofertilizer (e.g., polymer-coated nanofertilizer or nano-emulsified fertilizer) or nanopesticide (e.g., nanofertilizer or nanopesticide) boosts nutrient absorption and reduces off-target effect (Gill and Garg 2014). Nanoagrochemicals influence the biomass formation and rhizosphere-dominant microbiome and are especially significant in the cycles of nutrients and plant defenses (Ur Rahim et al. 2021). Nanoparticles can have different effects on microbes, ranging from disruption of membranes to

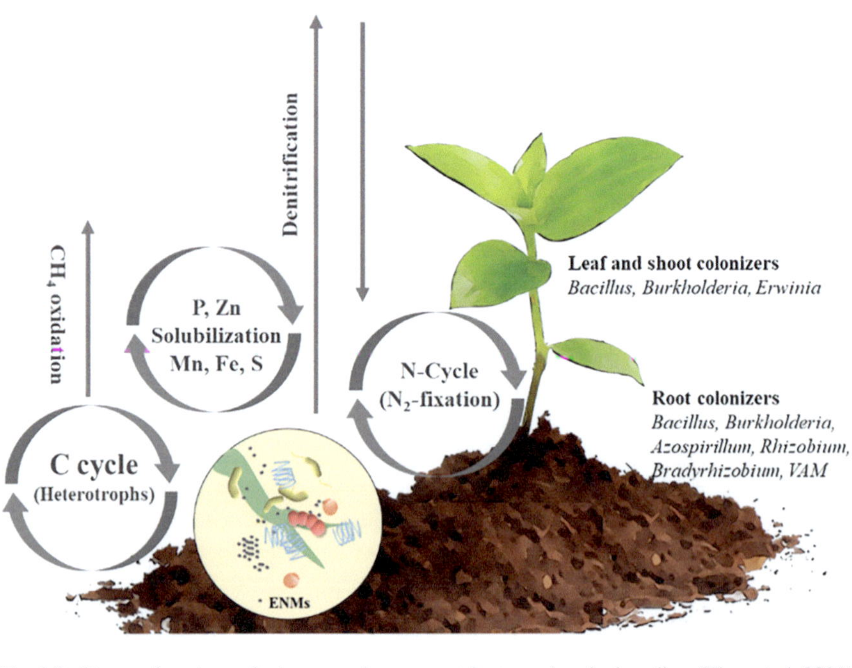

Fig. 2.2 Diverse functions of microorganisms on nutrient geochemical cycling (Khan et al. 2021)

oxidative stress (Noukelag et al. 2022; Rehman et al. 2023). However, more in-depth in vivo studies are still required.

2.4.1 *Enhancing Microbiome Efficiency Through Nanotechnology*

The plant microbiome, a form of microbe community, occupies different parts of the plant phyllosphere (leaf surfaces), rhizoplane (root surfaces), rhizosphere (soil in which roots are embedded), and endosphere (tissues inside the plant) (Kumar et al. 2022). These microbiomes are influenced by the composition of plant species, age, secretions, environmental conditions, and soil properties (Jansson and Hofmockel 2020). Selective pressure imposed on microbiomes by plants reselects favorable microorganisms that carry out a specific task (e.g., mobilization of nutrients, resistance to pathogens, toleration of abiotic stresses) (Deng et al. 2021).

Due to their exceptional surface properties and reactivity, nanomaterials (NMs) provide new avenues to the advancement of the functional dynamics of plant-microbiome interactions. An example is that NMs, including CuO, TiO_2, and ZnO, enhance root exudation and microbial activity as these NMs enhance nutrient mobilization and modify soil pH (Deng et al. 2021). One potential application of these NMs is to stimulate phytostimulation by increasing phytohormone biosynthesis and altering gene expression, thereby accelerating plant growth and enhancing resistance. In addition, secondary metabolite induction, such as siderophores, which benefits the solubilization of micronutrients such as Fe, Zn, and Cu, can also be modulated by microbial activity in the rhizosphere, facilitated by nano-enabled activity (Chen and Liu 2019; Layet et al. 2017). An improved synthesis of these compounds can boost nutrient bioavailability and reduce phytotoxicity of the metals, which leads to improved soil health and growth of plants.

Nonetheless, interactions of NMs with soil microbiota have both positive and negative effects. Several studies have shown that it is demonstrated that microbial diversity is disturbed, and the population of nitrogen fixing is reduced, in addition to the change of core enzyme activities in nutrient cycling, when exposed to Ag, TiO_2, and CeO_2 NMs (Spaepen and Vanderleyden 2011). These changes may have impacts on basal microbial communities, which are a key aspect of sustainable agriculture (Bray et al. 2019; Hunter 2016).

2.4.2 *Activity of ENMs Against Plant Pathogens*

ENMs are nanoscale materials (size 1–100 nm) with distinct physicochemical characteristics and have the prospect of being established as efficient agents in plant disease control. Novel avenues toward plant protection are available due to their

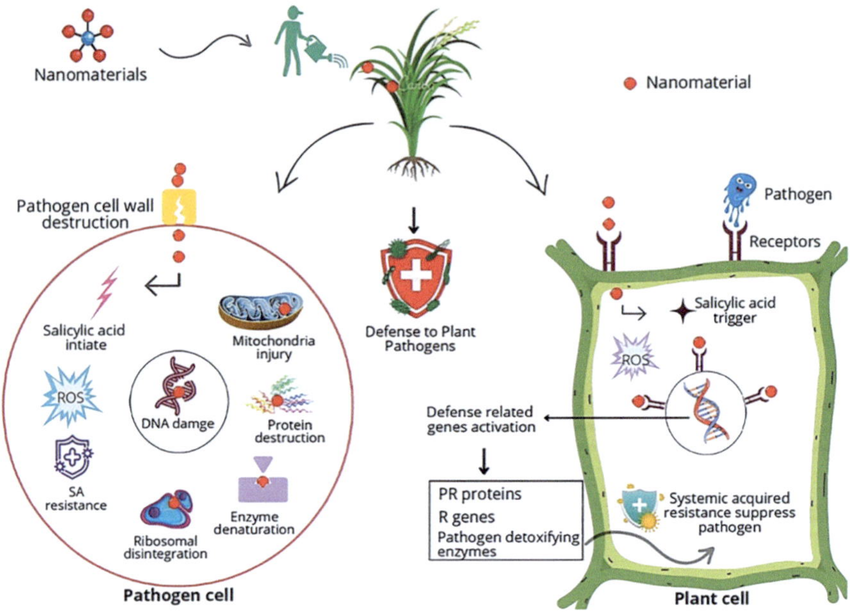

Fig. 2.3 The application of nanomaterials (NMs) is utilized for the diagnosis and protection against crop pathogens (Jabran et al. 2024)

large surface area and increased reactivity, and they can be tuned to react specifically with pathogens (Fig. 2.3) (Varma et al. 2025).

2.4.2.1 Against Bacteria

ENMs (primarily polymeric and metallic NPs) have been proven particularly effective against bacteria and phytopathogens. Cellulose nanocrystals and PLGA proved their efficacy in addressing *Pseudomonas syringae* infecting tomato plants without any phytotoxicity (Fortunati et al. 2016). Hyaluronan/chitosan nanofilms were also found to hinder *Xylella fastidiosa*. Several studies have reported that AgNPs bind with cysteine-rich proteins and enzymes and cause cell death (Avila-Quezada and Espino-Solis 2020). The ability of copper nanoparticles (CuNPs) to act as an environmentally safe alternative to bulk copper fungicides, minimizing the risk of resistance and environmental pollution, has been found (Bramhanwade et al. 2016).

2.4.2.2 Against Fungi

Antifungal property is also an attribute of ENMs. Conjugated carbendazim AgNPs (Cz-AgNPs) was effective in inhibiting anthracnose (*Colletotrichum gloeosporioides*) in crops (Shivamogga Nagaraju et al. 2020). NPs infiltrate the cell walls of fungal cells, interfere with the production of 2-glucan, and kill organelles (Arciniegas-Grijalba et al. 2017). Moreover, metal NPs trigger the production of reactive oxygen species (ROS), which causes DNA and protein damage to fungus cells, resulting in their death (Zhang et al. 2018a, b).

2.4.2.3 Against Viruses

About 50% of the plant diseases are caused by viruses (Vargas-Hernandez et al. 2020). ENMs, especially metallic nanoparticles (NPs), have displayed excellent antiviral activity. For example, silver nanoparticles (AgNPs) can kill viruses by interacting with surface glycoproteins, inhibiting genomic DNA binding, or destroying the viral capsid assembly (Galdiero et al. 2011). Au/CuSNPs have also been reported to kill norovirus capsid proteins. Also, ZnO and SiO_2 nanoparticles showed antiviral capacity by disrupting viral envelopes in tobacco mosaic virus (TMV) (Cai et al. 2019). Vectors (mainly insects causing ~70% of plant viruses) are also important to control (Cai et al. 2019).

2.4.3 Impacts of NPs on Plant-Microbiome Interaction

Biochemical, genetic, and metabolic interactions create a dynamic and complex network at the interface of the plant and microbiome. Such symbiosis is critical in nutrient acquisition, resisting diseases, inducing growth, as well as surviving abiotic stresses (Philippot et al. 2013). On the one hand, plants recruit helpful microorganisms using exudates such as organic compounds like flavonoids (Merten et al. 2020). On the other hand, they also affect microbial gene expression in response to quorum-sensing signals (Merten et al. 2020). In turn, the microbiota represents the possibility to influence the expression of the host genes that allows stating that the plant microbiome serves as the second genome, something which determines the plant phenotype. Nonetheless, nanoparticles (NPs), and hence deposition within the agricultural schemes, can interfere with this fine balance. TiO2 NPs, ZnO, CuO, and AgNPs are found to change the microbial diversity and, along with it, the enzyme activity in the rhizosphere (Singh et al. 2019; Rai-Kalal and Jajoo 2021). This disturbance is associated with to the capacity of NPs to produce reactive oxygen species (ROS), partake in crucial nutrient binding, and harm the architecture of microbial cells, inducing proscribed functionality or oblivion of the microorganism.

Soil factors such as porosity, organic matter, and pH determine the mobility and bioavailability of NP. Some studies have revealed minimal effects of NPs on the soil structure (Labud et al. 2007), whereas others have revealed that the microbial activity is reduced and that community composition is altered (Nyberg et al. 2008; Pradhan et al. 2011). Exposure to metal oxide nanoparticles leads to a decrease in litter degradation and microbial enzyme activity (Jeevanandam et al. 2018). Since it is expected that approximately 11 million tons of metal-based NPs can be created annually, the ecological significance of their production over many years must be addressed (Jeevanandam et al. 2018). With more and more research focusing on nanotechnology applications in agriculture, there must be some knowledge as to the possible, non-desirable consequences of such interactions on plant-microbe associations in a way that would guarantee sustainable and balanced soil health.

2.5 Applications of Nanotechnology in Plant-Microbiome Systems

Increasing world population and environmental concerns have necessitated the development of other models of using agrochemicals rather than the conventional ones. Nanoparticles (NPs) have an exciting potential in agriculture because of their tiny size, large surface area, and chemical stability. They can transport agrochemicals and stimulate the growth of plants as well as their endurance to stress. NPs have direct effects on agriculture by being directly taken up by roots or leaves as well as indirect effects by changing the soil microbiome (Etesami et al. 2021). For example, ZnO nanoparticles enhanced plant-microbe relationships and alleviated stress imposed by arsenic in contaminated soil (Fig. 2.4).

2.5.1 Nanoscale Monitoring of Plant Health and Microbiome Activity

Engineered nanaparticles are also used to track crop wellness and control diseases caused by bacteria, fungi, viruses, and nematodes. The most typical engineered nanaparticles are metallic oxides (Ag, Cu, Zn, Ti), metalloids (Si, B), carbon-based nanomaterials (graphene oxide, fullerenes, carbon nanotubes), etc., as well as liposomes and quantum dots (Elmer and White 2018). These substances have a direct antimicrobial effect or activate the plant defense systems. Nanoparticles of silica (SiO_2 NPs) are one of the most efficient—they penetrate into plant tissues, decrease the incidence of pathogens, enhance drought adaptation, and improve yield (Du et al. 2022). Their efficiency is due to silicic acid exudation that strengthens cell walls and inhibits the arrival of pathogens (Du et al. 2022). Other NPs (e.g., Cu, carbon NPs) also can induce SAR (systemic acquired resistance). This is related to the responses of innate immunity in animals (Elmer and White 2018).

Ag NPs have a wide spectrum of antifungal activity, and they also greatly inhibit infections in crops such as ryegrass and pepper (Jo et al. 2009). In the same manner, ZnO NPs alleviated *F. graminearum* in wheat without toxic accumulations of zinc. Material such as MWCNTs and GO are considered carbon nanomaterials, which have the ability to disrupt fungal growth (Wang et al. 2014). ENMs can enhance the

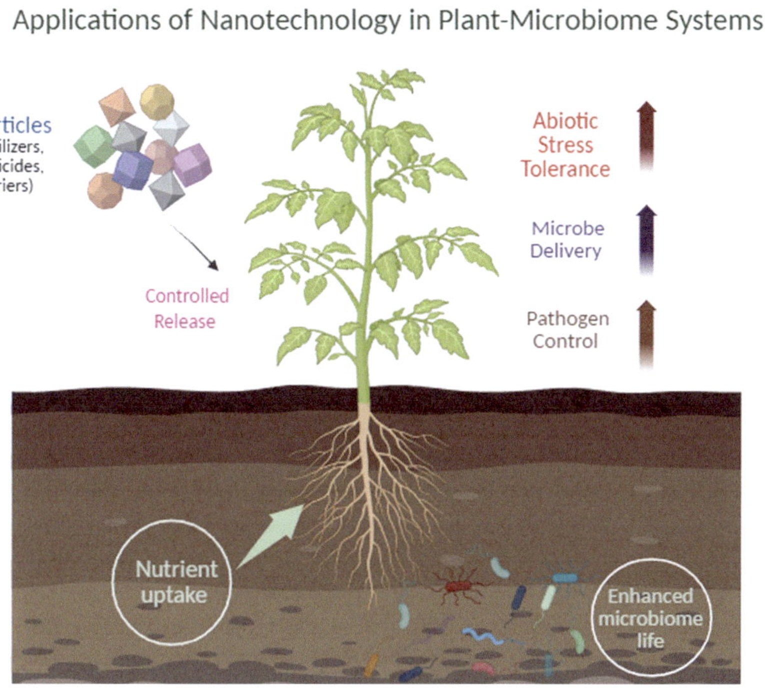

Fig. 2.4 Demonstration of how NPs (such as nanopesticides, nanofertilizers, and nanocarriers) improve stress tolerance, control infections, transport beneficial bacteria, and enhance nutrient uptake and how NPs minimize environmental damage and promote plant health through the regulated release of agrochemicals

activity of microbial biocontrol agents. For example, low-dose NPs increased the antifungal effect of *Pseudomonas protegens* against *Candida albicans* (Khan et al. 2018; Edgington et al. 1980). Targeted delivery and suppression of pathogens are also manifested by pesticides nano-formulated as conjugates of TiO_2/Ag with dimethomorph. Although ENMS are effective, they need to be evaluated critically in terms of dose and safety. Their combination with plant growth-promoting microbes also represents a potentially valuable course toward sustainable disease management (Berendsen et al. 2012; Sorty et al. 2016; Ali and Glick 2019).

2.5.2 *Nanoparticle-Mediated Alleviation of Abiotic Stress in Plants*

Non-biotic stress, including drought, salinity, heavy metal pollution, and extreme temperatures, greatly restricts the productivity and quality of plants (Sharma et al. 2020). Silicon nanoparticles (Si-NPs) have been proposed as one of the tools to mitigate the impacts of abiotic stresses because they can increase nutrient absorption and expression of stress-related genes and have increased antioxidant capacity. The stress caused by heavy metals, especially cadmium (Cd), arsenic (As), lead (Pb), and chromium (Cr), has adverse impacts on photosynthesis and plant growth. In *Brassica chinensis*, the observed effects were *Oryza sativa*, *P. sativum*, *Zea mays*, and many other species include a decrease in metal uptake and toxicity as well as an increase in photosynthetic efficiency and a reduction of the reactive oxygen species (ROS) (Hussain et al. 2020; Tripathi et al. 2021; Burachevskaya et al. 2021; Sharma et al. 2020).

Seed germination, assimilation of nutrients, and leaf gas exchange are also hampered by salinity stress. Si-NPs mitigate these situations in plants such as *Cucurbita*, *S. lycopersicum*, *Ocimum basilicum*, and *Lens culinaris* by controlling proline production and enhancing chlorophyll levels and morphological characteristics (Merwad et al. 2018; Ashkavand et al. 2015; Liu et al. 2014). The mitigation of drought stress, as indicated by findings showing the positive effect of Si-NPs on water uptake, stability of membranes, accumulation of osmolytes, and changes in the activities of antioxidant enzymes is mentioned in *Crataegus* sp., *Sorghum bicolor*, and wheat (Ashkavand et al. 2015; Hattori et al. 2005). ZnO and Fe NPs also enhance drought resistance by enhancing seed germination and decreasing chlorosis (Sedghi et al. 2013). Nanomaterials may also be nutrient carriers. However, the stabilization of Si-NPs is imperative, and in slow-release NPK fertilizers, they will regulate nutrient availability and enhance the resistance of plants to salinity and drought stress (Ashkavand et al. 2015).

2.5.3 Delivery of Beneficial Microbes Using Nanotechnology

Plant-hosting microbiome is a very complex and dynamic microbiome, which majorly determines plant physiology, nutrient uptake, stress tolerance, and plant growth (Ravanbakhsh et al. 2019). Recent technologies in nanotechnology have provided a new horizon for increasing this plant-microbe interaction by facilitating delivery mechanisms with precision and better microbial colonization. The microbiome can be adjusted with engineered nanoparticles (NPs) that manipulate the microbe communication and survival and enhance positive interactions (Cao et al. 2024). For example, nano-based materials (carbon dots, in particular) have demonstrated the possibility of enhancing nitrogen bioavailability and stimulating the growth of soybeans exposed to drought (Wang et al. 2022). In the same sense, the plant physiology may also be reprogrammed with the use of graphene and carbon-based nanomaterials, which alter the rhizosphere microbiome community, improving the plant health indirectly through stress (Hao et al. 2018; Cao et al. 2024).

Quorum sensing (QS) is one of the significant ways in which bacteria can effectively organize their activity in response to environmental conditions, including

NPs. This is a density-dependent communication, which entails autoinducers that are small signaling molecules. The existence of nanoparticles may disrupt QS, thus regulating microbial activities in biofilm formation, root association, and plant growth-promoting abilities (Turan et al. 2017). Nanoparticles may also augment the transport and adhesion processes of beneficial microbes and, hence, the rate of their uptake and colonization. As soon as they are applied in the environment (usually through a foliar spray or soil amendment), NP reach environmental processes of aggregation, oxidation, and corona formation, which modify their fate and interaction with the plant tissues (Fincheira et al. 2021). These modified NPs can enter the labile tissue through either apoplastic routes or symplastic routes, more likely through stomata, cuticular pores, or root epidermis. In other cases, such as *C. roseus*, endocytosis has been identified to play an important role in the uptake of carbon-based NPs (Fincheira et al. 2021).

2.6 Innovations and Future Prospects

One major promise of the potential of nanotechnology all over the world is to overcome some of the most significant challenges related to food security, assuming the world can overcome the problem of hunger, poverty, and food insecurity through soil health and more sustainable agriculture production (Doolette et al. 2016). It paves the way for a green revolution by reducing agricultural risks and enhancing resilience to environmental stressors. Nevertheless, nanotechnology in agriculture has not been well exploited despite having enormous potential (Berhe et al. 2018). Engineered nanomaterials (NMs) may facilitate plant-microbe interactions, reduce the stresses of soil (drought, salinity, and nutrient deficiency), and increase crop production. However, only few realistic studies concerning the effect of NMs on rhizospheric microbiomes have been conducted (Eivazi and Afrasiabi 2018).

To apply NMs sustainably, it is important to know how they will disintegrate in the environment, what their toxicity levels are, and what effects they will have on microbes (Priester et al. 2012). Incorrect (excessive) use may cause phytotoxicity and alter soil microbe processes, and the optimized use may minimize the use of chemical fertilizers (Zhao et al. 2020). It was revealed that NMs could reduce the effects of biotic and abiotic stress on crops (Zhang et al. 2018a, b), although there remain concerns regarding nanoparticle build-up and toxicity. When getting into agroecosystems, NMs can experience changes that will produce new, and possibly toxic, compounds (Amde et al. 2017). Therefore, extensive studies, risk analysis, and governance systems are necessary regarding the long-term effects that can only facilitate the equalization between the benefits of nanotechnology and environmental security (Xiao et al. 2017b).

2.6.1 Integration with Digital Agriculture and AI for Real-Time Monitoring and Control

Smart farming involves the proper combination of artificial intelligence (AI) and the Internet of Things (IoT) in cyber-physical systems to control farms comprehensively (Rosell-Polo et al. 2015). Applications of AI help assessment of soil health, monitoring crops and disease diagnosis, and weed management through mobile expert system programs and the Soil Risk Characterization Decision Support Systems (SRCDSS) provide farmers with support through smartphone (Khardia et al. 2022). Satellite imagery is another method that AI examines in real time, through which it can provide data-driven precision agriculture (Sharma and Shivandu 2024). The IoT bridges the gap linking sensors, UAVs, and ICT to gather and relay farm-related data. The 5G, LoRa, NB-IoT, ZigBee, and Wi-Fi technologies improve remote monitoring, data thickets, and low-latency communication (Bordolui 2024). LoRa in particular is popular due to the low cost, long distance, and efficiency. With the combination of cloud computing, IoT is used in support of smart applications in livestock and fisheries, and weather tracking. Robotics and Autonomous Systems (RAS), which is powered by AI, can mine data to provide information to predict yield, identify pests, and optimize fertilizers to increase precision and productivity (Villa-Henriksen et al. 2020).

2.6.2 Potential for Addressing Global Agricultural Challenges

Against the backdrop of the COVID-19 pandemic, climatic changes, and biodiversity losses, as well as geopolitical rivalries including the war in Ukraine, agrifood systems currently experience compounding pressures (Xie and Al-Aly 2022). The combination of these phenomena highlights the need for developing more sustainable and resilient systems that enhance productivity, minimize postharvest losses, and protect the environment, human health, and biodiversity (Banerjee et al. 2022). In the whole world, approximately three billion individuals do not have access to a healthy diet, which contributes to the development of multiple types of malnutrition, obesity, and non-communicable diseases (Naveed et al. 2023). The recently developed science of microbiomes provides an opportunity since it has been demonstrated that human gut microbiota is at the core of diet-related health, whereas soil, plant, and animal microbiomes are crucial to both environmental health and food production (Zhang et al. 2022). Strategies based on microbiomes can play a more decisive role in the future by improving the resilience and sustainability of agrifood systems. They are solutions that facilitate efficiency in the utilization of natural resources and preventive health practice throughout the One Health domain (McGurnaghan et al. 2021). Furthermore, due to new research findings, microbiomes can be enhanced in terms of structure and functionality, which hold the promise of being revolutionary throughout the food chain, including soils, crops, and humans (Holman et al. 2020). Incorporation of microbiome technology into different ecosystems can transform agrifood systems toward healthier, more productive, and crisis-resistant systems (Barron et al. 2020).

2.7 Challenges and Future Directions

Nanotechnology holds significant potential to provide organic farming, as it combines soil science and the ecology of microbes. Prospective developments are the biodegradable nanomaterials, multi-omics to investigate nanoparticle-microbe-plant interactions, synthetic microbiomes, and in situ biosensors (Kumar 2025). Nevertheless, there is still an issue with the toxicity of nanoparticles, in particular, the impact they have on non-target life, changes in the microbiome, and leaching (Kibbey and Strevett 2019). Toxicity is determined by the composition of particles, solubility, and charge. Nanoparticles that are biosynthesized have fewer hazards (Mgadi et al. 2024). Safe use entails biodegradable nanocarriers, microbial surveillance, and transparent policies. A synergetic work of agronomists, microbiologists, nanotechnologists, and policymakers is obligatory to extend laboratory-level studies into practical and sustainable field practices.

2.7.1 Environmental and Safety Considerations

Nanotechnology has a lot of potential in sustainable agriculture, though there is a need to study its impact on the environment, particularly non-target species and ecosystems. Although in experimental conditions engineered nanomaterials (ENMs) are frequently exhibiting low to moderate phytotoxicity (Moreno-Olivas et al. 2014), they are most of the time conducted under simplified conditions (hydroponics or sand culture), and they may not represent agricultural systems (Komire et al. 2024). Due to repeated field applications of ENMs, in the form of either biosolids or nanobiostimulants, there can be accumulation in the long run in the soil, which may affect the soil biota.

Environmental concerns are associated with the transformation of ENMs in natural environments, which can lead to the formation of toxic products through sedimentation, oxidation, dissolution, and agglomeration (Sambangi et al. 2022). The changes may be detrimental to positive soil microbes, including nitrogen-fixing and phosphate-solubilizing bacteria. TiO_2 and AgNPs nanoparticles, which are made of metals, have been demonstrated to weaken microbial cell structure, enzyme functioning, and nutrient supplies (Chavan et al. 2020; Upadhayay et al. 2019).

To overcome such risks, scientists have also proposed green nano approach, whereby nano-biofertilizers and nanobiostimulants, which promote microbe health without enhancing the dispersion of nanoparticles, are used. The behavior of nanoparticles can be controlled too, and it can be toxicity-reduced by the means of controlled release systems, as well as biodegradable coatings (Chhipa 2021). There is a need to carry out long-term field research to better evaluate the effect of ENMs on soil health, microbiomes, and ecosystem services.

2.7.2 Socioeconomic and Regulatory Challenges

Microbial nanobiotechnology has a transformative potential, particularly in developing countries, in agricultural applications, health care, and environmental sustainability. It is a combination of nanotechnology and microbiology with innovation, which enables one to foster amplification in productivity, and such innovation leads to growth in the economy based on knowledge (Aryal et al. 2025). However, problems exist. The marginalized group may find it hard to access; socioeconomic inequality may become an obstacle unless the policies are inclusive. Also, local innovation and autonomy can become a bottleneck because of being dependent on proprietary or foreign nanomaterials (Upadhyay et al. 2024). The problems of nanoomics tools, ownership of biological information, and environmental justice are also ethical issues related to environmental problems, which are comparable. Most countries lack appropriate legislation to regulate the use of nanoparticles, including bioaccumulation and impeding of the beneficial microorganisms (Raghib et al. 2020). Some of the strategies that should be used are those related to the production of biodegradable, low-toxicity nanomaterials, prolonged investigations of the environmental disorders, and unrestricted policy in the regulation of nanomaterials to ensure equal and safe application. Responsible innovations depend on establishing trust among the local community, in which they are involved in decision-making. The positive effects of nanobiotechnology would only be of value when it is regulated and incorporates more individuals in ensuring that the risks are minimal.

2.7.3 Policy and Regulatory Challenges

Microbial nanobiotechnology has a transformational potential for diverse industrial applications. The impact of nanobiotechnology using microbes remains transformational in developing countries, particularly in agriculture, health care, and environmental sustainability. It combines nanotechnology with microbiology, thus boosting its innovation, productivity, and knowledge-based economic development (Aryal et al. 2025). Nonetheless, there remain severe problems. Unless inclusive policy changes are introduced, socioeconomic inequality can restrict access for marginalized groups. The low risk of local innovation and autonomy reliance on proprietary or imported nanomaterials can be an obstacle (Upadhyay et al. 2024). Other ethical issues come into play, such as data ownership concerning biological characteristics and environmental fairness. Moreover, most countries do not have the regulations to deal with possible risks of nanoparticles, including bioaccumulation and inhibition of beneficial microbes (Raghib et al. 2020). Among approaches that can address the problem are the production of biodegradable, low-toxic nanoparticles, prolonged environmental evaluation, and regulatory oversight. Local community participation in decisions must be an important aspect in creating a trust relationship, and by being responsible. Finally, microbial nanobiotechnology can fully realize its potential benefits only when its governance structure is inclusive, ensuring that the knowledge is applied in ways that minimize environmental and social risks (Farooq et al. 2022; Singh et al. 2023).

2.8 Conclusion

The confluence of plant microbiome research and nanotechnology is an exciting field toward the development of sustainable, resilient, and high-performance agriculture systems. Plant microbiomes, particularly those in the rhizosphere, phyllosphere, and endosphere, play key roles in nutrient acquisition, stress alleviation, disease suppression, and plant development. The advent of multi-omics technologies has significantly enhanced our understanding of such microbial processes and their complex interactions with plant hosts and the environment. Simultaneously, nanotechnology has emerged as a revolutionizing factor in agriculture with new avenues for soil health improvement, activation of microbial activity, and crop productivity enhancement. Engineered nanomaterials have been demonstrated to be capable of modulating microbial communities, regulating plant metabolism, and improving nutrient delivery efficiency with very small environmental impact. Carefully deployed, nanotechnology can optimize and even supercharge the beneficial effects of the plant microbiome. The intersection of microbiome science with nanotechnology and AI- and IoT-based smart agriculture enables new opportunities for real-time monitoring, tailored interventions, and optimized resource allocation. Together, these technologies are capable of addressing some of the most pressing global agriculture problems, including soil erosion, climate stress, food insecurity, and reduced biodiversity. However, to unlock their full potential, it is essential to ensure safe, context-dependent use of nanomaterials, supported by long-term research, regulation, and multidisciplinary inquiry. Briefly, this chapter is about moving beyond traditional agriculture by embracing microbiome-based and nano-enabled technologies. These approaches, grounded in ecological awareness and technological progress, can result in a more sustainable, productive, and health-focused future for global food production.

References

Ahmed T, Noman M, Gardea-Torresdey JL, White JC, Li B (2023) Dynamic interplay between nano-enabled agrochemicals and the plant-associated microbiome. Trends Plant Sci 28(11):1310–1325

Alamri S, Nafady NA, El-Sagheer AM, El-Aal MA, Mostafa YS, Hashem M, Hassan EA (2022) Current utility of arbuscular mycorrhizal fungi and hydroxyapatite nanoparticles in suppression of tomato root-knot nematode. Agronomy 12(3):671

Ali S, Glick BR (2019) Plant–bacterial interactions in management of plant growth under abiotic stresses. In: New and future developments in microbial biotechnology and bioengineering. Elsevier, pp 21–45

Ali MH, Mandal S, Ghorai M, Lal MK, Tiwari RK, Kumar M et al (2023) Perspectives of omics and plant microbiome. In: Genomics, transcriptomics, proteomics and metabolomics of crop plants. Academic, pp 131–144

Amde M, Liu JF, Tan ZQ, Bekana D (2017) Transformation and bioavailability of metal oxide nanoparticles in aquatic and terrestrial environments. A review. Environ Pollut 230:250–267

Arciniegas-Grijalba PA, Patiño-Portela MC, Mosquera-Sánchez LP, Guerrero-Vargas JA, Rodríguez-Páez JE (2017) ZnO nanoparticles (ZnO-NPs) and their antifungal activity against coffee fungus *Erythricium salmonicolor*. Appl Nanosci 7(5):225–241

Argentel-Martínez L, Peñuelas-Rubio O, Herrera-Sepúlveda A, González-Aguilera J, Sudheer S, Salim LM et al (2025) Biotechnological advances in plant growth-promoting rhizobacteria for sustainable agriculture. World J Microbiol Biotechnol 41(1):21

Aryal S, Kumari P, Gautam CK, Gupta RK (2025) Impact of microbial Nanobiotechnology on the economy of developing countries. In: Nano-microbiology for sustainable development. Springer Nature Switzerland, Cham, pp 269–292

Ashkavand P, Tabari M, Zarafshar M, Tomaskova I, Struve D (2015) Effect of SiO2 nanoparticles on drought resistance in hawthorn seedlings. For Res Pap 76:350–359. https://doi.org/10.1515/frp-2015-0034

Avila-Quezada GD, Espino-Solis GP (2020) Silver nanoparticles offer effective control of pathogenic bacteria in a wide range of food. In: Pathogenic bacteria, vol 203

Badawy AA, Alotaibi MO, Abdelaziz AM, Osman MS, Khalil AM, Saleh AM et al (2021) Enhancement of seawater stress tolerance in barley by the endophytic fungus *Aspergillus ochraceus*. Metabolites 11(7):428

Banerjee M, Pal R, Dutta S (2022) Risk of incident diabetes post-COVID-19: a systematic review and meta-analysis. Prim Care Diabetes 16(4):591–593

Barron E, Bakhai C, Kar P, Weaver A, Bradley D, Ismail H et al (2020) Associations of type 1 and type 2 diabetes with COVID-19-related mortality in England: a whole-population study. Lancet Diabetes Endocrinol 8(10):813–822

Berendsen RL, Pieterse CM, Bakker PA (2012) The rhizosphere microbiome and plant health. Trends Plant Sci 17(8):478–486

Berg G, Rybakova D, Grube M, Köberl M (2016) The plant microbiome explored: implications for experimental botany. J Exp Bot 67(4):995–1002

Berhe AA, Barnes RT, Six J, Marín-Spiotta E (2018) Role of soil erosion in biogeochemical cycling of essential elements: carbon, nitrogen, and phosphorus. Annu Rev Earth Planet Sci 46(1):521–548

Bloch K, Webster TJ, Ghosh S (2021) Nanobiotechnology of the plant microbiome. In: Microbiome stimulants for crops. Woodhead Publishing, pp 349–367

Bodenhausen N, Horton MW, Bergelson J (2013) Bacterial communities associated with the leaves and roots of Arabidopsis thaliana. PLoS One 8(2):e56329

Bordolui SK (2024) Sensor based technology development: advancements in precision agriculture. Int J Adv Eng Sci Res. https://doi.org/10.53882/IJAESR.2023.1001004

Brader G, Compant S, Vescio K, Mitter B, Trognitz F, Ma LJ, Sessitsch A (2017) Ecology and genomic insights into plant-pathogenic and plant-nonpathogenic endophytes. Annu Rev Phytopathol 55(1):61–83

Bramhanwade K, Shende S, Bonde S, Gade A, Rai M (2016) Fungicidal activity of cu nanoparticles against fusarium causing crop diseases. Environ Chem Lett 14(2):229–235

Bray N, Kao-Kniffin J, Frey SD, Fahey T, Wickings K (2019) Soil macroinvertebrate presence alters microbial community composition and activity in the rhizosphere. Front Microbiol 10:256

Bulgarelli D, Schlaeppi K, Spaepen S, Van Themaat EVL, Schulze-Lefert P (2013) Structure and functions of the bacterial microbiota of plants. Annu Rev Plant Biol 64(1):807–838
Burachevskaya M, Minkina T, Fedorenko A, Fedorenko G, Chernikova N, Rajput VD, Mandzhieva S, Bauer T (2021) Accumulation, translocation, and toxicity of arsenic in barley grown in contaminated soil. Plant Soil. https://doi.org/10.1007/s11104-021-05067-9
Cai L, Liu C, Fan G, Liu C, Sun X (2019) Preventing viral disease by ZnONPs through directly deactivating TMV and activating plant immunity in *Nicotiana benthamiana*. Environ Sci Nano 6(12):3653–3669
Cao J, Chen Z, Wang L, Yan N, Lin J, Hou L et al (2024) Graphene enhances artemisinin production in the traditional medicinal plant Artemisia annua via dynamic physiological processes and miRNA regulation. Plant Commun 5(3):100742
Chauhan P, Sharma N, Tapwal A, Kumar A, Verma GS, Meena M et al (2023) Soil microbiome: diversity, benefits and interactions with plants. Sustainability 15(19):14643
Chavan S, Sarangdhar V, Nadanathangam V (2020) Toxicological effects of TiO2 nanoparticles on plant growth promoting soil bacteria. Emerg Contam 6:87–92
Chen Q, Liu S (2019) Identification and characterization of the phosphate-solubilizing bacterium *Pantoea* sp. S32 in reclamation soil in Shanxi, China. Front Microbiol 10:2171
Chen H, Song Y, Wang Y, Wang H, Ding Z, Fan K (2024) Zno nanoparticles: improving photosynthesis, shoot development, and phyllosphere microbiome composition in tea plants. J Nanobiotechnol 22(1):389
Chhipa H (2021) Nano-toxicity to microbes: potential implications of nanomaterials on microbial activity. Nanotoxicol Nanoecotoxicol 1:99–123
Chung H, Kim MJ, Ko K, Kim JH, Kwon HA, Hong I et al (2015) Effects of graphene oxides on soil enzyme activity and microbial biomass. Sci Total Environ 514:307–313
Compant S, Samad A, Faist H, Sessitsch A (2019) A review on the plant microbiome: ecology, functions, and emerging trends in microbial application. J Adv Res 19:29–37
Dastogeer KM, Li H, Sivasithamparam K, Jones MG, Wylie SJ (2018) Fungal endophytes and a virus confer drought tolerance to Nicotiana benthamiana plants through modulating osmolytes, antioxidant enzymes and expression of host drought responsive genes. Environ Exp Bot 149:95–108
Dastogeer KM, Tumpa FH, Sultana A, Akter MA, Chakraborty A (2020) Plant microbiome–an account of the factors that shape community composition and diversity. Curr Plant Biol 23:100161
Dastogeer KM, Kao-Kniffin J, Okazaki S (2022) Plant microbiome: diversity, functions, and applications. Front Microbiol 13:1039212
Deng X, Zhang N, Shen Z, Zhu C, Liu H, Xu Z, Li R, Shen Q, Salles JF (2021) Soil microbiome manipulation triggers direct and possible indirect suppression against *Ralstonia solanacearum* and *Fusarium oxysporum*. NPJ Biofilms Microbiomes 7:33
Doolette CL, Gupta VV, Lu Y, Payne JL, Batstone DJ, Kirby JK et al (2016) Quantifying the sensitivity of soil microbial communities to silver sulfide nanoparticles using metagenome sequencing. PLoS One 11(8):e0161979
Du J, Liu B, Zhao T, Xu X, Lin H, Ji Y et al (2022) Silica nanoparticles protect rice against biotic and abiotic stresses. J Nanobiotechnol 20(1):197
Edgington LV, Martin RA, Bruin GC, Parsons IM (1980) Systemic fungicides: a perspective after 10 years. Plant Dis 64(1):19–23
Eivazi F, Afrasiabi Z (2018) Effects of silver nanoparticles on the activities of soil enzymes involved in carbon and nutrient cycling. Pedosphere 28(2):209–214
Elmer W, White JC (2018) The future of nanotechnology in plant pathology. Annu Rev Phytopathol 56(1):111–133
Etesami H, Fatemi H, Rizwan M (2021) Interactions of nanoparticles and salinity stress at physiological, biochemical and molecular levels in plants: A review. Ecotoxicol Environ Saf 225:112769

Farhana, Munis MFH, Alamer KH, Althobaiti AT, Kamal A, Liaquat F et al (2022) ZnO nanoparticle-mediated seed priming induces biochemical and antioxidant changes in chickpea to alleviate Fusarium wilt. J Fungi 8(7):753

Farooq B, Anjum S, Farooq M, Rather GA, Nazir A, Nayak BK, Nanda A (2022) Role of nanotechnology in soil microbiome and agricultural development. In: Core microbiome: improving crop quality and productivity, pp 230–248

Fincheira P, Tortella G, Seabra AB, Quiroz A, Diez MC, Rubilar O (2021) Nanotechnology advances for sustainable agriculture: current knowledge and prospects in plant growth modulation and nutrition. Planta 254:1–25

Fortunati E, Rescignano N, Botticella E, La Fiandra D, Renzi M, Mazzaglia A et al (2016) Effect of poly (DL-lactide-co-glycolide) nanoparticles or cellulose nanocrystals-based formulations on Pseudomonas syringae pv. tomato (Pst) and tomato plant development. J Plant Dis Protect 123(6):301–310

Fotoukian H, Nazarpak MH, Solouk A, Bagheri F, Akbari S, Nouri A (2024) Surface modification of titanium implants via PLLA/HA fibrous composite coating to improve piezoelectric properties. Surf Interfaces 50:104501

Galdiero S, Falanga A, Vitiello M, Cantisani M, Marra V, Galdiero M (2011) Silver nanoparticles as potential antiviral agents. Molecules 16(10):8894–8918

Ghosh S, Webster TJ (eds) (2021) Nanobiotechnology: microbes and plant assisted synthesis of nanoparticles, mechanisms and applications. Elsevier

Gill HK, Garg H (2014) Pesticide: environmental impacts and management strategies. Pesticides-Toxic Aspects 8(187):10–5772

Gupta S, Schillaci M, Walker R, Smith PM, Watt M, Roessner U (2021) Alleviation of salinity stress in plants by endophytic plant-fungal symbiosis: current knowledge, perspectives and future directions. Plant Soil 461(1):219–244

Hacquard S (2016) Disentangling the factors shaping microbiota composition across the plant holobiont. New Phytol 209(2):454–457

Haider HI, Zafar I, Ain QU, Noreen A, Nazir A, Javed R et al (2023) Synthesis and characterization of copper oxide nanoparticles: its influence on corn (Z. mays) and wheat (*Triticum aestivum*) plants by inoculation of Bacillus subtilis. Environ Sci Pollut Res 30(13):37370–37385

Hakim S, Naqqash T, Nawaz MS, Laraib I, Siddique MJ, Zia R et al (2021) Rhizosphere engineering with plant growth-promoting microorganisms for agriculture and ecological sustainability. Front Sustain Food Syst 5:617157

Hao Y, Ma C, Zhang Z, Song Y, Cao W, Guo J et al (2018) Carbon nanomaterials alter plant physiology and soil bacterial community composition in a rice-soil-bacterial ecosystem. Environ Pollut 232:123–136

Hattori T, Inanaga S, Araki H, An P, Morita S, Luxová M et al (2005) Application of silicon enhanced drought tolerance in Sorghum bicolor. Physiol Plant 123:459–466. https://doi.org/10.1111/j.1399-3054.2005.00481.x

Holman N, Knighton P, Kar P, O'Keefe J, Curley M, Weaver A et al (2020) Risk factors for COVID-19-related mortality in people with type 1 and type 2 diabetes in England: a population-based cohort study. Lancet Diabetes Endocrinol 8(10):823–833

Hossain MS, Hossain MS, Ahmed S, Mobarak MB (2024) Characterization and adsorption performance of nano-hydroxyapatite synthesized from *Conus litteratus* waste seashells for Congo red dye removal. RSC Adv 14(52):38560–38577

Hunter P (2016) Plant microbiomes and sustainable agriculture. EMBO Rep 17:1696–1699

Hussain B, Lin Q, Hamid Y, Sanaullah M, Di L, Hashmi M, Khan MB, He Z, Yang X (2020) Foliage application of selenium and silicon nanoparticles alleviates cd and Pb toxicity in rice (*Oryza sativa* L.). Sci Total Environ 712:136497

Hussain M, Shakoor N, Adeel M, Ahmad MA, Zhou H, Zhang Z et al (2023) Nano-enabled plant microbiome engineering for disease resistance. Nano Today 48:101752

Jabran M, Ali MA, Muzammil S, Zahoor A, Ali F, Hussain S et al (2024) Exploring the potential of nanomaterials (NMs) as diagnostic tools and disease resistance for crop pathogens. Chem Biol Technol Agric 11(1):75

Jansson JK, Hofmockel KS (2020) Soil microbiomes and climate change. Nat Rev Microbiol 18:35–46

Jatav HS, Sharma LD, Sadhukhan R, Singh SK, Singh S, Rajput VD et al (2020) An overview of micronutrients: prospects and implication in crop production. In: Plant micronutrients: deficiency and toxicity management, pp 1–30

Jeevanandam J, Barhoum A, Chan YS, Dufresne A, Danquah MK (2018) Review on nanoparticles and nanostructured materials: history, sources, toxicity and regulations. Beilstein J Nanotechnol 9(1):1050–1074

Jo YK, Kim BH, Jung G (2009) Antifungal activity of silver ions and nanoparticles on phytopathogenic fungi. Plant Dis 93(10):1037–1043

Khan ST (2020) Interaction of engineered nanomaterials with soil microbiome and plants: their impact on plant and soil health. In: Sustainable agriculture reviews 41: nanotechnology for plant growth and development. Springer International Publishing, Cham, pp 181–199

Khan ST, Ahmad J, Ahamed M, Jousset A (2018) Sub-lethal doses of widespread nanoparticles promote antifungal activity in *Pseudomonas protegens* CHA0. Sci Total Environ 627:658–662

Khan MA, Sahile AA, Jan R, Asaf S, Hamayun M, Imran M et al (2021) Halotolerant bacteria mitigate the effects of salinity stress on soybean growth by regulating secondary metabolites and molecular responses. BMC Plant Biol 21(1):176

Khardia N, Sharma S, Kumawat H (2022) Precision farming. A monthly peer reviewed magazine for agriculture and allied sciences, 63

Kibbey TC, Strevett KA (2019) The effect of nanoparticles on soil and rhizosphere bacteria and plant growth in lettuce seedlings. Chemosphere 221:703–707

Komire K, Tiwari M, Gupta PC, Bahorun T, Sharma N (2024) Harnessing Nanobiostimulants for sustainable agriculture: unraveling plant-microbe synergies. In: Nanobiostimulants: emerging strategies for agricultural sustainability. Springer Nature Switzerland, Cham, pp 197–226

Kumar N (2025) Soil-root-microbe-Nano interactions: toward sustainable plant-microbiome systems. J Publ Int Res Eng Manag (JOIREM) 5(06)

Kumar A, Singh AK, Choudhary KK (eds) (2019) Role of plant growth promoting microorganisms in sustainable agriculture and nanotechnology. Woodhead Publishing

Kumar S, Sindhu SS, Kumar R (2022) Biofertilizers: An ecofriendly technology for nutrient recycling and environmental sustainability. Curr Res Microb Sci 3:100094

Kumari A, Rana V, Yadav SK, Kumar V (2023) Nanotechnology as a powerful tool in plant sciences: recent developments, challenges and perspectives. Plant Nano Biol 5:100046

Labud V, Garcia C, Hernandez T (2007) Effect of hydrocarbon pollution on the microbial properties of a sandy and a clay soil. Chemosphere 66(10):1863–1871

Lal R (2016) Soil health and carbon management. Food Energy Secur 5(4):212–222

Layet C, Auffan M, Santaella C, Chevassus-Rosset C, Montes M, Ortet P, Barakat M, Collin B, Legros S, Bravin MN et al (2017) Evidence that soil properties and organic coating drive the phytoavailability of cerium oxide nanoparticles. Environ Sci Technol 51:9756–9764

Lemanceau P, Blouin M, Muller D, Moënne-Loccoz Y (2017) Let the core microbiota be functional. Trends Plant Sci 22(7):583–595

Liu C, Wei L, Zhang S, Xu X, Li F (2014) Effects of nanoscale silica sol foliar application on arsenic uptake, distribution and oxidative damage defense in rice (*Oryza sativa* L.) under arsenic stress. RSC Adv 4:57227–57234

Mariyam S, Bhardwaj R, Khan NA, Sahi SV, Seth CS (2023) Review on nitric oxide at the forefront of rapid systemic signaling in mitigation of salinity stress in plants: crosstalk with calcium and hydrogen peroxide. Plant Sci 336:111835

McGurnaghan SJ, Weir A, Bishop J, Kennedy S, Blackbourn LA, McAllister DA et al (2021) Risks of and risk factors for COVID-19 disease in people with diabetes: a cohort study of the total population of Scotland. Lancet Diabetes Endocrinol 9(2):82–93

Meel S, Saharan BS (2024) Enhancing crop resilience towards drought: by integrating nanotechnology, microbiomes, and growth-promoting rhizobacteria. Discover Agric 2(1):112

Merten C, Schoonjans R, Di Gioia D, Peláez C, Sanz Y, Maurici D, Robinson T (2020) Exploring the need to include microbiomes into EFSA's scientific assessments. EFSA J 18(6):e18061

Merwad A-RMA, Desoky E-SM, Rady MM (2018) Response of water deficit-stressed Vigna unguiculata performances to silicon, proline or methionine foliar application. Sci Hortic 228:132–144

Mgadi K, Ndaba B, Roopnarain A, Rama H, Adeleke R (2024) Nanoparticle applications in agriculture: overview and response of plant-associated microorganisms. Front Microbiol 15:1354440

Moreno-Olivas F, Gant VU Jr, Johnson KL, Peralta-Videa JR, Gardea-Torresdey JL (2014) Random amplified polymorphic DNA reveals that TiO2 nanoparticles are genotoxic to Cucurbita pepo. J Zheijang Univ Sci A 15(8):618–623

Muhammad M, Wahab A, Waheed A, Mohamed HI, Hakeem KR, Li L, Li WJ (2024) Harnessing bacterial endophytes for environmental resilience and agricultural sustainability. J Environ Manag 368:122201

Naveed Z, García HAV, Wong S, Wilton J, McKee G, Mahmood B et al (2023) Association of COVID-19 infection with incident diabetes. JAMA Netw Open 6(4):e238866–e238866

Nawaz T, Gu L, Fahad S, Saud S, Bleakley B, Zhou R (2024) Exploring sustainable agriculture with nitrogen-fixing cyanobacteria and nanotechnology. Molecules 29(11):2534

Noukelag SK, Mewa-Ngongang M, Ngqoloda S, Kotsedi L, Razanamahandry LC, Ntwampe SKO et al (2022) Influence of synthesis method on structural, morphological, magnetic, and antimicrobial properties of Fe-Ag nanoparticles. J Inorg Organomet Polym Mater. https://doi.org/10.1007/s10904-022-02493-9

Nyberg L, Turco RF, Nies L (2008) Assessing the impact of nanomaterials on anaerobic microbial communities. Environ Sci Technol 42(6):1938–1943

Pallavi, Mishra RK, Sahu PK, Mishra V, Jamal H, Varma A, Tripathi S (2023) Isolation and characterization of halotolerant plant growth promoting rhizobacteria from mangrove region of Sundarbans, India for enhanced crop productivity. Front Plant Sci 14:1122347

Philippot L, Raaijmakers JM, Lemanceau P, Van Der Putten WH (2013) Going back to the roots: the microbial ecology of the rhizosphere. Nat Rev Microbiol 11(11):789–799

Power JF, Carere CR, Lee CK, Wakerley GL, Evans DW, Button M et al (2018) Microbial biogeography of 925 geothermal springs in New Zealand. Nat Commun 9(1):2876

Pradhan A, Seena S, Pascoal C, Cássio F (2011) Can metal nanoparticles be a threat to microbial decomposers of plant litter in streams? Microb Ecol 62(1):58–68

Priester JH, Ge Y, Mielke RE, Horst AM, Moritz SC, Espinosa K et al (2012) Soybean susceptibility to manufacture nanomaterials with evidence for food quality and soil fertility interruption. Proc Natl Acad Sci 109(37):E2451–E2456

Raghib F, Naikoo MI, Khan FA, Alyemeni MN, Ahmad P (2020) Interaction of ZnO nanoparticle and AM fungi mitigates Pb toxicity in wheat by upregulating antioxidants and restricted uptake of Pb. J Biotechnol 323:254–263

Rai-Kalal P, Jajoo A (2021) Priming with zinc oxide nanoparticles improve germination and photosynthetic performance in wheat. Plant Physiol Biochem 160:341–351

Rajput VD, Minkina T, Kumari A, Harish, Singh VK, Verma KK et al (2021) Coping with the challenges of abiotic stress in plants: new dimensions in the field application of nanoparticles. Plants 10(6):1221

Rajput VD, Faizan M, Upadhyay SK, Kumari A, Ranjan A, Sushkova S et al (2022) Influence of nanoparticles on the plant rhizosphere microbiome. In: The role of nanoparticles in plant nutrition under soil pollution: nanoscience in nutrient use efficiency. Springer International Publishing, Cham, pp 83–102

Rakgotho T, Ndou N, Mulaudzi T, Iwuoha E, Mayedwa N, Ajayi RF (2022) Green-synthesized zinc oxide nanoparticles mitigate salt stress in *Sorghum bicolor*. Agriculture 12(5):597

Ravanbakhsh M, Kowalchuk GA, Jousset A (2019) Root-associated microorganisms reprogram plant life history along the growth–stress resistance tradeoff. ISME J 13(12):3093–3101

Rehman H, Ali W, Khan NZ, Aasim M, Khan T, Khan AA (2023) Delphinium uncinatum mediated biosynthesis of zinc oxide nanoparticles and in-vitro evaluation of their antioxidant, cytotoxic, antimicrobial, anti-diabetic, anti-inflammatory, and anti-aging activities. Saudi J Biol Sci 30(1):103485

Reinhold-Hurek B, Bünger W, Burbano CS, Sabale M, Hurek T (2015) Roots shaping their microbiome: global hotspots for microbial activity. Annu Rev Phytopathol 53(1):403–424

Rich MK, Nouri E, Courty PE, Reinhardt D (2017) Diet of arbuscular mycorrhizal fungi: bread and butter? Trends Plant Sci 22(8):652–660

Rosell-Polo JR, Cheein FA, Gregorio E, Andújar D, Puigdomènech L, Masip J, Escolà A (2015) Advances in structured light sensors applications in precision agriculture and livestock farming. Adv Agron 133:71–112

Sabra MS, Allam EA, El-Aal MA, Hassan NH, Mostafa AHM, Ahmed AA (2025) A novel pharmacological strategy using nanoparticles with glutathione and virgin coconut oil to treat gentamicin-induced acute renal failure in rats. Naunyn Schmiedeberg's Arch Pharmacol 398(1):933–950

Samad A, Trognitz F, Compant S, Antonielli L, Sessitsch A (2017) Shared and host-specific microbiome diversity and functioning of grapevine and accompanying weed plants. Environ Microbiol 19(4):1407–1424

Sambangi P, Gopalakrishnan S, Pebam M, Rengan AK (2022) Nano-biofertilizers on soil health, chemistry, and microbial community: benefits and risks. Proc Indian Natl Sci Acad 88(3):357–368

Sasse J, Martinoia E, Northen T (2018) Feed your friends: do plant exudates shape the root microbiome? Trends Plant Sci 23(1):25–41

Schmid AK, Allers T, DiRuggiero J (2020) SnapShot: microbial extremophiles. Cell 180(4):818–818

Sedghi M, Hadi M, Toluie SG (2013) Effect of nano zinc oxide on the germination parameters of soybean seeds under drought stress. Ser Biol 16:73

Sharma K, Shivandu SK (2024) Integrating artificial intelligence and Internet of Things (IoT) for enhanced crop monitoring and management in precision agriculture. Sensors Int 5:100292

Sharma A, Kapoor D, Wang J, Shahzad B, Kumar V, Bali AS, Jasrotia S, Zheng B, Yuan H, Daoliang Y (2020) Chromium bioaccumulation and its impacts on plants: an overview. Plants 9:100

Shivamogga Nagaraju R, Holalkere Sriram R, Achur R (2020) Antifungal activity of Carbendazim-conjugated silver nanoparticles against anthracnose disease caused by Colletotrichum gloeosporioides in mango. J Plant Pathol 102(1):39–46

Shu WS, Huang LN (2022) Microbial diversity in extreme environments. Nat Rev Microbiol 20(4):219–235

Singh J, Vishwakarma K, Ramawat N, Rai P, Singh VK, Mishra RK et al (2019) Nanomaterials and microbes' interactions: A contemporary overview. 3 Biotech 9(3):68

Singh A, Agrawal S, Rajput VD, Ghazaryan K, Movsesyan HS, Minkina T et al (2023) Seed nanopriming: An innovative approach for upregulating seed germination and plant growth under salinity stress. In: Nanopriming approach to sustainable agriculture. IGI Global, pp 290–313

Sorty AM, Meena KK, Choudhary K, Bitla UM, Minhas P, Krishnani K (2016) Effect of plant growth promoting bacteria associated with halophytic weed (*Psoralea corylifolia* L) on germination and seedling growth of wheat under saline conditions. Appl Biochem Biotechnol 180:872–882. https://doi.org/10.1007/s12010-016-2139-z

Spaepen S, Vanderleyden J (2011) Auxin and plant-microbe interactions. Cold Spring Harb Perspect Biol 3:a001438

Srivastava A, Singh R (2021) Nanoparticles for sustainable agriculture and their effect on plants. Curr Nanosci 17(1):58–69

Tripathi DK, Rai P, Guerriero G, Sharma S, Corpas FJ, Singh VP (2021) Silicon induces adventitious root formation in rice under arsenate stress with involvement of nitric oxide and indole-3-acetic acid. J Exp Bot 72:4457–4471

Turan NB, Chormey DS, Büyükpınar Ç, Engin GO, Bakirdere S (2017) Quorum sensing: little talks for an effective bacterial coordination. TrAC Trends Anal Chem 91:1–11

Upadhayay VK, Khan A, Singh J, Singh AV (2019) Splendid role of nanoparticles as antimicrobial agents in wastewater treatment. In: Environmental contaminants: ecological implications and management, pp 119–136

Upadhyay SK, Srivastava AK, Rajput VD, Chauhan PK, Bhojiya AA, Jain D et al (2022) Root exudates: mechanistic insight of plant growth promoting rhizobacteria for sustainable crop production. Front Microbiol 13:916488

Upadhyay P, Navvuru S, Yadav PK, Lalotra S, Singh A, Rajput VD et al (2024) Nano-omics-based abiotic and biotic stresses management. In: Advances in environmental engineering and green technologies book series, pp 348–371

Ur Rahim H, Qaswar M, Uddin M, Giannini C, Herrera ML, Rea G (2021) Nano-enable materials promoting sustainability and resilience in modern agriculture. Nanomaterials 11(8):2068

Vandenkoornhuyse P, Quaiser A, Duhamel M, Le Van A, Dufresne A (2015) The importance of the microbiome of the plant holobiont. New Phytol 206(4):1196–1206

Vargas-Hernandez M, Macias-Bobadilla I, Guevara-Gonzalez RG, Rico-Garcia E, Ocampo-Velazquez RV, Avila-Juarez L, Torres-Pacheco I (2020) Nanoparticles as potential antivirals in agriculture. Agriculture 10(10):444

Varma KN, Vijaysri D, Aravindharajan STM, Kavya T, Govindasamy V, Hareesh V et al (2025) Long-term effects of ENM transformation on the agricultural ecosystem and the interconnections with soil quality and climate change. In: Nano-bioinoculants. Academic, pp 317–336

Venturi V, Keel C (2016) Signaling in the rhizosphere. Trends Plant Sci 21(3):187–198

Verma KK, Joshi A, Song XP, Singh S, Kumari A, Arora J et al (2024) Synergistic interactions of nanoparticles and plant growth promoting rhizobacteria enhancing soil-plant systems: a multigenerational perspective. Front Plant Sci 15:1376214

Villa-Henriksen A, Edwards GT, Pesonen LA, Green O, Sørensen CAG (2020) Internet of Things in arable farming: implementation, applications, challenges and potential. Biosyst Eng 191:60–84

Vishwakarma K, Kumar N, Shandilya C, Varma A (2020) Unravelling the role of endophytes in micronutrient uptake and enhanced crop productivity. In: Symbiotic soil microorganisms: biology and applications. Springer International Publishing, Cham, pp 63–85

Wang X, Liu X, Chen J, Han H, Yuan Z (2014) Evaluation and mechanism of antifungal effects of carbon nanomaterials in controlling plant fungal pathogen. Carbon 68:798–806

Wang C, Ji Y, Cao X, Yue L, Chen F, Li J et al (2022) Carbon dots improve nitrogen bioavailability to promote the growth and nutritional quality of soybeans under drought stress. ACS Nano 16(8):12415–12424

Xiao X, Chen W, Zong L, Yang J, Jiao S, Lin Y et al (2017a) Two cultivated legume plants reveal the enrichment process of the microbiome in the rhizo compartments. Mol Ecol 26(6):1641–1651

Xiao Y, Huang Q, Zheng Z, Guan H, Liu S (2017b) Construction of a Cordyceps sinensis exopolysaccharide-conjugated selenium nanoparticles and enhancement of their antioxidant activities. Int J Biol Macromol 99:483–491

Xie Y, Al-Aly Z (2022) Risks and burdens of incident diabetes in long COVID: a cohort study. Lancet Diabetes Endocrinol 10(5):311–321

Yaish MW, Kumar PP (2015) Salt tolerance research in date palm tree (Phoenix dactylifera L.), past, present, and future perspectives. Front Plant Sci 6:348

Yang Y, Wang N, Guo X, Zhang Y, Ye B (2017) Comparative analysis of bacterial community structure in the rhizosphere of maize by high-throughput pyrosequencing. PLoS One 12(5):e0178425

Yasmin H, Mazher J, Azmat A, Nosheen A, Naz R, Hassan MN et al (2021) Combined application of zinc oxide nanoparticles and biofertilizer to induce salt resistance in safflower by regulating ion homeostasis and antioxidant defence responses. Ecotoxicol Environ Saf 218:112262

Zhang D, Wei L, Zhong M, Xiao L, Li HW, Wang J (2018a) The morphology and surface charge-dependent cellular uptake efficiency of upconversion nanostructures revealed by single-particle optical microscopy. Chem Sci 9(23):5260–5269

Zhang L, Wu L, Si Y, Shu K (2018b) Size-dependent cytotoxicity of silver nanoparticles to Azotobacter vinelandii: growth inhibition, cell injury, oxidative stress and internalization. PLoS One 13(12):e0209020

Zhang T, Mei Q, Zhang Z, Walline JH, Liu Y, Zhu H, Zhang S (2022) Risk for newly diagnosed diabetes after COVID-19: a systematic review and meta-analysis. BMC Med 20(1):444

Zhao S, Su X, Wang Y, Yang X, Bi M, He Q, Chen Y (2020) Copper oxide nanoparticles inhibited denitrifying enzymes and electron transport system activities to influence soil denitrification and N_2O emission. Chemosphere 245:125394

Chapter 3
Synergistic Effect of Plant Growth–Promoting Rhizobacteria and Nanoparticles in Abiotic Stress Alleviation in Plants

B. Chithradevi, M. Petheeswaran, Kavya Bakka, and Dinakar Challabathula

Abstract Environmental stress factors such as salinity, drought, flood, high and low temperatures cause significant losses to agricultural crop productivity, thereby affecting global food security. Traditional agricultural practices hugely relied on chemical fertilizers, wherein although higher crop yield and productivity were observed, the soil health was seriously deteriorated. Further, the agrochemicals had a devastating effect on human health and environment. Under these circumstances, plant growth-promoting rhizobacteria (PGPR) were widely used as biofertilizers to improve the growth and productivity of crops not only in fertile soils but also in contaminated soils. Further PGPR were also employed to protect the plants from abiotic stress-induced damage. The usage of PGPR is a sustainable and environment friendly approach, which is hugely favoured due to varied benefits it offer, such as improved soil texture, decreased dependence on agrochemicals and ensured agricultural sustainability. In recent times, nanotechnology revolution in agriculture has gained significant attention due to the application of various nanomaterials or nano-based fertilizers to increase crop productivity. Nano particles were also added as amendments to the soil to enhance soil texture for improved agricultural yields. Since PGPR and nanomaterials offer varied benefits to soil and plants in isolation, it is expected that the combinatorial effect will be substantial than the effects observed by PGPRs or nanoparticles alone. In agreement, reports have indicated a significant increase in crop yields due to the combined application of PGPRs and nanoparticles than single treatments, advocating the usage of cocktail of PGPR's and nanomaterials as nanobiofertilizers for improved soil properties and increased crop yield. In the current chapter, functional aspects of combined treatments of

B. Chithradevi · K. Bakka
Department of Microbiology, School of Life Sciences, Central University of Tamil Nadu, Thiruvarur, Tamil Nadu, India

M. Petheeswaran · D. Challabathula (✉)
Department of Biotechnology, School of Integrative Biology, Central University of Tamil Nadu, Thiruvarur, Tamil Nadu, India
e-mail: dinakarc@cutn.ac.in

K. A. Abd-Elsalam (ed.), *Plant-Microbiome Nanotechnology*, Sustainability Sciences in Asia and Africa, https://doi.org/10.1007/978-981-95-5056-2_3

PGPRs and nanomaterials for sustainable crop productivity and abiotic stress alleviation will be discussed in detail, advocating their use as an ecofriendly and sustainable approach.

Keywords Abiotic stress · Biofertilizers · Nanobiofertilizers · Nanoparticles · PGPR

3.1 Introduction

Stress that is perceived by plants is defined as any change in the surrounding environment/substance/agents that negatively affect plant metabolism, growth and reproduction (Lichtenthaler 1996). Plants being sessile modify their physiological, biochemical and molecular mechanisms in order to adapt to unfavourable conditions; however, crop yield is often significantly affected by various stress factors. Both biotic and abiotic stresses adversely affect plant growth and development (Kumar and Nautiyal 2022). Abiotic stresses arising from global warming, anthropogenic activities and other environmental challenges are major contributors to reduced agricultural productivity. With the global population projected to reach nearly 10 billion by 2050 (United Nations 2019), the demand for food is expected to rise substantially. It is estimated that global food demand will grow by 60–70% in the next two decades (Duro et al. 2020; Gupta and Shrestha 2023). Alarmingly, the agricultural land is also shrinking due to urbanization, putting farmers under immense pressure to feed the expanding population with limited resources. To address this challenge, farmers have increasingly relied on synthetic chemical fertilizers, which enhance yields but pose long-term risks to both humans and soil health. Globally, around 5.2 billion pounds of pesticides are used annually, and such excessive dependence on pesticides and chemical fertilizers poses detrimental effects on living organisms and the environment (Mahmood et al. 2016; Khan et al. 2022). To overcome these challenges, both biological and non-biological amendments are being employed. Among biological amendments, plant growth-promoting rhizobacteria (PGPR) are well recognized for enhancing plant–soil interactions and supporting overall plant health. In recent years, non-biological amendments, such as nanoparticles, have also shown promising results in agriculture. PGPR promotes plant growth and helps to mitigate stress in plants through multiple mechanisms, including the production of extracellular polysaccharides and biofilms, increasing antioxidant enzymes, osmoprotectants, volatile organic compounds (VOCs), antibiotics, lytic enzymes and by inducing systemic resistance (ISR). Meanwhile, recent advances in nanotechnology have captured significant attention, particularly for their potential applications in agriculture to alleviate stress. The current chapter focuses on exploring the synergistic effects of combining both biological (PGPR) and non-biological amendments (nanoparticles) to mitigate abiotic stress in crops.

3.2 Nanomaterials for Sustainable Agriculture

Sustainability has emerged as a central focus across various sectors, including agriculture. In this context, sustainable agricultural practices are increasingly being adopted to enhance crop productivity with minimal environmental impact. Applications of nanotechnology in agriculture include nanofertilizers, nanopesticides, nanosensors and smart delivery systems, as well as roles in food processing, packaging and safety. With the advent of nanotechnology, nanoparticles are being utilized either as direct sources of micronutrients or as delivery platforms for bioactive agrochemicals, for enhanced crop growth, yield and quality (Prasad et al. 2017). In sustainable agriculture, nanoparticles (NPs) are increasingly being explored as innovative tools to improve crop resilience and productivity while reducing environmental harm. A wide range of metal-based NPs, such as caesium, cobalt, copper, copper oxide, titanium dioxide, calcium carbonate, magnesium, magnesium oxide, manganese, zinc, zinc oxide, selenium, silver, silicon, silicon oxide, iron oxide and molybdenum have been applied to plants under different stress conditions (Khalid et al. 2022). Many of these are formulated as nanofertilizers or nanopesticides, which not only boost plant growth but also lower the ecological footprint compared to traditional agrochemicals.

An especially promising approach is green synthesis, where microbes such as bacteria and fungi are harnessed to produce nanoparticles in an ecofriendly manner. While conventional chemical fertilizers and pesticides show accumulation in soil and deteriorate the soil quality along with environment, biologically derived nanomaterials are considered safe and are considered sustainable alternative to chemicals. Nanomaterials such as silica nanoparticles, quantum dots, carbon nanotubes, polymeric nanoparticles and liposome-based systems have been studied for their role in plant growth promotion, enhancement of nutrient efficiency, abiotic stress tolerance amelioration and optimizing resource use in agricultural farming (Balusamy et al. 2023; Choudhary et al. 2025). Nanoparticles classified as organic or inorganic based on the composition generally range from 1 to 100 nm in size. The nanoparticles render various advantages such as high mobility in plants, easy uptake through the root exudates, entry through roots, leaves or any other plant tissues directly influencing the photosynthesis, stress response and disease resistance (Fincheira et al. 2021). The mode of delivery of nanoparticles to plants varies ranging from seed priming, foliar spray to leaves and soil amendments, hydroponic studies and direct injection. Reports suggested improvement in stability and improved performance of nanoparticles when combined with carriers such as chitosan and biochar (Balusamy et al. 2023; Mi et al. 2023).

Nanomaterials for agricultural applications are considered as nanofertilizers since the nutrients are released to the plants in a slow and controlled manner for efficient nutrient use, resulting in modulation of plant hormone levels, activation of signalling pathways for stress tolerance and strengthening the plant defence system. The slow and steady release of nanoparticles not only leads to improved growth and tolerance to varied stress conditions but also reduces the risk of hyper-fertilization to plants. Although physical (top-down or bottom-up), chemical or biological methods are employed for the synthesis of nanomaterials, biological methods are considered environmentally safe and sustainable for improving plant growth and stress tolerance (Balusamy et al. 2023).

3.3 Mode of Application of Nanoparticles

3.3.1 Seed Priming

Seed priming is one of the widely used techniques facilitating nanoparticles penetration into seeds through passive water-mediated diffusion, demonstrating beneficial effects on seed performance with consistency. Seed priming with nanoparticles has been shown to break dormancy and mitigate the effects of seed ageing. Priming with different types of nanoparticles enhances the activity of key enzymes that facilitate early germination and contribute to activation of antioxidant defence. Seeds of *Brassica juncea* primed with CuO nanoparticles exhibited increased activity of carbonic anhydrase and nitrate reductase enzymes, which in turn promoted shoot and root elongation as well as improved photosynthetic efficiency (Faraz et al. 2023). Similar to zinc nanoparticles, other nanoparticles can penetrate the seed coat, stimulate key genes important for improved water uptake, enhance seed vigour, increase nutrient bioavailability and solubility in soil and provide a controlled, sustained release (Balusamy et al. 2023).

Foliar Application
Foliar application refers to the direct spraying of nanoparticles onto plant surfaces, primarily allowing their entry through stomata, after which they are translocated to different tissues via apoplastic and symplastic pathways (Hong et al. 2021). Foliar application of copper nanoparticles to leaves of tomato plants exposed to salinity stress conferred significant tolerance to salt stress with remarkable enhancement in the activity of several key antioxidant and defence-related enzymes, including glutathione peroxidase, catalase, superoxide dismutase, phenylalanine ammonia lyase and ascorbate peroxidase, thereby contributing to improved stress tolerance (Pérez-Labrada et al. 2019).

3.3.2 Soil Application

Synthesized nanoparticles can be directly applied to the soil at an optimum concentration, from where they are absorbed by the roots and subsequently utilized by the plant. In the case of soil application, careful consideration must be given to optimizing the concentration of nanoparticles, as excessive application can lead to phytotoxic effects and may also pose risks to soil health and the surrounding environment (Tourinho et al. 2012; Dimkpa et al. 2013; De Moraes et al. 2021).

3.4 Mitigation of Abiotic Stress in Plants by Employing NPs

3.4.1 Zinc Oxide Nanoparticles (ZnO-NPs)

Zinc (Zn) being an essential micronutrient is vital for plant growth and development. Although required in plants in minimal amounts and found to be toxic in excess quantities, its deficiency severely impairs carbohydrate, protein and chlorophyll synthesis. Being a crucial micronutrient, it is required in adequate amounts for plant growth, development and stress tolerance. However, lack of bioavailable zinc and usage of conventional sources of zinc such as zinc oxide (ZnO) and zinc sulfate ($ZnSO_4{\cdot}7H_2O$), which show low soil solubility renders limited bioavailability to plants causing Zn deficiency. Alternatively, zinc oxide nanoparticles (ZnO-NPs) are considered as suitable alternatives and a novel source of zinc fertilizer for agricultural applications for assessing the growth and development (Balusamy et al. 2023). The application of ZnO-NPs to sorghum plants under water-limiting conditions significantly improved grain yield along with improved acquisition of nitrogen, phosphorus and potassium through roots and shoots (Dimkpa et al. 2019). Similarly, compared to controls, increased levels of non-enzymatic antioxidants, including total flavonoid and total phenolic contents, were observed with simultaneous reduction in malondialdehyde (MDA) and hydrogen peroxide (H_2O_2) levels in potato plants treated with foliar application of ZnO-NPs (100 and 50 mg L^{-1}) and SiO_2-NPs (50 mg L^{-1}) under drought stress. The combined application not only improved yield but also enhanced several key biochemical attributes (Al-Selwey et al. 2023). Further, foliar application of ZnO-NPs (25 and 50 mg L^{-1} concentrations) synthesized from *Coleus forskohlii* leaf extracts effectively alleviated drought stress in tomato plants, by enhancing ascorbic acid and free phenol levels and boosting the activities of key antioxidant enzymes, including superoxide dismutase, catalase and ascorbate peroxidase (El-Zohri et al. 2021). In another study, while the *Leucaena leucocephala* seedlings exposed to cadmium and lead stress showed a decline in growth, plants treated with ZnO-NPs showed enhanced photosynthetic pigment levels, enhanced antioxidant levels and reduced MDA levels (Venkatachalam et al. 2017).

Among various environmental stresses encountered by plants, salinity is one of the most detrimental exerting negative effects from the earliest stages of seed germination and seedling establishment through to vegetative growth and reproduction. High salt concentrations disrupt osmotic balance, hinder water and nutrient uptake and trigger ionic toxicity, ultimately impairing photosynthesis, reducing biomass accumulation and significantly affecting crop yield and quality. Several reports emphasize the importance of initiating treatments at the earliest developmental stages, such as seed priming, to mitigate salinity-induced damage. Anwar et al. (2023a) demonstrated that seed priming with a combination of ZnO-NPs and melatonin significantly enhanced shoot growth, chlorophyll content and accumulation of total soluble sugars and proteins compared to untreated controls. Notably, the combined application of nanoparticles has been shown to exert greater benefits than individual treatments. Foliar application of both bulk and nano-sized ZnO and Fe_2O_3 nanoparticles to two wheat cultivars under salinity stress improved plant height, leaf area and shoot biomass (Fathi et al. 2017). Furthermore, the combined application of sugarcane bagasse with ZnO-NPs effectively mitigated cadmium and salinity stress in wheat plants, resulting in 75% increase in yield compared to 42% under individual treatments, while also contributing to soil remediation in contaminated sites (Hafez et al. 2025).

3.4.2 Titanium Dioxide Nanoparticles (TiO_2- NPs)

Titanium dioxide nanoparticles (TiO_2-NPs) exert significant biological effects on plants, showing huge beneficial effects at lower concentrations; however, at higher doses, they become extremely toxic. The influence of TiO_2-NPs on plants is highly context-dependent, varying with environmental conditions, plant species and the applied nanoparticle concentration. Although physical and chemical methods are widely used for synthesizing TiO_2 nanoparticles, plant extract–mediated synthesis has attracted increasing attention owing to its simplicity, cost-effectiveness, safety, non-toxicity and environmentally friendly nature. Leaves are a preferred material for green synthesis, as their metabolite-rich extracts serve as effective agents for the production of TiO_2 nanoparticles (Alabdallah et al. 2024). TiO_2 nanoparticles (TiO_2-NPs) have been reported to enhance photosynthetic rates in crops while also stimulating the activity of specific enzymes. TiO_2-NPs were synthesized using TiO_2 salt and *Buddleja asiatica* plant extract as a reducing and capping agent. The combined effect of TiO_2-NPs (40ppm) and $Ca_3(PO_4)_2$ (40ppm) on wheat under drought stress has led to increased root length, shoot length and fresh weight (Mustafa et al. 2021). In another study, increased antioxidant enzyme activities with lower MDA and H_2O_2 contents were observed in wheat plants treated with TiO_2 and ZnO-NPs under heat stress (Thakur et al. 2021). The transcriptome analysis of leaves of chickpea plants exposed to TiO_2-NPs under cold stress revealed active mitigation of oxidative damage by activation of distinct genes and pathways associated with cold stress tolerance mediated by NPs (Amini et al. 2017). A significant reduction in electrolyte leakage from leaves of cold-tolerant and cold-sensitive varieties of chick pea plants treated with TiO_2-NPs. A significant increase in the growth of *Oryza sativa* plants grown under varying soil conditions, along with increased nutrient uptake, increased root and shoot length and higher biomass accumulation was observed upon treatment with TiO_2-NPs (Arshad et al. 2021). These studies strongly advocate the positive effects of TiO_2-NPs on plant growth, development and stress tolerance.

3.4.3 Silica Nanoparticles (Si-NPs)

The role of silicon (Si) in biogeochemical processes and plant physiological functions cannot be underestimated. Being the second most abundant element in Earth's crust, it plays a key role in plant physiological functions, such as deposition in plant cell walls to increase physical strength, reinforcement of plant structure, improvement of stress tolerance and boosting disease resistance. Silicon and silicon nanoparticles (Si-NPs) are known to protect plants from oxidative damage by activating defence mechanisms upon exposure to biotic and abiotic stress conditions (Wang et al. 2022a, b). Remarkable differences were observed between untreated and Si and Si-NP-treated leaves of wheat plants under heat stress. While leaves of untreated plants showed plasma membrane disruption, leaves of NP treated plants showed cellular integrity (Younis et al. 2020). However, the responses with Si were more effective than Si-NP in wheat plants for improving thermotolerance. Increased expression levels of genes encoding *Triticum aestivum* plasma membrane intrinsic protein (TaPIP1) and *Triticum aestivum* nodulin 26-like intrinsic protein (TaNIP2) were observed in Si-treated plants correlating with improved relative water content confirming the effectiveness of Si over Si-NP in wheat thermotolerance (Younis et al. 2020). Species-specific and dose-dependent responses are crucial in the case of NPs; hence, utmost care is advocated for applying the correct dose and validating the responses of different species being tested. The foliar application of Si-NPs at 200 ppm to cucumber plants alleviated the salinity stress symptoms and showed increased germination rate, increased vigour and biomass (Alsaeedi et al. 2018). In a greenhouse study, using four Si-NPs concentrations under three irrigation regimes, it is observed that 200 mg/kg Si-NPs significantly improved the growth and nutrient uptake of cucumber plants, resulting in increased plant height, increased chlorophyll content and increased fruit yield not only under normal conditions but also under salinity stress conditions, highlighting the efficiency of SiNPs in mitigating salinity stress in cucumber plants (Alsaeedi et al. 2019). Application of Si-NPs to *Brassica oleracea* var. botrytis plants exposed to salinity stress showed improved root growth and increased antioxidant activity (Anwar et al. 2023b). Application of Si-NPs was also shown to mitigate the heavy metal tolerance in plants by displaying increased seed germination, increased shoot and root growth, higher chlorophyll content, reduced electrolyte leakage and higher antioxidant enzyme activities under heavy metal stress conditions (Ali et al. 2019). These findings strongly indicate the potential of Si-NPs in improving plants' tolerance to varied abiotic stress conditions.

3.4.4 Silver Nanoparticle (Ag-NPs)

Application of silver nanoparticles (Ag-NPs) to crop plants for improving productivity and stress tolerance is a promising approach, since AgNPs are reported to play a dual role in plants by restricting the attack of microbial pathogens and by promoting plant growth under different abiotic stress conditions. They have been widely used as antifungal agents and plant growth-promoting agents. Cucumber plants treated with Ag ions and Ag-NPs showed a significant increase in key amino-acids (cycloleucine, isoleucine, lysine, methionine, norleucine, ornithine, threonine and β-alanine and metabolites (arbazole, citraconic acid, raffinose, dithioerythritol, D-erythronolactone, lactulose and N-methyl-L-glutamic acid) contributing to increased antioxidant activity and plant growth promotion (Zhang et al. 2018). While application of AgNPs to *Satureja hortensis* plants showed improved germination efficiency, increased stem length and increased plant height under normal and salt stress conditions (Nejatzadeh 2021), AgNP treatment to *Pennisetum glaucum* L. plants has led to mitigation of oxidative damage and increased expression of genes encoding key antioxidant enzymes (superoxide dismutase, catalase and peroxidase) under salt stress conditions (Khan et al. 2023). In another study, application of Ag-NPs to wheat reduced the devastating effects of salinity stress by increasing the K^+ ion concentration in the cells, abscisic acid production, increasing the antioxidant activity, chlorophyll content, proline accumulation, nitrogen content along with increased activities of nitrogen assimilating enzymes nitrate reductase and nitrite reductase (Wahid et al. 2020). The results from these studies strongly demonstrate Ag-NP–mediated enhancement of growth and diversified responses in different plant species for enhanced stress tolerance through modulation in physiological, biochemical and molecular mechanisms for combating adverse environmental conditions.

3.4.5 Selenium Nanoparticles (Se-NPs)

The micronutrient selenium (Se) plays an important role in influencing various plant physiological and biochemical processes related to improvement of plant metabolism, enhancement of secondary metabolites and activation of biotic and abiotic stress tolerance mechanisms. Selenium nanoparticles (Se-NPs) were found to be more active than bulk Se for improving resistance to oxidative stresses, plant growth promotion, enhancement of soil nutrient status, enhancement of plant antioxidant levels and participation in the transpiration process. Tomato plants treated with Se - NPs have shown increased antioxidant activity and increased flavonoid and carotenoid contents under drought stress conditions (Ishtiaq et al. 2023). *Capsicum annum* plants treated with selenium, silicon and copper nanoparticles showed increased chlorophyll and lycopene content and enhanced activity of glutathione peroxidase, phenylalanine ammonia lyase, catalase and peroxidase in leaves, concomitant to elevated levels of phenols, β-carotene, flavonoids, yellow carotenoids and glutathione in fruit tissues (González García et al. 2021). Foliar application of Se-NPs in strawberry plants improved drought and salinity tolerance by showing increased plant growth, osmolyte accumulation and increased antioxidant activity (Zahedi et al. 2019). Foliar application of selenium (Se), silicon (Si) and zinc (Zn) nanoparticles has led to ameliorative effects of lead and cadmium heavy metal stress in *Salvia officinalis* L. plants. The treated plants showed increased biomass, yield and enhanced essential oil biosynthesis in addition to mitigation of heavy metal stress (Bakhtiari et al. 2023).

3.4.6 Cerium Oxide Nanoparticles (CeO_2-NPs)

Studies have indicated the beneficial effects of cerium oxide nanoparticles (CeO_2-NPs) in plant growth promotion and abiotic stress alleviation at low dose levels. However, higher doses are found to be toxic and detrimental to plants. CeO_2-NPs act as nanozymes and benefit plants by improving photosynthetic parameters and increasing antioxidants and morphological attributes. Plants treated with CeO_2-NPs show activation of stress signalling cascades to scavenge reactive oxygen species under different abiotic stress conditions. Significant improvement in photosynthesis and water-use efficiency was observed in *Glycine max* L. plants treated with cerium oxide nanoparticles (CeO_2-NPs) under drought stress. However, treatment with higher concentrations negatively affected Rubisco activity, possibly due to toxicity (Cao et al. 2018). Improvement of salinity tolerance in rice plants, along with increased chlorophyll content, shoot length and grain weight, was observed upon treatment with poly(acrylic acid)-coated cerium oxide nanoparticles (PNC) (Zhou et al. 2021). Significant reduction in lipid peroxidation and increase in photosynthetic carbon assimilation and yield were observed in *Sorghum bicolor* L. plants treated with CeO_2-NPs through foliar application (Djanaguiraman et al. 2018). Studies using CeO_2-NPs for the enhancement of plant growth, improved photosynthetic efficiency and stress tolerance reiterate the potential of nanoparticles for climate-resilient agriculture.

3.4.7 Fe_2O_3 Nanoparticles

Iron (Fe) as a vital macronutrient element is required for photosynthetic and respiratory electron transport chains in chloroplasts and mitochondria, and for the biosynthesis of chlorophyll and photosynthetic proteins. Iron nanoparticles are applied as fertilizers for the supply of Fe to plants, facilitating the solubility and dispersion of insoluble Fe, thereby increasing the bioavailability over a wide range of soil pH conditions. Foliar application of salicylic acid combined with Fe_2O_3 nanoparticles in Ajowan plants has led to improved H+ pump activity, increased ATP content in roots and leaves and increased essential oil production in vegetative and reproductive tissues, respectively (Ghassemi-Golezani and Abdoli 2021). Further, rice plants grown in arsenic-contaminated soil upon treatment with Fe-NPs enhanced glyoxylase and antioxidant enzyme activities, reduced electrolyte leakage and downregulated the arsenic transporter genes Lsi1 and Lsi2, thereby promoting plant growth (Bidi et al. 2021). Fe_2O_3-mediated graphene oxide mitigated the harmful effects of polystyrene nanoplastics by increasing antioxidant enzyme activity and chlorophyll fluorescence (Arikan et al. 2022). Fe_2O_3 nanoparticles promoted rice plant growth by reducing electrolyte leakage, reduced H_2O_2, and MDA levels, and enhanced antioxidant enzyme activities limiting cadmium uptake and accumulation mitigating cadmium toxicity by modulating cellular processes and regulating Cd and Fe transporter genes (Ghouri et al. 2025).

3.4.8 Copper-Based Nanoparticles (Cu-NPs)

Application of copper-based nanoparticles (Cu-NPs) has been shown to improve growth and increase tolerance of crop plants to varied stress conditions. Cu-NPs when applied alone or with plant growth regulators or plant beneficial microbes are known to mitigate impacts of abiotic and biotic stressors across diverse plant species by activating enzymatic and non-enzymatic antioxidant defence responses for reducing oxidative stress, membrane stability preventing cellular damage, promoting the synthesis of primary and secondary metabolites and improving photosynthetic efficiency. The effectiveness of Cu-NPs strongly depends on particle size, dosage, coating material, duration of exposure, target plant species and developmental stage, which need to be critically considered for designing nano-enabled agricultural strategies (Kaleem et al. 2024). Regardless of the benefits offered by Cu-NPs to host plants, studies have also reported negative impacts on plants under certain conditions. Increased oxidative damage and reduced nutrient uptake were observed in Brassica plants treated with Cu-NPs under cadmium stress (Wang et al. 2022a, b). Further, application of Cu-NPs suppressed the growth and increased oxidative stress markers, including ROS, H_2O_2, MDA and electrolyte leakage in *Cucumis melo* (Shah et al. 2023). Reports strongly suggest that Cu-NPs can boost plant growth and development and confer abiotic stress tolerance; however, effective application requires careful optimization for preventing phytotoxicity.

3.4.9 Other Nanoparticles Used for Improvement of Stress Tolerance

Manganese nanoparticles (Mn-NPs) applied to bell pepper plants through nanopriming under salinity stress improved seed germination and promoted root growth (Ye et al. 2020). Copper-induced metal toxicity in *Brassica napus* L. plants was significantly reduced upon application of sulfur nanoparticles (S-NPs) along with reduced oxidative damage and increased total chlorophyll and protein content, signifying the importance and potential of S-NPs to enhance tolerance to heavy metal stress (Yuan et al. 2023). Further, wheat plants treated with Astaxanthin nanoparticles (Ast-NPs) have shown cadmium stress mitigation by reducing cadmium translocation in the cells, maintaining ionic balance and enhancing antioxidant enzyme activities, leading to improved stress tolerance (Zeshan et al. 2022). Application of nano-zerovalent iron (nZVI) nanoparticles along with biochar to wheat plants significantly alleviated lead toxicity and showed increased root, shoot, grain biomass and increased activity of antioxidant enzymes (Nauman Mahamood et al. 2023). Further, water-soluble nanoparticles alleviated salinity stress in host plants by increasing chlorophyll content and by promoting root and shoot growth. Stimulation in lateral development while limiting the elongation of primary roots indicates a specific selective effect on root architecture (Baz et al. 2020). Taşkın et al. (2018) studied the effect of nano-hydroxyapatite (NHA) and soluble phosphorus in low- and high-calcareous soil where phosphorus availability is limited in soil. NHA is known to help plants to uptake more phosphorus and increase dry weight compared to soluble phosphorus. Krishnamoorthy and Rajiv (2016) studied the effect of a seed coated with cobalt nanoparticles and nutrient-rich hexaaminocyclotriphosphatriene (HACTP) compound in PVP nanofibers. Their conductivity studies showed that combinations of HACTP, PVP and Co-NPs show controlled release compared to other fibres. Chitosan nanoparticles (CH-NPs) are used to mitigate cold stress in banana plants by reducing reactive oxygen species and lipid peroxidation. Foliar application of different concentrations of CH-NPs resulted in increased photosynthetic pigments and antioxidant enzyme activity (Wang et al. 2021). Seed treatment and foliar application of chitosan nanofertilizer comprising copper and salicylic acid showed 2.5–3.5-fold increase in sucrose translocation between the internodes along with increased leaf chlorophyll content (Sharma et al. 2020). Phyto-nanoparticles promoted proline and phenol accumulation in plants and enhanced antioxidant activity, with green-synthesized nanoparticles exhibiting lower toxicity than chemically synthesized nanoparticles (Santos-Espinoza et al. 2021). Compared to untreated plants, chitosan-treated *Salvia abrotanoides* plants under drought stress showed enhanced antioxidant activity along with increased relative water content and chlorophyll content (Attaran Dowom et al. 2022). Abiotic stresses can severely impair plant growth and development across morphological, physiological, biochemical, metabolic, cellular and molecular levels, ultimately reducing agricultural productivity. The exogenous application of nanoparticles through seed priming, foliar sprays, root irrigation or soil amendments can alleviate these stresses, mitigate their harmful effects and enhance plant stress tolerance. Table 3.1 summarizes the effects of various nanoparticles on plants, highlighting the key physiological, biochemical and growth-related changes observed towards enhanced tolerance to different abiotic stresses.

3.5 Benefits of PGPR for Abiotic Stress Alleviation

Plant growth-promoting rhizobacteria (PGPR) possess diverse functional traits, including the presence of ACC deaminase activity, phosphate solubilization, siderophore production and the ability to synthesize phytohormones such as auxins,

Table 3.1 Effect of nanoparticles in enhancing plant growth under normal and different abiotic stress conditions

Nanoparticle	Stress	Key observations	Plant	References
ZnO-NP	Drought	Improved grain yield, root and shoot acquisition of nutrients P, N and K	*Sorghum bicolor*	Dimkpa et al. (2019)
ZnO and Fe_2O_3	Salinity	Enhanced plant growth, leaf area, fresh and dry biomass	*Triticum aestivum* (Kavir and Tajan)	Fathi et al. (2017)
ZnO-NPs	Drought	Improved shoot and root dry weight. Augmentation of ascorbic acid, free phenols and the activity of antioxidant enzymes	*Solanum lycopersicum*	El-zohri et al. (2021)
ZnO-NPs	Cd and Pb	Increased photosynthetic pigments and total soluble protein contents. Decreased lipid peroxidation	*Leucaena leucocephala*	Venkatachalam et al. (2017)
ZnO-NPs and Si-NPs	Drought	Increased yield correlating with increased antioxidants	*Solanum tuberosum*	Al-Selwey et al. (2023)
ZnO-NPs	Salinity	Higher shoot growth, increased chlorophyll content, total soluble sugar and protein	*Solanum melongena*	Anwar et al. (2023a)
Sugarcane bagasse and zinc oxide NPs	Cadmium and salinity stress	Increased wheat yield increased antioxidant levels	*Triticum aestivum*	Hafez et al. (2025)
TiO_2-NPs, ZnO - NPs	Heat	Higher antioxidant enzyme activities and lower H_2O_2 content and MDA levels	*Oryza sativa* L.	Thakur et al. (2021)
TiO_2 complexed with Calcium Phosphate	Drought	Increased root, shoot and dry weight, increase in chlorophyll content, membrane stability index, proline, yield and sugar content. Increase in antioxidant levels	*Triticum aestivum* L.	Mustafa et al. (2021)
TiO_2-NPs	Cold	Decrease in electrolyte leakage and oxidative stress	*Cicer arietinum* L.	Amini et al. (2017)
TiO_2-NPs	No stress (different soil is used)	Improvement in nutritional uptake, increase in the dry weight, biomass, root and shoot length	*Oryza sativa*	Arshad et al. (2021)
Chitosan - NPs	Drought	Increased antioxidant enzyme activities. Increased RWC and chlorophyll content	*Saliva abrotanoides*	Attaran Dowom et al. (2022)

(continued)

Table 3.1 (continued)

Nanoparticle	Stress	Key observations	Plant	References
Chitosan nanoparticles	Cold stress	Increased photosynthetic pigments and antioxidant enzyme activity	*Musa acuminata*	Wang et al. (2021)
Chitosan nanoparticles with Cu and SA	No stress	Increased photosynthetic pigments and sucrose translocation between internodes	*Triticum aestivum* L.	Sharma et al. (2020)
Nano silica	Salinity and water stress	Enhanced plant height, chlorophyll content and fruit yield	*Cucumis sativus*	Alsaeedi et al. (2019)
Si-NPs	Heavy metal	Improved seed germination, shoot and root length, Increased chlorophyll content and antioxidant enzyme activities	*Triticum aestivum* L. Wheat	Ali et al. (2019)
Si-NPs	Salinity	Increased root length and increased antioxidant enzyme activity	*Brassica oleracea var. botrytis*	Anwar et al. (2023b)
Silica-NP with Biochar	No stress	Increased nitrogen, phosphorous, chlorophyll and antioxidant	*Solanum lycopersicum* and *Citrullus lantatus*	Pavlicevic et al. (2022)
Si-NP	Heat stress	Disruption of plasma membrane in untreated plants upon heat stress, whereas in treated plants densely packed nucleus with an intact envelope is observed with downregulation of *TaPIP1* and *TaNIP2* gene expression	*Triticum aestivum* L.	Younis et al. (2020)
S-NPs	Metal (Cu) toxicity	Enhanced antioxidant defence system Reduced lipid peroxidation and increased nutrient accumulation	*Brassica napus* L.	Yuan et al. (2023)
Ag-NPs	No stress	Increased accumulation of metabolites like arbazole, raffinose, lactulose, citraconic acid, aspartic acid, dithioerythritol, D-erythronolactone and *N*-methyl-L-glutamic acid	*Cucumis sativus*	Zhang et al. (2018)
Ag-NPs	Salinity	Improved germination efficiency, stem length and plant height	*S. hortensis*	Nejatzadeh (2021)

(continued)

Table 3.1 (continued)

Nanoparticle	Stress	Key observations	Plant	References
Ag-NPs	Salt	Improved seed germination, antioxidant machinery and proline accumulation	*Triticum aestivum* L.	Wahid et al. (2020)
Ag-NPs	Salt	Decreased oxidative damage by increasing the antioxidant enzyme activities. Increased 500+ transcripts associated with various metabolic and molecular functions	*Pennisetum glaucum* L.	Khan et al. (2023)
Se-NPs	Drought	Decreased levels of H_2O_2 and MDA. Increased activity of antioxidant enzymes and increase in total flavonoids and carotenoids	*Solanum lycopersicum*	Ishtiaq et al. (2023)
Se-NPs, Si-NPs and Cu-NPs	Salt	Leaves: Increased lycopene, beta-carotene, chlorophyll and glutathione peroxidase Fruits: Increased enzyme activities. Decreased glutathione and flavonoid levels	*Capsicum annuum* L.	González García et al. (2021)
Se-NPs	Salinity	Enhanced osmolyte production, increased photosynthetic pigments and antioxidant enzyme activities	*Fragaria ananassa*	Zahedi et al. (2019)
Si-NPs, Se - NPs and Zn-NPs	Metal toxicity	Increased biomass, yield and essential oils (camphor, α-thujone, β-thujone and 1,8-cineole) and decreased Pb and Cd accumulation	*Salvia officinalis* L.	Bakhtiari et al. (2023)
Mn-NPs	Salt	Development of NP-corona complex, Enhancement in seed germination and root growth	Bell Peppers (*Capsicum annuum* L.)	Ye et al. (2020)
Ast-NPs (Astaxanthin NPs)	Metal (Cd) stress	Restricted translocation of Cd to aerial parts of the plant, improved length and nutrient profile	*Triticum aestivum* L.	Zeshan et al. (2022)
Nano zerovalent Iron (nZVI) NPs	Metal (Pb) toxicity	Increase in photosynthetic and antioxidant enzymatic activities	*Triticum aestivum* L.	Nauman Mahamood et al. (2023)

(continued)

Table 3.1 (continued)

Nanoparticle	Stress	Key observations	Plant	References
CuO-NPs	Negative impact	Suppressed the growth rate and affected physiological and biochemical activities	Melon (*Cucumis melo* L.)	Shah et al. (2023)
Cu-NPs and CuO-NPs	Metal toxicity (Cd)	Decreased nutritional value and increased oxidative damage	Brassica (*Brassica campestris* L.)	Wang et al. (2022a, b)
Water-soluble carbon nano particles (CNP)	Salinity	Alleviated the salt stress and improved chlorophyll content, root and shoot length. Promotion in lateral root growth	*Lactuca sativa*	Baz et al. (2020)
Fe_2O_3-mediated graphene oxide (FGO)	Nanoplastics	Increased in chlorophyll fluorescence	*Triticum aestivum* L.	Arikan et al. (2022)
Nano Fe_2O_3 and salicylic acid	Salt	Improved plant growth Enhanced activities of H+-ATPase and H+ PPase	*Trachyspermum ammi* L.	Ghassemi-Golezani and Abdoli (2021)
Fe-NPs	Heavy metal (Arsenic)	Nourishment of glyoxalase system and antioxidant enzymes. Reduction in electrolyte leakage and downregulate the expression of gene Lsi1 and Lsi2 (involved in Arsenic uptake)	*Oryza sativa* L.	Bidi et al. (2021)
CeO_2-NPs	Salt	Elevated chlorophyll content, shoot length and grain weight	Rice (*Oryza sativa* L.)	Zhou et al. (2021)
Cerium oxide (CeO_2)-NPs	Drought	Huge impact on the photosynthesis and water use efficiency as well as high concentration of CeO_2 NPs affected the Rubisco activity.	Soybean (*Glycine max* (L.) Merr.)	Cao et al. (2018)
Cerium oxide (CeO_2)-NPs	Drought	Decreased membrane lipid peroxidation and elevated the carbon assimilation rates	*Sorghum bicolor* (L.) Moench	Djanaguiraman et al. (2018)

gibberellins, cytokinins, salicylic acid and jasmonic acid. These attributes not only enhance plant growth and development but also help plants withstand and adapt to various abiotic stresses. Crops cultivated under open-field natural conditions are constantly exposed to multiple environmental stresses such as drought, salinity, nutrient scarcity and elevated temperatures, which often occur simultaneously and intensify the severity of stress. Further, the accumulation of heavy metals like Mn, Pb, Cd and Cr in agricultural soils, largely driven by industrial expansion and population growth, poses another major constraint for sustainable crop production. Most of the crop species are sensitive to these stress conditions and lack the adaptive capacity to tolerate such harsh environments, resulting in significant loss in crop productivity and yield, with environmental stresses alone contributing to nearly 50% of global yield losses (Shah et al. 2021; National Climate Assessment-USDA). To address these challenges, a sustainable and ecofriendly approach that enhances resource-use efficiency while minimizing reliance on chemical inputs is required. In this context, plant growth-promoting rhizobacteria (PGPR), first described in the early 1980s (Kloepper et al. 1980), have emerged as a promising biological tool, with numerous strains now isolated and characterized for their potential to improve plant growth and resilience under stress-prone conditions.

Soil salinity, largely driven by urbanization and industrial activities, poses a serious challenge to agricultural productivity. Salinity is a critical abiotic stress that deteriorates soil quality and severely hampers crop performance at all stages of growth, posing a global threat to agriculture with an estimated annual loss of nearly US$12 billion (FAO 2020). Soil quality is important for the efficient growth of plants. In case of salinity stress, plants are exposed to high concentrations of salt, which affects plants in all developmental stages. Both physiological and biochemical characteristics are altered like water status, photosynthesis, protein synthesis, leaf area and lipid metabolism (Saleem et al. 2021; Hmaeid et al. 2019). To overcome these changes, an ecofriendly approach of employing plant growth-promoting rhizobacteria is a viable option. These beneficial bacteria help plants to ameliorate adverse conditions by employing various stress tolerance mechanisms. Also they are genetically flexible and release plant growth-promoting hormones (Dodd and Pérez-Alfocea 2012; Hanin et al. 2016; Meena et al. 2017; Bakka and Challabathula 2020). PGPR can be introduced into plants through different application methods such as seed coating, soil drenching with bacterial cultures or foliar sprays. Somna et al. (2025) demonstrated that combining seed inoculation with subsequent soil drenching significantly enhanced root and shoot development, showing clear differences in growth between PGPR inoculated plants and the uninoculated plants. Neshat et al. (2022) investigated the role of *Pseudomonas* sp. C16-2O and *Enterobacter* sp. S16-3 in supporting canola plants under saline conditions. Their study revealed that PGPR inoculation enhanced antioxidant enzyme activity (SOD, CAT, APX, PPO), improved root surface development, reduced MDA levels and effectively mitigated salt-induced damage compared to untreated controls. Similarly, salicylic acid, a key plant signalling molecule, is known for its role in stress tolerance. Ali et al. (2023b) reported that the combined application of salt-tolerant *Pseudomonas aeruginosa* and exogenous salicylic acid in salt-sensitive maize not only alleviated salinity stress but also resulted in a 25% increase in photosynthetic pigments and a 27% improvement in grain yield, outperforming their individual applications. Lu et al. (2024) discovered a novel *S. lasalocidi* JCM 3373 bacteria, which alleviates salinity stress in soybean. It does not colonize roots but secretes chemical compound indole-3-carboxaldehyde (ICA1d), which alters root architecture. Exogenous application of ICA1d (50 μM) showed increased expression of stress-tolerant genes and also augmentation of root volume, diameter and fresh weight in treated plants compared to control.

3.6 Production of Volatile Organic Compounds (VOCs) by Microbes

Volatile organic compounds (VOCs) released by microorganisms have been shown to enhance plant growth even under stress conditions. In a study by Cappellari et al. (2023), *Bacillus amyloliquefaciens* GB03 served as the source of VOCs, supplied through sterile vials containing bacterial inoculum. *Mentha piperita* seedlings exposed to both salt stress and VOCs exhibited a 20% increase in shoot fresh weight and higher antioxidant activity when compared to salt-stressed plants not exposed to mVOCs. Drought stress triggers excessive accumulation of reactive oxygen species and osmolytes in plants, which in turn alters key physiological processes such as photosynthesis, transpiration, flowering time, stomatal behaviour and cell turgor. Prolonged exposure damages cellular structures by impairing membrane integrity, protein synthesis and other vital metabolic pathways (Zia et al. 2021; Wasaya et al. 2022; Gupta et al. 2024). Arora et al. (2023) observed that drought-induced oxidative stress can inhibit the growth of drought-tolerant plant growth-promoting bacteria (PGPB), such as *Enterobacter bugandensis* WRS7. Metabolic profiling showed accumulation of protective osmolytes like proline, betaine, trehalose and exopolysaccharides, providing valuable insights into microbial drought tolerance mechanisms. Upon inoculation, drought-tolerant bacterium enhanced plant growth not only under normal conditions but also under water-limited conditions. *Microbacterium paraoxydans*, which was isolated from arsenic-contaminated environment, was able to show plant growth promotion by significantly improving plant height, biomass, branch number, root and shoot development, as well as chlorophyll content (Mandal et al. 2024). Similarly, *Acinetobacter oleivorans* IRS14, which was identified as a psychrotrophic bacterium due to its higher growth rate at 10 °C, exhibited multiple PGPR traits, including siderophore production, phosphate solubilization, IAA synthesis and ACC deaminase activity. Inoculation of wheat seeds with IRS14 showed enhancement in seed germination under chilling stress to 80%, in contrast to only 40% in untreated controls. Further, plants treated with IRS14 showed increased levels of photosynthetic pigments, antioxidants and proline, promoting its role in mitigating cold stress (Ali et al. 2023a).

Co-inoculation of the arbuscular mycorrhizal fungus *Rhizophagus irregularis* and the plant growth-promoting bacterium *Bacillus megaterium* to maize plants has shown enhanced protection from drought and high-temperature stress by exhibiting improved photosystem II efficiency, higher stomatal conductance, enhanced photosynthetic performance and root hydraulic conductivity through the regulation of aquaporin genes. These kinds of synergistic interactions are proven to be more effective than single-microbe inoculations, advocating their potential for improving crop resilience under changing environmental conditions (Romero-Munar et al. 2023).

3.7 Soil Quality Improvement by PGPR and Nanoparticle Interactions

Contamination of soil is a critical environmental issue primarily caused due to industrial effluents, agricultural runoff and extended use of conventional chemical fertilizers and pesticides. Extensive use and application of conventional fertilizers to agricultural fields destroy soil fertility and pose risk to human health and the environment. These are considered chemical pollutants that deteriorate soil health, reduce crop productivity and pose long-term risks to agricultural fields. Studies have suggested an integrated approach combining nanoparticles with plant growth-promoting rhizobacteria (PGPR) as a promising solution for the remediation of soil. While PGPR serve as biofertilizers to enhance nutrient availability, promote plant growth and improve stress tolerance, nanoparticles function in immobilization or degradation of toxic compounds, leading to soil quality restoration. This synergistic approach is advantageous with remarkable potential for sustainable agriculture and environmental management, serving as a simple and viable alternative to conventional remediation methods.

The quality of soil is crucial in shaping the structure and function of microbial communities, which is important for assessing any fertilizer or natural amendment before soil application. The combined use of PGPR and nanoparticles benefits plants by decreasing sodium uptake and increasing the K^+/Na^+ ratio and maintaining ionic balance under salt stress (Al-Turki et al. 2023). The combined use of PGPR and nanoparticles significantly improves soil properties, water-holding capacity, pH, electrical conductivity, phosphorus, nitrate content and other organic compounds, predominating observed effects with individual applications (Akhtar et al. 2021; Rajput et al. 2023). The interaction between nanoparticles and PGPR under different stress conditions helps them to alleviate the stress and show improved plant growth promotion compared to untreated plants. The combined effects of bulk and nano forms of ZrO_2 (zirconia) and TiO_2 with PGPR on properties of soil and nutrient dynamics were evaluated for understanding the potential of nanomaterials in sustainable nutrient management (Karunakaran et al. 2013). Nanofertilizers in combination with PGPR were found to minimize nutrient losses, reduce environmental contamination and improve plant growth. Cadmium contamination in agricultural soils is a serious concern, wherein it can accumulate in agricultural crops and enter the food chain, posing serious risk to human health. The use of ZnO-NPs has been explored to reduce cadmium accumulation and increase zinc uptake in wheat plants. The experiments conducted in soils treated with 25 mg/kg of cadmium across clay loam, sandy clay loam and sandy loam revealed the effectiveness of application of 300 mg/kg of ZnO-NPs in clay loam soils in significantly reducing cadmium concentrations in roots, shoots, husks and grains with increased Zn content in plants (Usman et al. 2023). Further, evaluation of the effects of CuO and ZnO nanoparticles on soil microbial communities revealed enhanced microbial activity and promotion in soil biodiversity upon treatment of soil with low doses of CuO and ZnO nanoparticles (Liu et al. 2021). Analysing the effect of silica-containing nanoparticles and the biochar produced by different methods and their efficacy as carriers of PGPR in two different plant species has been carried out. Utilizing biochar and nanoparticles as carriers of PGPR followed by inoculation to tomato and watermelon plants has led to increased nitrogen, phosphorous, chlorophyll and antioxidant potential along with increase in phosphate-solubilizing and nitrogen-fixing bacteria (Pavlicevic et al. 2022). Investigation on the combined use of nano-sized zero-valent iron (nZVI) and rhizosphere microorganisms for remediating heavy metal-contaminated soil revealed the effectiveness of low doses of combined fertilizers as potential growth promoters (Mokarram-Kashtiban et al. 2019). Similarly, application of TiO_2 and Fe_3O_4 nanoparticles to soil increased phosphorus availability and soil moisture leading to increased phosphorus accumulation in shoots and roots of *Lactuca sativa* plants (Zahra et al. 2015). Green synthesis of nanoparticles is the process of using biologically active compounds like enzymes, proteins, flavonoids, polyphenols and terpenoids, which can act as catalysts, reducing agents, stabilizers or capping agents. Combined effect of green-synthesized nanoparticle and PGPR showed a promising role not only aiding in the remediation of contaminated soils but also improving soil fertility, thereby supporting sustainable agricultural practices.

3.8 PGPR and Nanoparticles Synergism for Abiotic Stress Alleviation

Continuous usage of chemical fertilizers and chemical pesticides poses a serious threat to soil health, environment and human beings. In recent times, the synergistic use of PGPR and NPs has emerged as a sustainable strategy to enhance plant growth and development by mitigating various abiotic stresses such as heavy metal toxicity, drought and salinity. Experiments in greenhouse and field have conclusively shown the efficiency of copper oxide (CuO) nanoparticles used as foliar treatment in the improvement of disease resistance and increase in fresh weight of watermelon plants (Elmer et al. 2018). Combined application of silver nanoparticles and PGPR (*Pseudomonas* sp.) as foliar spray on two cucumber varieties significantly increased flavonoid and sugar content, chlorophyll levels, antioxidant activity and defence-related enzymes, demonstrating the potential of integrating PGPR with nanoparticles for improved plant growth, stress tolerance and productivity (Nawaz and Bano 2020) Further, combined use of Fe_3O_4 nanoparticles and *Rhizobium* in bean plants has led to enhanced nodule formation, greater nodule dry weight and improved symbiotic nitrogen fixation, as indicated by increased nitrogenase activity and leghaemoglobin content (De Souza-Torres et al. 2021). Chitosan-based nanoparticles are particularly advantageous due to their biodegradability, non-toxic nature and antimicrobial properties. Khati et al. (2017) investigated the combined effect of nanochitosan and maize-derived PGPR, which led to higher levels of organic acids (including acid esters, aldehydes and alcohols), increased activity of soil health-promoting enzymes (alkaline phosphatase, fluorescein diacetate hydrolysis and dehydrogenase) and improvements in physiological parameters, such as seed germination, plant height and leaf area. In another study, Kumar et al. (2020) isolated PGPR from spinach and synthesized silver nanoparticles using *Tagetes erecta* (marigold) leaf and flower extracts. The effects of PGPR, silver nanoparticles and their combination were tested on *Zea mays*. Interestingly, seeds treated solely with silver nanoparticles exhibited better growth responses than those receiving the combined treatment, highlighting that the efficacy of nanoparticle-microbe interactions may vary depending on the crop and treatment conditions. Table 3.2 summarizes the effects of combined application of nanoparticle and PGPR in enhancing plant growth under different abiotic stress conditions. Figure 3.1 summarizes the effect of various abiotic stresses on plant growth and physiological responses in control

Table 3.2 Effect of combined application of nanoparticles and PGPR towards enhancement of plant growth under different abiotic stress conditions

Stress	Nanoparticle	PGPR	Key observations	Plant	References
Salinity	Si-NP (Foliar application)	*Pseudomonas koreensis MG209738* and *Bacillus coagulans NCAIM B.01123*	Reduction in oxidative stress. Increased SOD, CAT and peroxidase activities	*Beta vulgaris*	Alharbi et al. (2022)
Salinity	Silver (Ag)	*Pseudomonas moraviensis*	Increased chlorophyll, carotenoids, sugars, phenolic contents in onion leaves and bulb	*Allium cepa*	Jahangir et al. (2020)
Salinity	Ag NPs	*Comamonas testosteroni*	Increase in total sugar, protein, proline and total chlorophyll content. Increased antioxidant enzyme activities	*Linum usitatissimum*	Khalofah et al. (2021)
Salinity	Si-Zn-NPs	*Bradyrhizobium japonicum* (USDA 110) and *Trichoderma harizianum*	Increase in root length, leaf area, activity of antioxidant enzymes and heavier dry weight of nodules	*Glycine max* L.	Osman et al. (2021)

(continued)

Table 3.2 (continued)

Stress	Nanoparticle	PGPR	Key observations	Plant	References
Salinity	Fe_3O_4-NPs	*Rhizobium* sp. (MK3588), *Stenotrophomonas maltophilia*, mycorrhizal fungus	Increased nutrient uptake, antioxidant enzyme activities, soluble carbohydrate and protein	Chickpea (*Cicer arietinum* L.)	Abd-Alla et al. (2019)
Salinity	Graphene oxide nanoparticles	*Cyanothece* sp., *Rhizobacterium*	Increased antioxidant enzymes and reduced electrolyte leakage	*Hordeum vulgare*, *Vicia faba*	El Semary et al. (2020)
Salinity	SiO_2-NPs	*Pseudomonas*, *Bacillus* and *Variovorax*	The foliar spray improved leaf surface area and recruitment of beneficial microbes in the root rhizosphere	*Glycine max* L.	Wang et al. (2024)
Salinity	ZnO-NPs	*Azospirillum lipoferum* SP2, *Bacillus coagulans* NCAIM B.01123, *Bacillus circulance* NCAIM B.02324 and *Bacillus subtilis* MF497446	Increased antioxidant enzyme activities. Decreased electrolyte leakage H_2O_2 and MDA levels.	*Triticum aestivum* L.	Alharbi et al. (2023)
Metal toxicity (Cd)	ZnO-NPs and TiO_2-NPs	*Bacillus pumilus*	Increased seed germination rate	*Zea mays*	Shafiq et al. (2022)
Metal toxicity	Ag-NPs	*Bacillus mojavensis* BZ-13	Increased yield, dry biomass, root and shoot length and leaf area	*Withania somnifera* L.	Danish et al. (2022)
Metal toxicity (Chromium)	ZnO-NPs	*Staphylococcus aureus* strain K1	Increased growth and antioxidant defence	*Triticum aestivium* L.	Ahmad et al. (2022)

(continued)

Table 3.2 (continued)

Stress	Nanoparticle	PGPR	Key observations	Plant	References
Metal toxicity (Chromium)	Fe-NPs	*S. aureus*	Increased biomass, chlorophyll content and photosynthetic rates	*Oryza sativa*	Alharby and Ali (2022)
Metal toxicity (Arsenic)	Mo-NPs	*Bacillus* sp. strain ZH16	Improved plant growth and development	*Triticum aestivum* L.	Ahmed et al. (2022)
Metal toxicity (Arsenic)	ZnO-NPs	*Providencia vermicola*	Enhanced proline, photosynthetic pigments, total sugars and proteins Decreased Arsenic accumulation in shoots and roots	*Luffa acutangula*	Tanveer et al. (2022)
Metal toxicity (Cd)	TiO_2-NPs	*Pseudomonas fluorescens*	Reduction in uptake of Cd from soil	*Trifolium repens* L.	Zand et al. (2020)
Metal toxicity (Cd)	TiO_2-NPs	*Bacillus mycoides*	Alleviation of Cd toxicity and plant growth promotion	*H. vulgare*	Ma et al. (2023)
Metal toxicity (Chromium)	Ag-NPs	*HASR1*	Augmentation of plant growth, antioxidant enzyme and nutrient uptake	*Hordeum vulgare L*	Zhu et al. (2023)
Drought, heat	ZnO-NPs	Rhizobacteria, Pseudomonas sp.	Production of proline and antioxidant enzymes, reduced electrolyte leakage	*Triticum aestivum* L.	Azmat et al. (2022)
Drought	ZnO	*Azospirillum brasilense*	Enhanced antioxidant levels and augmentation of photosynthetic pigments	*Triticum aestivum* L.	Muhammad et al. (2022)

(continued)

Table 3.2 (continued)

Stress	Nanoparticle	PGPR	Key observations	Plant	References
Drought	Nano zinc (Zn) oxide, nano iron (Fe) oxide and nano Zn-Fe oxide	*Azotobacter*, *Azosperilium* and *Pseudomonas*	Decreased electrical conductivity. Increased peroxidase activity, polyphenol oxidase activity and yield	Wheat (*Triticum aestivum* L.)	Seyed Sharifi et al. (2020)
Drought	Siliconoxide-NPs (SiO_2)	*Bacillus* sp. *Azospirillum lipoferum* and *Azospirillum brasilenss* (MT482404, MT742664 and MT 742,666)	Improved biomass and chlorophyll content as well as enhancing the relative water content, nutrient uptake, osmolytes, gas exchange and antioxidant activities	Wheat (*Triticum aestivum* L.)	Akhtar et al. (2021)
Excess Mn and Fe	Ag-NPs	*Pseudomonas stutzeri*	Increased activity of antioxidant enzymes. Increased production of carotenoids, flavonoids and proline content	*Momordica charantia* L.	Tariq and Bano (2023)
Galaxolide-contaminated soil	Se-NPs	*Actinobacterium* sp.	Increased chlorophyll content, maximum efficiency of photosystem II, stomatal conductance and photorespiratory metabolism as well as accumulation of phytochelatins and metallothioneins	*Glycine max* L.	Halawani and Aloufi (2023)
Drought, salinity	TiO_2-NPs	*Bacillus thuringiensis* AZP2, *Paenibacillus polymyza* A26	Increased fresh and dry weight	*Triticum aestivum* cv. *Stava*	Timmusk et al. (2018)

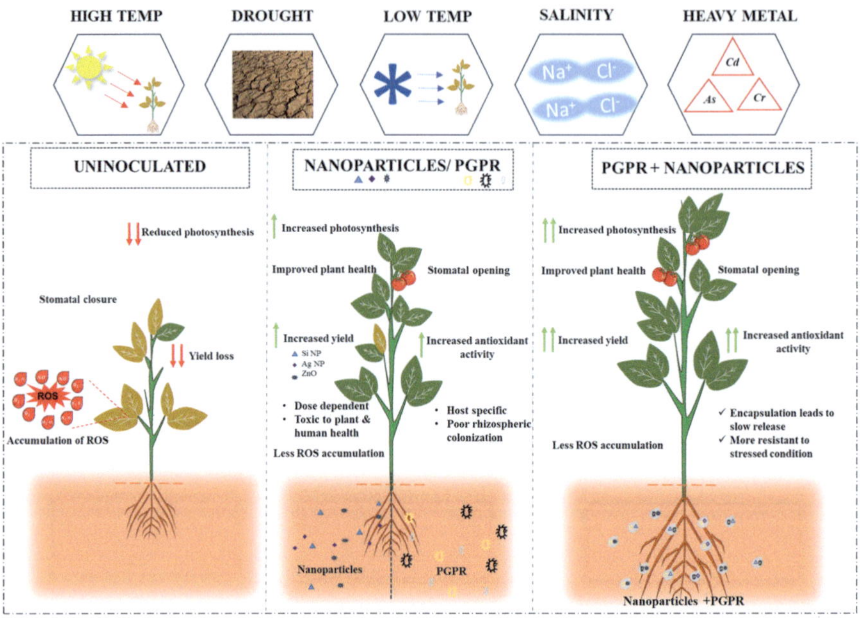

Fig. 3.1 Effect of various abiotic stresses on plant growth and physiological responses in control plants, plants inoculated with nanoparticles or PGPR alone, and plants receiving a combined treatment of PGPR and nanoparticles

uninoculated plants, plants inoculated with nanoparticles or PGPR alone and plants receiving a combined treatment of PGPR and nanoparticles. It illustrates differences in growth parameters, stress tolerance and biochemical markers, highlighting the enhanced mitigation of stress in plants treated with the combined PGPR-nanoparticle approach compared to single treatments or untreated controls. The data represent the synergistic effects of nanoparticles and PGPR in alleviating stress-induced damage, improving photosynthetic efficiency, antioxidant activity and overall plant health. Figure 3.2 depicts the schematic representation of the mechanism of action of nanoparticles (NPs) in combination with plant growth-promoting rhizobacteria (PGPR) in enhancing plant growth and stress tolerance. The diagram depicts the combined application of PGPR and nanoparticles towards enhancement of nutrient availability, improvement of soil physicochemical properties and elevation of antioxidant levels in plants. This synergistic interaction enables plants to better withstand abiotic stresses such as salinity, drought and heavy metal toxicity by enhancing antioxidant enzyme activity, stabilizing cell membranes, increasing photosynthetic pigments and maintaining ionic balance. As a result, plant growth is promoted, biomass accumulation is improved, and overall plant health is supported under stressful environmental conditions.

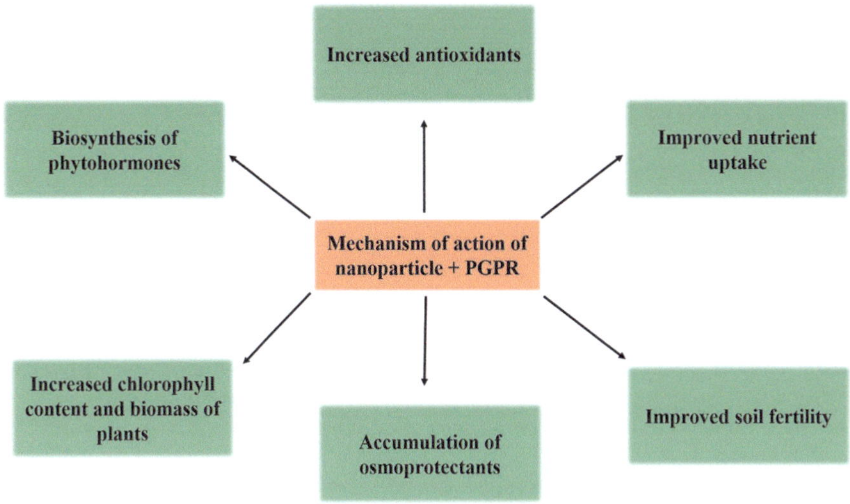

Fig. 3.2 Schematic representation of the mechanism of action of nanoparticles (NPs) in combination with plant growth-promoting rhizobacteria (PGPR) in enhancing plant growth and stress tolerance

3.9 Synergism Between PGPR and Nanoparticles for Alleviating Heavy Metal Stress

Despite the heavy reliance of modern agriculture on chemical fertilizers, surprisingly, plants absorb only a small fraction of the applied nutrients. The excessive use of fertilizers often leads to the accumulation of heavy metals in soil, with chromium being one of the major contaminants. Chromium exposure induces oxidative stress and disrupts the activity of antioxidant enzymes. Studies have shown that treating *Hordeum vulgare* L. with silver nanoparticles (Ag-NPs) in combination with the PGPR strain HAS31 helps mitigate chromium toxicity while enhancing plant growth, antioxidant enzyme activity and nutrient uptake (Zhu et al. 2023). Similarly, Tanveer et al. (2022) evaluated the combined effect of plant growth-promoting bacteria (PGPB, *Providencia vermicola*), zinc oxide nanoparticles (ZnO-NPs) and oxalic acid on *Luffa acutangula* grown in arsenic-contaminated soils and observed a significant reduction in electrolyte leakage and lipid peroxidation concomitant to activation of plant defence mechanisms, improved relative water content, increased total sugar content, increased photosynthetic pigment, IAA production and antioxidant enzyme activities. The co-application of *Bacillus* sp. and molybdenum nanoparticles (Mo-NPs) to wheat plants significantly reduced arsenic contamination by decreasing arsenic translocation in tissues, increased antioxidant enzyme activities, promoted IAA production and enhanced phosphate solubilization (Ahmed et al. 2022). In another study, the bioremediation potential of PGPR *Pseudomonas stutzeri* and silver nanoparticles was evaluated in *Momordica charantia*

L. Surface-sterilized seeds treated with PGPR and silver nanoparticles were irrigated with water containing high concentrations of iron and manganese from industrial effluents. While the PGPR treatment significantly reduced chemical oxygen demand and total dissolved solids, silver nanoparticles were effective in decreasing the oxidative stress by stimulating the antioxidant enzymes, thus highlighting the importance of synergistic effect of PGPR and nanoparticles for plant growth promotion and mitigation of heavy metal stress tolerance (Tariq and Bano 2023). In another study, the responses of rice plants grown in chromium-contaminated soils with iron nanoparticles (Fe-NPs) and *Staphylococcus aureus* treatment have shown marked differences between treated and untreated plants. While the rice plants treated with combined application of Fe NPs and *S. aureus* effectively alleviated chromium stress, the Fe NP-treated plants exhibited significant improvement in biomass, yield, chlorophyll content and photosynthetic rates, with significantly reduced oxidative damage induced by chromium stress compared to untreated plants. The inoculation with chromium-resistant *S. aureus* enhanced plant metal tolerance by converting toxic Cr^{6+} into the less harmful Cr^{3+} form (Alharby and Ali 2022). Similarly, co-application of *S. aureus* and zinc oxide nanoparticles (ZnO NPs) to wheat plants has shown increased chlorophyll content, increased antioxidant enzyme activity and improved nutrient uptake, enabling wheat plants to survive and grow in chromium-contaminated soils (Ahmad et al. 2022). Anthropogenic activities have led to the accumulation of several heavy metal pollutants in the environment, such as cadmium, which is released due to industrial water discharge, solid waste disposal, waste water irrigation, etc. Plants grown in Cd-polluted soil/water show a significant decrease in growth, biomass, photosynthetic pigments, nutrient content in leaves and roots. The combined effect of PGPR (*Bacillus mycoides* PM35) and TiO_2 nanoparticle was found to be effective in alleviation of cadmium toxicity and growth promotion of *Hordeum vulgare* (Ma et al. 2023). Combined application of TiO_2-NPs along with *Pseudomonas fluorescens* to *Trifolium repens* L. plants under cadmium stress reduced cadmium uptake from the soil and restricted the accumulation to the roots, thereby allowing the plant to develop adaptive tolerance mechanisms (Zand et al. 2020). The combined treatment of selenium nanoparticles and *Actinobacterium* sp. in soybean plants grown in soils contaminated with galaxolide has shown enhanced chlorophyll content, photosystem II efficiency, stomatal conductance and photorespiratory metabolism along with accumulation of phytochelatins and metallothioneins (Halawani and Alotfi 2023). These studies strongly suggest beneficial effects in plants upon integration of nanoparticles with PGPR for mitigating heavy metal stress and for enhancing plant growth. Evaluation of combined effect of *Bacillus mojavensis* and silver nanoparticles on Ashwagandha grown in nanoparticle-contaminated soil revealed significant improvement in plant health by increasing the Soil Plant Analysis Development (SPAD) index, photochemical and non-photochemical quenching, PSII quantum yield, total chlorophyll, flavonoids, phenols and tannin content, along with a significant reduction in lipid peroxidation (Danish et al. 2022). Similarly, evaluation of the synergistic effect of titanium dioxide (TiO_2) and zinc oxide (ZnO) nanoparticles in combination with PGPR (*Bacillus pumilus*) on maize plants exposed to metal toxicity revealed improved relative water content, proline, protein, total sugar and photosynthetic pigment levels, with decreased lipid peroxidation indicating the immense potential of PGPR-nanoparticle synergism for mitigating abiotic stress responses and improvement of plant growth (Shafiq et al. 2022).

3.10 Synergism Between PGPR and Nanoparticles for Alleviating Salinity Stress

Nanoparticles in combination with plant growth-promoting rhizobacteria (PGPR) have been effectively used in many studies to mitigate salt stress in various crops. The effect of silver nanoparticles alone and in combination with PGPR used to mitigate salinity stress responses in onion seedlings has shown significantly increased chlorophyll content, carotenoid contents, sugars and phenols in both leaves and bulbs. While two PGPR strains were tested, *Bacillus pumilus* performed better under non-stress conditions, and *Pseudomonas moraviensis* was more effective under salt stress (Jahangir et al. 2020). Sugar beet plants treated with PGPR alone performed better than those treated with PGPR and foliar spray with silica nanoparticles under salinity conditions (Alharbi et al. 2022). In another study, the combined application of Ag-NPs and *Comamonas testosteroni* in *Linum usitatissimum* enhanced antioxidant activity and osmoprotectant levels, such as proline, resulting in improved growth and chlorophyll content under high salinity stress conditions (Khalofah et al. 2021). Co-application of *Bradyrhizobium japonicum* (USDA 110), *Trichoderma harzianum* and silica nanoparticles to *Glycine max* plants under salt stress conditions enhanced root length, leaf area, antioxidant enzyme activity, total soluble sugars and nodule dry weight (Osman et al. 2021). Furthermore, triple microbial inoculation combined with Fe_3O_4 nanoparticles to plants has shown increased nutrient uptake, including potassium, phosphorus, protein and carbohydrate content, highlighting its potential for biofertilizer development (Abd-Alla et al. 2019). Encapsulation of biofertilizers has also yielded positive results in mitigating abiotic stress conditions. Encapsulation of *Cyanothece* sp. and *Enterobacter cloacae* with nanoparticles (graphene oxide and carbon nanotubes), combined with methyl salicylate treatment to chickpea plants, has shown enhanced antioxidant enzyme activity and reduced electrolyte leakage, alleviating salinity-induced inhibition (El Semary et al. 2020). Conventional silicon and silicon nanoparticles (SiNPs) were compared for salinity stress amelioration in soybean plants. While SiNPs not only improved plant growth more effectively than conventional silicon, they also enhanced beneficial microbial populations in the soil as well as endospores in leaves and roots. Further, foliar application of SiNPs increased leaf surface area and reduced the negative effects of salinity, suggesting the importance of providing the nutrients in the form of nanomaterials for visualizing beneficial effects (Wang et al. 2024). The combined effect of ZnO nanoparticles and PGPR on wheat under salinity stress showed significant enhancement in the activity of antioxidant enzymes and reduction in electrolyte leakage, lipid peroxidation and H_2O_2 levels, highlighting the potential of synergistic action of nanoparticles and PGPR or microbial consortia for improvement of plant growth, improved stress tolerance and maintenance of soil health (Alharbi et al. 2023).

3.11 Synergism Between PGPR and Nanoparticles for Alleviating Drought Stress

Drought is one of the major abiotic factors limiting crop productivity. Combined application of zinc oxide nanoparticles (ZnO NPs) and the plant growth-promoting rhizobacterium *Azospirillum brasilense* in wheat has been shown to mitigate drought stress while enhancing antioxidant levels and increasing photosynthetic pigment content (Muhammad et al. 2022). The simultaneous exposure of wheat plants to heat and drought stress was alleviated through treatment with both ZnO NPs and *Pseudomonas* sp., which promoted the production of photosynthetic pigments, proline, soluble sugars, proteins, IAA, antioxidant enzymes such as superoxide dismutase, peroxidase, catalase, ascorbate peroxidase, glutathione reductase, dehydroascorbate reductase and the stress hormone abscisic acid (Azmat et al. 2022). Drought-tolerant PGPR strains isolated from wheat rhizosphere and screened for high production of phytohormones, including indole acetic acid (IAA), cytokinin, abscisic acid and osmolytes, have demonstrated that their combined application with silicon nanoparticles (SiNPs) significantly increases biomass and photosynthetic pigment levels even under drought conditions (Akhtar et al. 2021). Further, application of nano iron (Fe) oxide along with *Azotobacter*, *Azospirillum* and *Pseudomonas* under drought stress has been reported to enhance peroxidase and polyphenol oxidase activities, increase yield and improve overall plant performance. The study also suggested that dual inoculation with *Pseudomonas* and mycorrhizal biofertilizers could further enhance drought mitigation (Seyed Sharifi et al. 2020). Co-application of titanium dioxide (TiO_2) nanoparticles with *Bacillus thuringiensis* AZP2 and *Paenibacillus polymyxa* A26 under drought and salinity stress conditions in wheat plants improved plant growth, including fresh and dry weight, demonstrating the effectiveness of combining nanoparticles with beneficial microbial consortia to alleviate abiotic stress conditions (Timmusk et al. 2018).

3.12 Conclusions and Future Perspectives

Nanotechnology, representing a rapidly emerging field, has diverse applications in different fields of science and technology. In agriculture, nanoparticles in combination with PGPR are promising candidates for sustainable crop production and abiotic stress tolerance. Although PGPR or nanoparticles in isolation can show plant growth promotion and development in addition to possessing different PGPR traits, reports have indicated that the combined effect is stronger than the isolated effect. The synergistic action of NPs with PGPRs has shown significant potential in enhancing plant resilience and mitigating various abiotic stresses. This combined treatment to plants improves nutrient uptake, stimulates antioxidant activity and enhances photosynthetic efficiency, providing an effective biological alternative to conventional chemical fertilizers. Despite their uses in agriculture, the widespread application of nanoparticles in soils raises environmental concerns along with phytotoxicity symptoms seldom observed in host plants. Due to their small size and high reactivity, NPs can accumulate in the soil, have the potential to enter the food chain and pose risks to consumers. Long-term exposure to NPs may alter soil microbial communities, causing disruption to ecological balance, changing soil characteristics and contributing to environmental pollution. Excessive application of nanoparticles can negatively impact plant growth, soil health and the environment, highlighting the need for careful regulation and precise management. In spite of known beneficial effects, the mechanisms underlying the synergistic effects of NPs and PGPR remain poorly understood. A deeper understanding of the synergy between the PGPR and NPs is essential to optimize nanoparticle dosages, ensuring maximum plant benefits with minimal toxicity. Emphasis should be laid on elucidating the physiological, biochemical and molecular mechanisms driving the interactions between nanoparticles, PGPR and different plant species across diverse soil types. A detailed understanding of the mechanisms will allow researchers to fine-tune nanoparticle formulations and application strategies, potentially replacing or reducing the reliance on conventional agrochemicals. The knowledge gained could revolutionize sustainable agriculture by providing a safer, more efficient and environmentally friendly alternatives for improving crop productivity along with mitigation of abiotic stresses.

References

Abd-Alla MH, Nafady NA, Bashandy SR, Hassan AA (2019) Mitigation of effect of salt stress on the nodulation, nitrogen fixation and growth of chickpea (*Cicer arietinum* L.) by triple microbial inoculation. Rhizosphere 10:100148

Ahmad S, Mfarrej MF, El-Esawi MA, Waseem M, Alatawi A, Nafees M, Saleem MH, Rizwan M, Yasmeen T, Anayat A, Ali S (2022) Chromium-resistant *Staphylococcus aureus* alleviates chromium toxicity by developing synergistic relationships with zinc oxide nanoparticles in wheat. Ecotoxicol Environ Saf 230:113142

Ahmed T, Noman M, Rizwan M, Ali S, Ijaz U, Nazir MM, AL Haithloul HAS, Alghanem SM, Abdulmajeed AM, Li B (2022) Green molybdenum nanoparticles-mediated bio-stimulation of *Bacillus sp.* strain ZH16 improved the wheat growth by managing in planta nutrients supply, ionic homeostasis and arsenic accumulation. J Hazard Mater 423:127024

Akhtar N, Ilyas N, Hayat R, Yasmin H, Noureldeen A, Ahmad P (2021) Synergistic effects of plant growth promoting rhizobacteria and silicon dioxide nano-particles for amelioration of drought stress in wheat. Plant Physiol Biochem 166:160–176

Alabdallah NM, Alluqmani SM, Almarri HM, AL-Zahrani AA (2024) Physical, chemical, and biological routes of synthetic titanium dioxide nanoparticles and their crucial role in temperature stress tolerance in plants. Heliyon 10(4):e26537

Alharbi K, Hafez E, Omara AE, Awadalla A, Nehela Y (2022) Plant growth promoting rhizobacteria and silica nanoparticles stimulate sugar beet resilience to irrigation with saline water in salt-affected soils. Plants (Basel) 11(22):3117

Alharbi K, Hafez EM, Omara AED, Rashwan E, Alshaal T (2023) Zinc oxide nanoparticles and PGPR strengthen salinity tolerance and productivity of wheat irrigated with saline water in sodic-saline soil. Plant Soil 493:475–495

Alharby HF, Ali S (2022) Combined role of Fe nanoparticles (Fe NPs) and *Staphylococcus aureus* L. in the alleviation of Chromium stress in rice plants. Life (Basel) 12:338

Ali S, Rizwan M, Hussain A, Zia Ur Rehman M, Ali B, Yousaf B, Wijaya L, Alyemeni MN, Ahmad P (2019) Silicon nanoparticles enhanced the growth and reduced the cadmium accumulation in grains of wheat (*Triticum aestivum* L.). Plant Physiol Biochem 140:1–8

Ali A, Dindhoria K, Kumar R (2023a) *Acinetobacter oleivorans* IRS14 alleviates cold stress in wheat by regulating physiological and biochemical factors. J Appl Microbiol 134(8):176

Ali Q, Ahmad M, Kamran M, Ashraf S, Shabaan M, Babar BH, Zulfiqar U, Haider FU, Ali MA, Elshikh MS (2023b) Synergistic effects of rhizobacteria and salicylic acid on maize salt-stress tolerance. Plants (Basel) 12(13):2519

Alsaeedi A, El-Ramady H, Alshaal T, El-Garawany M, Elhawat N, Al-Otaibi A (2018) Exogenous nanosilica improves germination and growth of cucumber by maintaining K^+/Na^+ ration under elevated Na^+ stress. Plant Physiol Biochem 125:164–171

Alsaeedi A, El-Ramady H, Alshaal T, El-Garawany M, Elhawat N, Al-Otaibi A (2019) Silica nanoparticles boost growth and productivity of cucumber under water deficit and salinity stresses by balancing nutrients uptake. Plant Physiol Biochem 139:1–10

Al-Selwey WA, Alsadon AA, Alenazi MM, Tarroum M, Ibrahim AA, Ahmad A, Osman M, Seleiman MF (2023) Morphological and biochemical response of potatoes to exogenous application of ZnO and SiO_2 nanoparticles in a water deficit environment. Horticulturae 9(8):883

Al-Turki A, Murali M, Omar AF, Rehan M, Sayyed RZ (2023) Recent advances in PGPR-mediated resilience toward interactive effects of drought and salt stress in plants. Front Microbiol 14:1214845

Amini S, Maali-Amiri R, Mohammadi R, Shahandashti SS (2017) cDNA-AFLP analysis of transcripts induced in chickpea plants by TiO2 nanoparticles during cold stress. Plant Physiol Biochem 111:39–49

Anwar T, Qureshi H, Fatimah H, Siddiqi EH, Anwaar S, Moussa IM, Adil MF (2023a) Elucidating effect of ZnO-nanoparticles and melatonin on physiological adjustments and growth of *Solanum melongena* under salinity stress. Scientia Hortic 322:112455

Anwar T, Qureshi H, Fatimah H, Siddiqi EH, Anwaar S, Moussa IM, Adil MF (2023b) Improvement of physio-biochemical attributes and mitigation of salinity stress by combined application of melatonin and silicon nanoparticles in *Brassica oleracea* var. *botrytis*. Sci Hortic 322:112456

Arikan B, Alp FN, Ozfidan-Konakci C, Balci M, Elbasan F, Yildiztugay E, Cavusoglu H (2022) Fe_2O_3-modified graphene oxide mitigates nanoplastic toxicity via regulating gas exchange, photosynthesis, and antioxidant system in *Triticum aestivum*. Chemosphere 307(4):136048

Arora S, Babele PK, Jha PN (2023) Biochemical and metabolic signatures are fundamental to drought adaptation in PGPR *Enterobacter bugandensis* WRS7. Mol Omics 19(8):640–652

Arshad M, Nisar S, Gul I, Nawaz U, Irum S, Ahmad S, Sadat H, Mian IA, Ali S, Rizwan M, Alsahli AA, Alyemeni MN (2021) Multi-element uptake and growth responses of Rice (Oryza sativa L.) to TiO2 nanoparticles applied in different textured soils. Ecotoxicol Environ Saf 215:112149

Attaran Dowom S, Karimian Z, Mostafaei Dehnavi M, Samiei L (2022) Chitosan nanoparticles improve physiological and biochemical responses of *Salvia abrotanoides* (Kar.) under drought stress. BMC Plant Biol 22(1):364

Azmat A, Tanveer Y, Yasmin H, Hassan MN, Shahzad A, Reddy M, Ahmad A (2022) Coactive role of zinc oxide nanoparticles and plant growth promoting rhizobacteria for mitigation of synchronized effects of heat and drought stress in wheat plants. Chemosphere 297:133982

Bakhtiari M, Raeisi Sadati F, Raeisi Sadati SY (2023) Foliar application of silicon, selenium, and zinc nanoparticles can modulate lead and cadmium toxicity in sage (*Salvia officinalis* L.) plants by optimizing growth and biochemical status. Environ Sci Pollut Res 30:54223–54233

Bakka K, Challabathula D (2020) Amelioration of salt stress tolerance in plants by plant growth-promoting rhizobacteria: insights from "Omics" approaches. In: Varma A et al (eds) Plant microbe symbiosis, pp 303–330

Balusamy SR, Joshi AS, Perumalsamy H et al (2023) Advancing sustainable agriculture: a critical review of smart and eco-friendly nanomaterial applications. J Nanobiotechnol 21:372

Baz H, Creech M, Chen J, Gong H, Bradford K, Huo H (2020) Water-soluble carbon nanoparticles improve seed germination and post-germination growth of lettuce under salinity stress. Agronomy 10(8):1192

Bidi H, Fallah H, Niknejad Y, Barari Tari D (2021) Iron oxide nanoparticles alleviate arsenic phytotoxicity in rice by improving iron uptake, oxidative stress tolerance and diminishing arsenic accumulation. Plant Physiol Biochem 163:348–357

Cao Z, Rossi L, Stowers C, Zhang W, Lombardini L, Ma X (2018) The impact of cerium oxide nanoparticles on the physiology of soybean (*Glycine max* (L.) *Merr.*) under different soil moisture conditions. Environ Sci Pollut Res Int 25(1):930–939

Cappellari LDR, Gil SS, Palermo TB, Palermo JS, Meneguzzi R, Giordano W, Banchio E (2023) A practical method to investigate the effect of volatile organic compounds emitted by rhizobacteria on plant growth under conditions of salt stress. MethodsX 10:102099

Choudhary P, Sharma N, Pandey VK, Garg A, James J, Rustagi S (2025) Synergistic role of microbiologically synthesized nanoparticles and PGPR in sustainable agriculture: a critical review. Z Naturforsch C. J Biosci. https://doi.org/10.1515/znc-2025-0087. Epub ahead of print. PMID: 40854010

Danish M, Shahid M, Zeyad MT, Bukhari NA, Al-Khattaf FS, Hatamleh AA, Ali S (2022) *Bacillus mojavensis*, a metal-tolerant plant growth-promoting bacterium, improves growth, photosynthetic attributes, gas exchange parameters, and alkalo-polyphenol contents interaction between metal nanoparticles and PGPR on the plant in silver nanoparticle (Ag-NP)-treated *Withania somnifera* L.(Ashwagandha). ACS Omega 7(16):13878–13893

De Moraes ACP, Ribeiro L, de Camargo ER, Lacava PT (2021) The potential of nanomaterials associated with plant growth-promoting bacteria in agriculture. 3 Biotech 11:318

De Souza-Torres A, Govea-Alcaide E, Gómez-Padilla E, Masunaga SH, Effenberger FB, Rossi LM, López-Sánchez R, Jardim RF (2021) Fe_3O_4 nanoparticles and Rhizobium inoculation enhance nodulation, nitrogen fixation and growth of common bean plants grown in soil. Rhizosphere 17:100275

Dimkpa CO, Latta DE, McLean JE, Britt DW, Boyanov MI, Anderson AJ (2013) Fate of CuO and ZnO nano-and microparticles in the plant environment. Environ Sci Technol 47:4734–4742

Dimkpa CO, Singh U, Bindraban PS, Elmer WH, Gardea-Torresdey JL, White JC (2019) Zinc oxide nanoparticles alleviate drought induced alterations in *Sorghum* performance, nutrient acquisition, and grain fortification. Sci Total Environ 688:926–934

Djanaguiraman M, Nair R, Giraldo JP, Prasad PVV (2018) Cerium Oxide nanoparticles decrease drought-induced oxidative damage in *Sorghum* leading to higher photosynthesis and grain yield. ACS Omega 3(10):14406–14416

Dodd IC, Pérez-Alfocea F (2012) Microbial amelioration of crop salinity stress. J Exp Bot 63:3415–3428

Duro JA, Lauk C, Kastner T, Erb KH, Haberl H (2020) Global inequalities in food consumption, cropland demand and land-use efficiency: a decomposition analysis. Glob Environ Change 64:102124

El Semary NAH, Alouane MHH, Nasr O, Aldayel MF, Alhaweti FH, Ahmed F (2020) Salinity stress mitigation using encapsulated biofertilizers for sustainable agriculture. Sustainability 12:9218

Elmer W, De La Torre-Roche R, Pagano L, Majumdar S, Zuverza-Mena N, Dimkpa C, Gardea-Torresdey J, White JC (2018) Effect of metalloid and metal oxide nanoparticles on fusarium wilt of watermelon. Plant Dis 102(7):1394–1401

El-Zohri M, Al-Wadaani NA, Bafeel SO (2021) Foliar sprayed green zinc oxide nanoparticles mitigate drought-induced oxidative stress in tomato. Plants (Basel) 10(11):2400

FAO (2020) Salt-affected soils [Online]. FAO, Rome

Faraz A, Faizan MD, Rajput V, Minkina T, Hayat S, Faisal M et al (2023) CuO nanoparticle-mediated seed priming improves physio-biochemical and enzymatic activities of *Brassica juncea*. Plants (Basel) 12(4):803

Fathi A, Zahedi M, Torabian S, Khoshgoftar A (2017) Response of wheat genotypes to foliar spray of ZnO and Fe_2O_3 nanoparticles under salt stress. J Plant Nutr 40:1376–1385

Fincheira P, Tortella G, Seabra AB, Quiroz A, Diez MC, Rubilar O (2021) Nanotechnology advances for sustainable agriculture: current knowledge and prospects in plant growth modulation and nutrition. Planta 254(4):66

Ghassemi-Golezani K, Abdoli S (2021) Improving ATPase and PPase activities, nutrient uptake and growth of salt stressed ajowan plants by salicylic acid and iron-oxide nanoparticles. Plant Cell Rep 40:559–573

Ghouri F, Shahid MJ, Ali S, Ashraf H, Alomrani SO, Liu J, Alshehri MA, Shahid MQ (2025) Tetraploidy and Fe2O3 nanoparticles: dual strategy to reduce the Cd-induced toxicity in rice plants by ameliorating the oxidative stress and downregulation of metal transporters. Environ Sci Nano 12:634–646

González García Y, Cárdenas Álvarez C, Cadenas Pliego G, Benavides-Mendoza A, Cabrera de la Fuente M, Sandoval-Rangel A, Valdés Reyna J, Juárez Maldonado A (2021) Effect of three nanoparticles (Se, Si and Cu) on the bioactive compounds of bell pepper fruits under saline stress. Plants 10(2):217

Gupta B, Shrestha J (2023) Editorial: abiotic stress adaptation and tolerance mechanisms in crop plants. Front Plant Sci 14:1278895

Gupta R, Kumar S, Jain D, Saini RP, Halli HM, Yadav VK, Singh AK, Tomar M, Singh P (2024) Drought stress effects on Lathyrus seedlings: a morpho-physiological and molecular study. J Food Legumes 37(1):53–71

Hafez EM, Alharbi K, Gharib HS, Omara AED, Elatafi E, Hamada MM, Rashwan E, Alshaal T (2025) Synergistic effect of sugarcane bagasse and zinc oxide nanoparticles on eco-remediation of cadmium-contaminated saline soils in wheat cultivation. Plants (Basel) 14(1):85

Halawani RF, Aloufi FA (2023) Galaxolide-contaminated soil and tolerance strategies in soybean plants using biofertilization and selenium nanoparticle supplementation. Front Plant Sci 14:1221780

Hanin M, Ebel C, Ngom M, Laplaze L, Masmoudi K (2016) New insights on plant salt tolerance mechanisms and their potential use for breeding. Front Plant Sci 7:1787

Hmaeid N, Wali M, Mahmoud OMB, Pueyo JJ, Ghnaya T, Abdelly C (2019) Efficient rhizobacteria promote growth and alleviate NaCl-induced stress in the plant species *Sulla carnosa*. Appl Soil Ecol 133:104–113

Hong J, Wang C, Wagner DC, Gardea-Torresdey JL, He F, Rico CM (2021) Foliar application of nanoparticles: mechanisms of absorption, transfer, and multiple impacts. Environ Sci Nano 8(5):1196–1210

Ishtiaq M, Mazhar MW, Maqbool M, Hussain T, Hussain SA, Casini R, Abd-ElGawad AM, Elansary HO (2023) Seed priming with the selenium nanoparticles maintains the redox status in the water stressed tomato plants by modulating the antioxidant defense enzymes. Plants (Basel) 12(7):1556

Jahangir S, Javed K, Bano A (2020) Nanoparticles and plant growth promoting rhizobacteria (PGPR) modulate the physiology of onion plant under salt stress. Pak J Bot 52:1473–1480

Kaleem Z, Xu W, Ulhassan Z, Shahbaz H, He D, Naeem S, Ali S, Shah AM, Sheteiwy MS, Zhou W (2024) Harnessing the potential of copper-based nanoparticles in mitigating abiotic and biotic stresses in crops. Environ Sci Pollut Res 31:59727–59748

Karunakaran G, Suriyaprabha R, Manivasakan P, Yuvakkumar R, Rajendran V, Kannan N (2013) Impact of nano and bulk ZrO_2, TiO_2 particles on soil nutrient contents and PGPR. J Nanosci Nanotechnol 13(1):678–685

Khalid MF, Iqbal Khan R, Jawaid MZ, Shafqat W, Hussain S, Ahmed T, Rizwan M, Ercisli S, Pop OL, Alina Marc R (2022) Nanoparticles: the plant saviour under abiotic stresses. Nanomaterials (Basel) 6:3915

Khalofah A, Kilany M, Migdadi H (2021) Phytostimulatory influence of *Comamonas testosterone* and silver nanoparticles on *Linum usitatissimum* L. under salinity stress. Plants 10:790

Khan AR, Mustafa A, Hyder S, Valipour M, Rizvi ZF, Gondal AS, Yousuf Z, Iqbal R, Daraz U (2022) *Bacillus* spp. as bioagents: uses and application for sustainable agriculture. Biology (Basel) 11(12):1763

Khan I, Awan SA, Rizwan M, Akram MA, Zia Ur Rehman M, Wang X, Zhang X, Huang L (2023) Physiological and transcriptome analyses demonstrate the silver nanoparticles mediated alleviation of salt stress in pearl millet (*Pennisetum glaucum* L). Environ Pollut 318:120863

Khati P, Chaudhary P, Gangola S, Bhatt P, Sharma A (2017) Nanochitosan supports growth of *Zea mays* and also maintains soil health following growth. 3 Biotech 7:81

Kloepper JW, Leong J, Teintze M, Schroth MN (1980) Enhanced plant growth by siderophores produced by plant growth-promoting rhizobacteria. Nature 286:885–886

Krishnamoorthy V, Rajiv S (2016) Potential seed coatings fabricated from electrospinning hexaaminocyclotriphosphazene and cobalt nanoparticles incorporated polyvinylpyrrolidone for sustainable agriculture. ACS Sustain Chem Eng 5(1):146–152

Kumar V, Nautiyal CS (2022) Plant abiotic and biotic stress alleviation: from an endophytic microbial perspective. Curr Microbiol 79(10):311

Kumar P, Pahal V, Gupta A, Vadhan R, Chandra H, Dubey RC (2020) Effect of silver nanoparticles and *Bacillus cereus* LPR2 on the growth of *Zea mays*. Sci Rep 10(1):1

Lichtenthaler HK (1996) Vegetation stress: An introduction to the stress concept in plants. J Plant Physiol 148:4–14

Liu Y, Li Y, Pan B, Zhang X, Zhang H, Steinberg CEW, Qiu H, Vijver MG, Peijnenburg WJGM (2021) Application of low dosage of copper oxide and zinc oxide nanoparticles boosts bacterial and fungal communities in soil. Sci Total Environ 757:143807

Lu L, Liu N, Fan Z, Liu M, Zhang X, Tian J, Yu Y, Lin H, Huang Y, Kong Z (2024) A novel PGPR strain, *Streptomyces lasalocidi* JCM 3373^T, alleviates salt stress and shapes root architecture in soybean by secreting indole-3-carboxaldehyde. Plant Cell Environ 47(6):1941–1956

Ma J, Li Y, Chen F, Sun Y, Zhu Y, Wang L (2023) *Bacillus mycoides* PM35 in combination with titanium dioxide (TiO_2) - nanoparticles enhanced morpho-physio-biochemical attributes in Barley (*Hordeum vulgare* L.) under cadmium stress. Chemosphere 323:138224

Mahmood I, Imadi SR, Shazadi K, Gul A, Hakeem KR (2016) Effects of pesticides on environment. Plant Soil Microbes 1:253–269

Mandal D, Das SK, Adhikari J, Chatterjee D, Bandyopadhyay TK, Basu A (2024) Genome sequencing, annotation and application of a strain of *Microbacterium paraoxydans*–A bacterium with arsenic bioremediation and plant growth promoting potential. Microbe 4:100132

Meena KK, Sorty AM, Bitla UM, Choudhary K, Gupta P, Pareek A, Singh DP, Prabha R, Sahu PK, Gupta VK, Singh HB, Krishnani KK, Minhas PS (2017) Abiotic stress responses and microbe mediated mitigation in plants: the omics strategies. Front Plant Sci 8:172

Mi K, Yuan X, Wang Q, Dun C, Wang R, Yang S, Yang Y, Zhang H, Zhang H (2023) Zinc oxide nanoparticles enhanced rice yield, quality, and zinc content of edible grain fraction synergistically. Front Plant Sci 14:1196201

Mokarram-Kashtiban S, Hosseini SM, Kouchaksaraei MT, Younesi H (2019) The impact of nanoparticles zero-valent iron (nZVI) and rhizosphere microorganisms on the phytoremediation ability of white willow and its response. Environ Sci Pollut Res Int 26(11):10776–10789

Muhammad F, Raza MAS, Iqbal R, Zulfiqar F, Aslam MU, Yong JWH, Altaf MA, Zulfiqar B, Amin J, Ibrahim MA (2022) Ameliorating drought effects in wheat using an exclusive or co-applied rhizobacteria and ZnO nanoparticles. Biology (Basel) 11(11):1564

Mustafa H, Ilyas N, Akhtar N, Raja NI, Zainab T, Shah T, Ahmad A, Ahmad P (2021) Biosynthesis and characterization of titanium dioxide nanoparticles and its effects along with calcium phosphate on physicochemical attributes of wheat under drought stress. Ecotoxicol Environ Saf 15:112519

Nauman Mahamood M, Zhu S, Noman A, Mahmood A, Ashraf S, Aqeel M, Ibrahim M, Ashraf S, Liew RK, Lam SS, Irshad MK (2023) An assessment of the efficacy of biochar and zero-valent iron nanoparticles in reducing lead toxicity in wheat (*Triticum aestivum* L.). Environ Pollut 319:120979

Nawaz S, Bano A (2020) Effects of PGPR (*Pseudomonas* sp.) and Ag-nanoparticles on enzymatic activity and physiology of cucumber. Recent Patents Food Nutr Agric 11(2):124–136

Nejatzadeh F (2021) Effect of silver nanoparticles on salt tolerance of *Satureja hortensis* L. during in vitro and in vivo germination tests. Heliyon 7(2):e05981

Neshat M, Abbasi A, Hosseinzadeh A, Sarikhani MR, Dadashi Chavan D, Rasoulnia A (2022) Plant growth promoting bacteria (PGPR) induce antioxidant tolerance against salinity stress through biochemical and physiological mechanisms. Physiol Mol Biol Plants 28(2):347–361

Osman HS, Gowayed SM, Elbagory M, Omara AE, El-Monem AM, Abd El-Razek UA, Hafez EM (2021) Interactive impacts of beneficial microbes and Si-Zn nanocomposite on growth and productivity of soybean subjected to water deficit under salt-affected soil conditions. Plants (Basel) 10:1396

Pavlicevic M, Abdelraheem W, Zuverza-Mena N, O'Keefe T, Mukhtar S, Ridge G, Ranciato J, Haynes C, Elmer W, Pignatello J, Pagano L, Caldara M, Marmiroli M, Maestri E, Marmiroli N, White JC (2022) Engineered nanoparticles, natural nanoclay and biochar, as carriers of plant-growth promoting bacteria. Nanomaterials 12(24):4474

Pérez-Labrada F, López-Vargas ER, Ortega-Ortiz H, Cadenas-Pliego G, Benavides-Mendoza A, Juárez-Maldonado A (2019) Responses of tomato plants under saline stress to foliar application of copper nanoparticles. Plants (Basel) 8(6):151

Prasad R, Bhattacharyya A, Nguyen QD (2017) Nanotechnology in sustainable agriculture: recent developments, challenges, and perspectives. Front Microbiol 20:1014

Rajput VD, Kumari A, Upadhyay SK, Minkina T, Mandzhieva S, Ranjan A, Sushkova S, Burachevskaya M, Rajput P, Konstantinova E, Singh J, Verma KK (2023) Can nanomaterials improve the soil microbiome and crop productivity? Agriculture 13(2):231

Romero-Munar A, Aroca R, Zamarreño AM, García-Mina JM, Perez-Hernández N, Ruiz-Lozano JM (2023) Dual inoculation with Rhizophagus irregularis and bacillus megaterium improves maize tolerance to combined drought and high temperature stress by enhancing root hydraulics, photosynthesis and hormonal responses. Int J Mol Sci 24(6):5193

Saleem S, Iqbal A, Ahmed F, Ahmad M (2021) Phytobeneficial and salt stress mitigating efficacy of IAA producing salt tolerant strains in *Gossypium hirsutum*. Saudi J Biol Sci 28(9):5317–5324

Santos-Espinoza AM, González-Mendoza D, Ruiz-Valdiviezo VM, Luján-Hidalgo MC, Jonapa-Hernández F, Valdez-Salas B, Gutiérrez-Miceli FA (2021) Changes in the physiological and biochemical state of peanut plants (*Arachis hypogaea* L.) induced by exposure to green metallic nanoparticles. Int J Phytoremediation 23(7):747–754

Seyed Sharifi R, Khalilzadeh R, Pirzad A, Anwar S (2020) Effects of biofertilizers and nano zinc-iron oxide on yield and physicochemical properties of wheat under water deficit conditions. Commun Soil Sci Plant Anal 51(19):2511–2524

Shafiq T, Yasmin H, Shah ZA, Nosheen A, Ahmad P, Kaushik P, Ahmad A (2022) Titanium oxide and zinc oxide nanoparticles in combination with cadmium tolerant *Bacillus pumilus* ameliorates the cadmium toxicity in maize. Antioxidants (Basel) 11(11):2156

Shah A, Nazari M, Antar M, Msimbira LA, Naamala J, Lyu D et al (2021) PGPR in agriculture: a sustainable approach to increasing climate change resilience. Front Sustain Food Syst 5:1–22

Shah IH, Manzoor MA, Sabir IA, Ashraf M, Liaquat F, Gulzar S, Chang L, Zhang Y (2023) Phytotoxic effects of chemically synthesized copper oxide nanoparticles induce physiological, biochemical, and ultrastructural changes in *Cucumis melo*. Environ Sci Pollut Res Int 30(18):51595–51606

Sharma G, Kumar A, Devi KA, Prajapati D, Bhagat D, Pal A, Raliya R, Biswas P, Saharan V (2020) Chitosan nanofertilizer to foster source activity in maize. Int J Biol Macromol 145:226–234

Somna G, Sangwan S, Baskar C, Bakka K, Challabathula D (2025) Protocol for inoculation of PGPR *staphylococcus sciuri* to seeds and seedlings of rice and tomato plants for increased root and shoot growth. Bio Protoc 15(6):e5241

Tanveer Y, Yasmin H, Nosheen A, Ali S, Ahmad A (2022) Ameliorative effects of plant growth promoting bacteria, zinc oxide nanoparticles and oxalic acid on *Luffa acutangula* grown on arsenic enriched soil. Environ Pollut 300:118889

Tariq S, Bano A (2023) Role of PGPR and silver nanoparticles on the physiology of *Momordica charantia* L. irrigated with polluted water comprising high Fe and Mn. Int J Phytoremediation 25:1643–1655

Taşkın MB, Şahin Ö, Taskin H, Atakol O, Inal A, Gunes A (2018) Effect of synthetic nano-hydroxyapatite as an alternative phosphorus source on growth and phosphorus nutrition of lettuce (*Lactuca sativa* L.) plant. J Plant Nutr 41(9):1148–1154

Thakur S, Asthir B, Kaur G, Kalia A, Sharma A (2021) Zinc oxide and titanium dioxide nanoparticles influence heat stress tolerance mediated by antioxidant defense system in wheat. Cereal Res Commun 50:385–396

Timmusk S, Seisenbaeva G, Behers L (2018) Titania (TiO2) nanoparticles enhance the performance of growth-promoting rhizobacteria. Sci Rep 8(1):617

Tourinho PS, Van Gestel CA, Lofts S, Svendsen C, Soares AM, Loureiro S (2012) Metal-based nanoparticles in soil: fate, behavior, and effects on soil invertebrates. Environ Toxicol Chem 31:1679–1692

United Nations (2019) Department of Economic and Social Affairs, population division. World population prospects: highlights. ST/ESA/SER. A/423

Usman M, Zia-Ur-Rehman M, Rizwan M, Abbas T, Ayub MA, Naeem A, Alharby HF, Alabdallah NM, Alharbi BM, Qamar MJ, Ali S (2023) Effect of soil texture and zinc oxide nanoparticles on growth and accumulation of cadmium by wheat: a life cycle study. Environ Res 216(1):114397

Venkatachalam P, Jayaraj M, Manikandan R, Geetha N, Rene ER, Sharma NC, Sahi SV (2017) Zinc oxide nanoparticles (ZnONPs) alleviate heavy metal-induced toxicity in *Leucaena leucocephala* seedlings: A physiochemical analysis. Plant Physiol Biochem 110:59–69

Wahid I, Kumari S, Ahmad R, Hussain SJ, Alamri S, Siddiqui MH, Khan MIR (2020) Silver nanoparticle regulates salt tolerance in wheat through changes in aba concentration, ion homeostasis, and defense systems. Biomolecules 10(11):1506

Wang A, Li J, Al-Huqail AA, Al-Harbi MS, Ali EF, Wang J, Ding Z, Rekaby SA, Ghoneim AM, Eissa MA (2021) Mechanisms of chitosan nanoparticles in the regulation of cold stress resistance in banana plants. Nanomaterials (Basel) 11(10):2670

Wang L, Ning C, Pan T, Cai K (2022a) Role of silica nanoparticles in abiotic and biotic stress tolerance in plants: a review. Int J Mol Sci 9:1947

Wang S, Fu Y, Zheng S, Xu Y, Sun Y (2022b) Phytotoxicity and accumulation of copper-based nanoparticles in *brassica* under cadmium stress. Nanomaterials (Basel) 12(9):1497

Wang P, Zhang H, Hu X, Xu L, An X, Jin T, Ma R, Li Z, Chen S, Du S, Wei G (2024) Comparing the potential of silicon nanoparticles and conventional silicon for salinity stress alleviation in soybean (*Glycine max* L.): growth and physiological traits and rhizosphere/endophytic bacterial communities. J Agric Food Chem 72(19):10781–10793

Wasaya A, Hassan J, Yasir TA, Ateeq M, Raza MA (2022) Foliar application of silicon improved physiological indicators, yield attributes, and yield of pearl millet (*pennisetum glaucum* l.) under terminal drought stress. J Soil Sci Plant Nutr 22(4):4458–4472

Ye Y, Cota-Ruiz K, Hernandez-Viezcas JA, Valdes C, Medina-Velo IA, Turley RS, Gardea-Torresdey JL (2020) Manganese nanoparticles control salinity-modulated molecular responses in *Capsicum annuum* L. through priming: a sustainable approach for agriculture. ACS Sustain Chem Eng 8:1427–1436

Younis A, Khattab H, Emam MM (2020) Impacts of silicon and silicon nanoparticles on leaf ultrastructure and TaPIP1 and TaNIP2 gene expressions in heat stressed wheat seedlings. Biol Plant 64:343–352

Yuan H, Liu Q, Fu J, Wang Y, Zhang Y, Sun Y, Tong H, Dhankher OP (2023) Co-exposure of sulfur nanoparticles and Cu alleviate Cu stress and toxicity to oilseed rape *Brassica napus* L. J Environ Sci (China) 124:319–329

Zahedi SM, Abdelrahman M, Hosseini MS, Hoveizeh NF, Tran LP (2019) Alleviation of the effect of salinity on growth and yield of strawberry by foliar spray of selenium-nanoparticles. Environ Pollut 253:246–258

Zahra Z, Arshad M, Rafique R, Mahmood A, Habib A, Qazi IA, Khan SA (2015) Metallic nanoparticle (TiO_2 and Fe_3O_4) application modifies rhizosphere phosphorus availability and uptake by *Lactuca sativa*. J Agric Food Chem 63:6876–6882

Zand AD, Tabrizi AM, Heir AV (2020) Application of titanium dioxide nanoparticles to promote phytoremediation of Cd-polluted soil: contribution of PGPR inoculation. Biorem J 2(3):171–189

Zeshan A, Abdullah M, Adil MF, Wei D, Noman M, Ahmed T, Sehar S, Ouyang Y, Shamsi IH (2022) Improvement of morpho-physiological, ultrastructural and nutritional profiles in wheat seedlings through astaxanthin nanoparticles alleviating the cadmium toxicity. J Hazard Mater 424(PtC):126511

Zhang H, Du W, Peralta-Videa JR, Gardea-Torresdey JL, White JC, Keller A, Guo H, Ji R, Zhao L (2018) Metabolomics reveals how cucumber (*cucumis sativus*) reprograms metabolites to cope with silver ions and silver nanoparticle-induced oxidative stress. Environ Sci Technol 52(14):8016–8026

Zhou H, Wu H, Zhang F, Su Y, Guan W, Xie Y, Giraldo JP, Shen W (2021) Molecular basis of cerium oxide nanoparticle enhancement of rice salt tolerance and yield. Environ Sci Nano 8(11):3294–3311

Zhu Y, Wang L, Ma J, Li Y, Chen F, Peijnenburg W (2023) Comparative physiological and metabolomics analyses using Ag-NPs and HAS31 (PGPR) to alleviate Cr stress in barley (*Hordeum vulgare* L.). Environ Pollut 333:122010

Zia R, Nawaz MS, Siddique MJ, Hakim S, Imran A (2021) Plant survival under drought stress: implications, adaptive responses, and integrated rhizosphere management strategy for stress mitigation. Microbiol Res 242:126626

Chapter 4
Nanofertilizers: Application, Slow-Release Formulations, and Safety Considerations in Agricultural Systems

Heriberto Fortino Ramírez-Cariño, Paula Cecilia Guadarrama-Mendoza, and Rogelio Valadez-Blanco

Abstract An increase in global population necessitates increased food production. However, drought, soil degradation, and environmental factors have reduced agricultural land. Strategies such as organic amendments, biofertilizers, and nanofertilizers have been explored to promote sustainable agriculture. Nanofertilizers (NFs) are formulations with distinctive properties, including large surface area that enhances nutrient assimilation efficiency, and encapsulated structures that enable sustained release, thereby reducing losses through leaching or volatilization. In addition, NFs act as signaling molecules, strengthening plant defense systems and mitigating stress caused by drought, extreme temperatures, and heavy-metal toxicity. Research on NF formulations has focused primarily on cereals, followed by legumes, and other vegetables and has expanded rapidly with applications through both foliar and soil routes. Among these formulations, dual or multi-nanonutrient types, particularly those containing Zn and Fe, are the most prevalent, followed by macronutrient-based NFs incorporating N, P, and Ca. Bottom-up chemical synthesis remains the preferred method due to its simplicity and reproducibility. Furthermore, advanced NF designs featuring slow-release multinutrients embedded in hydroxyapatite or chitosan matrices show promising results for yield improvement and nutrient uptake. However, the commercial application of nanofertilizers is not yet readily established, mainly due to unresolved safety concerns. Rigorous evaluation, which includes toxicity, dosage, timing, and crop-specific responses, is essential. Future research must value null and adverse findings to provide a balanced understanding of risks for plants, ecosystems, and human health. In conclusion, nanofertilizers represent a promising tool in agricultural systems, whose long-term success

H. F. Ramírez-Cariño
Universidad de la Sierra Juárez, Avenida Universidad S/N, Ixtlán de Juárez, Oaxaca, Mexico

P. C. Guadarrama-Mendoza · R. Valadez-Blanco (✉)
Instituto de Agroindustrias, Universidad Tecnológica de la Mixteca, Oaxaca, Mexico
e-mail: rvaladez@mixteco.utm.mx

K. A. Abd-Elsalam (ed.), *Plant-Microbiome Nanotechnology*, Sustainability Sciences in Asia and Africa, https://doi.org/10.1007/978-981-95-5056-2_4

will depend on the integration of technological advances with comprehensive assessments of their safety, efficacy, and sustainability.

Keywords Agronanotechnology · Crop productivity · Environmental safety · Food security · Nanofertilizers · Nutrient use efficiency · Sustainable agriculture

4.1 Introduction

The steadily growing global population demands intensified food production to ensure food security and support the sustained development of society—especially within the agricultural sector. However, modern agriculture is facing several complex challenges which are driven by social, economic, and environmental pressures (Basavegowda and Baek 2021). Among the environmental factors, climate change and global warming have resulted in water scarcity and soil depletion; in addition, social changes have caused the reduction of agricultural land and labor, whereas economic forces drive current agricultural practices almost exclusively toward crop productivity without considering their environmental impact.

In this regard, conventional fertilizers (CFs), also called chemical fertilizers, for many decades have been extremely helpful in increasing crop productivity to meet the growing food demands for the world's population. Therefore, intensive use of these agrochemicals has been justified to achieve global food security and good nutritional value of the harvested products. However, in recent years, there is an increasing social awareness of the environmental impact generated by human activities, which has been prompted by current evidences of climate change and global warming. The latter have exacerbated the agricultural challenges mentioned above. In this regard, the problems related to the excessive use of chemical fertilizers, which are oftentimes indiscriminate, are well known (Al Tawaha et al. 2024a; Okey-Onyesolu et al. 2021): (i) low nutrient utilization efficiency—several chemical fertilizers are immobile and are not readily available for plants, while others are easily soluble in water and are not retained by plants, thereby resulting in overall low utilization of nutrients such as N (30–35%), P (18–20%), and K (35–40%) (Haydar et al. 2024); (ii) the nutrients that are not utilized by the plant remain in the soil or dissolved in the irrigation runoff; (iii) these nutrients that are not utilized result in pollution of water bodies and soil, where nutrient leachate causes eutrophication, and soil contamination increases soil salinity and eventually leads to soil depletion in terms of mineral and organic nutrients, microbiota, and general biota. The latter changes also favor pathogen and pest resistance, thereby ultimately resulting in soil infertility and the need to increase the amounts of chemical fertilizers to maintain crop productivity.

To mitigate or overcome the issues posed by CFs, recent trends have focused on the use of more environmentally friendly approaches, in which the goal is not only to sustain the productivity of the crops but also to preserve soil richness, thus contributing to its fertility, while reducing the environmental impact caused by the agricultural practice. The main strategies implemented in this regard are the following: (i) sustainable agriculture management (Hobbs et al. 2008); (ii) soil amendments such as composts (Aguilar-Paredes et al. 2023); (iii) biofertilizers (Ramírez-Cariño et al. 2024); and (iv) nanofertilizers (Escribà-Gelonch et al. 2023). All these strategies contribute to the development of sustainable agriculture (Basavegowda and Baek 2021; Haydar et al. 2024), which not only supports global food security but also restores and maintains soil fertility, improves the efficiency of resource utilization, and reduces the environmental footprint of agriculture activities. In this chapter, we critically examine recent advancements in the development and application of engineered nanomaterials used as crop fertilizers—commonly referred to as nanofertilizers. Our analysis considers their use across different food crops, relevant agronomic factors, and the influence of synthesis and formulation methods on advanced crop production. In addition, we provide a thorough assessment of current health and safety concerns associated with the large-scale deployment of nanofertilizers in modern agriculture.

4.2 Nanomaterials as Efficient Fertilizers: Nanofertilizers

In the last decade, the application of nanotechnology in agriculture has experienced a great boost, thus generating a remarkable volume of research and practical opportunities within the sector. Among the relatively new applications, nanofertilizers have gained notable relevance due to the critical status of modern agriculture and, as outlined above, there is a need to produce large amounts of nutritious foods while preserving soil fertility, conserving natural resources, and minimizing environmental impact. Nanofertilizers (NFs) can be defined as formulations of plant nutrients, which include one or more nanoscale materials or nanomaterials (NM) with at least one of its dimensions having a size lesser than 100 nm. The NFs possess specific physicochemical properties that make them a good alternative over conventional fertilizers: i) they have a large surface area/volume ratio, which allows these materials to penetrate into plants and be readily available for plant nutrition, thus improving nutrient use efficiency (NUE); ii) NFs can be formulated in an encapsulated or embedded form, by which these products have the property of slow or controlled nutrient delivery to the plant, which allows for the reduction of nutrient losses by runoff or volatilization, thus, improving nutrition efficiency and reducing the environmental impact of crop fertilization. The slow or controlled release characteristics of these NFs have led their classification as "precision delivery" and "targeted delivery" systems, as part of the smart agriculture practices, by which many of the CF´s negative effects are circumvented achieving comparable crop productivities over extended periods of time, while minimizing nutrient resources usage. Moreover, the slow or controlled release of the NFs can drastically reduce nutrient losses to the environment, thereby preventing soil depletion and water eutrophication.

The effectiveness of NFs in crop production has been well documented in the literature, showing improvements in plant growth, crop yields, and nutrient accumulation in harvested products, in comparison to conventional fertilizers. In addition, a number of studies have evidenced the slow- and controlled-released effect of specific NFs that results in minimizing nutrient runoff and waste, which further complement the beneficial attributes that these novel fertilizing products can bring to modern agriculture. It is also noteworthy that there is growing evidence about the ameliorating effect of NFs on both biotic and abiotic stress factors in agricultural systems. In this regard, much of the attention in the literature has been focused on the effect of NFs on abiotic stress factors such as water drought, high salinity, extreme temperatures, and soils contaminated with heavy metals. The main mechanism proposed for the stress-mitigation activity of NFs is that these nanomaterial formulations act as stress-signaling molecules that trigger the upregulation of genes related to defense and antioxidant system in plants. This helps plants cope with environmental stress and continue its growth and development (Singh et al. 2023).

However, the nanometric scale of these materials also poses some safety, environmental, and health risks since these NMs can enter the living cells without surveillance, which can potentially cause undesirable reactions or cell responses not only in the plant cells but also in the soil biota and ultimately in human cellular systems. Exposure of humans to nanomaterials may occur at manufacturing centers, in agriculture fields, or during consumption of food products. The environmental fate of these nanomaterials is still largely unknown, so the safety issues of NFs have a tremendous impact on the consolidation of this technology as part of the global agriculture (Haydar et al. 2024; Singh et al. 2023; Yadav et al. 2023).

The efficiency of NFs in agricultural systems is very complex since various factors need to be taken into account, of which some cannot be controlled. Here, we have categorized these factors into two groups: (i) characteristics of agricultural system and (ii) traits of NF formulations. Some characteristics of agricultural system include the type of crop and cultivar, soil characteristics, agronomical managing practices, and environmental conditions. Among these factors, particular interest is focused on the crop-fertilization system: foliar, soil-irrigated, or hydroponics (Ding et al. 2023; Singh et al. 2023). On the other hand, NF characteristics can be classified based on the following factors: nanomaterial size, concentration, chemical properties, nutrient type (macro or micro), single or multiple nutrient, nanomaterial synthesis, NF product formulation or delivery system, and toxicological aspects (Nongbet et al. 2022).

4.3 Application of NFs in Agricultural Systems

Below we have listed the most relevant reports of NFs in agricultural systems in the past five years and have discussed the main factors influencing their efficiency in terms of crop productivity and sustainability. A comprehensive list of the works related to NFs utilization in agricultural systems for different crops is summarized in Table 4.1. The classification of the types of vegetables and fruits was based on

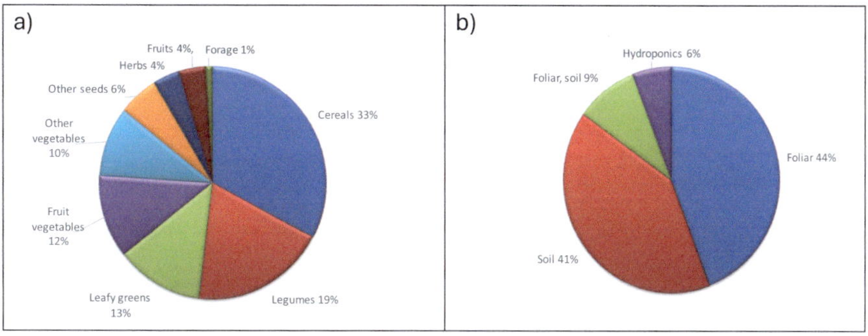

Fig. 4.1 Bibliometric analysis of nanofertilizers applied to different food crops according to the characteristics of the agricultural system: (**a**) vegetable and fruit botanical category ($n = 95$); (**b**) fertilization regime ($n = 84$)

their primary botanical attributes. In the table it can be noted that almost all vegetable groups are represented, including herbs and forage, with spices as the only missing group. Although forage is not produced directly for human consumption, it constitutes an important agro-industrial product, which serves as the basis for the whole food chain, toward which many important NF advancements and applications have been reported (Mejias et al. 2021).

A bibliometric analysis was conducted encompassing 95 articles focused on NF applications in the cultivation of vegetable and fruit crops. Figure 4.1a depicts the relative rate of research contributions in the NF literature for the different crop types according to their botanical classification (Table 4.1), in which it can be seen that cereals are the most researched food crop, with wheat being the most predominant crop, followed by rice and corn, with an equal number of reports. Legumes are the second most researched crops with more NF applications, with green bean cultivation as the most prominent crop. The groups following in importance are leafy greens and fruit vegetables, with lettuce and tomato being the most represented crops, respectively. Which is to be expected due to the commercial global importance of these vegetables products.

4.3.1 Foliar Versus Soil Fertilization

NFs can be applied in crop cultivation mainly by three different methods: hydroponics, foliar fertilization, and fertigation, the latter two being the most commonly used in large-scale agriculture. In the case of soil application, the nanoparticles enter the plant through different mechanisms depending on their size: particles smaller than 5 nm are absorbed through the pores of the root's cell wall, whereas larger particles require the activation of specific transporter proteins. Similarly, in foliar fertilization, small nanomaterials enter the leaves through the cuticle and the larger ones through the stomata. After entering the plant, NFs are transported to plant's organs

Table 4.1 Application of nanofertilizers in food crop products classified according to their primary botanical traits

Type of vegetable/fruit	Crop products
Cereals	
	**Wheat** (Beig et al. 2023; Chamani et al. 2023; Huang et al. 2024; Ibrahim et al. 2022; Jacobo and Murillo 2025; Kumar et al. 2022a, b; Lian et al. 2024; Noruzi et al. 2023; Ramírez-Rodríguez et al. 2020; Rizk et al. 2025; Saad et al. 2022; Sharma et al. 2022; Sheoran et al. 2021; Svietlova et al. 2021), **corn** (Al-Aaraji and Al-Rubaye 2020; de França Bettencourt et al. 2020; Jacobo and Murillo 2025; Kumaraswamy et al. 2021; Neto et al. 2020; Salama and Badry 2020; Sharma et al. 2024; Shoukat et al. 2025; Yoon et al. 2020), **rice** (Baral et al. 2020; Dey et al. 2023; Koley and Mondal 2024; Mehmood et al. 2024; Naseem et al. 2020; Pongpeera and Maensiri 2023; Sehgal et al. 2024; Tahir et al. 2021; Yuvaraj et al. 2023), ***Sorghum*** (Dimkpa et al. 2022), **pearl millet** (Kumar et al. 2022a)
Legumes	
	**Green bean** (Amaya-Olivas et al. 2023; Franco-Lagos et al. 2023; Márquez-Prieto et al. 2022; Muñoz-Márquez et al. 2022, 2024; Ponce-García et al. 2023; Salama et al. 2021, 2022), **soybean** (Batsmanova et al. 2020; Ernst et al. 2024; Li et al. 2023), **mungbean** (Motlhalamme et al. 2023; Ram et al. 2024; Rani et al. 2023; Ukidave and Ingale 2022), **broad bean** (Mahmoud et al. 2022), **moth bean** and **chick pea** (Ukidave and Ingale 2022), **grass pea** (Arfaoui et al. 2021), **lupin** (Murgueitio-Herrera et al. 2022)
Other seeds	
	**Flax plant** (Yaseen and Amin 2021), **mustard** and **sesame** (Kumar et al. 2022a), **cocoa** (Gómez-Vera et al. 2021), ***Amaranthus*** (Francis et al. 2022), **sunflower** (Ďurišová et al. 2023)
Fruit vegetables	
	**Tomato** (Almendros et al. 2022; Chen et al. 2025; Ekanayake and Godakumbura 2021; Gudkov et al. 2020; Helal et al. 2023), **cucumber** (Carmona et al. 2021; Modi et al. 2021), **eggplant** (Elbasuney et al. 2024; Gudkov et al. 2020), **zucchini** (Meier et al. 2020), **peppers** (Abdel-Aziz et al. 2021), **okra** (Tarafder et al. 2020)
Leafy greens	
	**Lettuce** (Garza-Alonso et al. 2023b; Meier et al. 2020; Pandorf et al. 2020; Ren et al. 2025; Sakhno et al. 2023; Salama et al. 2024b; Song and Kim 2020), **cabagge** (Abdulhameed et al. 2021; Murgueitio-Herrera et al. 2022; Yang et al. 2022), **pak choi** (Guardiola-Márquez et al. 2024), **alfalfa** (El-Shal et al. 2022), **ming** (Le et al. 2022)

(continued)

Table 4.1 (continued)

Type of vegetable/fruit	Crop products
Root vegetables	
	**Potato** (Abd El-Azeim et al. 2020), **radish** (Garza-Alonso et al. 2023a), **carrot** (Song and Kim 2020), **beet** (Kandil et al. 2020)
Flower vegetables	
	**Broccoli** (Guardiola-Márquez et al. 2023a, b), **cauliflower** (Saurabh et al. 2024)
Bulb and stem vegetables	
	**Onion** (Coelho et al. 2025; Salama et al. 2024a), ***Asparagus*** (Le et al. 2022)
Herbs	
	**Basil** (Fortis-Hernández et al. 2023), **coriander** (Fahad et al. 2022), **rosemary** (Elsayed et al. 2022), **sage** (El-Mahrouk et al. 2024)
Fruits	
	**Grape** (Gaiotti et al. 2021)**, orange** (El-Gioushy et al. 2021)**, strawberry** (El-Bialy et al. 2023)**, cherries**, **apples**, **peaches**, **pears**, **tomatoes**, **grapes**, **paraguayan fruits,** and **apricots** (Carrasco-Correa et al. 2023)
Other vegetables	
	**Forage** (Reis et al. 2022)

via the xylem or the phloem, from the soil or the leaves, respectively (Singh et al. 2023).

Fertilization regime has an effect on the efficacy of the NF treatment. In soil fertilization, NFs are prone to aggregation and degradation due to their large surface area, which also can cause nanotoxicity toward the biota present in the soil. Thus, in general, soil nanofertilizers may have similar drawbacks to conventional fertilizers such as reduced efficiency and losses through the fertigation runoff. Considering the above, in principle, foliar fertilization can prevent NF degradation and provide a more efficient and direct application of the nutrients into the plant organs (Ding et al. 2023; Pradhan et al. 2024). However, it has to be considered that the foliar application is likely to pose greater risks of exposure to the farmers and to the above-the-ground biota. These particles can also become airborne, leading to the uncontrolled release of nanomaterials into the environment. In contrast, the soil environment acts as a buffered system for NFs, significantly reducing the risk of nanoparticle dispersal by the wind.

To compile bibliometric analysis with respect to fertilization methods, germination studies were discarded, which resulted in a total of 84 articles. The distribution of NF studies according fertilization methods is depicted in Fig. 4.1b, in which the number of publications is similar for foliar and soil fertilization, with only 6% reports for hydroponics cultivation. This indicates that currently there is similar research interest in both methods of fertilization. In addition, in 9% of the articles, a comparative study of both foliar and soil methods for crop fertilization was performed, whose main findings are discussed in the following.

In a study on potato cultivation, an NPK nanofertilizer was applied in field potato plantations using both foliar and soil fertilization regimes (Abd El-Azeim et al. 2020). The authors reported that foliar fertilization using the NF formulation at 50% of the recommended NPK full dose resulted in higher economic yield, starch rates, NPK nutrient use efficiency, and harvest index, thus increasing the overall potato yield while conserving the product quality. In conclusion, foliar application of NF formulation resulted in the highest economic profit as compared with both NF soil treatment and control (100% dose conventional fertilization). Dey et al. (2023) studied the effect of a Cu-Fe bimetallic NF formulation on the growth, biochemical, and toxicological parameters of rice plantlets using both foliar and soil fertilization. The authors reported that the foliar treatments increased plant growth compared to both the control (with no NF application) and the soil NF treatment; although for the latter, greater Fe and Cu contents were quantified in the plant tissues. In a similar study, Gaiotti et al. (2021) investigated the effect of a nanocomposite formulation composed of urea and amorphous calcium phosphate (ACP) on plant growth and fruit nutritional quality of grapevines cultivated under foliar and soil fertilization, reporting that when using only 80% of the nitrogen dose of the conventional fertilization treatment, the plant growth, yield, nitrogen status, and grape nutritional qualities were comparable with the non-nanofertilized control, whereas no significant differences were found between the studied fertilization regimes.

A study on radish cultivation using Fe_2O_3 nanofertilizers applied via foliar and soil fertilization method was conducted by Garza-Alonso et al. (2023a). NF treatments via both fertilization modes resulted in major increases in the nutritional quality, productivity, and biochemical parameters evaluated. In addition, an improvement in Fe utilization efficiency of up to 43% in comparison with the treatments without NF application was reported. The authors concluded that the use of Fe nanofertilizers can fully substitute conventional fertilization by improving agricultural parameters such as plant growth, nutritional quality, and nutrient utilization efficiency. In a similar report, the same research group (Garza-Alonso et al. 2023b) evaluated the application of ZnO nanofertilizers on the cultivation of lettuce under foliar and soil fertilization and reported that foliar NF application resulted in better overall performance compared to both the soil NF treatment and control. The plant parameters evaluated in the study were plant biomass, chlorophylls, antioxidant status (metabolites and enzymatic activities), and micronutrients. From these results, it was concluded that Zn nanofertilizer can be used as an alternative to conventional fertilizers with improved performance on plant growth, nutritional quality, and phytochemical quality of lettuce crops.

Guardiola-Márquez et al. (2024) studied the effect of a bimetallic Zn-Fe nanofertilizer on the growth and micronutrient uptake of pak choi plants cultivated under foliar and soil fertilization. The NF formulation was comprised of ZnO and Fe_2O_3 nanoparticles that were either functionalized or left unmodified, using extract of the algae *Arthrospira platensis*. The authors reported a synergic effect of the bimetallic NFs with the algae extract used as capping agent on increasing plant growth (32%), as well as the contents of Zn (sixfold) and Fe (twofold) in comparison with the plants without NF treatment and the uncapped NF formulation. No differences in the performance of the foliar and soil fertilization regimes were reported. In summary, from the studies reviewed, it can be concluded that despite that in some reports fertilization presented better results in terms of plant growth and productivity than soil fertilization, these beneficial effects may depend on many other different factors such as crop type, agronomical parameters, soil quality, as well as on the characteristics of the NF formulation.

4.3.2 Nutrient Components in Nanofertilizers

Plant nutrition in large-scale agriculture requires the application of both macro (N, P, K, Ca, Mg, S) and micro nutrients (Fe, Zn, B, Mn, Cu, Cl, Mo, among others) in the right amounts to foster the development and health of the plant. The literature survey (Fig. 4.2a) shows that most of the NF research works focused on the study of micronutrient nanofertilizers (56%), whereas only 26% focused on macronutrient fertilization. In addition, 16% of the studies included both micro and macro nutrients in their formulations. A few studies employed carbon (Pandorf et al. 2020; Yang et al. 2022) and chitosan (Saad et al. 2022; Sharma et al. 2024) nanoparticles without including a nanoscale readily available nutrient. Interestingly, these studies showed increases in plant growth and crop yield, which was most probably due to the very high surface area per unit volume which gives these nanomaterials the

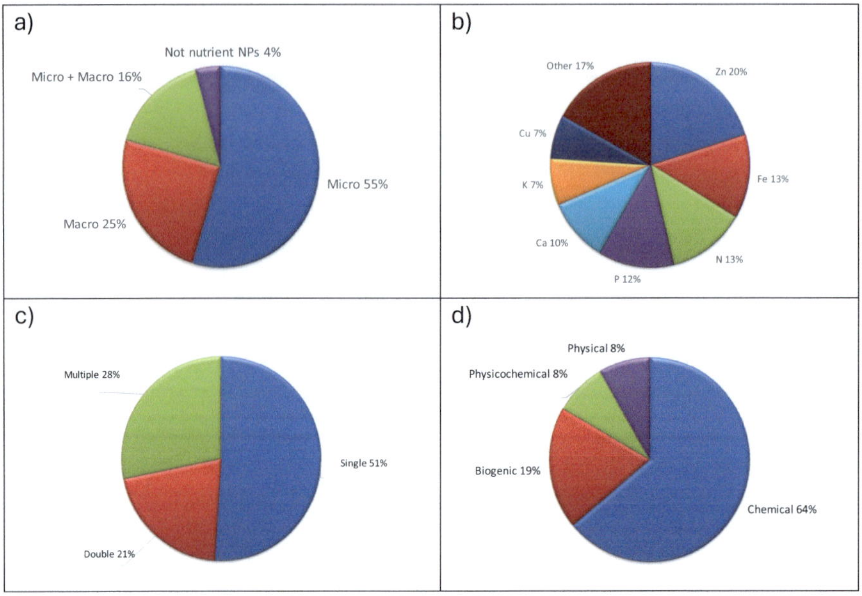

Fig. 4.2 Bibliometric analysis of nanofertilizers (NFs) applied to different food crops according to the characteristics of the NF formulation: (**a**) nutrient type ($n = 96$); (**b**) nanoscale nutrient in NF formulation ($n = 191$); (**c**) number of nanoparticle components ($n = 96$); (**d**) nanoparticle's synthetizing method ($n = 72$). NP nanoparticle

capacity to improve nutrient uptake in the plant (G. Sharma et al. 2024), thus minimizing nutrient losses through the leachate (Pandorf et al. 2020).

In addition, the nanoscale nutrients that have been mostly studied in NF research (Fig. 4.2b) comprise the micronutrients Zn and Fe, followed by the macronutrients N (with the same number of reports than Fe), P, Ca, and K. The high number of reports employing Zn as the basis of NF formulations derives from the importance of this micronutrient in the regulation of several physiological functions in plants. Zn is a critical cofactor of several enzymes related to the plant´s growth and maintenance. This metal participates in the metabolism of carbohydrates and proteins and also is a key component of plants defense system as it synthesizes auxins (Nongbet et al. 2022). Similarly, Fe is an essential micronutrient for plants growth and development and, although required in very small quantities, it acts as a cofactor or component in a number of physiological process such as chlorophyll synthesis, cellular respiration, photosynthesis, energy production, defense mechanisms, and nitrogen metabolism (Liang 2022).

In addition, another important aspect of nanofertilizers is the number of nanonutrients contained in the NF formulation; based on this, we classified NF formulations as single-, double-, or multiple-component nanofertilizers (Fig. 4.2c). The amount of NF reports utilizing two or more nanonutrients are approximately the same than those employing single-nutrient formulations. Moreover, multiple nutrient NF formulations are very significant (30%), which implies that current technology in this area is focusing on the development of multicomponent nanofertilizers. In order to further pursue this goal, research and development focuses on the synthesis of nanocomposite fertilizers with multiple nutrient components, as well as the delivery system for slow- or controlled-release of nanonutrients (Yadav et al. 2023).

4.3.3 Methods for the Synthesis of Nanoparticles and Nanomaterials

The formulation of NFs requires the selection of correct synthetizing method depending on the final agricultural application, the NFs physicochemical properties, and the characteristics of the NF formulation. As it has been widely documented in the literature, synthesis of nanoparticles (NPs) can be achieved using a wide number of techniques, which may be also combined, thus providing a complex area for NF development since there are a large number of factors and variables that affect the final formulated product and its fertilization efficiency (Yadav et al. 2023). The synthetizing methods can be generally classified into top-down (from macro to nanoparticles) and bottom-up (from single molecules to nanoparticles) methods. Most of the top-down methods involve physical techniques among which the most used are ball milling, evaporation-condensation, high electrical discharges, and laser ablation. On the other hand, chemical, biogenic, and physicochemical methods are commonly used in the bottom-up NP synthesis approach.

A total of 96 articles were analyzed for the method of synthesis of nanomaterials related to NF formulations, from which 26% were rejected since the works did not report the synthesis method or the nanoparticles were purchased from external suppliers. As depicted in Fig. 4.2d, chemical methods, particularly co-precipitation, have been the most utilized techniques for NP synthesis. Recently, special attention has been put on biogenic NF synthesis considering the need of reducing the use of chemicals as well as the implementation of synthetizing methods which are carried out under mild physicochemical conditions. This approach provides an environmentally friendly alternative to the chemical and physical methods, a characteristic that has portrayed the products thus obtained as green nanofertilizers (Al Tawaha et al. 2024b). However, these methods very often lead to physicochemical heterogeneity in the biogenic extracts, which may result in low efficiencies in NF synthesis. The low efficiency of these methods can be due to the low-specificity of the reduction and capping agents, as well as the presence of impurities and non-desired reactions in the synthesizing media. These factors negatively affect the reproducibility and scalability of the biogenic methods, thus reducing its commercial viability. On the other hand, some authors in the literature have described this technique as capable of producing nanoparticles that are safer for plants, the environment, and human health; however, this appears to be a misjudgment of the physicochemical properties of the final product, which consists merely of metallic nanoparticles. Therefore, the authors believe that such judgments should be carefully reconsidered based on experimental evidence regarding the safety of the formulated NF products.

The chemical co-precipitation method has been widely used because it is a relatively simple process, provides good control on the formation of nanoparticles in a reproducible and scalable manner, and allows the synthesis of both single- and multiple-component nanomaterials for the formulation of multinutrient NFs. This technique also allows for the development of efficient delivery systems which allows the production of slow- or controlled-release NF formulations. On the other hand, it is noteworthy that the reported use of physical methods for nanoparticle formation, such as ball milling (with only 3% of reports), has had very little application in current research works. This result is interesting since the ball milling method, although energetically intensive, provides a technique that is easy to scale-up, reproducible, and economically viable for NP production. It would be interesting to determine whether the current research inclination toward chemical methods stems from their ease of implementation at the laboratory level and the limited availability of equipment required for physical approaches, particularly the ball-milling technique, or if this trend is influenced by technological, economic, or environmental considerations within industry-driven research strategies.

It would be interesting to determine whether the current research preference for chemical methods stems from their ease of implementation at the laboratory level and the limited availability of equipment required for physical approaches—particularly the ball-milling technique—or if this trend is influenced by technological, economic, or environmental considerations within industry-driven research strategies.

4.3.4 Formulation of NFs and Delivery Systems

A critical stage in the development of NFs is the formulation of the final product, in which it is important to take into account the efficient delivery of nutrient nanomaterials to the plant (Yadav et al. 2023). Figure 4.3 shows a schematic representation of the different NF formulation systems that are utilized in current research. The simplest NF formulation is the application of single-nutrient NPs to the plant via foliar or soil fertilization (Fig. 4.3a). This delivery method is simple and straightforward and allows the direct application of the NFs to the soil or leaves. However, the main drawback of these formulations is that nanofertilizers can be lost through the leachate or may fail to reach the roots of the plant in soil fertilization (Haydar et al. 2024), or may be disperse through the air during foliar application, potentially causing airborne nanoparticle contamination. Therefore, the effectiveness and safety of this delivery system is undermined by the aforementioned considerations.

Based on the bibliometric analysis (Fig. 4.2c), about 50% of the research works focused on single-nutrient NF formulations, which implies that this delivery method is practical, affordable, and produces effective fertilization results. Interestingly, a few authors utilized single NPs as fertilizers and reported a reduction in nutrient

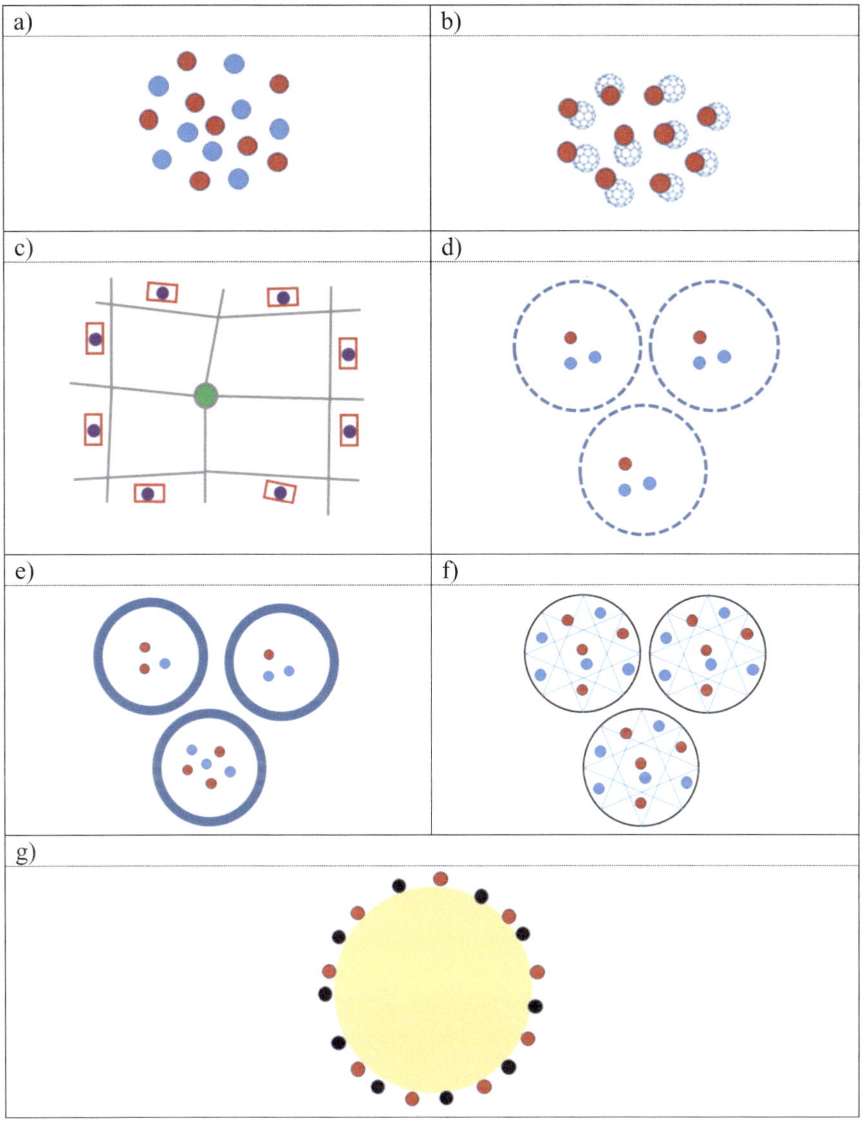

Fig. 4.3 Schematic representations of nanofertilizer (NF) formulations: (**a**) NFs made of only nanoparticles (NPs); (**b**) NPs capped with NPs; (**c**) NP matrix capped with NPs; (**d**) NPs coated with large capping agents; (**e**) NPs encapsulated by polymeric materials; (**f**) NPs embedded in polymeric matrices; (**g**) non-nanoscale nutrients coated with NPs

losses through the fertilization leachate (Mehmood et al. 2024; Sheoran et al. 2021; Yuvaraj et al. 2023) probably due to the large surface area of the nanoparticles. Other authors have reported the coating of granular nutrients in the micro or macroscale with metallic nanoparticles (Fig. 4.3g), in which case the granular material acts as the carrier of the nanonutrient (Beig et al. 2023). The formulation of NFs with double or multiple nanonutrient components can be achieved by using several physicochemical methods in which multiple nanonutrients are combined, usually through weak physicochemical forces, to constitute multinutrient nanocomposites (Fig. 4.3b, c). However, these formulations may have the same drawbacks of single-NP formulations.

To overcome the limitations of NF formulations comprising only nanoscale materials, larger molecules have been used as capping agents of the nutrient nanoparticles to formulate functionalized nanofertilizers (Fig. 4.3d), in which the coating materials are regularly comprised of organic materials such as humic acids (Yoon et al. 2020) and molasses (Beig et al. 2023), as well as biological substrates, such as bacterial and microalgal cells (Guardiola-Márquez et al. 2023a, b). The function of coating materials is to stabilize and protect the NPs in the nutrient solution through their transport to the plant's cells, as well as to serve as a vehicle for nanoparticle delivery. Nanoparticle functionalization can also provide additional physicochemical or biological properties to the NF product during its interactions with the plant. In addition, this type of formulation can retard the release of nutrients from the coated nanofertilizer, thereby imparting slow-release properties to the NF product.

At the front core of NF formulations is the development of slow- or controlled-release nanofertilizers, in which the nutrient nanoparticles are encapsulated or embedded in a polymeric or hydrogel structure that has the following main purposes: (i) protect the NPs from direct interaction with biotic or abiotic agents of the crop fertilization system; (ii) prevent losses of NFs through the runoff; (iii) provide a slow or controlled release of nanonutrients to the plant (Haydar et al. 2024). The above attributes allow the NFs thus formulated to achieve greater effectiveness (plant growth and yield) while maximizing nutrient usage and reducing losses by runoff, thus yielding efficient and sustainable NF products with the potential to become commercially viable. In this regard, two types of NF formulations have been used: (i) nutrient nanoparticles encapsulated usually in an organic polymer, in which a continuous film-coating of the NPs is attained (Fig. 4.3e), which has been termed in the literature as controlled-release NFs and (ii) nutrient nanoparticles embedded in a polymeric hydrogel (Fig. 4.3f), termed in the literature as slow-release NFs (Haydar et al. 2024). In this review, no distinction is made between slow and controlled release NFs; therefore, only the slow-release term (SRNF) is used, since we consider that generally the delivery mechanisms are too complex to regard the encapsulated NFs as controlled-release delivery systems. The synthesis of slow-release NF is diverse and has been reported in good detail elsewhere (Haydar et al. 2024; Yadav et al. 2023).

The main materials used in the formulation of slow-release NF can be divided into inorganic and polymeric. Polymeric organic materials have been used for the elaboration of slow-release NFs, such as chitosan (Gómez-Vera et al. 2021; Kumaraswamy et al. 2021; Saad et al. 2022; Sharma et al. 2024), alginate (Ekanayake and Godakumbura 2021; Le et al. 2022), lignin (Helal et al. 2023), and starch (Reis et al. 2022). Among the inorganic materials used for the production of SRNP are aluminum silicates which constitute a mesoporous silica matrix (Naseem et al. 2020) and clays (Reis et al. 2022). However, clearly the most used materials to synthetize SRNFs are hydroxyapatite (HAP) and amorphous calcium phosphate (ACP), which apart from acting as embedding inorganic matrices provide a source of P and Ca. These inorganic matrices have been the basis of a wide range of NF formulations which can comprise single or multiple nanoscale components, and both micro and/or micronutrients.

4.4 Slow-Release Nanofertilizers

As discussed above, an important number of nanofertilizer formulations based on HAP and ACP have resulted in an efficient slow release of nanonutrients. In addition, these formulations offer many advantages to the NF product, as they can incorporate multiple nanoparticle components, both micro- and macronutrients, and also provide a source of P and Ca to the fertilization system. In a forage agronomical system, Reis et al. (2022) tested a nanocomposite material based on HAP nanoparticles embedded in a starch/bentonite matrix added to bulk urea and reported an improvement in P efficiency as compared with the conventional P fertilizer triple superphosphate (TSP). In another study, Sehgal et al. (2024) applied a urea-HAP nanocomposite as fertilizer in rice cultivation reporting an increase in plant growth and yield, as well as an improvement in the plant and soil chemical properties. In an okra (*Abelmoschus esculentus*) cultivation experiment (Tarafder et al. 2020), a multinutrient formulation comprising a urea-HAP nanocomposite added with Cu, Fe, and Zn nanoparticles was applied as fertilizer at a 100-fold lower dose than the conventional fertilizer. The authors reported that the nanofertilizer reduced nutrient losses in the leachate and improved plant's uptake of Cu, Fe, and Zn, thus demonstrating the slow-release capacity of the NF formulation. Similarly, in a wheat cultivation experiment, Sharma et al. (2022) synthetized a multinutrient nanofertilizer comprised of urea-HAP nanocomposite doped with Zn or Mg. The authors reported that the nanofertilizer formulated with half of the recommended fertilization dose resulted in a slow release of urea and similar wheat yield as compared with the control treatment with full fertilization dose. A corn fertilization study was performed by Yoon et al. (2020) utilizing HAP coated with humic acids, in which the application of the nanofertilizer resulted in higher plant yields as compared with conventional fertilizers, whereas the rhizospheric bacteria were not affected by the application of the nanohybrid product.

In another type of SRNF formulation, Helal et al. (2023) reported the application of a nanohybrid comprised of urea embedded in a lignin-bentonite matrix as a nanofertilizer in tomato cultivation. The study showed that the NF formulation had a nitrogen uptake efficiency much greater (47–88%) than that of the conventional fertilizer (33%). In addition, the NF treatment applied at 25% of the recommended fertilization dose resulted in higher growth, yield, and fruit quality than the control. In a corn cultivation study (Kumaraswamy et al. 2021), a Si nanohybrid was formulated using a chitosan-tripolyphosphate (TPP) encapsulating matrix, resulting in 45% higher yield as compared to the control treatment fertilized with bulk SiO_2. Naseem et al. (2020) carried out a rice cultivation experiment applying a nanohybrid product comprising urea embedded in an aluminosilicate mesoporous matrix. The authors reported higher crop yields in the nanofertilizer treatment in comparison with the application of a commercial urea fertilizer. This result was attributed to a slow-release effect of the NF formulation which resulted in better nutrient use efficiency of the urea-Zn nanofertilizer as compared to the control.

4.5 Safety Considerations

One of the main factors that hamper the commercial consolidation of nanofertilizers is their safety toward plant crops, the environment, and human health. These concerns are substantial and warrant careful consideration. The safety and health issues related to the application of nanofertilizers at a large scale have been discussed comprehensively elsewhere (Haydar et al. 2024; Singh et al. 2023; Yadav et al. 2023). In general, from this literature review, it can be implied that the general effects of nanofertilizers on crop plants are encouraging and favor the growth and health of the treated plants. On the other hand, the toxicity of nanofertilizers on the plants and/or soil environment has been addressed in some of the articles surveyed in this review. We discuss some important aspects of these reports in the following.

In the article previously discussed by Dey et al. (2023), the authors reported the use of Cu-Fe nanofertilizers on rice cultivation at two different concentrations (5 and 10 mg/L). Whereas the low concentration of the nanofertilizer had a positive effect on plant growth, the 10 mg/L NF dose caused an increase in the oxidative stress of the plant, thus indicating NF toxicity at this concentration. Similarly, Arfaoui et al. (2021) reported that ZnO nanofertilizers applied in green pea cultivation inhibited root and seedling development at high NF concentrations (0.1 and 1 g/L), whereas at lower NF doses (10 and 100 mg/L), the embryos manage to cope with NF toxicity by inducing secretion of hydrogen peroxide scavenging capacity to modulate the oxidative stress caused by the Zn nanoparticles.

In another study, Ibrahim et al. (2022) carried out the application of CuO nanoparticles in wheat cultivation in order to compare its performance with a conventional fertilizer, reporting that both the nanofertilizer and bulk Cu treatments caused a drastic reduction in root growth. The authors measured the effective toxicity dose concentration for the fertilizers concluding that bulk Cu was 2.5 times more toxic than the Cu nanofertilizer (0.94 mg/L). In a similar study, Noruzi et al. (2023) reported on the effect of different HAP nanoparticle formulations on wheat cultivation and observed acceptable fertilization performance in acidic media. However, the authors reported that both the nanofertilizer and the conventional fertilizer inhibited seed germination at high P concentrations, with a lower toxicity of the NF treatments in comparison with that of the conventional P fertilizer (TPP).

A comprehensive toxicological study was performed by Ďurišová et al. (2023), who studied the effect of different single-NP nanofertilizers (ZnO, $ZnSO_4$, Fe_3O_4 and TiO_2) on pollen and seed yield of sunflower plants. Despite the fact that NP treatments changed pollen morphology, seed yields were not affected; on the contrary, they even increased in the Zn and Fe treatments as compared with the control. The authors attributed these results to the low NP concentrations used in the study: 10 and 76 mg/L for the Zn and Fe nanoparticles, respectively. In a study previously mentioned, Sharma et al. (2022) studied the effect of urea-HAP capped with Zn/Mg on the growth of wheat, finding that the NF treatments enhanced enzymatic activity in the soil in comparison with the control, thus confirming the innocuous effect of the nanofertilizer on the soil environment. Similarly, Salama et al. (2024a, b) reported the application of a Ca nanofertilizer on lettuce cultivation, observing significant improvements in plant growth and yield, as well as in the nutritional and phytochemical quality of products compared to the conventional fertilizer. In addition, the authors reported that the NF-treated lettuces were safe for human consumption by measuring the effective toxic concentrations (EC_{50}) of the vegetable products using a Microtox 500 analyzer. In a study that investigated the toxic concentrations of Zn nanofertilizers, Song and Kim (2020) applied ZnO nanoparticles at concentrations of 0, 1, 5, 20, 100, 1000 mg/kg for lettuce and carrot fertilization. The authors reported that the NF treatments exhibited excellent fertilization performance in terms of plant growth and nutrient content as compared to the control. The only toxic concentration of the NF treatment was 1000 mg/kg, which can be regarded as an excessive concentration since NF concentrations lower than 100 mg/kg resulted in efficient fertilization of the crop products.

In a study on flax plants fertilization, Yaseen et al. (2021) assessed the effect of mycosynthesized P nanoparticles on crop cultivation and determined the safe dose of the NF on plants as 12.5 mg/L. In addition, the best plant growth results were obtained using a 1:1 combination of the nanofertilizer with the conventional P fertilizer (CF). The authors reported that NF treatment had detrimental effect on the structure of leaf cells, while no damage was observed in the combined NF-CF treatment. Sehgal et al. (2024) studied the effect of urea-HAP nanofertilizer on rice cultivation and reported that NF treatment had no negative effects on the soil microbiota. On the contrary, the application of nanofertilizer was associated with an increase in nitrogen-fixing bacteria.

From this review, we can conclude that NF formulations have been generally reported to be relatively safe for plants and soil microbiota at the dose ranges typically used (approximately 5–50 ppm). Moreover, in most of the studies reviewed, nanofertilizer improved plant growth and crop yield as well as the nutritional and phytochemical quality of the crop products. However, before any large-scale implementation of a specific nanofertilizer, it is strongly recommended to determine its effective toxic concentration (EC_{50}) for the target crop. This requirement arises from a key limitation: the toxicity levels of NF formulations depend on multiple interacting factors, including genotypic and phenotypic traits of the plant, soil characteristics, and the physicochemical properties of the NFs themselves. Therefore, toxicological assessments must be carried out on a case-by-case basis, considering the specific crop–NF system. In this context, it is crucial to evaluate the toxicological profile of nanofertilizers in a more systematic manner not only through in vitro and in planta assays but also by incorporating in vivo tests on soil macrobiota and animal models (e.g., murine systems), in order to assess their potential implications for the environment and human health.

Although some studies have investigated the effects of NFs on soil microbial communities using traditional microbiological approaches, there is a clear need for more comprehensive research addressing both soil micro- and macrobiota. Ideally, such studies should integrate omics-based approaches—including genomics, transcriptomics, and metabolomics—to unravel the complex interactions between nanofertilizers and the biological components of the soil–plant environment. Finally, it is important to note that the limited number of reports describing NF toxicity may not necessarily indicate a low toxic potential. Instead, this could reflect a publication bias, as research on nanofertilizers often falls within the realm of applied science and technology, where academic publishing tends to favor studies with positive or "successful" outcomes. As a result, negative or null findings are frequently underreported or excluded. Given the critical importance of safety in the large-scale application of NFs, we believe that the current publication paradigm should shift toward the standards used in medical, toxicological, and pharmacological sciences, where neutral and adverse findings are given equal importance. This shift is essential to ensure a more balanced and comprehensive understanding of NF risks toward crop plants, the environment, and human health.

4.6 Conclusions and Future Prospects

This chapter critically examined recent advancements in engineering-synthesized nanomaterials used as crop fertilizers. Nanoparticle technologies have demonstrated high efficacy in boosting crop yields, enhancing nutrient-use efficiency (via nanofertilizers (NFs)), and alleviating both biotic and abiotic stress. Research on NFs has expanded rapidly across diverse food crop groups, with balanced application through both foliar and soil routes. Among these formulations, dual or multi-nanonutrient types, particularly those containing Zn and Fe, are the most prevalent, followed by macronutrient-based NFs incorporating N, P, and Ca. Bottom-up chemical synthesis remains the preferred method due to its simplicity and reproducibility. Additionally, advanced NF designs featuring slow-release multinutrients embedded in hydroxyapatite or chitosan matrices show promising results for both yield enhancement and nutrient uptake.

In addition, this review addresses safety considerations regarding the use of NFs in large-scale agriculture. Common application levels (5–50 ppm) are generally considered safe; however, comprehensive toxicological evaluations, including EC_{50} determinations, remain essential. Toxicity is influenced by plant and soil traits as well as nanoparticle characteristics, requiring case-specific assessments. Omics approaches may provide valuable insights into the complex interactions within the soil–plant–NF systems. It is also important to acknowledge that publication bias toward "successful results" may obscure potential risks of the use of NFs for plants, the environment, and human health. Given the critical importance of health and safety in the implementation of large-scale NF systems, unbiased reporting aligned with toxicological standards is vital for responsible application of NF products.

Acknowledgments Ramírez-Cariño thanks Secretaria de Ciencia, Humanidades, Tecnología e Innovación (SECIHTI), Mexico, for the financial support of his postdoctoral fellowship (No. 644716).

Declaration of Competing Interests The authors declare that they have no known competing financial interests or personal relationships that could have appeared to influence the work reported in this chapter.

References

Abd El-Azeim MM, Sherif MA, Hussien MS, Tantawy IAA, Bashandy SO (2020) Impacts of nano- and non-nanofertilizers on potato quality and productivity. Acta Ecol Sin 40(5):388–397. https://doi.org/10.1016/j.chnaes.2019.12.007

Abdel-Aziz HMM, Soliman MI, Abo Al-Saoud AM, El-Sherbeny GA (2021) Waste-derived npk nanofertilizer enhances growth and productivity of capsicum annuum l. Plants 10(6):1144. https://doi.org/10.3390/plants10061144

Abdulhameed MF, Taha AA, Ismail RA (2021) Improvement of cabbage growth and yield by nanofertilizers and nanoparticles. Environ Nanotechnol Monitor Manag 15(January):100437. https://doi.org/10.1016/j.enmm.2021.100437

Aguilar-Paredes A, Valdés G, Araneda N, Valdebenito E, Hansen F, Nuti M (2023) Microbial community in the composting process and its positive impact on the soil biota in sustainable agriculture. Agronomy 13(2):542. https://doi.org/10.3390/agronomy13020542

Al Tawaha ARM, Singh A, Rajput VD, Varshney A, Agrawal S, Ghazaryan K, Minkina T, Al Zoubi OM, Habeeb T, Dionis L, Hasan HA, Shawaqfeh S (2024a) Green Nanofertilizers—the need for modern agriculture, intelligent, and environmentally-friendly approaches. Ecol Eng Environ Technol 25(1):1–21. https://doi.org/10.12912/27197050/172946

Al Tawaha ARM, Singh A, Rajput VD, Varshney A, Agrawal S, Ghazaryan K, Minkina TM, Al Zoubi OM, Habeeb T, Dionis L, Hasan HA, Shawaqfeh S (2024b) Green Nanofertilizers: the need for modern agriculture, intelligent, and environmentally-friendly approaches. Ecol Eng Environ Technol 25(1):1–21. https://doi.org/10.12912/27197050/172946

Al-Aaraji HTM, Al-Rubaye MA (2020) Effect of nanoparticles and ordinary zinc oxide fertilizer and humic acid on zinc and phosphorous availably in the soil. Plant Arch 20(1):2403–2410

Almendros P, González D, Fernández MD, García-Gomez C, Obrador A (2022) Both Zn biofortification and nutrient distribution pattern in cherry tomato plants are influenced by the application of ZnO nanofertilizer. Heliyon 8(3):e09130. https://doi.org/10.1016/j.heliyon.2022.e09130

Amaya-Olivas NI, Sánchez E, Hernández-Ochoa L, Ojeda-Barrios DL, Ávila-Quezada GD, Flores-Córdova MA, Chávez-Flores D, Ayala-Soto JG, Salcido-Martínez A, Ramírez-Estrada CA (2023) Biofortification with magnesium nanofertilizer on bioactive compounds and antioxidant capacity in green beans. Notulae Botanicae Horti Agrobotanici Cluj-Napoca 51(1):1–11. https://doi.org/10.15835/nbha51112830

Arfaoui H, Karmous I, Mahjoubi Y, Kharbech O, Tlahig S, Loumerem M (2021) Screening of the effects of zinc oxide based nanofertilizers on the germination of Lathyrus sativa L. seeds. J Appl Bot Food Qual 94:53–60. https://doi.org/10.5073/JABFQ.2021.094.007

Baral K, Kumar D, Shlvay YS (2020) Effect of zinc oxide nanoparticles embedded npk fertilizer on growth, yield and zinc concentration and uptake of aromatic rice (oryza sativa). Indian J Agric Sci 90(8):1626–1630. https://doi.org/10.56093/ijas.v90i8.105980

Basavegowda N, Baek KH (2021) Current and future perspectives on the use of nanofertilizers for sustainable agriculture: the case of phosphorus nanofertilizer. *3*. Biotech 11(7):357. https://doi.org/10.1007/s13205-021-02907-4

Batsmanova L, Taran N, Konotop Y, Kalenska S, Novytska N (2020) Use of a colloidal solution of metal and metal oxide-containing nanoparticles as fertilizer for increasing soybean productivity. J Cent Eur Agric 21(2):311–319. https://doi.org/10.5513/JCEA01/21.2.2414

Beig B, Niazi MBK, Jahan Z, Haider G, Zia M, Shah GA, Iqbal Z, Hayat A (2023) Development and testing of zinc sulfate and zinc oxide nanoparticle-coated urea fertilizer to improve N and Zn use efficiency. Front Plant Sci 13(January):1058219. https://doi.org/10.3389/fpls.2022.1058219

Carmona FJ, Dal Sasso G, Ramírez-Rodríguez GB, Pii Y, Delgado-López JM, Guagliardi A, Masciocchi N (2021) Urea-functionalized amorphous calcium phosphate nanofertilizers: optimizing the synthetic strategy towards environmental sustainability and manufacturing costs. Sci Rep 11(1):1–14. https://doi.org/10.1038/s41598-021-83048-9

Carrasco-Correa EJ, Mompó-Roselló Ò, Simó-Alfonso EF (2023) Calcium oxide nanofertilizer as alternative to common calcium products for the improvement of the amount of peel fruit calcium. Environ Technol Innov 31:103180. https://doi.org/10.1016/j.eti.2023.103180

Chamani M, Naseri B, Rafiee-Dastjerdi H, Emaratpardaz J, Ebadollahi A, Palla F (2023) Some physiological effects of Nanofertilizers on wheat-aphid interactions. Plants 12(14):2602. https://doi.org/10.3390/plants12142602

Chen W, Hu Z, Liu D (2025) Compared to MgSO4, the use of magnesium Nanofertilizer alleviates potassium-magnesium antagonism in tomato roots. Agriculture 15(4):368. https://doi.org/10.3390/agriculture15040368

Coelho A, Andréia Cavalari A, Haddad P, do Nascimento AN (2025) Nanofertilizers for enhancing food production: A case study on microgreens enrichment using superparamagnetic iron oxide nanoparticles (SPIONs). Food Chem 463(September):141364. https://doi.org/10.1016/j.foodchem.2024.141364

de França Bettencourt GM, Degenhardt J, Zevallos Torres LA, de Andrade Tanobe VO, Soccol CR (2020) Green biosynthesis of single and bimetallic nanoparticles of iron and manganese using bacterial auxin complex to act as plant bio-fertilizer. Biocatal Agric Biotechnol 30:101822. https://doi.org/10.1016/j.bcab.2020.101822

Dey S, Nath S, Alam Ansari T, Biswas A, Barman F, Mukherjee S, Gopal G, Bhattacharyya A, Mukherjee A, Kundu R, Paul S (2023) Application of green synthesized bimetallic nZVI-Cu nanoparticle as a sustainable alternative to chemical fertilizers to enhance growth and photosynthetic efficiency of rice seedlings. Plant Physiol Biochem 201(February):107837. https://doi.org/10.1016/j.plaphy.2023.107837

Dimkpa CO, Campos MGN, Fugice J, Glass K, Ozcan A, Huang Z, Singh U, Santra S (2022) Synthesis and characterization of novel dual-capped Zn–urea nanofertilizers and application in nutrient delivery in wheat. Environ Sci: Adv 1(1):47–58. https://doi.org/10.1039/d1va00016k

Ding Y, Zhao W, Zhu G, Wang Q, Zhang P, Rui Y (2023) Recent trends in foliar nanofertilizers: a review. Nanomaterials 13(21):2906. https://doi.org/10.3390/nano13212906

Ďurišová Ľ, Ďúranová H, Kšiňan S, Ernst D, Šebesta M, Žitniak Čurná V, Eliáš P, Qian Y, Straka V, Feng H, Tomovičová L, Kotlárová N, Kratošová G, Kolenčík M (2023) Exploring the impact of metal-based Nanofertilizers: a case study on sunflower pollen morphology and yield in field conditions. Agronomy 13(12):2922. https://doi.org/10.3390/agronomy13122922

Ekanayake SA, Godakumbura PI (2021) Synthesis of a dual-functional Nanofertilizer by embedding ZnO and CuO nanoparticles on an alginate-based hydrogel. ACS Omega 6(40):26262–26272. https://doi.org/10.1021/acsomega.1c03271

Elbasuney S, El-Sayyad GS, Abdelaziz AM, Rizk SH, Tolba MM, Attia MS (2024) Stable colloidal iron oxide nanoparticles: A new Green Nanofertilizer and therapeutic nutrient for eggplant immune response against fusarium wilt disease. J Clust Sci 35(4):983–997. https://doi.org/10.1007/s10876-023-02527-3

El-Bialy SM, El-Mahrouk ME, Elesawy T, Omara AED, Elbehiry F, El-Ramady H, Áron B, Prokisch J, Brevik EC, Solberg S (2023) Biological nanofertilizers to enhance growth potential of strawberry seedlings by boosting photosynthetic pigments, plant enzymatic antioxidants, and nutritional status. Plants 12(2):302. https://doi.org/10.3390/plants12020302

El-Gioushy SF, Ding Z, Bahloul AME, Gawish MS, Abou El Ghit HM, Abdelaziz AMRA, El-Desouky HS, Sami R, Khojah E, Hashim TA, Kheir AMS, Zewail RMY (2021) Foliar application of nano, chelated, and conventional iron forms enhanced growth, nutritional status, fruiting aspects, and fruit quality of washington navel orange trees (Citrus sinensis l. osbeck). Plants 10(12):2577. https://doi.org/10.3390/plants10122577

El-Mahrouk ESM, Atef EAM, Gabr MK, Aly MA, Mohamed AE, Eisa EA, Gururani MA (2024) Response of Salvia officinalis to zinc and silicon nanoparticles and pollen extract as alternates to traditional fertilizers. Front Plant Sci 15(November):1469691. https://doi.org/10.3389/fpls.2024.1469691

Elsayed AAA, EL-Gohary A, Taha ZK, Farag HM, Hussein MS, AbouAitah K (2022) Hydroxyapatite nanoparticles as novel nano-fertilizer for production of rosemary plants. Sci Hortic 295(November 2021):110851. https://doi.org/10.1016/j.scienta.2021.110851

El-Shal RM, El-Naggar AH, El-Beshbeshy TR, Mahmoud EK, El-Kader NIA, Missaui AM, Du D, Ghoneim AM, El-Sharkawy MS (2022) Effect of Nano-fertilizers on alfalfa plants grown under different salt stresses in hydroponic system. Agriculture 12(8):1113. https://doi.org/10.3390/agriculture12081113

Ernst D, Kolenčík M, Kupec M, Šebesta M, Qian Y, Straka V, Černý I, Soulange JG, Ducsay L (2024) Effects of zinc oxide and zinc–silica-based Nanofertilizers with yeasts on selected components of soybean in the central European agronomic region: a short-term study. Agronomy 14(9):2138. https://doi.org/10.3390/agronomy14092138

Escribà-Gelonch M, Butler GD, Goswami A, Tran NN, Hessel V (2023) Definition of agronomic circular economy metrics and use for assessment for a nanofertilizer case study. Plant Physiol Biochem 196(December 2022):917–924. https://doi.org/10.1016/j.plaphy.2023.02.042

Fahad, Balouch A, Agheem MH, Memon SA, Baloch AR, Tunio A, Abdullah, Pato AH, Jagirani MS, Panah P, Gabole AA, Qasim S (2022) Efficient mitigation of cadmium and lead toxicity in coriander plant utilizing magnetite (Fe3O4) nanofertilizer as growth regulator and antimicrobial agent. Int J Environ Anal Chem 102(16):3868–3879. https://doi.org/10.1080/03067319.2020.1776861

Fortis-Hernández M, De La Rosa-Reta CV, Galindo-Guzmán M, Preciado-Rangel P, Galindo-Guzmán AP, Flores-Loyola E (2023) Biofortification with ZnO NPs as nanofertilizers to improve sustainable commercial and phytochemical quality in basil plants. Notulae Botanicae Horti Agrobotanici Cluj-Napoca 51(2):13078. https://doi.org/10.15835/nbha51213078

Francis DV, Sood N, Gokhale T (2022) Biogenic CuO and ZnO nanoparticles as nanofertilizers for sustainable growth of Amaranthus hybridus. Plants 11(20):2776. https://doi.org/10.3390/plants11202776

Franco-Lagos CL, Sánchez E, Palacio-Márquez A, Pérez-Álvarez S, Terrazas-Gómez M, Villalobos-Cano O, Ramírez-Estrada CA (2023) Efficacy of the application of boron nanofertilizer on biomass, yield, nitrogen assimilation and photosynthetic activity in green beans. Notulae Botanicae Horti Agrobotanici Cluj-Napoca 51(1):12795. https://doi.org/10.15835/nbha51112795

Gaiotti F, Lucchetta M, Rodegher G, Lorenzoni D, Longo E, Boselli E, Cesco S, Belfiore N, Lovat L, Delgado-López JM, Carmona FJ, Guagliardi A, Masciocchi N, Pii Y (2021) Urea-doped calcium phosphate nanoparticles as sustainable nitrogen nanofertilizers for viticulture: implications on yield and quality of pinot gris grapevines. Agronomy 11(6):1026. https://doi.org/10.3390/agronomy11061026

Garza-Alonso CA, Cadenas-Pliego G, Juárez-Maldonado A, González-Fuentes JA, Tortella G, Benavides-Mendoza A (2023a) Fe2O3 nanoparticles can replace Fe-EDTA fertilizer and boost the productivity and quality of Raphanus sativus in a soilless system. Sci Hortic 321(April):112261. https://doi.org/10.1016/j.scienta.2023.112261

Garza-Alonso CA, Juárez-Maldonado A, González-Morales S, Cabrera-De la Fuente M, Cadenas-Pliego G, Morales-Díaz AB, Trejo-Téllez LI, Tortella G, Benavides-Mendoza A (2023b) ZnO nanoparticles as potential fertilizer and biostimulant for lettuce. Heliyon 9(1):e12787. https://doi.org/10.1016/j.heliyon.2022.e12787

Gómez-Vera P, Blanco-Flores H, Francisco AM, Castillo J, Tezara W (2021) Silicon dioxide nanofertilizers improve photosynthetic capacity of two Criollo cocoa clones (Theobroma cacao L.). Exp Agric 57(2):85–102. https://doi.org/10.1017/S0014479721000065

Guardiola-Márquez CE, García-Sánchez CV, Sánchez-Arellano ÓA, Bojorquez-Rodríguez EM, Jacobo-Velázquez DA (2023a) Biofortification of broccoli microgreens (Brassica oleracea var. italica) with Glucosinolates, zinc, and iron through the combined application of bio- and Nanofertilizers. Foods 12:3826. https://doi.org/10.3390/foods12203826

Guardiola-Márquez CE, López-Mena ER, Segura-Jiménez ME, Gutierrez-Marmolejo I, Flores-Matzumiya MA, Mora-Godínez S, Hernández-Brenes C, Jacobo-Velázquez DA (2023b) Development and evaluation of zinc and iron nanoparticles functionalized with plant growth-promoting Rhizobacteria (PGPR) and microalgae for their application as bio-Nanofertilizers. Plants 12(20):3657. https://doi.org/10.3390/plants12203657

Guardiola-Márquez CE, del Martínez-Ballesta MC, García-Sánchez CV, Bojorquez-Rodríguez EM, Jacobo-Velázquez DA, Martínez-Hernández GB (2024) Effects of foliar and root application of Zn and Fe bio-nanofertilizers on the glucosinolate, Zn and Fe contents of Pak choi (Brassica rapa Subsp. Chinensis) grown under hydroponic and pot cultivation. Sci Hortic 338(May):113654. https://doi.org/10.1016/j.scienta.2024.113654

Gudkov SV, Shafeev GA, Glinushkin AP, Shkirin AV, Barmina EV, Rakov II, Simakin AV, Kislov AV, Astashev ME, Vodeneev VA, Kalinitchenko VP (2020) Production and use of selenium nanoparticles as fertilizers. ACS Omega 5(28):17767–17774. https://doi.org/10.1021/acsomega.0c02448

Haydar MS, Ghosh D, Roy S (2024) Slow and controlled release nanofertilizers as an efficient tool for sustainable agriculture: recent understanding and concerns. Plant Nano Biol 7(January):100058. https://doi.org/10.1016/j.plana.2024.100058

Helal MID, El-Mogy MM, Khater HA, Fathy MA, Ibrahim FE, Li YC, Tong Z, Abdelgawad KF (2023) A controlled-release Nanofertilizer improves tomato growth and minimizes nitrogen consumption. Plants 12(10):1978. https://doi.org/10.3390/plants12101978

Hobbs PR, Sayre K, Gupta R (2008) The role of conservation agriculture in sustainable agriculture. Philos Trans R Soc B: Biol Sci 363(1491):543–555. https://doi.org/10.1098/rstb.2007.2169

Huang X, Wang X, Liu X, Cheng L, Pan J, Yang X (2024) Nanotechnology in agriculture: manganese ferrite nanoparticles as a micronutrient fertilizer for wheat. Plants 13(10):1395. https://doi.org/10.3390/plants13101395

Ibrahim AS, Ali GAM, Hassanein A, Attia AM, Marzouk ER (2022) Toxicity and uptake of CuO nanoparticles: evaluation of an emerging nanofertilizer on wheat (Triticum aestivum L.) Plant. Sustainability 14(9):4914. https://doi.org/10.3390/su14094914

Jacobo MYA, Murillo JML (2025) Evaluation of the effect of ZnO nanoparticles on the germination process of wheat (Triticum) and maize (Zea mays) seeds and their possible application as a fertilizer. MRS Adv 10(3):364–369. https://doi.org/10.1557/s43580-025-01111-3

Kandil EE, Abdelsalam NR, El Aziz AAA, Ali HM, Siddiqui MH (2020) Efficacy of nanofertilizer, fulvic acid and boron fertilizer on sugar beet (Beta vulgaris L.) yield and quality. Sugar Tech 22(5):782–791. https://doi.org/10.1007/s12355-020-00837-8

Koley R, Mondal NK (2024) Synthesis of calcium-based nanofertilizer and its efficacy towards reduction of oxidative stress and fluoride uptake in rice (Oryza sativa L.). Plant Nano Biol 9(December 2023):100087. https://doi.org/10.1016/j.plana.2024.100087
Kumar A, Singh K, Verma P, Singh O, Panwar A, Singh T, Kumar Y, Raliya R (2022a) Effect of nitrogen and zinc nanofertilizer with the organic farming practices on cereal and oil seed crops. Sci Rep 12(1):6938. https://doi.org/10.1038/s41598-022-10843-3
Kumar R, Dhiman M, Sharma L, Dadhich A, Kaushik P, Sharma MM (2022b) Nanofertilizers: the targeted nutrient supplier and enhance nutrients uptake by pearl millets (Pennisetum glaucum). Biocatal Agric Biotechnol 45(October):102524. https://doi.org/10.1016/j.bcab.2022.102524
Kumaraswamy RV, Saharan V, Kumari S, Chandra Choudhary R, Pal A, Sharma SS, Rakshit S, Raliya R, Biswas P (2021) Chitosan-silicon nanofertilizer to enhance plant growth and yield in maize (Zea mays L.). Plant Physiol Biochem 159:53–66. https://doi.org/10.1016/j.plaphy.2020.11.054
Le TTH, Mai TTT, Phan KS, Nguyen TM, Tran TLA, Dong TN, Tran HC, Ngo TTH, Hoang PH, Ha PT (2022) Novel integrated Nanofertilizers for improving the growth of Polyscias fruticosa and Asparagus officinalis. J Nanomater 2022:5791922. https://doi.org/10.1155/2022/5791922
Li M, Zhang P, Guo Z, Cao W, Gao L, Li Y, Tian CF, Chen Q, Shen Y, Ren F, Rui Y, White JC, Lynch I (2023) Molybdenum Nanofertilizer boosts biological nitrogen fixation and yield of soybean through delaying nodule senescence and nutrition enhancement. ACS Nano 17(15):14761–14774. https://doi.org/10.1021/acsnano.3c02783
Lian J, Cheng L, Huang X, Wang X, Wang Y, Deng C, Xin X, Zou T, Chen Y, Yu H, Liu W, Pan J, He Z, Yang X, White JC (2024) Foliar application of iron-based nanofertilizers to wheat grown in a Cd-contaminated field: implications for food safety and biofortification. Environ Sci Nano 11(6):2577–2590. https://doi.org/10.1039/d3en00958k
Liang G (2022) Iron uptake, signaling, and sensing in plants. Plant Commun 3(5):100349. https://doi.org/10.1016/j.xplc.2022.100349
Mahmoud AWM, Ayad AA, Abdel-Aziz HSM, Williams LL, El-Shazoly RM, Abdel-Wahab A, Abdeldaym EA (2022) Foliar application of different iron sources improves Morpho-physiological traits and nutritional quality of broad bean grown in Sandy soil. Plants 11(19):2599. https://doi.org/10.3390/plants11192599
Márquez-Prieto AK, Palacio-Márquez A, Sánchez E, Macias-López BC, Pérez-Álvarez S, Villalobos-Cano O, Preciado-Rangel P (2022) Impact of the foliar application of potassium nanofertilizer on biomass, yield, nitrogen assimilation and photosynthetic activity in green beans. Notulae Botanicae Horti Agrobotanici Cluj-Napoca 50(1):12569. https://doi.org/10.15835/nbha50112569
Mehmood S, Kumar N, Mansoori A, Mohan M, Kumar A, Ghorai TK (2024) Effect of ZnMgO2 nanoparticles used as a nanofertilizer: promoting the growth activities of rice seedlings. Environ Sci Nano 11(4):1571–1581. https://doi.org/10.1039/d3en00770g
Meier S, Moore F, Morales A, González ME, Seguel A, Meriño-Gergichevich C, Rubilar O, Cumming J, Aponte H, Alarcón D, Mejías J (2020) Synthesis of calcium borate nanoparticles and its use as a potential foliar fertilizer in lettuce (Lactuca sativa) and zucchini (Cucurbita pepo). Plant Physiol Biochem 151(March):673–680. https://doi.org/10.1016/j.plaphy.2020.04.025
Mejias JH, Salazar F, Pérez Amaro L, Hube S, Rodriguez M, Alfaro M (2021) Nanofertilizers: A cutting-edge approach to increase nitrogen use efficiency in grasslands. Front Environ Sci 9(March):635114. https://doi.org/10.3389/fenvs.2021.635114
Modi S, Kumar S, Dubey PK (2021) Dynamics of chitosan based NPK-nanofertilizers in greenhouse cucumber production system. J Environ Biol 42(January):162–168. https://doi.org/10.22438/jeb/42/1/MRN-1251
Motlhalamme T, Mohamed H, Kaningini AG, More GK, Thema FT, Mohale KC, Maaza M (2023) Bio-synthesized calcium carbonate (CaCO3) nanoparticles: their anti-fungal properties and application as nanofertilizer on Lycopersicon esculentum growth and gas exchange measurements. Plant Nano Biol 6(October):100050. https://doi.org/10.1016/j.plana.2023.100050

Muñoz-Márquez E, Soto-Parra JM, Noperi-Mosqueda LC, Sánchez E (2022) Application of molybdenum Nanofertilizer on the nitrogen use efficiency, growth and yield in Green beans. Agronomy 12(12):3163. https://doi.org/10.3390/agronomy12123163

Muñoz-Márquez E, Soto-Parra JM, Pérez-Leal R, Sanchez E (2024) New nitrogen use efficiency indices for biomass formation and productivity in Green beans under foliar fertilization with molybdenum Nanofertilizer. Nitrogen 5(3):667–687. https://doi.org/10.3390/nitrogen5030044

Murgueitio-Herrera E, Falconí CE, Cumbal L, Gómez J, Yanchatipán K, Tapia AD, Martínez K, Sinde-Gonzalez I, Toulkeridis T (2022) Synthesis of iron, zinc, and manganese Nanofertilizers, using Andean blueberry extract, and their effect in the growth of cabbage and Lupin plants. Nanomaterials 12(11):1921. https://doi.org/10.3390/nano12111921

Naseem F, Zhi Y, Farrukh MA, Hussain F, Yin Z (2020) Mesoporous ZnAl2Si10O24 nanofertilizers enable high yield of Oryza sativa L. Sci Rep 10(1):10841. https://doi.org/10.1038/s41598-020-67611-4

Neto ME, Britt DW, Jackson KA, Braccini AL, Inoue TT, Batista MA (2020) Early development of corn seedlings primed with synthetic tenorite nanofertilizer. J Seed Sci 42:e202042040. https://doi.org/10.1590/2317-1545v42240979

Nongbet A, Mishra AK, Mohanta YK, Mahanta S, Ray MK, Khan M, Baek KH, Chakrabartty I (2022) Nanofertilizers: a smart and sustainable attribute to modern agriculture. Plants 11(19):2587. https://doi.org/10.3390/plants11192587

Noruzi M, Hadian P, Soleimanpour L, Ma'mani L, Shahbazi K (2023) Hydroxyapatite nanoparticles: an alternative to conventional phosphorus fertilizers in acidic culture media. Chem Biol Technol Agric 10(1):71. https://doi.org/10.1186/s40538-023-00437-0

Okey-Onyesolu CF, Hassanisaadi M, Bilal M, Barani M, Rahdar A, Iqbal J, Kyzas GZ (2021) Nanomaterials as nanofertilizers and nanopesticides: an overview. ChemistrySelect 6(33):8645–8663. https://doi.org/10.1002/slct.202102379

Pandorf M, Pourzahedi L, Gilbertson L, Lowry GV, Herckes P, Westerhoff P (2020) Graphite nanoparticle addition to fertilizers reduces nitrate leaching in growth of lettuce (: Lactuca sativa). Environ Sci Nano 7(1):127–138. https://doi.org/10.1039/c9en00890j

Ponce-García OC, Noperi-Mosqueda LC, Soto-Parra JM, Yáñez-Muñoz RM, Pérez-Leal R, Navarro-León E, Sánchez E (2023) Assaying the efficiency of sulfate, chelate and zinc nanoparticle fertilizers in green bean grown in alkaline soil. J Plant Nutr 46(5):653–664. https://doi.org/10.1080/01904167.2022.2067062

Pongpeera S, Maensiri D (2023) Application of hydroxyapatite nanoparticles as fertilizer improved yield quality of Rice (Oryza sativa L.). Chiang Mai J Sci 50(5):1–12. https://doi.org/10.12982/CMJS.2023.053

Pradhan SR, Chinnusamy T, Thiruvenkatasamy S, Govindaraj S, Rajendran P, Pandian K, Chandrasekaran IR (2024) Harnessing the power of foliar nanofertilizers to enhance vegetable crop performance. Plant Sci Today 11(sp4):01–11. https://doi.org/10.14719/pst.5742

Ram B, Senthivelu M, Prabukumar G, Gnanachitra M, Prasad VBR, Subramanian KS (2024) Physiological, biochemical and soil microbial responses of green gram (Vigna radiata L.) to foliar nutrition of nanofertilizers. Plant Sci Today 11(sp4):01–09. https://doi.org/10.14719/pst.5326

Ramírez-Cariño HF, Morales I, Guadarrama-Mendoza PC, Tlilayatzi-Bello AA, Valadez-Blanco R (2024) Beneficial traits of crop-associated rhizobacteria toward plant growth promotion, biofertilization, and nutrient enhancement of harvested agro-products. In: Abd-Elsalam KA, Mohamed HI (eds) Bacterial secondary metabolites. Elsevier, pp 299–334. https://doi.org/10.1016/B978-0-323-95251-4.00005-3

Ramírez-Rodríguez GB, Miguel-Rojas C, Montanha GS, Carmona FJ, Dal Sasso G, Sillero JC, Pedersen JS, Masciocchi N, Guagliardi A, Pérez-De-luque A, Delgado-López JM (2020) Reducing nitrogen dosage in triticum durum plants with urea-doped nanofertilizers. Nanomaterials 10(6):10061043. https://doi.org/10.3390/nano10061043

Rani S, Kumari N, Sharma V (2023) Biosynthesized zinc oxide nanoparticles as a novel Nanofertilizer influence growth, yield, antioxidant enzymes, and arsenic accumulation in

Mungbean (Vigna radiata L.) under arsenic stress. J Soil Sci Plant Nutr 23(2):2360–2380. https://doi.org/10.1007/s42729-023-01188-5

Reis HPG, Giroto AS, Guimarães GGF, Putti FF, Pavinato PS, Teles AP, Ribeiro C, Fernandes DM (2022) Role of slow-release phosphate Nanofertilizers in forage nutrition and phosphorus lability. ACS Agric Sci Technol 2(3):564–572. https://doi.org/10.1021/acsagscitech.2c00012

Ren Y, Zhang D, Cheng B, Chen B, Yue L, Cao X, Wang C, Wang Z (2025) Foliar spraying zinc-carbon dot Nanofertilizer promotes yield and quality of lettuce (Lactuca sativa L.) through leaf-root regulation. *ACS agricultural*. Sci Technol 5(3):371–380. https://doi.org/10.1021/acsagscitech.4c00651

Rizk R, Ahmed M, Abdul-Hamid D, Zedan M, Tóth Z, Decsi K (2025) Resulting key physiological changes in Triticum aestivum L. plants under drought conditions after priming the seeds with conventional fertilizer and greenly synthesized zinc oxide nanoparticles from corn wastes. Agronomy 15(1):211. https://doi.org/10.3390/agronomy15010211

Saad AM, Alabdali AYM, Ebaid M, Salama E, El-Saadony MT, Selim S, Safhi FA, ALshamrani SM, Abdalla H, Mahdi AHA, El-Saadony FMA (2022) Impact of Green chitosan nanoparticles fabricated from shrimp processing waste as a source of Nano nitrogen fertilizers on the yield quantity and quality of wheat (Triticum aestivum L.) cultivars. Molecules 27(17):5640. https://doi.org/10.3390/molecules27175640

Sakhno Y, Degli Esposti L, Adamiano A, Borgatta J, Cahill M, Vaidya S, White JC, Iafisco M, Jaisi DP (2023) Citrate-stabilized amorphous calcium phosphate nanoparticles doped with micronutrients as a highly efficient nanofertilizer for environmental sustainability. ACS Agric Sci Technol 3(10):845–854. https://doi.org/10.1021/acsagscitech.3c00117

Salama HSA, Badry HH (2020) Effect of partial substitution of bulk urea by nanoparticle urea fertilizer on productivity and nutritive value of teosinte varieties. Agron Res 18(4):2568–2580. https://doi.org/10.15159/AR.20.189

Salama DM, Abd El-Aziz ME, El-Naggar ME, Shaaban EA, Abd El-Wahed MS (2021) Synthesis of an eco-friendly nanocomposite fertilizer for common bean based on carbon nanoparticles from agricultural waste biochar. Pedosphere 31(6):923–933. https://doi.org/10.1016/S1002-0160(21)60024-3

Salama DM, Abd El-Aziz ME, Shaaban EA, Osman SA, Abd El-Wahed MS (2022) The impact of nanofertilizer on agro-morphological criteria, yield, and genomic stability of common bean (Phaseolus vulgaris L.). Sci Rep 12(1):18552. https://doi.org/10.1038/s41598-022-21834-9

Salama DM, Khater MA, Abd El-Aziz ME (2024a) The influence of potassium nanoparticles as a foliar fertilizer on onion growth, production, chemical content, and DNA fingerprint. Heliyon 10(11):e31635. https://doi.org/10.1016/j.heliyon.2024.e31635

Salama DM, Osman SA, Mahmoud SH, El-Tanahy AMM, Abd El-Aziz ME (2024b) Improving the productivity and physiological characteristics of lettuce plants using spraying calcium as a Nanofertilizer. Horticulturae 10(11):1157. https://doi.org/10.3390/horticulturae10111157

Saurabh A, Kaur M, Khan R, Guleria G, Shandilya M, Thakur S (2024) Foliar application of Fe2O3 nanofertilizer on growth and yield of cauliflower (Brassica oleracea var. Botrytis L.) cv. Pusa Snowball K-1. Int J Phytoremediation 26(7):993–1002. https://doi.org/10.1080/15226514.2023.2288894

Sehgal Y, Kalia A, Dhillon BS, Dheri GS (2024) Effect of a slow-release urea Nanofertilizer on soil microflora and yield of direct seeded Rice (Oryza sativa L.). Nitrogen 5(4):1074–1091. https://doi.org/10.3390/nitrogen5040069

Sharma B, Shrivastava M, Afonso LOB, Soni U, Cahill DM (2022) Zinc- and magnesium-doped hydroxyapatite nanoparticles modified with urea as smart nitrogen fertilizers. ACS Appl Nano Mater 5(5):7288–7299. https://doi.org/10.1021/acsanm.2c01192

Sharma G, Saharan V, Pal A, Sharma SS (2024) Chitosan nanofertilizer to strengthen sink strength and provide resistance against PFSR (post flowering stalk rot) disease in maize. Biocatal Agric Biotechnol 60:103303. https://doi.org/10.1016/j.bcab.2024.103303

Sheoran P, Grewal S, Kumari S, Goel S (2021) Enhancement of growth and yield, leaching reduction in Triticum aestivum using biogenic synthesized zinc oxide nanofertilizer. Biocatal Agric Biotechnol 32(February):101938. https://doi.org/10.1016/j.bcab.2021.101938

Shoukat A, Saqib ZA, Nawaz A, Amir KZ, Ahmad I, Hamza A, Mühling KH (2025) Nanofertilizers benefited maize to cope oxidative stress under saline environment. Plant Nano Biol 11(August 2024):100141. https://doi.org/10.1016/j.plana.2025.100141

Singh A, Rajput VD, Tawaha ARM, Al Z, Al OM, Habeeb T, Rawat S, Ghazaryan K, Minkina T (2023) A review on crop responses to Nanofertilizers for mitigation of multiple environmental stresses. Ecol Eng Environ Technol 24(7):280–296. https://doi.org/10.12912/27197050/169313

Song U, Kim J (2020) Zinc oxide nanoparticles: a potential micronutrient fertilizer for horticultural crops with little toxicity. Hortic Environ Biotechnol 61(3):625–631. https://doi.org/10.1007/s13580-020-00244-8

Svietlova N, Topchiy N, Storozhenko V, Smirnov O, Kovalenko M, Batsmanova L, Taran N (2021) Photosynthetic response of some triticum cultivars to the combined influence of nanofertilizers and water deficit. J Cent Eur Agric 22(3):539–545. https://doi.org/10.5513/JCEA01/22.3.3174

Tahir M, Imran M, Nawaz F, Shahid M, Naeem MA, Ahmad I, Akram M, Khalid U, Farooq ABU, Bakhat HF, Kamran M, Shah ZA (2021) Effects of Bacillus sp. MR-1/2 and magnetite nanoparticles on yield improvement of rice by urea fertilizer under different watering regimes. J Appl Microbiol 131(5):2433–2447. https://doi.org/10.1111/jam.15110

Tarafder C, Daizy M, Alam MM, Ali MR, Islam MJ, Islam R, Ahommed MS, Aly Saad Aly M, Khan MZH (2020) Formulation of a hybrid Nanofertilizer for slow and sustainable release of micronutrients. ACS Omega 5(37):23960–23966. https://doi.org/10.1021/acsomega.0c03233

Ukidave VV, Ingale LT (2022) Green synthesis of zinc oxide nanoparticles from Coriandrum sativum and their use as fertilizer on Bengal gram, Turkish gram, and green gram plant growth. Int J Agron 2022:8310038. https://doi.org/10.1155/2022/8310038

Yadav A, Yadav K, Abd-Elsalam KA (2023) Exploring the potential of nanofertilizers for a sustainable agriculture. Plant Nano Biol 5(February):100044. https://doi.org/10.1016/j.plana.2023.100044

Yang R, Shen J, Zhang Y, Jiang L, Sun X, Wang Z, Tang B, Shen Y (2022) The role of biochar nanoparticles performing as Nanocarriers for fertilizers on the growth promotion of Chinese cabbage (Brassica rapa (Pekinensis Group)). Coatings 12(12):1984. https://doi.org/10.3390/coatings12121984

Yaseen R, Amin BH (2021) The use of phosphorus nanoparticles synthesized by rhizospheric fungus Aspergillius fumigatus as a nanofertilizer for flax plant. Malays J Microbiol 17(3):244–253. https://doi.org/10.21161/mjm.200852

Yoon HY, Lee JG, Esposti LD, Iafisco M, Kim PJ, Shin SG, Jeon JR, Adamiano A (2020) Synergistic release of crop nutrients and stimulants from hydroxyapatite nanoparticles functionalized with humic substances: toward a multifunctional nanofertilizer. ACS Omega 5(12):6598–6610. https://doi.org/10.1021/acsomega.9b04354

Yuvaraj M, Subramanian KS, Cyriac J (2023) Efficiency of zinc oxide nanoparticles as controlled release nanofertilizer for rice (Oryza sativa L). J Plant Nutr 46(18):4477–4493. https://doi.org/10.1080/01904167.2023.2233561

Chapter 5
Positive and Negative Effects of Metal Nanoparticles on Bacterial and Fungal Biocontrol Agents

Seweta Srivastava, Aspak, Barnik Debnath, A. K. Malavika Ram, and Meenakshi Rana

Abstract Recently, nanoparticles (NPs) have drawn more interest because of their special qualities. In the agricultural and related sectors, metal-based NPs are frequently utilized. Examining the potential advantages and risks that different NPs may pose to microbial diversity and plants is the goal of this chapter. By directly influencing growth of the plant and indirectly the surroundings of ecological environment, nanoparticles can support the green revolution and ensure the sustainable growth of agriculture. Although the beneficial impacts of nanoparticles on the diversity of endophytic microbial communities in plants and around the rhizosphere have been well documented by several studies, some research continues to indicate that the effects of nanoparticles on microorganisms and plants may be dose dependent. The equilibrium of soil ecosystems may be upset by the overuse of nanoparticles. The process of interaction between nanoparticles and microbes must be thoroughly studied, and appropriate utilization techniques and dosages must be investigated, in order to maximize the advantages of nanoparticles and minimize any potential hazards. To guarantee the safe use of nanoparticles, it is also essential to improve the monitoring and assessment of potential environmental risks.

Keywords Biocontrol · Metal · Microbes · Nanoparticles · Plant disease management

---

S. Srivastava (✉) · Aspak · M. Rana
School of Agriculture, Lovely Professional University, Phagwara, Punjab, India
e-mail: seweta.21896@lpu.co.in

B. Debnath
ICAR-Central Institute for Cotton Research, Nagpur, Maharashtra, India

A. K. Malavika Ram
Hawkesbury Institute for the Environment, Western Sydney University, Sydney, NSW, Australia

K. A. Abd-Elsalam (ed.), *Plant-Microbiome Nanotechnology*, Sustainability Sciences in Asia and Africa, https://doi.org/10.1007/978-981-95-5056-2_5

5.1 Introduction

The study and manipulation of materials with sizes approximately between 1 and 100 nm is known as nanotechnology. It is regarded as a nanoparticle if one of the dimensions falls within this range (Klaine et al. 2008). Norio Taniguchi of the University of Tokyo first used the word *nanotechnology* in a report that was published in 1974. Eric Drexler coined the word "nanotechnology" in his book *Engines of Creation* (Drexler 1986). Because of their special qualities that fall in between those of individual molecules and bulk materials, nanoparticles (NPs) have drawn more attention recently. Depending on their primary material, nanoparticles can be classified as either natural or manufactured. While these compounds and carbon nanotubes are classified as organic NPs, inorganic NPs are classified as metal (Ag, Al, Au, Bi, Co, Cu, Fe, In, Mo, Ni, Si, Sn, Ti, W, Zn), metal oxide (Al_2O_3, CeO_2, CuO, Cu_2O, In_2O_3, La_2O_3, MgO, NiO, SiO_2, SnO_2, TiO_2, ZnO, ZrO_2), and quantum dots. The detrimental impacts of metal-based nanoparticles (NPs) (Ag, CuO, FeO, TiO_2, and ZnO) on the abundance, diversity, and activity of plants and animals are extensively studied (Rajput et al. 2018).

The usage of artificial fertilizers, the shrinking amount of arable land, the growing global population, and the new issues brought about by climate change are all contributing factors to the demand for sustainable agriculture (Joel et al. 2023). Furthermore, unpredictable climate conditions that result in heat waves, droughts, and floods have had a negative impact on plant health and output. Moreover, diseases, pests, and weeds all reduce plant productivity. Together, these factors drastically lower agricultural species' productivity, health, adaptability, and yield, which has a negative impact on livestock and the global human population (Meel and Saharan 2024). Coordinated efforts are therefore needed to preserve plant health without threatening the environment or the climate. Plant components, both above and below ground, are connected to a wide variety of microbial species collectively known as the microbiome (Li et al. 2024). The subject of nanoparticles has drawn special attention recently from a variety of sectors. The word "nano" refers to a measurement on the scale of one-billionth (10^9) of a meter in size and is derived from the Greek word "nanos," which means dwarf (Pantidos and Horsfall 2014). The primary characteristic of nanoparticles that makes it easier for them to interact with other particles is their surface area to volume aspect ratio (Banik and Pérez-de-Luque 2017). For example, reducing the size of the particles is an effective method of enhancing their antibacterial activity and improving their biocompatibility as well as their ability to penetrate cell membranes (Sirelkhatim et al. 2015). Currently, a variety of metal and metal oxide nanoparticles (NPs) are manufactured, including titanium (Ti), copper oxide (CuO), zinc oxide (ZnO), silver (Ag), and gold (Au). Numerous studies have demonstrated the effectiveness of metal nanoparticles (MNPs) against a variety of food pollutants and fungal infections (Sardella et al. 2019).

Because they lessen the need for chemical pesticides, biocontrol agents like beneficial bacteria and fungi in particular are crucial to organic farming and integrated pest management (Ramezani et al. 2019). However, a number of environmental factors can affect how successful they are. Because metal nanoparticles can interact with both plant pathogens and biocontrol agents, they have the potential to increase the activity and survival of these microorganisms, which could increase their effectiveness in controlling diseases and pests (Ali et al. 2020). Notwithstanding the potential, metal nanoparticles and biocontrol agents interact in a complicated fashion, and the results are not always predictable (Adamecz 2023). Although some research indicates that nanoparticles might promote microbial growth and improve their capacity for biocontrol, other studies draw attention to the possibility that these nanoparticles could be harmful to the very species they are meant to help by naturally suppressing or controlling plant diseases (Fig. 5.1), BCAs which include bacteria and fungi help to lessen the need for chemical pesticides and promote more environmentally friendly farming methods (Tariq et al. 2022). However, environmental stressors, competition with other bacteria, and the presence of pollutants like metal nanoparticles can all affect how effective BCAs (Mahmood et al. 2024). MNPs have the potential to improve these beneficial species' biocontrol capabilities, but they also carry some hazards, such as the potential for toxicity, growth inhibition, or disruption of important biological processes (Padmakumar et al. 2023). The objectives of this chapter are to: explore the interaction of metal nanoparticles (MNPs) with bacterial and fungal biocontrol agents at the physiological, biochemical, and molecular levels; study the potential role of MNPs in enhancing the efficacy of biocontrol agents by improving their antagonistic properties or stress resistance; and examine the detrimental effects of MNPs on biocontrol agents, including their development, metabolic processes, and survival.

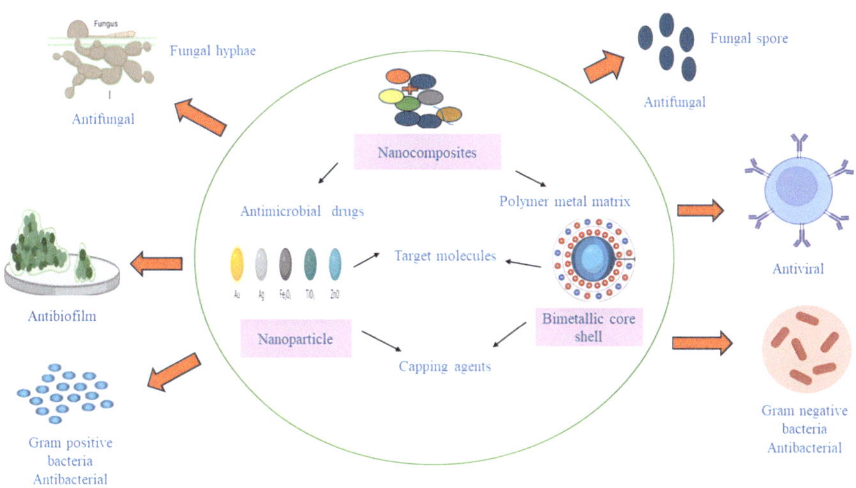

Fig. 5.1 Application of nanoparticle-based nanocomposites in controlling phytopathogens. (Source: Made by authors)

5.2 Microbe-Mediated Metallic Nanoparticles

In contrast to conventional physical and chemical methods, farmers are now interested in using microbe-mediated nanoparticles (NPs) as nanopesticides, primarily for plant disease control (Ahmed et al. 2023). Nanoparticles are generated using a variety of traditional chemical and physical techniques, such as solvent evaporation, vapor condensation, physical fragmentation, sol–gel, precipitation from microemulsion, and interferometric lithography (Agarwal et al. 2018). These methods use poisonous and potentially hazardous materials, which add to pollution in the atmosphere. Additionally, by binding with plants, these harmful compounds may cause NPs to build up in the food chain through food consumption, endangering the health of human beings (Jovanović et al. 2016). Nevertheless, compared to conventional chemical procedures, the microbe-mediated synthesis of NPs has a number of advantages (Fig. 5.2), including potentially biocompatible, economical, and ecologically benign (Shinde et al. 2022). Because of its innate capacity to generate highly resilient nanoparticles, microbe-mediated synthesis of NPs using a variety of microorganisms, including bacteria, fungi, yeast, and microalgae, presents a potential strategy (Ali et al. 2020). Furthermore, microbial production may precisely regulate the shape, size, and content of NPs that can significantly effect their properties and actual applications (Kaur et al. 2022). The effectiveness of biogenic NPs is largely dependent on their stability, dispersion, and particle size (Kumar et al. 2021). The importance of these features has been emphasized by numerous investigations, which demonstrate how microbial synthesis techniques can be exploited to modulate and modify particle size (Sharma et al. 2023). Furthermore, surface changes

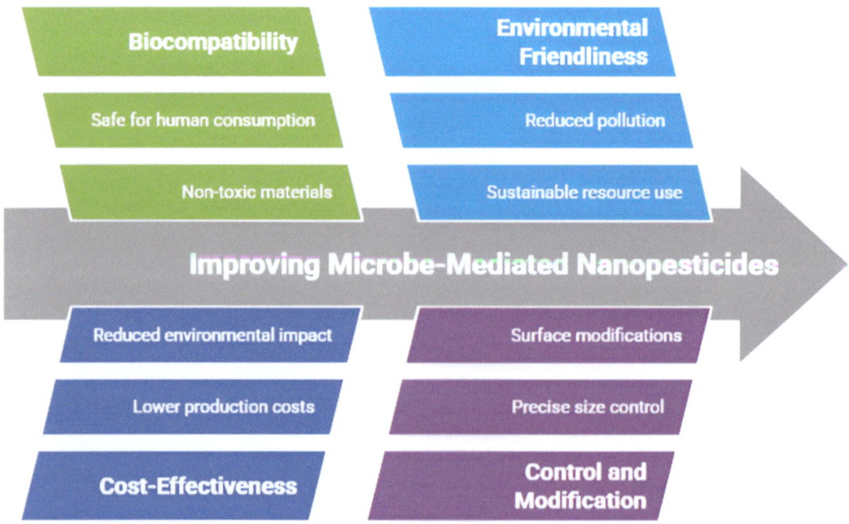

Fig. 5.2 Improving the synthesis and functional efficiency of microbe-mediated metallic nanoparticles (MM-MNPs). (Source: Made by authors)

might enhance the spreading of microbially generated nanoparticles, enabling better interactions with target pathogens (Osonga et al. 2020).

5.2.1 Synthesis of Nanoparticles Using Bacteria

A viable and eco-friendly process of producing NPs with a range of uses in the agricultural industry is the utilization of bacteria (Hulkoti and Taranath 2014). Because bacteria may create extracellular enzymes which may convert metal ions to their equivalent NPs, they have been employed to synthesize NPs (Ibrahim et al. 2019). Numerous processes, including intracellular biosynthesis, extracellular biosynthesis, and bioaccumulation, are able to synthesize NPs utilizing bacteria (Alfryyan et al. 2022; Yang et al. 2022). Inside their cells, bacteria produce NPs through intracellular biosynthesis. In this process, metal ions are taken in by the bacterial cells and reduced to nanoparticles (NPs) by intracellular enzymes. Following cell lysis, the NPs are subsequently discharged into the extracellular milieu (Dağlıoğlu and Yılmaz Öztürk 2019; Öcal et al. 2020). Bacteria can produce NPs outside of their cells through a process called extracellular biosynthesis. Throughout the process, extracellular enzymes that can change metal ions into their corresponding NPs are released. Following their release, the NPs can then be collected and purified in the extracellular environment (Spagnoletti et al. 2019; Taha et al. 2019). Numerous research have documented the environmentally friendly production of metallic nanoparticles employing bacterial microorganisms in the past (Noman et al. 2020). In order to manage the rice bacterial disease, Ahmed et al. (2020a) reported the production of silver nanoparticles using *Bacillus cereus*. Similarly, Varshney et al. (2010) used *Pseudomonas stutzeri*, which was isolated from wastewater, to demonstrate the environmentally friendly creation of copper nanoparticles. *Acinetobacter calcoaceticus*, *P. stutzeri*, *Escherichia coli*, *B. amyloliquefaciens*, and *Lactobacillus* sp. were among the bacterial strains that were previously employed for microbe-mediated NP production (Ahmed et al. 2023). AgNPs were recently developed using bacterial strains from the following species: *Pseudomonas stutzeri*, *Escherichia coli*, *Lactobacillus* sp., *Bacillus licheniformis*, *Bacillus megaterium*, *Bacillus amyloliquefaciens*, and *Acinetobacter calcoaceticus* (Prasad et al. 2013; Ali et al. 2020). Both intracellular and extracellular biosynthesis are capable of creating silver nanoparticles (NPs), which have demonstrated antibacterial efficacy against a variety of pathogenic pathogens (Saratale et al. 2018). AgNPs and AuNPs were greenly synthesized using a unique bacteria called *Stenotrophomonas*. In this case, the low molecular weight secretory proteins found in the supernatant are essential for the production of AgNPs and AuNPs (Malhotra et al. 2013). The reductase enzyme is essential for reducing metal ions, according to the green synthesis of copper, silver, and zinc nanoparticles from *Streptomyces* sp. (Karthik et al. 2014). Copper nanoparticles (NPs) ranging in size from 5 to 30 nm were synthesized by the marine bacterium strain *Kocuria flava* (Kaur et al. 2015). Using *Shewanella loihica* PV-4, ultra-small palladium and platinum nanoparticles

(NPs) with sizes ranging from 2 to 7 nm were synthesized (Ahmed et al. 2018). *Ochrobactrum* sp. was utilized to create tellurium nanoparticles, and it may be a valuable nanofactory for transforming the hazardous tellurite oxyanions into beneficial NPs (Zonaro et al. 2017). The biosynthesis of NPs from bacteria is a practical and ecologically safe way to produce NPs when all the variables are considered. Table 5.1 enlists the effective bacterial species employed in the production of nanoparticles.

Synthesis of Nanoparticles Using Fungi

Because they can produce a variety of metabolites and extracellular enzymes that can minimize metal ions to their respective NPs, due to the diverse nature of fungi, they have been popularly exploited for NP synthesis (Rai et al. 2021; Yadav et al. 2023; Zhao et al. 2024). To produce NPs from fungi, a variety of methods have been employed, such as fungal biomass and intracellular and extracellular biosynthesis (Mistry et al. 2021). Metal ions are taken in by fungal cells during intracellular biosynthesis, and various intracellular enzymes can then convert them to NPs. Following cell lysis, the NPs are subsequently discharged into the extracellular milieu (Tyagi et al. 2019; Rajeshkumar and Sivapriya 2020; Sidhu et al. 2025). Extracellular biosynthesis comprises the production of extracellular enzymes that are capable of converting metallic ions to their equivalent NPs outside the fungal cells (Spagnoletti et al. 2019; Zhang et al. 2019). Adjusting certain abiotic factors, such as agitation, temperature, light, culture, and synthesis times, can yield the envisioned properties of NPs from various fungal species. To regulate the size and form of NPs, these criteria should be sustained throughout the culture of fungi and synthesis of NPs (Birla et al. 2013; Nawawi et al. 2020). Additionally, it was discovered that physicochemical properties of NPs can be determined by variations in pH, temperature, growth medium, biomass amount, and concentration of the metal precursor (Saxena et al. 2016; Liang et al. 2017). *Rhizopus stolonifer* filtrate was used to create AgNPs, which had NP sizes of 2.86, 25.89, and 48.43 nm at 40, 20, and 60 °C, respectively. However, neither 10 °C nor 80 °C was used. Enzymes and active molecules required for AgNPs biogenesis may become denatured or inactivated at extremely low or high temperatures (AbdelRahim et al. 2017).

Fusarium oxysporum was used in the biosynthesis of AgNPs by Husseiny et al. (2015), who also documented the impact of substrate concentration and incubation temperature. AgNPs were biosynthesised via biotransformation by using both the dried and fresh culture of *Saccharomyces cerevisiae* as the biocatalyst, and it was found that freshly cultivated yeast produced more AgNPs than dried culture. AgNPs were biosynthesised in a cell-associated way by using a strain of the marine yeast *Yarrowia lipolytica*. According to this study, the biomineralization of metallic ions may be caused by the brown pigment melanin (Apte et al. 2013). AuNPs with a diameter of 1–100 nm were synthesized intracellularly using *Pichia jadinii*. In order to manage the size and shape of AuNPs, the growth and cellular activities of *P. jadinii* were readily regulated in this study (Gericke and Pinches 2006). *Candida utilis* NCIM 3469, which has a size range of 20 to 80 nm, was used in an extracellular method to produce the green synthesis AgNPs (Waghmare et al. 2015). More than

Table 5.1 Bacterial-mediated green synthesis of nanoparticles (NPs)

S. No.	Bacteria	Nanoparticle	Size (mm)	Morphology	Application	References
1.	*Aeromonas* sp. SH10	Silver (Ag)	6.4	Spherical	Antimicrobial activity	Singh et al. (2017)
2.	*Bacillus cereus*	Silver (Ag)	20–40	Spherical	Antimicrobial and antifungal activity	Sunkar and Nachiyar (2012), Zehra et al. (2025)
3.	*Bacillus subtilis* 168	Gold (Au)	5–25	Octahedral	Enhancing plant growth	Bremer et al. (2023)
4.	*Bacillus subtilis*	Silver (Ag)	5–50	Spherical and triangular	Enhancing plant growth	Vanlalveni et al. (2024)
5.	*Clostridium thermoaceticum*	Cadmium sulfide (CdS)	5–50	Amorphus	Improving plant growth, providing pest control, or enhancing soil properties	Roy and Pal (2025)
6.	*Corynebacterium* sp. SH09	Silver (Ag)	10–15	Amorphus	Used as nano-pesticides and nano-fertilizers	Roy and Pal (2025)
7.	*Desulfobacteraceae*	Zinc sulfide (ZnS)	2–5	Spherical	Used as nano-fertilizers	Roy and Pal (2025)
8.	*Geobacillus* sp.	Gold (Au)	5–50	Quasi-hexagonal	Antifungal and antibacterial activity	Correa-Llantén et al. (2013)
9.	*Klebsiella aerogenes*	Cadmium sulfide (CdS)	20–200	Crystalline	Used as sensors for precision farming and crop protection	Babu et al. (2022)
10.	*Klebsiella pneumonia*	Silver (Ag)	28.2–122 (avg 52.2)	Spherical	Nano-pesticides, nano-fertilizers, and antifungal/antibacterial agents	Franci et al. (2015)
11.	*Lactobacillus* strains	Gold (Au)	20–50 or above	Hexagonal, crystalline, triangular, cluster	Antifungal and antibacterial agents, and as delivery vehicles for fertilizers and pesticides	Bahrulolum et al. (2021)
12.	*Lactobacillus* strains	Silver (Ag)	15–500	Crystalline, hexagonal, triangular, cluster	Nano-pesticides, nano-fertilizers, or in crop protection	Ayenew et al. (2024)
13.	*Lactobacillus* strains	Titanium (Ti)	40–60	Spherical	Enhancing growth and yield parameters of the plant	Chandoliya et al. (2024)

(continued)

Table 5.1 (continued)

S. No.	Bacteria	Nanoparticle	Size (mm)	Morphology	Application	References
14.	*Lactobacillus casei* subsp. *casei*	Silver (Ag)	25–50	Spherical	Antibacterial activity	Korbekandi et al. (2012), Roy and Pal (2025)
15.	*Nocardiopsis* sp. MBRC-1	Silver (Ag)	~45	Spherical	Used as nano-pesticides, nano-fertilizers, and in food packaging and protection	Manivasagan et al. (2013), Roy and Pal (2025)
16.	*Plectonema boryanum* UTEX 485	Gold (Au)	10–25 and ~1–10 and 10–6000	Cubic and octahedral platelet	Used as nano-fertilizers, plant growth regulators, and in pest control	Koul et al. (2021)
17.	*Pseudomonas aeruginosa*	Gold (Au)	15–30	-	Soil improvement	Singh et al. (2023)
18.	*Pseudomonas aeruginosa*	Lanthanum (La)	15–30	Crystalline and needle-like	Enhancing plant growth and soil health	Korbekandi et al. (2012), Roy and Pal (2025)
19.	*Pseudomonas fluorescens*	Gold (Au)	50–70	Spherical	Used as antagonists to control plant pathogens, as well as in biosensors for detecting soil quality, plant pathogens, and heavy metals	Dutta et al. (2023)
20.	*Pseudomonas putida* NCIM 2650	Silver (Ag)	~70	Spherical	Plant growth enhancement and pest control	Thamilselvi and Radha (2013), Roy and Pal (2025)
21.	*Pseudomonas stutzeri* AG259	Silver (Ag)	35–46 and up to 200	Hexagonal, equilateral triangle, crystalline silver, and monoclinic silver sulfide acanthite	Antimicrobial agents and promoters of plant growth	Krishnaraj et al. (2012), Zehra et al. (2025)
22.	*Rhodobacter sphaeroides*	Zinc sulfide (ZnS)	Average diameter of 8	Spherical	Used as nanofertilizers, agrochemical carriers, and can also help in targeted nutrient delivery to plants	Bai et al. (2006), Roy and Pal (2025)

23.	*Rhodopseudomonas capsulata*	Gold (Au)	10–20	Nanoplate and Spherical	Can improve seed germination and the interaction between soil and plant roots	He et al. (2007); Roy and Pal (2025)
24.	*Rhodopseudomonas palustris*	Cadmium sulfide (CdS)	8.01+_0.25	Crystalline, face-centered cubic	Used in various agricultural applications, such as for biosensors and potentially in fertilizers or to stimulate plant growth	Bai et al. (2009), Yen et al. (2022)
25.	*Serratia nematodiphila*	Silver (Ag)	10–31	Spherical and crystalline	Used as nano-pesticides, nano-fertilizers, and for improving overall plant health	Malarkodi et al. (2013)
26.	*Shewanella algae* strain BRY	Gold (Au)	Varied sizes, changed with pH	–	Used in phytoremediation and as plant growth regulators	Wu and Ng (2017)

70% of the copper nanoparticles (CuNPs) produced by *Saccharomyces cerevisiae* were between 10 and 12 nm in size, according to a different study (do Nascimento et al. 2018) (Table 5.2).

According to numerous studies, the synthesis of NPs using fungi offers a number of benefits, including low cost, high production rate, and simplicity of cultivation (Qu et al. 2019; Qamar and Ahmad 2021; Srivastava et al. 2024). As an example, using the cell-free filtrate of *Trichoderma virens* HZA14, Tomah et al. (2020) showed how to biosynthesize silver nanoparticles to combat the *Sclerotria sclerotiorum* infection. Additionally, Singh et al. (2022) demonstrated that *Aspergillus niger* was used to generate zinc oxide nanoparticles (NPs) that protect potato (*Solanum tuberosum* L.) plants from the *Alternaria solani* pathogen, which can cause early blight disease. In a comparable manner *Aspergillus aeneus*-mediated zinc oxide nanoparticle production was described by Jain et al. (2013). Like other microorganisms, yeasts have also been widely used for the large-scale production of NPs (Karthik et al. 2014). When combined, fungal-mediated NP synthesis provides a flexible, affordable, and sustainable method for producing green NPs with special qualities and potential uses in agriculture. New, environmentally friendly technology with a variety of uses could result from more study in this field.

Properties of Metal Nanoparticles

Metal-based nanoparticles (MNPs) possess unique physical, chemical, and biological properties due to their huge surface area and small size. These characteristics make them appealing for a range of uses, such as environmental science, agriculture, and medicine. An outline of the main characteristics of MNPs is provided in Table 5.3.

Positive and Negative Effects of Metal Nanoparticles on Bacterial and Fungal Biocontrol Agents

The potential of metal nanoparticles (MNPs) as substitutes for chemical fertilizers and pesticides has drawn a lot of interest in the field of agriculture (Sidhu et al. 2025). Their interactions with bacterial and fungal biocontrol agents (BCAs), however, can affect the effectiveness of biocontrol techniques in both good and negative ways. A summary of these consequences is provided below (Table 5.4).

Nanoparticles are capable of interacting with plants, root microbes, and soil organic matter through a variety of physical, chemical, and biologic mechanisms, including vulcanization, flocculation, precipitation, and adsorption (Chang et al. 2024). In order to regulate the adaptability of the plant surroundings, nanoparticles were initially transported to xylem tissue before being transported to various other tissues. They begin observing the physiological, biochemical, and genetic processes of soil microorganisms as soon as they enter the soil system (Ahmed et al. 2023). Here, we mainly examine how nanoparticles can improve the rhizosphere soil micro-environment and promote interactions between microbial consortia.

The substantial nutrients requirement during the period of rapid plant growth attracts a greater concentration of rhizosphere microorganisms, which are essential for the plants' absorption of nutrients (Chaparro et al. 2014; Zhang et al. 2024). Concurrently, the interaction of rhizosphere microbes and nanoparticles may

Table 5.2 Potential role of fungal-mediated metal/metal oxide nanoparticles in agricultural sector

S. No.	Fungi	NPs	Size (nm)	Shape	Applications	References
1.	*Aspergillus fumigatus*	Zinc oxide (ZnO)	1.2–6.8	Spherical and hexagonal	Improving plant growth, enhancing nutrient uptake, and potentially controlling plant diseases	Aftab et al. (2025)
2.	*Aspergillus oryzae*	Iron (III) chloride ($FeCl_3$)	10–24.6	Spherical	Plant growth promotion, nutrient delivery, and potential biocontrol agents	Tarafdar and Raliya (2013)
3.	*Aspergillus tubingensis*	Tricalcium phosphate ($Ca_3P_2O_8$)	28.2	Spherical	Used as nano-fertilizers to improve plant growth and nutrient uptake	Tarafdar et al. (2012)
4.	*Rhizopus oryzae*	Gold (Au)	10	Nanocrystalline	Promoting plant growth, improving nutrient uptake, and acting as antifungal and antibacterial agents	Das et al. (2009)
5.	*Rhizopus stolonifer*	Gold (Au)	1–5	Irregularly (Uniform)	Promoting plant growth, improving nutrient uptake, and acting as antifungal and antibacterial agents	Sarkar et al. (2012)
6.	*Aspergillus niger*	Gold (Au)	12.79 ± 5.61	Spherical	Enhancing plant growth, improving soil fertility, and also acting as nano-fertilizers or nano-pesticides	Bhambure et al. (2009), Sadaf et al. (2024)

(continued)

Table 5.2 (continued)

S. No.	Fungi	NPs	Size (nm)	Shape	Applications	References
7.	*Aureobasidium pullulans*	Gold (Au)	29 ± 6	Spherical	Improving nutrient uptake and enhancing plant growth	Zhang et al. (2011)
8.	*Colletotrichum* spp.	Gold (Au)	20–40	Decahedral and icosahedral	Disease management	Shankar et al. (2003), dos Santos et al. (2024)
9.	*Fusarium oxyporum*	Silver (Ag)	20–50	Spherical	Antibacterial activity	Durán et al. (2005)
10.	*Aspergillus fumigatus*	Silver (Ag)	5–25	Mostly spherical, some triangular	Enhancing plant growth and improving crop production	Bhainsa and D'Souza (2006)
11.	*Aspergillus niger*	Silver (Ag)	3–30	Spherical	Antibacterial and antifungal activity	Jaidev and Narasimha (2010)
12.	*Pleurotus sajor caju*	Silver (Ag)	30.5	Spherical	Antibacterial activity	Vigneshwaran and Kathe (2007)
13.	*Aspergillus flavus*	Silver (Ag)	8.92	Spherical	Antifungal activity	Vigneshwaran et al. (2007)
14.	*Aspergillus niger*	Silver (Ag)	5–35	Spherical	Antimicrobial	Kathiresan et al. (2009)
15.	*Trichoderma asperellum*	Silver (Ag)	13–18	Noncrystalline	Plant protection and plant growth promotion	Ahmed and Dutta (2019)
16.	*Phoma glomerata*	Silver (Ag)	60–80	Spherical	Antibiotic	Birla et al. (2009)
17.	*Coriolus versicolor*	Silver (Ag)	25–75, 444–491	Spherical	Enhancing seed germination, plant growth, and pest management	Sanghi and Verma (2009)
18.	*Trichoderma viride*	Silver (Ag)	5–40	Spherical, rod like	Antibacterial activity, synergistic effect with antibiotics	Fayaz et al. (2010), Li et al. (2025)
19.	*Amylomyces rouxii* KSU-09	Silver (Ag)	5–27	Spherical	Used as nano-pesticides and nano-fertilizers	Musarrat et al. (2010)

(continued)

Table 5.2 (continued)

S. No.	Fungi	NPs	Size (nm)	Shape	Applications	References
20.	*Aspergillus clavatus*	Silver (Ag)	550–650	Spherical	Antimicrobial activity	Saravanan and Nanda (2010)
21.	*Aspergillus flavus* NJP08	Silver (Ag)	17	Spherical	Plant growth enhancement and disease management	Jain et al. (2011)
22.	*Rhizopus stolonifer*	Silver (Ag)	25–30	Quasi-spherical	Used as nano-pesticides and nano-fertilizers	Binupriya et al. (2010)
23.	*Aspergillus terreus* CZR-1	Silver (Ag)	2.5	Spherical	Antifungal activity	Raliya and Tarafdar (2012)
24.	*Aspergillus flavus*	Titanium dioxide (TiO_2)	62–74	Spherical	Antimicrobial	Rajakumar et al. (2012)
25.	*Aspergillus flavus TFR7*	Titanium dioxide (TiO_2)	12–15	Spherical	Plant nutrient	Raliya et al. (2015)
26.	*Fusarium oxysporum*	Barium Titanate ($BaTiO_3$)	4–5	Quasi-spherical	Plant growth enhancement and antifungal activity	Bansal et al. (2006)
27.	*Fusarium oxysporum*	Platinum (Pt)	70–180	Rectangular, triangular, spherical, and aggregates	Improve seed germination, nutrient uptake, and plant disease management	Govender et al. (2009)
28.	*Fusarium* spp.	Zinc (Zn)	100–200	Irregular, some spherical	Antifungal activity	Velmurugan et al. (2010)
29.	*Aspergillus versicolor mycelia*	Mercury (Hg)	20.5 ± 1.82	Alteration	Antifungal and growth-promoting	Das et al. 2008

Table 5.3 Properties of metal nanoparticles (MNPs)

S. No.	Property	Description	Applications	References
1.	Surface area and size	The size of nanoparticles is 1–100 nm. A high ratio of surface to volume	Reactivity is high. Improved communication with chemicals and biological systems	Albanese et al. (2012)
2.	Quantum effects	Electrons have distinct energy states at the nano scale. Special optical, magnetic, and electrical characteristics	Applications in optics and electronics, such as surface plasmon resonance (SPR)	Philip and Kumar (2022)
3.	Surface chemistry	Because of the exposed atoms, it has a high surface energy and can be functionalized with organic compounds or polymers	Functionalization enhances target-specific behavior and biocompatibility	Rai et al. (2022)
4.	Shape and morphology	Nanoparticles can be spherical, rod-shaped, or any combination of these. Surface chemistry and characteristics are influenced by shape	Customized characteristics for particular uses (e.g., medicine delivery, sensors)	Yang et al. (2019)
5.	Charge and dipole interactions	Positive, negative, and neutral surface charges affect interactions	Impacts biological interactions, stability, and dispersion	Barroso da Silva et al. (2014)
6.	Optical properties (SPR)	Metal nanoparticles can be made to exhibit surface plasmon resonance (SPR), which results in intense light absorption and scattering	SPR is utilized in target molecule detection, imaging, biosensing, and diagnostics	Fathi et al. (2019)
7.	Catalytic properties	Catalytic processes are facilitated by high surface area. The surface of the nanoparticle contains active sites	Chemical reactions, such as oxidation and hydrogenation, can be catalyzed	Zhu and Xu (2016)
8.	Color effects	Because of their size and surface plasmon resonance, nanoparticles change color	This makes them useful for biological target detection and diagnostic applications	Zhang et al. (2016)
9.	Antimicrobial activity	Elevated surface energy can lead to collisions that disrupt metabolic processes, create reactive oxygen species, and harm cell membranes	Management of fungal and bacterial infections	Wang et al. (2017)
10.	Magnetic properties	Some nanoparticles (like iron oxide) have superparamagnetic properties	These are employed in medication delivery, magnetic targeting, and bio-separation	Shabatina et al. (2020)

(continued)

Table 5.3 (continued)

S. No.	Property	Description	Applications	References
11.	Bio-compatibility	Numerous MNPs can safely interact with biological systems because they are biocompatible	Applications in tissue engineering, medication delivery, and medical imaging	Malehmir et al. (2023)
12.	Toxicity considerations	At large quantities, ion release and high surface energy can be hazardous	May damage environmental bacteria or organisms, both helpful and harmful	Sajid et al. (2015)
13.	Stability and aggregation	Without stability, nanoparticles are prone to aggregating. Reactivity may be decreased by aggregation	For features (such surface charge and dispersion) to be maintained, adequate stabilization is required	Shrestha et al. (2020)
14.	Degradation and environmental fate	Metal ions may be released when nanoparticles break down over time. The size and content have an impact on the environment	Possible hazards to the environment and bio-accumulation	Zhang et al. (2018)
15.	Colloidal stability	Aggregation is stopped via stabilization methods (such as surface charge or polymers)	Assures efficacy and uniformity in environmental and biomedical applications	Gambinossi et al. (2015)

promote soil health and plant growth (Tripathi et al. 2024). According to the study, selenium nanoparticles (Se NPs) increase the number of *Sphingomonas* and various other bacterial species throughout the rice seedlings' developing stage and promote the release of root exudates (Jiao et al. 2023). According to Jiao et al. (2023), these factors worked in concert to adjust nutrient intake and promote rice plant growth. A similar regulating role is present throughout the growth phase in *Brassica chinensis* L. (Wang et al. 2022). Plants and rhizosphere microorganisms coexisted in a symbiotic relationship that improved the bioavailability of nutrients in the rhizosphere soil, increasing the plants' capacity to absorb nutrients (Xu et al. 2023). Applying zinc oxide quantum dots (ZnO QDs) to pumpkins during their growth phase helps to improve the endophytic and rhizosphere habitats for beneficial bacteria, which in turn promotes plant growth and nutrient uptake (Xu et al. 2023). It has been demonstrated that treating tomatoes with 50 mg/kg of $Fe_7(PO_4)_6$ nanoparticles will boost the relative abundance of helpful bacteria linked to nutrient accumulation, hence hastening nutrient accumulation (Jiao et al. 2024).

The effective and sustained supply of nitrogen in the rhizosphere soil is mainly due to certain microbial communities in the root zone that convert inert nitrogen into nitrogen compounds. Thus, one important factor influencing plant nitrogen intake was nitrogen transformation fueled by rhizosphere microbial populations (Moreau et al. 2019). For instance, applying FeNPs to alfalfa improved the rhizosphere microbial community's diversity, which in turn improved the roots' capacity to fix nitrogen (Zhang et al. 2024). The number of rhizospheric bacteria, including

Table 5.4 Positive impact of metal nanoparticles (MNPs) on fungal and bacterial biocontrol agents

Mechanism	Impact	Example	References
Antimicrobial synergistic action	Improves the inhibition of pathogens when paired with BCAs	*Pseudomonas fluorescens* combined with AgNPs to combat *Fusarium*	Afshan (2023)
Supply of micro-nutrients	Encourages the growth of advantageous microorganisms (low dosages of Fe, Zn, and Cu NPs)	Phosphate-solubilizing bacteria are enhanced by FeO_4 NPs	Vasant and Raghav (2024)
Support for biofilm and colonization	Rarely, certain NPs increase microbial adherence to roots	ZnO NPs promote the colonization of *Bacillus* in roots	Ahmed et al. (2020b)
Enhancement of stress tolerance	Makes people more resilient to environmental stressors like drought and UV light	*Trichoderma* is shielded from UV damage by TiO_2 NPs	Thabet and Alqudah (2024)
Controlled biocontrol agent delivery	NPs transport microbial metabolites, such as enzymes and antibiotics	AgNPs coated with chitosan deliver antifungal substances	Cota-Arriola et al. (2013)
Generation of free radicals	Nanoparticles interact negatively with the fungal cell wall and cause intracellular components to leak out	*Carica papaya*-based Fe_3O_4 magnetic nanoparticles were able to inhibit *Fusarium oxysporum* f.sp. *ciceris* by up to 87 % causing wilt of chickpea	Patel et al. (2024)
Enhanced enzymatic activity	Boosted activities of peroxidase (POD), polyphenol oxidase (PPO), catalase (CAT), and chitinase enzymes	$Ca_3(PO_4)_2$, SiO_2, and CuO NPs controlling black scurf disease of potato caused by *Rhizoctonia solani*	El-Shewy (2019)
Inhibition of appressorium formation	Upregulation of *OsNAC4*, *OsPRO10*, *OsKSL4*, and *OsPR1b* genes involved in resistance	ZnO nanoparticles effectively protect rice from rice blast caused by *Magnaporthe oryzae*	Qiu et al. (2023)
Increased phenolic activity	Chitinases, peroxidases, and polyphenol oxidases are among the responsive enzymes and phenolic compounds that are enhanced by the silica-mediated defense mechanism against fungi	Nanosilica (Si NPs) treated maize plants express more resistance to fungal biotic stresses caused by *Aspergillus* spp., *Fusarium* spp., etc.	Suriyaprabha et al. (2014)
Decreased heavy metal accumulation	Root length, plant height, hormone level, antioxidant enzyme activity, and other physiological parameters were improved after exposure of TiO_2 NPs	TiO_2 NPs decreased Cd accumulation along with increased activity of antioxidant system in maize	Zhou et al. (2021)
Activation of phytohormones	Fe_2O_3 NPs affected the levels of phytohormones such as ABA, brassinolide, and zeatin riboside. Such effects may play a significant role in stress response and plant development	The foliar application of Fe_2O_3 NPs increased the growth of tobacco (*N. benthamiana*) plants infected with Turnip mosaic virus as well as the growth of healthy plants	Hao et al. (2018)

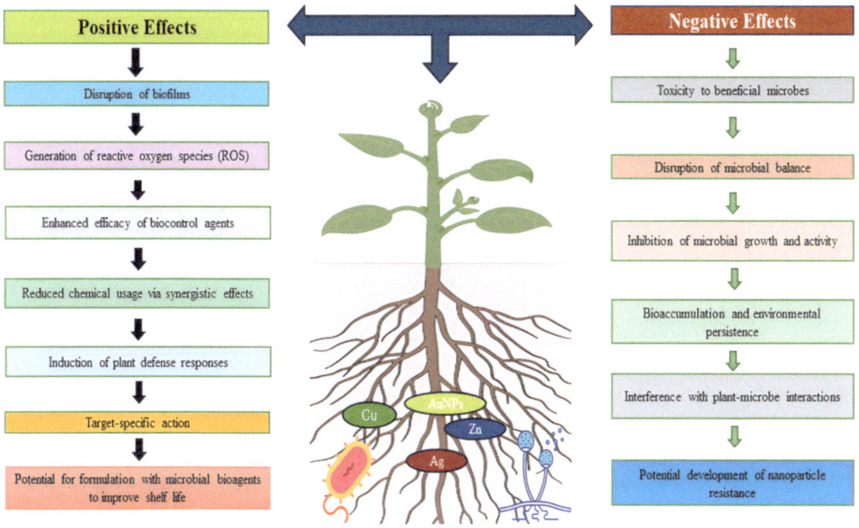

Fig. 5.3 An illustration showing the advantages and disadvantages of metal nanoparticles for bacterial and fungal biocontrol agents. (Source: Made by authors)

Saccharimonadia (called plant growth-promoting rhizobacteria, or PGPR), was increased by silver nanoparticles (AgNPs). Some strains of these bacteria assist in fixing nitrogen in the atmosphere, whereas others help plants to grow and adapt to nutrients (Sellstedt et al. 2013; Francioli et al. 2021) (Fig. 5.3).

Even though several studies have documented the beneficial impact of nanoparticles on the diversity of plant and rhizosphere microbial communities, some research still indicates that the impacts of NPs on microorganisms and plants may show dose-dependent results, particularly at high concentrations, where the NPs may have negative impacts on certain rhizosphere microbial communities and plant growth (Saghaï et al. 2022; Ren et al. 2024). For instance, seedlings of *Vigna radiata* and *Cicer arietinum* grew more when exposed to low concentrations of ZnO NPs in plant growth trials. Plant development would be inhibited by treatment doses of nanoparticles greater than 20 and 1 ppm, respectively (Mahajan et al. 2011). Simultaneously, *Solanum lycopersicum* germination and metabolic activity can be enhanced by 1.2 mM low concentration ZnO nanoparticles, whereas plants exceeding this concentration experience inhibition of germination and metabolic activity (Singh et al. 2016). Excessive nanoparticle concentrations alter the amount of edible nutrients in addition to seed germination and plant development rate. Low concentrations of TiO_2 nanoparticles (50 mg/L) can increase the nutrient contents in *Coriandrum sativum* L., whereas high concentrations can reduce the edible nutrient content and also inhibit growth. According to Yang et al. (2021) and Lin et al. (2022), microbial activity typically exhibits a dose-dependent pattern that is also associated with the kind of nanoparticles. By encouraging metabolism and energy conversion, small doses of nanoparticles may benefit microorganisms in the soil

Table 5.5 Negative impacts of metal nanoparticles on bacterial and fungal biocontrol agents

Mechanism	Impact on BCAs	Example	References
Growth and metabolism suppression	Microbial growth is inhibited by high NP concentrations	CuNPs stop *Pseudomonas* from growing	Ghasemian et al. (2015)
Direct toxicity	Destroys or inhibits BCAs by DNA disruption, ROS, and membrane damage	*Bacillus subtilis* viability is decreased by AgNPs	Do et al. (2025)
Unbalanced microbial community	Beneficial microorganisms are harmed by nontarget toxicity	Rhizosphere symbionts are depleted by broad-spectrum AgNPs	Sillen (2020)
Decreased colonization	Interferes with root adherence and biofilm production	CuNPs hinder the synthesis of *Pseudomonas* biofilms	LewisOscar et al. (2015)
Pathogens' induced resistance	Extended exposure could favor microorganisms that are resistant to MNPs	AgNP tolerance is gradually developed by *Fusarium* strains	Jian et al. (2022)
Production of reactive oxygen species (ROS)	ROS induce acute oxidative stress, inhibit enzymes, and cause damage to lipids, proteins, and DNA/RNA	Gold (Au), magnesium oxide (MgO), and zinc oxide (ZnO) NPs promote ROS formation through increased catalytic activity by producing H_2O_2 from glucose oxidase	Zheng et al. (2017)
Dysfunction of cytosolic proteins	Protein dysfunction is another mode of action by which multimetallic NPs induce their antimicrobial response by binding to cytosolic proteins, such as DNA and enzymes	Silver (Ag) and copper (Cu) NPs inhibit DNA replication, cell division, and DNA degradation, whereas gold NPs interact with DNA within the cell by upregulating genes	Chatterjee et al. (2014), Durán et al. (2016) and Roy et al. (2019)

rhizosphere. For instance, the kinds and total number of soil rhizosphere microorganisms were greatly enhanced when *Medicago truncatula* was treated with 5 mg/kg of Ag and 50 mg/kg of Zn and Ti nanoparticles (Chen et al. 2017). In the rhizosphere of *Lactuca sativa* L., low dose (10 mg/kg) ZnO nanoparticles stimulate the growth of Cyanobacteria; nevertheless, under the same treatment, 100 mg/kg ZnO has no discernible effect on this colony (Xu et al. 2018) (Table 5.5).

Beneficial microbial communities (fungi and bacteria) may be threatened by the release of NPs into the environment. By measuring soil respiration and enzymatic processes, one could ascertain how NPs affect microbial activity (Simonin and Richaume 2015). TiO_2 and CuO NPs reduced soil microbial biomass and enzymatic activities, which had an impact on their community patterns in flooded paddy soil, according to a study by Xu et al. (2015). You et al. (2018) found similar outcomes when they investigated the impact of ZnO, TiO_2, CeO_2, and Fe_3O_4 NPs on the bacterial populations of saline-alkali and black soils as well as soil enzymatic activity (invertase, urease, catalase, and phosphatase). The findings indicated an imminent threat to biological nitrogen fixation, alterations in the soil bacterial ecology, and an impact on soil enzyme activity. According to Jiling et al. (2016), a high concentration of Fe3O4 NPs dramatically reduced the amount of bacteria present in the soil. According to Chai et al. (2015), zinc oxide and CeO_2 NPs decrease enzymatic activities and have an impact on the plate counts of Azotobacter, P-solubilizing, and K-solubilizing bacteria. Titanium dioxide nanoparticles have a negative impact on microbial activity, variety, and abundance along with decreasing the number of functional soil bacteria and enzymatic activity (Buzea et al. 2007; Solanki et al. 2008). Excessive consumption of metal-based fertilizers and agricultural pesticides in crops poses a serious risk to the ecology. Furthermore, inappropriate agricultural methods and significant harm to soil life have been brought on by a rising need for food. Due to its critical role in biogeochemical cycles and significant contribution to the effective cultivation of food, soil microbiota (SM) is a decisive element in the overall sustainability of the natural environment. Despite being more effective than conventional agrochemicals, the consumption of metallic nanoparticles (MNPs), which are occasionally misused in agriculture, has had a variety of effects on SM, mostly on microbial diversity and structure. This viewpoint thus sought to deal with the effects of MNP utilization in agriculture, primarily on the soil microbiota, and broaden the conversation to include the threats to the equilibrium of the ecosystem.

5.3 Conclusion and Future Prospects

Technologies for microbial synthesis offer a more eco-friendly, economic, efficient, and nontoxic way to produce metal nanoparticles than other conventional methods, including physical and chemical ones. Numerous metals and metal oxides can be produced by microbes such as bacteria, fungi, actinomycetes, and others. The size, shape, and morphology of bio-synthesized nanoparticles are being identified through a variety of characterization techniques. Experts from all around the world aim to create microbial nanoparticles because they are easy to utilize, biocompatible, and possess many applications. Researchers can modify already-existing microorganisms and develop metabolic pathways to produce nanoparticles with adjustable sizes and shapes for a range of applications. Despite the numerous advantages of using recombinant organisms to bio-fabricate NPs, their use is still relatively uncommon because of concerns about NP synthesis's viability, economic viability, environmental safety, and reproducibility. While using GMOs to produce NPs, researchers must also consider concerns like public acceptability, biosafety, biosecurity, etc. Nevertheless is expected that in the upcoming years, the bio-fabrication of nanoparticles utilizing genetically modified organisms would attract a lot of interest and emerge as the favored technique for the feasible and practical manufacturing of nanoparticles. Difficult purification procedures for the recovery of intracellular nanoparticles are a major obstacle in the manufacture of nanoparticles using microorganisms. In agriculture, NPs are used extensively for crop protection, crop enhancement, and other purposes. Although lab-level methods based on nanoparticles have been established, a complete investigation of nanotoxicology for harmful health consequences in humans and the environment has not yet been conducted.

Another interesting area of study is how the generated nanoparticles interact with nutrients that are essential for plants and other materials. Therefore, precise evaluation of exposure duration, application technique, ambient conditions, nanoparticle form, and concentration is necessary. Frameworks should be established by the government to ensure the appropriate and safe use of nanomaterials in agricultural activities.

The aforementioned background suggests that nanoparticles are crucial for enhancing rhizosphere microbial populations and plant nutrient absorption. The two are closely related and have a synergistic effect that influences the development of plants in addition to ecosystem stability and sustainability. By loading nutrients and precisely delivering them to various plant sections, nanoparticles can increase the efficiency of nutrient usage and encourage plant nutrient uptake. At the same time, nanoparticles are essential for regulating rhizosphere microbes. In addition to encouraging plant growth and yield, combating pathogenic microbial strains further modifies the composition of the rhizospheric microbial community by changing the physical and chemical properties of the soil and root exudates, which impacts plant development and growth.

In order to guarantee sustainable evolution of agriculture and promote the green revolution, nanoparticles possess the capacity to both directly and indirectly impact the ecological environment around plants. At the same time, more attention should be paid to how rhizospheric microbial populations, soil, NPs, and plants interact. To determine whether NPs are useful, more research is required. Furthermore, to demonstrate relationships between NP dose, soil type, and ecological implications, long-term studies of cereals and certain other important cultivated crops are necessary. Such studies are essential to check the sustainable growth of agriculture in future decades and to lessen dependency on chemical fertilizers and pesticides. In conclusion, this chapter summarizes the results of nanoparticles in improving plant growth and modifying the rhizosphere microbiome through a variety of mechanisms, providing a basis for the cooperative collaboration between future studies in the fields of environmental science, sustainable agriculture, and nanoscience.

References

AbdelRahim K, Mahmoud SY, Ali AM, Almaary KS, Mustafa AEZMA, Husseiny SM (2017) Extracellular biosynthesis of silver nanoparticles using *Rhizopus stolonifer*. Saudi J Biol Sci 24:208–216

Adamecz DI (2023) The possible modulatory effect of metallic nanoparticles on macrophage polarization. Acta Biol Szeged 67:145–166

Afshan NUS (2023) Recent advancement in fungal biocontrol agents. In: Plant Mycobiome: diversity, interactions and uses. Springer International Publishing, Cham, pp 203–223

Aftab ZH, Mirza FS, Anjum T, Rizwana H, Akram W, Aftab M, Ali MD, Li G (2025) Antifungal potential of biogenic zinc oxide nanoparticles for controlling Cercospora leaf spot in mung bean. Nanomaterials 15:143. https://doi.org/10.3390/nano15020143

Agarwal H, Menon S, Kumar SV, Rajeshkumar S (2018) Mechanistic study on antibacterial action of zinc oxide nanoparticles synthesized using green route. Chem Biol Interact 286:60–70

Ahmed AA, Dutta P (2019) *Trichoderma asperellum* mediated synthesis of silver nanoparticles: characterization and its physiological effects on tea [*Camellia sinensis* (L.) Kuntze var. *assamica* (J. Masters) Kitam.]. Int J Curr Microbiol App Sci 8(4):1215–1229

Ahmed E, Kalathil S, Shi L, Alharbi O, Wang P (2018) Synthesis of ultra-small platinum, palladium and gold nanoparticles by *Shewanella loihica* PV-4 electrochemically active biofilms and their enhanced catalytic activities. J Saudi Chem Soc 22:919–929

Ahmed T, Shahid M, Noman M, Niazi MBK, Mahmood F, Manzoor I, Zhang Y, Li B, Yang Y, Yan C (2020a) Silver nanoparticles synthesized by using *Bacillus cereus* SZT1 ameliorated the damage of bacterial leaf blight pathogen in rice. Pathogens 9(3):160

Ahmed B, Ameen F, Rizvi A, Ali K, Sonbol H, Zaidi A, Khan MS, Musarrat J (2020b) Destruction of cell topography, morphology, membrane, inhibition of respiration, biofilm formation, and bioactive molecule production by nanoparticles of Ag, ZnO, CuO, TiO_2, and Al_2O_3 toward beneficial soil bacteria. ACS Omega 5(14):7861–7876

Ahmed T, Luo J, Noman M, Ijaz M, Wang X, Masood HA, Manzoor N, Wang Y, Li B (2023) Microbe-mediated nanoparticle intervention for the management of plant diseases. Crop Health 1:3

Albanese A, Tang PS, Chan WC (2012) The effect of nanoparticle size, shape, and surface chemistry on biological systems. Ann Rev Biomed Eng 14(1):1–16

Alfryyan N, Kordy MG, Abdel-Gabbar M, Soliman HA, Shaban M (2022) Characterization of the biosynthesized intracellular and extracellular plasmonic silver nanoparticles using *Bacillus cereus* and their catalytic reduction of methylene blue. Sci Rep 12(1):12495

Ali M, Ahmed T, Wu W, Hossain A, Hafeez R, Islam Masum M, Wang Y, An Q, Sun G, Li B (2020) Advancements in plant and microbe-based synthesis of metallic nanoparticles and their antimicrobial activity against plant pathogens. Nanomaterials 10(6):1146

Apte M, Sambre D, Gaikawad S, Joshi S, Bankar A, Kumar AR, Zinjarde S (2013) Psychrotrophic yeast *Yarrowia lipolytica* NCYC 789 mediates the synthesis of antimicrobial silver nanoparticles via cell-associated melanin. AMB Express 3:32

Ayenew BM, Satheesh N, Zegeye ZB, Kassie DA (2024) A review on the production of nanofertilizers and its application in agriculture. Heliyon 11(1):e41243. https://doi.org/10.1016/j.heliyon.2024.e41243

Babu PJ, Saranya S, Longchar B, Rajasekhar A (2022) Nanobiotechnology-mediated sustainable agriculture and post-harvest management. Curr Res Biotechnol 4:326–336

Bahrulolum H, Nooraei S, Javanshir N, Tarrahimofrad H, Mirbagheri VS, Easton AJ, Ahmadian G (2021) Green synthesis of metal nanoparticles using microorganisms and their application in the agrifood sector. J Nanobiotechnol 19:86. https://doi.org/10.1186/s12951-021-00834-3

Bai HJ, Zhang ZM, Gong J (2006) Biological synthesis of semiconductor zinc sulfide nanoparticles by immobilized *Rhodobacter sphaeroides*. Biotechnol Lett 28(14):1135–1139. https://doi.org/10.1007/s10529-006-9063-1

Bai HJ, Zhang ZM, Guo Y, Yang GE (2009) Biosynthesis of cadmium sulfide nanoparticles by photosynthetic bacteria *Rhodopseudomonas palustris*. Colloids Surf B: Biointerfaces 70(1):142–146

Banik S, Pérez-de-Luque A (2017) In vitro effects of copper nanoparticles on plant pathogens, beneficial microbes and crop plants. Span J Agric Res 15(2):e1005–e1005

Bansal V, Poddar P, Ahmad A, Sastry M (2006) Room-temperature biosynthesis of ferroelectric barium Titanate nanoparticles. J Am Chem Soc 128:11958–11963

Barroso da Silva FL, Bostrom M, Persson C (2014) Effect of charge regulation and ion–dipole interactions on the selectivity of protein–nanoparticle binding. Langmuir 30(14):4078–4083

Bhainsa KC, D'Souza SF (2006) Extracellular biosynthesis of silver nanoparticles using the fungus *Aspergillus fumigatus*. Colloids Surf B Biointerfaces 47:160–164

Bhambure R, Bule M, Shaligram N, Kamat M, Singhal R (2009) Extracellular biosynthesis of gold nanoparticles using Aspergillus Niger—its characterization and stability. Chem Eng Technol 32:1036–1041

Binupriya AR, Sathishkumar M, Yun SI (2010) Biocrystallization of silver and gold ions by inactive cell filtrate of *Rhizopus stolonifer*. Colloids Surf B Biointerfaces 79:531–534

Birla SS, Tiwari VV, Gade AK, Ingle AP, Yadav AP, Rai MK (2009) Fabrication of silver nanoparticles by *Phoma glomerata* and its combined effect against *Escherichia coli, Pseudomonas aeruginosa* and *Staphylococcus aureus*. Lett Appl Microbiol 48(2):173–179

Birla SS, Gaikwad SC, Gade AK, Rai MK (2013) Rapid synthesis of silver nanoparticles from fusarium oxysporum by optimizing Physicocultural conditions. Sci World J 2013:796018

Bremer E, Calteau A, Danchin A, Harwood C, Helmann JD, Médigue C, Palsson BO, Sekowska A, Vallenet D, Zuniga A, Zuniga C (2023) A model industrial workhorse: Bacillus subtilis strain 168 and its genome after a quarter of a century. Microb Biotechnol 16(6):1203–1231. https://doi.org/10.1111/1751-7915.14257

Buzea C, Pacheco II, Robbie K (2007) Nanomaterials and nanoparticles: sources and toxicity. Biointerphases 2:17. https://doi.org/10.1116/1.2815690

Chai H, Yao J, Sun J, Zhang C, Liu W, Zhu M, Ceccanti B (2015) The effect of metal oxide nanoparticles on functional bacteria and metabolic profiles in agricultural soil. Bull Environ Contam Toxicol 94(4):490–495. https://doi.org/10.1007/s00128-015-1485-9

Chandoliya R, Sharma S, Sharma V, Joshi R, Sivanesan I (2024) Titanium dioxide nanoparticle: A comprehensive review on synthesis, applications and toxicity. Plants 13:2964. https://doi.org/10.3390/plants13212964

Chang M, Liu Y, Xu M, Li H, Li SW (2024) Particle morphology and soil properties affect the retention of copper oxide nanoparticles in agricultural soils. Environ Geochem Health 46:281 https://doi.org/10.1007/s10653-024-02057-5

Chaparro JM, Badri DV, Vivanco JM (2014) Rhizosphere microbiome assemblage is affected by plant development. ISME J 8:790–803. https://doi.org/10.1038/ismej.2013.196

Chatterjee AK, Chakraborty R, Basu T (2014) Mechanism of antibacterial activity of copper nanoparticles. Nanotechnol 25:135101

Chen C, Tsyusko OV, McNear DH, Judy J, Lewis RW, Unrine JM (2017) Effects of biosolids from a wastewater treatment plant receiving manufactured nanomaterials on *Medicago truncatula* and associated soil microbial communities at low nanomaterial concentrations. Sci Total Environ 609:799–806. https://doi.org/10.1016/j.scitotenv.2017.07.188

Correa-Llantén DN, Muñoz-Ibacache SA, Castro ME, Muñoz PA, Blamey JM (2013) Gold nanoparticles synthesized by *Geobacillus* sp. strain ID17 a thermophilic bacterium isolated from Deception Island, Antarctica. Microb Cell Factories 12:75. https://doi.org/10.1186/1475-2859-12-75

Cota-Arriola O, Onofre Cortez-Rocha M, Burgos-Hernández A, Marina Ezquerra-Brauer J, Plascencia-Jatomea M (2013) Controlled release matrices and micro/nanoparticles of chitosan with antimicrobial potential: development of new strategies for microbial control in agriculture. J Sci Food Agric 93(7):1525–1536

Dağlıoğlu Y, Yılmaz Öztürk B (2019) A novel intracellular synthesis of silver nanoparticles using *Desmodesmus* sp. (Scenedesmaceae): different methods of pigment change. Rend Lincei Sci Fis Nat 30:611–621

Das S, Das A, Guha A (2008) Adsorption behavior of mercury on functionalized *Aspergillus versicolor* mycelia: atomic force microscopic study. Langmuir 25:360–366

Das SK, Das AR, Guha AK (2009) Gold nanoparticles: microbial synthesis and application in water hygiene management. Langmuir 25:8192–8199

do Nascimento JM, de Oliveira JD, de Lima Rizzo AC, Leite SGF (2018) Biogenic production of copper nanoparticles by *Saccharomyces cerevisiae*. J Bionanosci 12:689–693

Do HTT, Nguyen NPU, Saeed SI, Dang NT, Doan L, Nguyen TTH (2025) Advances in silver nanoparticles: unraveling biological activities, mechanisms of action, and toxicity. Appl Nanosci 15(1):1

dos Santos OAL, dos Santos MS, Filho SA, Backx BP (2024) Nanotechnology for the control of plant pathogens and pests. Plant Nano Biol 8:100080

Drexler KE (1986) Engines of creation: the coming era of nanotechnology. Anchor Books Edition

Durán N, Marcato PD, Alves OL, de Souza GIH, Esposito E (2005) Mechanistic aspects of biosynthesis of silver nanoparticles by several Fusarium oxysporum strains. J Nanobiotechnol 3:8. https://doi.org/10.1186/1477-3155-3-8

Durán N, Durán M, de Jesus MB, Seabra AB, Fávaro WJ, Nakazato G (2016) Silver nanoparticles: A new view on mechanistic aspects on antimicrobial activity. Nanomed Nanotechnol Biol Med 12:789–799

Dutta P, Kumari A, Mahanta M, Upamanya GK, Heisnam P, Borua S, Kaman PK, Mishra AK, Mallik M, Muthukrishnan G, Sabarinathan KG, Puzari KR, Vijayreddy D (2023) Nanotechnological approaches for management of soil-borne plant pathogens. Front Plant Sci 14:1136233. https://doi.org/10.3389/fpls.2023.1136233

El-Shewy E (2019) The efficacy of copper oxide, tri-calcium phosphate and silicon dioxide nanoparticles in controlling black scurf disease of potato. Ann Agric Sci Moshtohor 57:129–138

Fathi F, Rashidi MR, Omidi Y (2019) Ultra-sensitive detection by metal nanoparticles-mediated enhanced SPR biosensors. Talanta 192:118–127

Fayaz AM, Balaji K, Girilal M, Yadav R, Kalaichelvan PT, Venketesan R (2010) Biogenic synthesis of silver nanoparticles and their synergistic effect with antibiotics: a study against gram-positive and gram-negative bacteria. Nanomed Nanotechnol Biol Med 6(1):103–109

Franci G, Falanga A, Galdiero S, Palomba L, Rai M, Morelli G, Galdiero M (2015) Silver nanoparticles as potential antibacterial agents. Molecules 20(5):8856–8874. https://doi.org/10.3390/molecules20058856

Francioli D, Cid G, Kanukollu S, Ulrich A, Hajirezaei MR, Kolb S (2021) Flooding causes dramatic compositional shifts and depletion of putative beneficial bacteria on the spring wheat microbiota. Front Microbiol 12:773116. https://doi.org/10.3389/fmicb.2021.773116

Gambinossi F, Mylon SE, Ferri JK (2015) Aggregation kinetics and colloidal stability of functionalized nanoparticles. Adv Colloid Interf Sci 222:332–349

Gericke M, Pinches A (2006) Microbial production of gold nanoparticles. Gold Bull 39:22–28

Ghasemian E, Naghoni A, Rahvar H, Kialha M, Tabaraie B (2015) Evaluating the effect of copper nanoparticles in inhibiting Pseudomonas aeruginosa and Listeria monocytogenes biofilm formation. Jundishapur J Microbiol 8(5):e17430

Govender Y, Riddin T, Gericke M, Whiteley CG (2009) Bioreduction of platinum salts into nanoparticles: a mechanistic perspective. Biotechnol Lett 31:95–100

Hao Y, Yuan W, Ma C, White JC, Zetian Z, Adeel M, Zhou T, Rui Y, Xing B (2018) Engineered nanomaterials suppress turnip mosaic virus infection in tobacco (Nicotiana benthamiana). Environ Sci Nano 5:1685–1693

He S, Guo Z, Zhang Y, Zhang S, Wang J, Gu N (2007) Biosynthesis of gold nanoparticles using the bacteria *Rhodopseudomonas capsulata*. Mater Lett 61(18):3984–3987

Hulkoti NI, Taranath T (2014) Biosynthesis of nanoparticles using microbes—a review. Colloids Surf B Biointerfaces 121:474–483

Husseiny SM, Salah TA, Anter HA (2015) Biosynthesis of size controlled silver nanoparticles by *Fusarium oxysporum,* their antibacterial and antitumor activities. Beni-Suef Univ J Basic Appl Sci 4:225–231

Ibrahim E, Fouad H, Zhang M, Zhang Y, Qiu W, Yan C, Li B, Mo J, Chen J (2019) Biosynthesis of silver nanoparticles using endophytic bacteria and their role in inhibition of rice pathogenic bacteria and plant growth promotion. RSC Adv 9(50):29293–29299

Jaidev LR, Narasimha G (2010) Fungal mediated biosynthesis of silver nanoparticles, characterization and antimicrobial activity. Colloids Surf B Biointerfaces 81:430–433

Jain N, Bhargava A, Majumdar S, Tarafdar JC, Panwar J (2011) Extracellular biosynthesis and characterization of silver nanoparticles using *Aspergillus flavus* NJP08: a mechanism perspective. Nanoscale 3:635–641

Jain N, Bhargava A, Tarafdar JC, Singh SK, Panwar J (2013) A biomimetic approach toward synthesis of zinc oxide nanoparticles. Appl Microbiol Biotechnol 97(2):859–869

Jian Y, Chen X, Ahmed T, Shang Q, Zhang S, Ma Z, Yin Y (2022) Toxicity and action mechanisms of silver nanoparticles against the mycotoxin-producing fungus Fusarium graminearum. J Adv Res 38:1–12

Jiao L, Cao X, Wang C, Chen F, Zou H, Yue L, Wang Z (2023) Crosstalk between *in situ* root exudates and rhizobacteria to promote rice growth by selenium nanomaterials. Sci Total Environ 878:163175. https://doi.org/10.1016/j.scitotenv.2023.163175

Jiao L, Yue L, Le X, Cao X, Chen F, Wang C, Zhang X, Zou H, Wang Z (2024) Dynamic microbial regulation of triiron tetrairon phosphate nanomaterials in the tomato rhizosphere. Environ Sci Nano 11:1157–1169. https://doi.org/10.1039/D3EN00797A

Jiling C, Youzhi F, Xiangui L, Junhua W (2016) Arbuscular mycorrhizal fungi alleviate the negative effects of iron oxide nanoparticles on bacterial community in rhizospheric soils. Front Environ Sci 4:10

Joel JM, Johnson R, Puthur JT (2023) Co-application of arbuscular mycorrhizal fungi and engineered nanomaterials: a promising strategy for crop resilience against abiotic stresses. S Afr J Bot 162:314–323

Jovanović B, Bezirci G, Çağan AS, Coppens J, Levi EE, Oluz Z, Tuncel E, Duran H, Beklioğlu M (2016) Food web effects of titanium dioxide nanoparticles in an outdoor freshwater mesocosm experiment. Nanotoxicology 10(7):902–912

Karthik L, Kumar G, Kirthi AV, Rahuman A, Rao KB (2014) *Streptomyces* sp. LK3 mediated synthesis of silver nanoparticles and its biomedical application. Bioprocess Biosyst Eng 37(2):261–267

Kathiresan K, Manivannan S, Nabeel M, Dhivya B (2009) Studies on silver nanoparticles synthesized by a marine fungus, *Penicillium fellutanum* isolated from coastal mangrove sediment. Colloids Surf B Biointerfaces 71:133–137

Kaur H, Dolma K, Kaur N, Malhotra A, Kumar N, Dixit P, Sharma D, Mayilraj S, Choudhury AR (2015) Marine microbe as nano-factories for copper biomineralization. Biotechnol Bioprocess Eng 20:51–57

Kaur M, Gautam A, Guleria P, Singh K, Kumar V (2022) Green synthesis of metal nanoparticles and their environmental applications. Curr Opin Environ Sci Health 29:100390

Klaine SJ, Alvarez PJ, Batley GE, Fernandes TF, Handy RD, Lyon DY, Mahendra S, McLaughlin MJ, Lead JR (2008) Nanomaterials in the environment: behavior, fate, bioavailability, and effects. Environ Toxicol Chem 27:1825–1851

Korbekandi H, Iravani S, Abbasi S (2012) Optimization of biological synthesis of silver nanoparticles using *Lactobacillus casei* subsp. *Casei*. J Chem Technol Biotechnol 87(7):932–937

Koul B, Poonia AK, Yadav D, Jin JO (2021) Microbe-mediated biosynthesis of nanoparticles: applications and future prospects. Biomolecules 11:886. https://doi.org/10.3390/biom11060886

Krishnaraj C, Jagan EG, Rajasekar S, Selvakumar P, Kalaichelvan PT, Mohan N (2012) Optimization for rapid synthesis of silver nanoparticles and its effect on phytopathogenic fungi. Spectrochim Acta A Mol Biomol Spectrosc 93:95–99

Kumar P, Pandhi S, Mahato DK, Kamle M, Mishra A (2021) Bacillus-based nanobioformulations for phytopathogens and insect–pest management. Egypt J Biol Pest Control 31:1–11

LewisOscar F, MubarakAli D, Nithya C, Priyanka R, Gopinath V, Alharbi NS, Thajuddin N (2015) One pot synthesis and anti-biofilm potential of copper nanoparticles (CuNPs) against clinical strains of Pseudomonas aeruginosa. Biofouling 31(4):379–391

Li K, Li H, ur Rahman S, Rehman A, Khalid M (2024) Nanoparticle-mediated modulation of plant performance and microbiome dynamics: insights into interplay mechanisms. Biocatal Agric Biotechnol 61:103366

Li T, Tao R, Zhong Z, Liu X, Gao Z (2025) Combining *Trichoderma* sp. and biogenic AgNPs from *Trichoderma* strains as a synergistic control complex to improve the growth of muskmelon and suppress *Fusarium oxysporum* f. sp. *melonis*. Environ Sci Nano 12:2034–2049

Liang M, Wei S, Jian-Xin L, Xiao-Xi Z, Zhi H, Wen L, Liu ZC, Tang JX (2017) Optimization for extracellular biosynthesis of silver nanoparticles by *Penicillium aculeatum* Su1 and their antimicrobial activity and cytotoxic effect compared with silver ions. Mater Sci Eng C77:963–971

Lin J, Ma K, Chen H, Chen Z, Xing B (2022) Influence of different types of nanomaterials on soil enzyme activity: a global meta-analysis. Nano Today 42:101345. https://doi.org/10.1016/j.nantod.2021.101345

Mahajan P, Dhoke SK, Khanna AS (2011) Effect of Nano-ZnO particle suspension on growth of mung (*Vigna radiata*) and gram (*Cicer arietinum*) seedlings using plant agar method. J Nanotechnol 2011:696535. https://doi.org/10.1155/2011/696535

Mahmood I, Ansari RA, Rizvi R (eds) (2024) Nanotechnology and plant disease management. CRC Press

Malarkodi C, Rajeshkumar S, Paulkumar K, Vanaja M, Jobitha GDG, Annadurai G (2013) Bactericidal activity of bio mediated silver nano-particles synthesized by Serratia nematodiphila. Drug Invent Today 5:119–125

Malehmir S, Esmaili MA, Khaksary Mahabady M, Sobhani-Nasab A, Atapour A, Ganjali MR et al (2023) A review: hemocompatibility of magnetic nanoparticles and their regenerative medicine, cancer therapy, drug delivery, and bioimaging applications. Front Chem 11:1249134

Malhotra A, Dolma K, Kaur N, Rathore YS, Ashish MS, Choudhury AR (2013) Biosynthesis of gold and silver nanoparticles using a novel marine strain of Stenotrophomonas. Bioresour Technol 142:727–731

Manivasagan P, Venkatesan J, Senthilkumar K, Sivakumar K, Kim S (2013) Biosynthesis, antimicrobial and cytotoxic effect of silver nanoparticles using a novel *Nocardiopsis* sp. MBRC-1. BioMed Res Int 2013:287638

Meel S, Saharan BS (2024) Enhancing crop resilience towards drought: by integrating nanotechnology, microbiomes, and growth-promoting rhizobacteria. Discover Agric 2(1):112

Mistry H, Thakor R, Patil C, Trivedi J, Bariya H (2021) Biogenically proficient synthesis and characterization of silver nanoparticles employing marine procured fungi Aspergillus brunneoviolaceus along with their antibacterial and antioxidative potency. Biotechnol Lett 43:307–316

Moreau D, Bardgett RD, Finlay RD, Jones DL, Philippot L (2019) A plant perspective on nitrogen cycling in the rhizosphere. Funct Ecol 33(4):540–552

Musarrat J, Dwivedi S, Singh BR, Al-Khedhairy AA, Azam A, Naqvi A (2010) Production of antimicrobial silver nanoparticles in water extracts of the fungus *Amylomyces rouxii* strain KSU-09. Bioresour Technol 101:8772–8776

Nawawi WMFBW, Jones M, Murphy RJ, Lee KY, Kontturi E, Bismarck A (2020) Nanomaterials derived from fungal sources—is it the new hype? Biomacromolecules 21(1):30–55

Noman M, Shahid M, Ahmed T, Niazi MBK, Hussain S, Song F, Manzoor I (2020) Use of biogenic copper nanoparticles synthesized from a native Escherichia sp. as photocatalysts for azo dye degradation and treatment of textile effluents. Environ Pollut 257:113514

Öcal N, Ceylan A, Duman F (2020) Intracellular biosynthesis of PbS quantum dots using Pseudomonas aeruginosa ATCC 27853: evaluation of antibacterial effects and DNA cleavage activities. World J Microbiol Biotechnol 36(10):147

Osonga FJ, Akgul A, Yazgan I, Akgul A, Eshun GB, Sakhaee L, Sadik OA (2020) Size and shape-dependent antimicrobial activities of silver and gold nanoparticles: a model study as potential fungicides. Molecules 25(11):2682

Padmakumar A, Pavani C, Eswar K, Kong L, Yang W, Gopalakrishnan S et al (2023) Bacteria-premised nanobiopesticides for the management of phytopathogens and pests. ACS Agric Sci Technol 3(5):370–388

Pantidos N, Horsfall LE (2014) Biological synthesis of metallic nanoparticles by bacteria, fungi and plants. J Nanomed Nanotechnol 5(5):1

Patel D, Rathod K, Parekh K, Thakker JN (2024) NanoBioProtectors: role of plant derived magnetic nanoparticles as a potent biocontrol agent against *Fusarium oxysporum ciceris*. Plant Nano Biol 9:100086

Philip A, Kumar AR (2022) The performance enhancement of surface plasmon resonance optical sensors using nanomaterials: A review. Coord Chem Rev 458:214424

Prasad TNVKV, Kambala VSR, Naidu R (2013) Phyconanotechnology: synthesis of silver nanoparticles using brown marine algae Cystophoramoniliformis and their characterisation. J Appl Phycol 25:177–182

Qamar SUR, Ahmad JN (2021) Nanoparticles: mechanism of biosynthesis using plant extracts, bacteria, fungi, and their applications. J Mol Liq 334:116040

Qiu J, Chen Y, Liu Z, Wen H, Jiang N, Shi H, Kou Y (2023) The application of zinc oxide nanoparticles: An effective strategy to protect rice from rice blast and abiotic stresses. Environ Pollut 331:121925

Qu Y, Li X, Lian S, Dai C, Jv Z, Zhao B, Zhou H (2019) Biosynthesis of gold nanoparticles using fungus Trichoderma sp. WL-Go and their catalysis in degradation of aromatic pollutants. IET Nanobiotechnol 13(1):12–17

Rai M, Bonde S, Golinska P, Trzcińska-Wencel J, Gade A, Abd-Elsalam KA, Shende S, Gaikwad S, Ingle AP (2021) *Fusarium* as a novel fungus for the synthesis of nanoparticles: mechanism and applications. J Fungi 7(2):139

Rai S, Guin M, De A, Singh NB (2022) Functionalized nanomaterials: basics, properties and applications. In: Functionalized nanomaterials for corrosion mitigation: synthesis, characterization, and applications. American Chemical Society, pp 27–66

Rajakumar G, Rahuman A, Roopan SM, Khanna VG, Elango G, Kamaraj C, Zahir AA, Velayutham K (2012) Fungus-mediated biosynthesis and characterization of TiO2 nanoparticles and their activity against pathogenic bacteria. Spectrochim Acta A Mol Biomol Spectrosc 91:23–29

Rajeshkumar S, Sivapriya D (2020) Fungus-mediated nanoparticles: characterization and biomedical advances. Nanoparticles Med 2020:185–199

Rajput VD, Minkina T, Sushkova S, Tsitsuashvili V, Mandzhieva S, Gorovtsov A, Nevidomskyaya D, Gromakova N (2018) Effect of nanoparticles on crops and soil microbial communities. J Soils Sediments 18:2179–2187

Raliya R, Tarafdar JC (2012) Novel approach for silver nanoparticle synthesis using *Aspergillus terreus* CZR-1: mechanism perspective. J Bionanosci 6:12–16

Raliya R, Biswas P, Tarafdar JC (2015) TiO_2 nanoparticle biosynthesis and its physiological effect on mung bean (*Vigna radiata* L.). Biotechnol Rep 5:22–26

Ramezani M, Ramezani F, Gerami M (2019) Nanoparticles in pest incidences and plant disease control. In: Nanotechnology for agriculture: crop production & protection, pp 233–272

Ren Y, Wang G, Su Y, Li J, Zhang H, Ma G, Han J (2024) Effect of CeO_2, TiO_2 and SiO_2 nanoparticles on the growth and quality of model medicinal plant *Salvia miltiorrhiza* by acting on soil microenvironment. Ecotoxicol Environ Saf 280:116552. https://doi.org/10.1016/j.ecoenv.2024.116552

Roy B, Pal AK (2025) Nano-technology as an eco-friendly approach in agriculture. Plant Sci Today 12(1):1–7. https://doi.org/10.14719/pst.3027

Roy A, Bulut O, Some S, Mandal AK, Yilmaz MD (2019) Green synthesis of silver nanoparticles: biomolecule-nanoparticle organizations targeting antimicrobial activity. RSC Adv 9:2673–2702

Sadaf A, Afolayan JS, Perry CC (2024) Developing gold nanoparticle mycelial composites: effect of nanoparticle surface functionality on *Aspergillus niger* viability and cell wall biochemistry. Curr Res Biotechnol 7:100185

Saghaï A, Wittorf L, Philippot L, Hallin S (2022) Loss in soil microbial diversity constrains microbiome selection and alters the abundance of N-cycling guilds in barley rhizosphere. Appl Soil Ecol 169:104224. https://doi.org/10.1016/j.apsoil.2021.104224

Sajid M, Ilyas M, Basheer C, Tariq M, Daud M, Baig N, Shehzad F (2015) Impact of nanoparticles on human and environment: review of toxicity factors, exposures, control strategies, and future prospects. Environ Sci Pollut Res 22:4122–4143

Sanghi R, Verma P (2009) Biomimetic synthesis and characterization of protein capped silver nanoparticles. Bioresour Technol 100:501–504

Saratale RG, Karuppusamy I, Saratale GD, Pugazhendhi A, Kumar G, Park Y, Ghodake GS, Bharagava RN, Banu JR, Shin HS (2018) A comprehensive review on green nanomaterials using biological systems: recent perception and their future applications. Colloids Surf B Biointerfaces 170:20–35

Saravanan M, Nanda A (2010) Extracellular synthesis of silver bionanoparticles from Aspergillus clavatus and its antimicrobial activity against MRSA and MRSE. Colloids Surf B Biointerfaces 77:214–218

Sardella D, Gatt R, Valdramidis VP (2019) Metal nanoparticles for controlling fungal proliferation: quantitative analysis and applications. Curr Opin Food Sci 30:49–59

Sarkar J, Ray S, Chattopadhyay D, Laskar A, Acharya K (2012) Mycogenesis of gold nanoparticles using a phytopathogen *Alternaria alternata*. Bioprocess Biosyst Eng 35:637–643

Saxena J, Sharma PK, Sharma MM, Singh A (2016) Process optimization for green synthesis of silver nanoparticles by *Sclerotinia sclerotiorum* MTCC 8785 and evaluation of its antibacterial properties. Springerplus 5:861

Sellstedt A, Richau K, Sellstedt A, Richau KH (2013) Aspects of nitrogen-fixing Actinobacteria, in particular free-living and symbiotic Frankia. FEMS Microbiol Lett 342:179–186. https://doi.org/10.1111/1574-6968.12116

Shabatina TI, Vernaya OI, Shabatin VP, Melnikov MY (2020) Magnetic nanoparticles for biomedical purposes: modern trends and prospects. Magnetochemistry 6(3):30

Shankar SS, Ahmad A, Pasricha R, Sastry M (2003) Bioreduction of chloroaurate ions by geranium leaves and its endophytic fungus yields gold nanoparticles of different shapes. J Mater Chem 13:1822–1826

Sharma B, Tiwari S, Kumawat KC, Cardinale M (2023) Nanobiofertilizers as bioemerging strategies for sustainable agriculture development: potentiality and their limitations. Sci Total Environ 860:160476

Shinde BH, Inamdar SN, Nalawade SA, Chaudhari SB (2022) A systematic review on antifungal and insecticidal applications of bio-synthesized metal nanoparticles. Mater Today Proc 73:412–417

Shrestha S, Wang B, Dutta P (2020) Nanoparticle processing: understanding and controlling aggregation. Adv Colloid Interf Sci 279:102162

Sidhu AK, Agrawal SB, Verma N, Kaushal P, Sharma M (2025) Fungal-mediated synthesis of multimetallic nanoparticles: mechanisms, unique properties, and potential applications. Front Nanotechnol 7:1549713. https://doi.org/10.3389/fnano.2025.1549713

Sillen WMA, Thijs S, Abbamondi GR, Marzorati M, Weyens N (2020) Nanoparticle treatment of maize analyzedthrough the metatranscriptome: Compromised nitrogen cycling, possible phytopathogen selection, and planthormesis. Microbiome 8:127 https://doi.org/10.1186/s40168-020-00904-y

Simonin M, Richaume A (2015) Impact of engineered nanoparticles on the activity, abundance, and diversity of soil microbial communities: a review. Environ Sci Pollut Res 22:13710–13723. https://doi.org/10.1007/s11356-015-4171-x

Singh A, Singh NB, Hussain I, Singh H, Yadav V, Singh SC (2016) Green synthesis of nano zinc oxide and evaluation of its impact on germination and metabolic activity of *Solanum lycopersicum*. J Biotechnol 233:84–94. https://doi.org/10.1016/j.jbiotec.2016.07.010

Singh H, Du J, Yi TH (2017) Biosynthesis of silver nanoparticles using Aeromonas sp. THG-FG1.2 and its antibacterial activity against pathogenic microbes. Artif Cells Nanomed Biotechnol 45(3):584–590

Singh A, Gaurav SS, Shukla G, Rani P (2022) Assessment of mycogenic zinc nanofungicides against pathogenic early blight (*Alternaria solani*) of potato (*Solanum tuberosum* L.). Mater Today Proc 49:3528–3537

Singh NA, Narang J, Garg D, Jain V, Payasi D, Suleman S, Swami RK (2023) Nanoparticles synthesis via microorganisms and their prospective applications in agriculture. Plant Nano Biol 5:100047

Sirelkhatim A, Mahmud S, Seeni A, Kaus NHM, Ann LC, Bakhori SKM et al (2015) Review on zinc oxide nanoparticles: antibacterial activity and toxicity mechanism. Nano-Micro Lett 7:219–242

Spagnoletti FN, Spedalieri C, Kronberg F, Giacometti R (2019) Extracellular biosynthesis of bactericidal Ag/AgCl nanoparticles for crop protection using the fungus Macrophomina phaseolina. J Environ Manag 231:457–466

Srivastava S, Mehta CM, Rana M, Banik D, Rajput VD, Konstantinova E (2024) Chapter 2: Nanotechnology and soil health: current updates and future perspective. In: Nanotechnology applications and innovations for improved soil health. IGI Global, Rajpur, pp 24–42

Solanki A, Kim JD, Lee KB (2008) Nanotechnology for regenerative medicine: nanomaterials for stem cell imaging. Nanomedicine 3:567–578. https://doi.org/10.2217/17435889.3.4.567

Sunkar S, Nachiyar CV (2012) Biogenesis of antibacterial silver nanoparticles using the endophytic bacterium Bacillus cereus isolated from *Garcinia xanthochymus*. Asian Pac J Trop Biomed 2(12):953–959

Suriyaprabha R, Karunakaran G, Kavitha K, Yuvakkumar R, Rajendran V, Kannan N (2014) Application of silica nanoparticles in maize to enhance fungal resistance. IET Nanobiotechnol 8:133–137

Taha ZK, Hawar SN, Sulaiman GM (2019) Extracellular biosynthesis of silver nanoparticles from Penicillium italicum and its antioxidant, antimicrobial and cytotoxicity activities. Biotechnol Lett 41:899–914

Taniguchi N (1974) On the basic concept of nanotechnology. In: Proceedings of the international conferenceon production engineering, Tokyo, pp 18–23

Tarafdar JC, Raliya R (2013) Rapid, low-cost, and ecofriendly approach her for iron nanoparticle synthesis using *Aspergillus oryzae* TFR9. J Nanopart 2013:141274. https://doi.org/10.1155/2013/141274

Tarafdar JC, Raliya R, Rathore I (2012) Microbial synthesis of phosphorous nanoparticle from tricalcium phosphate using *Aspergillus tubingensis* TFR-5. J Bionanosci 6:84–89

Tariq M, Mohammad KN, Ahmed B, Siddiqui MA, Lee J (2022) Biological synthesis of silver nanoparticles and prospects in plant disease management. Molecules 27(15):4754

Thabet SG, Alqudah AM (2024) Unraveling the role of nanoparticles in improving plant resilience under environmental stress condition. Plant Soil 503:1–18

Thamilselvi V, Radha KV (2013) Synthesis of silver nanoparticles from *Pseudomonas putida* NCIM 2650 in silver nitrate supplemented growth medium and optimization using response surface methodology. Dig J Nanomater Biostruct 8:1101–1111

Tomah AA, Alamer ISA, Li B, Zhang J-Z (2020) Mycosynthesis of silver nanoparticles using screened *Trichoderma* isolates and their antifungal activity against *Sclerotinia sclerotiorum*. Nanomaterials. 10(10):1955

Tripathi S, Tiwari K, Mahra S, Victoria J, Rana S, Tripathi DK, Sharma S (2024) Nanoparticles and root traits: mineral nutrition, stress tolerance and interaction with rhizosphere microbiota. Planta 260:34. https://doi.org/10.1007/s00425-024-04409-y

Tyagi S, Tyagi PK, Gola D, Chauhan N, Bharti RK (2019) Extracellular synthesis of silver nanoparticles using entomopathogenic fungus: characterization and antibacterial potential. SN Appl Sci 1:1–9

Vanlalveni C, Ralte V, Zohmingliana H, Das S, Anal JMH, Lallianrawna S, Rokhum SL (2024) Review of microbes mediated biosynthesis of silver nanoparticles and their enhanced antimicrobial activities. Heliyon 10:e32333

Varshney R, Bhadauria S, Gaur MS, Pasricha R (2010) Characterization of copper nanoparticles synthesized by a novel microbiological method. JOM 62(12):102–104

Vasant GH, Raghav R (2024) Co-application of metal oxide nanoparticle (s) and plant growth promoting Rhizobacteria on the growth of groundnut plant (Arachis hypogaea L.). Doctoral dissertation, Department of Microbiology Faculty of Science Atmiya University

Velmurugan P, Shim J, You Y, Choi S, Kamala-Kannan S, Lee KJ, Kim HJ, Oh BT (2010) Removal of zinc by live, dead, and dried biomass of *Fusarium* spp. isolated from the abandoned-metal mine in South Korea and its perspective of producing nanocrystals. J Hazard Mater 182:317–324

Vigneshwaran N, Kathe A (2007) Silver-protein (core-shell) nanoparticle production using spent mushroom substrate. Langmuir 23:7113–7117

Vigneshwaran N, Ashtaputre NM, Varadarajan PV, Nachane RP, Paralikar KM, Balasubramanya RH (2007) Biological synthesis of silver nanoparticles using the fungus *Aspergillus flavus*. Mater Lett 61:1413–1418

Waghmare SR, Mulla MN, Marathe SR, Sonawane KD (2015) Ecofriendly production of silver nanoparticles using Candida utilis and its mechanistic action against pathogenic microorganisms. 3 Biotech 5:33–38

Wang L, Hu C, Shao L (2017) The antimicrobial activity of nanoparticles: present situation and prospects for the future. Int J Nanomedicine 12:1227–1249

Wang Q, Marconi M, Guan C, Wabnik K, Jiao Y (2022) Polar auxin transport modulates early leaf flattening. Proceedings of the National Academy of Sciences of the United States of America 119(50):e2215569119. https://doi.org/10.1073/pnas.2215569119

Wu JW, Ng IS (2017) Biofabrication of gold nanoparticles by *Shewanella* species. Bioresour Bioprocess 4:50. https://doi.org/10.1186/s40643-017-0181-5

Xu C, Peng C, Sun L, Zhang S, Huang H, Chen Y, Shi J (2015) Distinctive effects of TiO_2 and CuO nanoparticles on soil microbes and their community structures in flooded paddy soil. Soil Biol Biochem 86:24–33. https://doi.org/10.1016/j.soilbio.2015.03.006

Xu J, Luo X, Wang Y, Feng Y (2018) Evaluation of zinc oxide nanoparticles on lettuce (*Lactuca sativa* L.) growth and soil bacterial community. Environ Sci Pollut Res 25:6026–6035. https://doi.org/10.1007/s11356-017-0953-7

Xu X, Liang A, Li H, Shang H, Qian K, Jia W, White JC, Ma C, Xing B (2023) Foliar applied ZnO quantum dots boost pumpkin (*Cucurbita moschata* duch.) growth and positively alter endophytic and rhizosphere microbial communities. ACS Sustain Chem Eng 11:8503–8516. https://doi.org/10.1021/acssuschemeng.3c00954

Yadav SA, Suvathika G, Alghuthaymi MA, Abd-Elsalam KA (2023) Fungal-derived nanoparticles for the control of plant pathogens and pests. In: Fungal cell factories for sustainable nanomaterials productions and agricultural applications, pp 755–784

Yang L, Zhou Z, Song J, Chen X (2019) Anisotropic nanomaterials for shape-dependent physicochemical and biomedical applications. Chem Soc Rev 48(19):5140–5176

Yang H, Zhang X, Dang D, Zheng L, Yang Q, Yu Y (2021) Effects of iron oxide nanoparticles on CH_4 and N_2O emissions and microbial communities in two typical paddy soils. Chin J Appl Environ Biol 27:725–733. https://doi.org/10.19675/j.cnki.1006-687x.2020.07031

Yang Y, Waterhouse GI, Chen Y, Sun-Waterhouse D, Li D (2022) Microbial-enabled green biosynthesis of nanomaterials: current status and future prospects. Biotechnol Adv 55:107914

Yen KS, Sundar LS, Chao YY (2022) Foliar application of *Rhodopseudomonas palustris* enhances the rice crop growth and yield under field conditions. Plants (Basel) 11(19):2452. https://doi.org/10.3390/plants11192452

You T, Liu D, Chen J, Yang Z, Dou R, Gao X, Wang L (2018) Effects of metal oxide nanoparticles on soil enzyme activities and bacterial communities in two different soil types. J Soil Sediment 18(1):211–221. https://doi.org/10.1007/s11368-017-1716-2

Zehra SH, Ramzan K, Viskelis J, Viskelis P, Balciunaitiene A (2025) Advancements in Green synthesis of silver-based nanoparticles: antimicrobial and antifungal properties in various films. Nanomaterials (Basel). 15(4):252. https://doi.org/10.3390/nano15040252

Zhang X, He X, Wang K, Yang X (2011) Different active biomolecules involved in biosynthesis of gold nanoparticles by three fungus species. J Biomed Nanotechnol 7:245–254

Zhang Y, McKelvie ID, Cattrall RW, Kolev SD (2016) Colorimetric detection based on localized surface plasmon resonance of gold nanoparticles: merits, inherent shortcomings and future prospects. Talanta 152:410–422

Zhang J, Guo W, Li Q, Wang Z, Liu S (2018) The effects and the potential mechanism of environmental transformation of metal nanoparticles on their toxicity in organisms. Environ Sci Nano 5(11):2482–2499

Zhang H, Zhou H, Bai J, Li Y, Yang J, Ma Q, Qu Y (2019) Biosynthesis of selenium nanoparticles mediated by fungus Mariannaea sp. HJ and their characterization. Colloids Surf A Physicochem Eng Asp 571:9–16

Zhang H, Zheng T, Wang Y, Li T, Chi Q (2024) Multifaceted impacts of nanoparticles on plant nutrient absorption and soil microbial communities. Front Plant Sci 15:1497006. https://doi.org/10.3389/fpls.2024.1497006

Zhao X, Hou N, Wan C, Zhang L, Liu X (2024) Gold nanoparticles synthesis mediated by fungus isolated from aerobic granular sludge: process and mechanisms. Heliyon 10(6):e28281. https://doi.org/10.1016/j.heliyon.2024.e28281

Zheng K, Setyawati MI, Leong DT, Xie J (2017) Antimicrobial gold nanoclusters. ACS Nano 11:6904–6910

Zhou P, Adeel M, Shakoor N, Guo M, Hao Y, Azeem I, Li M, Liu M, Rui Y (2021) Application of nanoparticles alleviates heavy metals stress and promotes plant growth: an overview. Nanomaterials 11(1):26

Zhu QL, Xu Q (2016) Immobilization of ultrafine metal nanoparticles to high-surface-area materials and their catalytic applications. Chem 1(2):220–245

Zonaro E, Piacenza E, Presentato A, Monti F, Dell'Anna R, Lampis S, Vallini G (2017) *Ochrobactrum* sp. MPV1 from a dump of roasted pyrites can be exploited as bacterial catalyst for the biogenesis of selenium and tellurium nanoparticles. Microb Cell Factories 16:215

Chapter 6
Metal-Based Nanoparticles and Their Toxicity to Beneficial Soil Bacteria and Fungi

Olja Šovljanski, Tara Budimac, Anja Vučetić, Teodora Marić, and Ana Tomić

Abstract Metal-based nanoparticles (MNPs) are now widely used across industry, agriculture, and healthcare sectors thanks to their distinctive antimicrobial, catalytic, and physical-chemical characteristics. Nevertheless, their rising presence in land-based ecosystems—especially soil—has sparked concerns over potential environmental consequences. The most vulnerable targets are beneficial soil microorganisms, including bacteria and fungi, fundamental to nutrient cycling, organic matter decomposition, plant symbiosis, and overall soil health. This chapter comprehensively evaluates the ecotoxicological effects of MNPs on beneficial soil microbiota, emphasising the underlying mechanisms of toxicity, such as oxidative stress induction, membrane disruption, enzymatic inhibition, and genetic alterations. Special focus is given to key soil microbial groups that play pivotal roles in sustainable agriculture and ecosystem functioning. In addition to discussing the specific impacts of common MNPs such as AgNPs, CuNPs, ZnONPs, TiO_2NPs, and Fe_3O_4NPs, the chapter explores recent advances in mitigation strategies, including phytoremediation, bioremediation, and green nanoparticle synthesis. The role of regulatory frameworks and environmental risk assessment protocols is also examined, highlighting the need for more robust guidelines to ensure the safe integration of nanotechnologies into agricultural and environmental practices. Through this synthesis, the chapter contributes to a growing understanding of nanoparticle–microbe interactions and the need for environmentally conscious nanomaterial design.

Keywords Metal-based nanoparticles (MNPs) · Sustainable nanotechnology · Soil microbiota toxicity · Beneficial soil bacteria · Beneficial soil fungi

O. Šovljanski (✉) · T. Budimac · A. Vučetić · T. Marić · A. Tomić
University of Novi Sad, Faculty of Technology Novi Sad, Novi Sad, Serbia
e-mail: oljasovljanski@uns.ac.rs

K. A. Abd-Elsalam (ed.), *Plant-Microbiome Nanotechnology*, Sustainability Sciences in Asia and Africa, https://doi.org/10.1007/978-981-95-5056-2_6

6.1 Introduction

Nanotechnology has emerged as a transformative force across numerous industries, including agriculture, food processing, biomedical sciences, and environmental management. Among the most widely utilised nanomaterials are metal-based nanoparticles (MNPs). These nanoparticles possess unique physicochemical, catalytic, and antimicrobial properties, making them valuable in applications ranging from nano-fertilisers and pesticides to antimicrobial coatings and wastewater treatment agents (Fernandez-Triana 2024). However, the increasing production and use of MNPs raise significant environmental concerns, particularly regarding accumulation in soil ecosystems. Once in the soil, these materials can persist, interact with soil particles, and become bioavailable to microorganisms. Soil microbiota, especially beneficial bacteria and fungi, is critical for maintaining soil fertility, structure, and biological functions. They drive key ecological processes such as nutrient cycling, organic matter decomposition, symbiotic nitrogen fixation, and plant-microbe communication. The introduction of MNPs into the soil can adversely affect microbial communities through a variety of mechanisms. Moreover, prolonged exposure to MNPs can alter microbial diversity and community structure, potentially leading to ecological imbalance and reduced soil functionality (Abbas et al. 2024).

6.2 Types and Sources of Metal-Based Nanoparticles in Soil

MNPs enter soil ecosystems from various anthropogenic and natural sources. Their quantities in soil are difficult to estimate; however, as the production volume of metal nanoparticles reaches approximately 15,000–50,000 tons annually, their presence in soil is expected (Fernandez-Triana 2024). The massive production of MNPs will unavoidably lead to their accumulation in the environment, and their stability, reactivity, and mobility will determine their impact on microbial communities. Numerous additional parameters, such as charge, solubility, and binding affinity towards a biological target, can also influence their toxicity (Sharma et al. 2023b; Fernandez-Triana 2024). The antifungal, anticandidal, and antibacterial properties of MNPs and their derivatives indicate their highly hazardous nature and toxicity towards microorganisms. Further, MNPs show indirect toxicity due to their interaction with natural organic substances, which leads to the promotion of their toxicity (Sharma et al. 2023b). Still, numerous types of metal-based nanoparticles have been synthesised due to their benefits to the agricultural, electronic, medical, pharmaceutical, and chemical industries. Several prominent MNPs are widely used in nanotechnology, including silver, titanium, aluminium, silica, nickel, gold, zinc, and copper. Additionally, metal oxides have been used to produce nanoparticles, the most common being titanium dioxide, copper oxide, zinc oxide, iron oxide, silicon dioxide, and aluminium oxide, among others (Ameen et al. 2021). Representative metal-based nanoparticles are discussed in more depth below.

This chapter explores the environmental impact of metal-based nanoparticles (MNPs) on beneficial soil microorganisms. It highlights the sources, types, and widespread use of MNPs, particularly in agriculture, and examines their toxic effects on key soil bacteria and fungi. The chapter discusses mechanisms of toxicity, including oxidative stress, membrane damage, and genetic disruption, supported by recent experimental data. It also outlines mitigation strategies such as nanoparticle surface modification, phytoremediation, and sustainable nanotechnology. Finally, it emphasises the need for improved regulatory frameworks and long-term studies to ensure the safe application of MNPs in environmental and agricultural contexts.

6.2.1 *Types of Metal-Based Nanoparticles*

Silver nanoparticles (AgNPs) are used in textile, agricultural, electronics, and biomedical sectors due to their remarkable physicochemical properties (Abbas et al. 2024). Further, AgNPs are well known for their wide range of biocidal activity against Gram-positive and Gram-negative bacteria, fungi, and mycobacteria, and their combination with antibiotics even increases their efficacy. These characteristics originate from various complex reactions, such as producing reactive oxygen species (ROS), DNA damage, bacterial cell membrane degradation, and protein and amino acids synthesis suppression (Noga et al. 2023). However, AgNPs are unavoidably present in the environment due to their wide range of applications and growing market share. They are released not only during the use of products like cosmetics, wound dressings, food packaging, textiles, and personal care items but also during their production processes (Boersma et al. 2023; Rodrigues et al. 2024).

Copper nanoparticles (CuNPs) are inexpensive, readily available, and have characteristics like those of noble metals, supporting their enormous utilisation in various applications, such as in construction materials, electrical equipment, industrial processes, antibacterial agents, and the formation of composites with other metals (Din et al. 2017; Malhotra et al. 2020). CuONPs stand out due to their physical and chemical characteristics. However, they are regarded as dangerous and highly harmful pollutants, with water being one of the primary locations for accumulation in the natural environment. CuONPs in water have a greater transmission capacity and have wider effects than those in soil, especially when organic matter is present, as their stability and dispersibility in water can be further enhanced (Ban et al. 2021). In environmental settings, sulfidation of CuONPs changes their characteristics and may raise their apparent solubility, increasing their bioavailability and, consequently, their ecotoxicity (Rajput et al. 2020). Mechanisms of expressing ecotoxicology are related to the generation of ROS, which increases oxidative stress and affects cell viability by leaking intracellular proteins and sugars (Šovljanski et al. 2025).

Zinc oxide nanoparticles (ZnONPs) were primarily applied in the rubber industry, followed by uses in the cosmetic and medical sector, textile, agricultural, and electrotechnology industries. It possesses outstanding physical and chemical properties, such as high electrical, thermal conductivity and stability, UV blocking, antibacterial, and antimicrobial qualities, that justify its many utilisations (Al Jabri et al. 2022). Because of their many applications, 31,500–34,000 tons of ZnONPs are expected to be produced annually worldwide (Sun et al. 2021). The assumed toxicity mechanisms of ZnONPs include contact with biological targets, production of ROS, and dissolution of ZnONPs to zinc ions (Shen et al. 2016).

Titanium dioxide nanoparticles (TiO_2NPs) have extensive applications in the industry due to their exceptional optical, electrical, and antibacterial qualities and their combustible and chemically stable characteristics. Extensive use is notable in the following sectors: paint (51% of total production), plastic (19%), and paper (17%), as well as in food, agriculture, leather, cosmetic, medical, and pharmaceutical industries (Baranowska-Wójcik et al. 2020; Sungur 2021).

Iron oxide nanoparticles (Fe_3O_4NPs) can be employed in many biomedical applications, including imaging, drug administration, diagnostics, etc. Furthermore, Fe_3O_4NPs were identified as promising candidates for biological and technical applications because of their possible polymorphism and electron-hopping characteristics (Vallabani and Singh 2018). Besides biomedical applications, iron oxide nanoparticles can be used in agriculture, electronics, energy, and animal biotechnology (Ajinkya et al. 2020). The improvements in their utilisation of cutting-edge applications and technology were achieved by formulating reinforced active layers with polymers, inorganic metals, metal oxides, or bioactive compounds (Vallabani and Singh 2018). Their primary mechanism of Fe_3O_4NPs is the production of ROS, which raises lipid peroxidation, lowers antioxidant enzymes, and can cause protein aggregation (Gudkov et al. 2021).

Although the harmful effects of metal nanoparticles in the soil environment have not been thoroughly studied, individual studies of metal nanoparticles can be found (Table 6.1).

6.2.2 *Sources of MNPs in Soil*

One of the ways MNP contamination in soil comes from agricultural runoffs that can carry different MNPs used in the field (Shah et al. 2025b). According to Abbas et al. (2024), one of the possible sources of MNPs in soil is from industrial processes and streams, including solid and liquid waste from companies that use or produce nanoparticles. In water waste, the stability, solubility, and bioavailability of MNPs are impacted by their interactions with colloids such as humic substances and a wide range of ions and cations. The behaviours of MNPs may be significantly mediated by these interactions, which may impact their mobility, dissolution, and aggregation (Shah et al. 2025a). Moreover, treated sewage sludge contains residual nanoparticles with estimated concentrations of less than 10 mg/kg, which is also a substantial sink for MNPs in this respect.

Table 6.1 Comparative toxicity of various metal-based nanoparticles

Type of MNPs	Beneficial soil microorganism	Exposure	Toxicity	References
AgNPs	*Azotobacter vinelandii*	AgNPs: 10 and 50 nm	Apoptosis by 20.23% and 3.14%; Decreased cell number inhibited biological nitrogen fixation and ROS production	Zhang et al. (2018)
CuONPs	*B. megaterium AGNTL MM7*	CuONPs: approx. 188 nm, 0.1, 1, 5, 10, and 50 mg/ml	The minimal inhibitory concentration was 4 mg/ml	Tripathi et al. (2024)
CuNPs	Nitrification-, denitrification-, and C respiration-related soil microbiota	CuNPs: 100 mg/kg for 90 days	The denitrifier was the most susceptible to CuONPs	Simonin et al. (2018)
	B. subtilis MTCC 441 *P. fluorescens* MTCC 1749	CuNPs: 11–14 nm	The MBCs for *B. subtilis*: 40–70 μg/ml; The MBCs for *P. fluorescens*: 40–70 μg/ml	Sharma et al. (2023b)
TiO_2NPs	*Rhizobium leguminosarum*	TiO_2NPs: 250 mg/L	Morphological damage and nitrogen fixation were impeded	Fan et al. (2014)
	Rhizobium trifolii	TiO_2NPs: 100 nm, 18–24 mg/L	Reduced nitrogen fixation by 54%	Moll et al. (2016)
AgNPs	*Bradyrhizobium japonicum USDA110*	AgNPs: 10 μg/mL	No cell grows in 4 days of exposure	Boersma et al. (2023)
	Bradyrhizobium spp.	AgNPs: 20 nm in 10 mg/kg	Reduction of cell number	Peixoto et al. (2022)
	Arbuscular Mycorrhizal fungi (AMF) and *Rhizobium*	AgNPs: 8.1 nm, 1, 2.5, and 5 mM	1 mM AgNPs: AMF decreased; *Rhizobium* increased; 2.5 mM AgNPs: The harmful effects were not detected; 5 mM of AgNPs: AMF and Rhizobium decreased	Rahman et al. (2023)

(continued)

Table 6.1 (continued)

Type of MNPs	Beneficial soil microorganism	Exposure	Toxicity	References
ZnONPs	Soil microbial biomass	ZnONPs: 65.82 nm, 2.5 mg/kg	Reduce the soil bacterial count (1.2×10^6 CFU) compared to the control (2.1×10^6 CFU)	Verma et al. (2022)
		ZnONPs: 10–80 nm, 1 and 10 mg/g fresh soil	Inhibited microbial respiration, ammonification, dehydrogenase, and fluorescent diacetate hydrolase	Shen et al. (2015)
Fe_2O_3NPs		Fe_2O_3NPs: 2000 mg/kg of soil	After 180 days, nitrogen mineralisation efficiency decreased to 60%; bacterial and fungal CFU decreased to 79% and 89%, respectively	Rashid et al. (2017a, b)
		Fe_2O_3NPs: <50 nm, 1g/kg of soil	The bacterial concentration decreased by 42% and 19% in poultry and farmyard manure soil	Kamran et al. (2020)
ZnONPs, SiO_2NPs, TiO_2NPs CeO_2NPs	*Azotobacter,* P- and K-solubilising bacteria	NPs: 1 mg/g	TiO_2NPs decreased bacterial and enzymatic activity ZnONPs and CeO_2NPs suppressed thermogenic metabolism, decreased bacterial population, and enzymatic activity SiO_2NPs minimally increased bacterial activity	Chai et al. (2015)

Moreover, the estimated MNP concentrations in sludge-treated soils were 42, 662, and 31 μg/kg, respectively, for TiONPs, AgNPs, or ZnNPs (Fernandez-Triana 2024). Further, MNPs are detectable in the atmosphere due to emissions from the production of nanomaterials, construction and agricultural activities, and vehicular sources (Abbas et al. 2024; Singh et al. 2024). The MNP's small size, less than 100 nm, makes it hard to remove from the air. Utilisation of MNPs in the textile industry is also a potential source of soil contamination due to leaking from materials during washing, especially silver and titanium dioxide nanoparticles (Pacheco and Buzea 2020). Also, MNPs in soil can be found due to waste leaking from landfills and other waste disposal locations (Shah et al. 2025). Even though the air and water are carriers of MNPs, the principal recipient in the environment is the soil, as migration of MNPs in the soil is limited (Sun et al. 2022). Penetration of MNPs in the soil allows interactions with other solid constituents, including clay particles, aggregates, and soil organic matter, via adsorption, cation exchange, and fixation processes. However, such interactions limit the MNPs' impact on soil organisms and their functions by reducing their mobility in the soil matrix (Shah et al. 2025a).

6.3 Beneficial Soil Bacteria and Fungi Affected by MNPs

Beneficial soil bacteria and fungi (Fig. 6.1) are vital for maintaining soil health, nutrient cycling, organic matter decomposition, and plant productivity. These microbes form intricate symbiotic relationships with plants, contributing to natural pest control and stress tolerance (Chen et al. 2024a, b). Key microbial groups' loss or functional inhibition can cascade into larger ecosystem dysfunctions, including reduced nutrient availability, impaired plant growth, compromised soil resilience, etc. (Markowicz 2023).

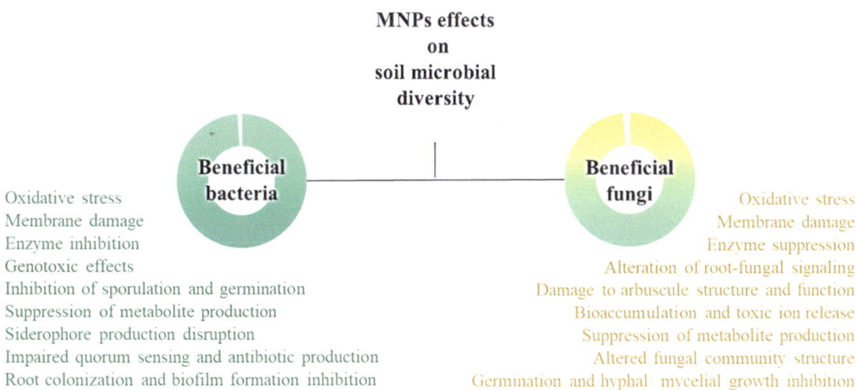

Fig. 6.1 Metal-based nanoparticles (MNPs) impact on soil microbial diversity

6.3.1 Effects on Beneficial Bacteria

The functional capacity of beneficial soil bacteria, such as *Rhizobium*, *Pseudomonas*, and *Bacillus* (Fig. 6.2), is highly susceptible to nanoparticle-induced stress (Mgadi et al. 2024). Across multiple studies, MNPs have been shown to exert toxic effects that compromise bacterial cell integrity, reduce enzymatic activity, disrupt biofilm formation, and inhibit key symbiotic processes. Suppressing these beneficial bacterial populations due to MNP contamination can result in a cascade of adverse outcomes, including reduced soil fertility, increased vulnerability to plant pathogens, and lower crop yields.

Rhizobacteria convert atmospheric nitrogen into ammonia through biological nitrogen fixation, making nitrogen bioavailable to plants. In exchange, the host plant supplies carbohydrates and niche protection. This mutualistic relationship significantly reduces the need for synthetic nitrogen fertilisers and enhances soil fertility (Fahde et al. 2023; Zhu et al. 2023). The rhizobia disruption by MNPs begins at the early stages of rhizobia-legume communication. Nanoparticles can interfere with the flavonoid-nod factor signalling cascade, which is crucial for initiating root hair curling and nodule formation. As a result, reduced or malformed nodules are commonly observed in MNP-contaminated soils (Haskett et al. 2025). The underlying mechanisms by which MNPs affect rhizobia include the induction of ROS in rhizobial cells, damaging membranes, proteins, and nucleic acids (Yeremko et al. 2025).

Additionally, nanoparticles adhere to and penetrate bacterial membranes (More et al. 2023). Metal ions released from MNPs interact with key functional groups in bacterial enzymes, leading to their deactivation or reduced expression (Slavin et al. 2017). MNPs have also been associated with DNA fragmentation and mutagenic alterations in rhizobia, potentially affecting long-term adaptation and symbiotic competence (Hunter et al. 2007; Wang et al. 2024).

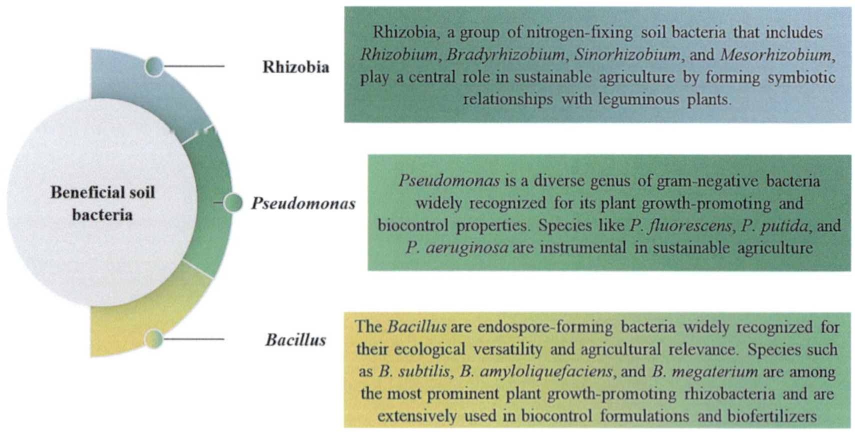

Fig. 6.2 Representatives of beneficial soil bacteria

The beneficial traits connected to *Pseudomonas* species (produce antibiotics, secrete siderophores to sequester iron, solubilise phosphate and produce plant hormones, trigger systemic resistance in plants, and suppress soil-borne pathogens through competition and antibiosis) are significantly compromised when bacteria are exposed to MNPs. Namely, siderophores are high-affinity iron-chelating compounds secreted by *Pseudomonas* to survive in iron-limited environments and outcompete pathogens. AgNPs and CuNPs inhibit siderophore biosynthesis by interfering with iron homeostasis and redox signalling pathways (Xie et al. 2024). Quorum sensing is vulnerable to oxidative stress and membrane damage induced by MNPs. CuNPs and AgNPs have been shown to downregulate QS genes in *P. aeruginosa*, leading to a loss of antibiotic production and reduced suppression of fungal pathogens like *Fusarium* and *Rhizoctonia* (Abo-Zaid et al. 2020).

Furthermore, nanoparticle-induced morphological alterations in *Pseudomonas* cells, including membrane perforations, cytoplasmic leakage, and cell shrinkage, are primarily caused by ROS and nanoparticle adhesion to the bacterial envelope, leading to loss of cell viability (Parra-Ortiz and Malmsten 2022). Effective plant root colonisation by *Pseudomonas* involves robust biofilm formation and motility. AgNPs and TiO_2NPs can inhibit these processes by disrupting flagellar motility, altering EPS (extracellular polymeric substances) composition, and downregulating genes involved in adhesion (García et al. 2025).

The ability of *Bacillus* bacteria to form robust biofilms, secrete a wide array of enzymes and metabolites, and survive in harsh conditions through sporulation makes them key players in maintaining their multiple roles in ecological niches (Šovljanski et al. 2021). Recent evidence indicates that MNPs can negatively impact *Bacillus* populations by interfering with critical physiological functions, potentially diminishing their role in agroecosystems. Key functional roles of *Bacillus* in soil health include facilitating root colonisation, enhancing stress tolerance, creating protective niches for other beneficial microbes, increases nutrient availability in the rhizosphere, as well as lipopeptides suppress pathogens like *Fusarium*, *Rhizoctonia*, and *Pythium*. Additionally, it triggers plant defence pathways, improving tolerance to both biotic and abiotic stressors (Russo et al. 2023). Biofilms are structured microbial communities embedded in extracellular polymeric substances (EPS), crucial for microbe communication (Šovljanski et al. 2022a, b; Dujković et al. 2023). AgNPs and TiO_2NPs significantly impair *Bacillus* biofilm development by disrupting quorum sensing and EPS secretion.

On the other hand, sporulation is a key survival mechanism for *Bacillus* in adverse environments (Šovljanski et al. 2022a, b, 2024a, b), but MNPs can interfere with sporulation pathways, rendering the bacteria more susceptible to stress. Like other soil bacteria, *Bacillus* species experience nanoparticle-induced oxidative stress, resulting in ROS accumulation and subsequent damage to DNA, proteins, and lipids. Many analyses have revealed cellular deformation, membrane disintegration, and cytoplasmic shrinkage in *Bacillus* cells exposed to CuNPs and AgNPs. MNPs also inhibit the biosynthesis of secondary metabolites, including the antimicrobial lipopeptides that contribute to *Bacillus*'s biocontrol efficacy (Montes de Oca-Vásquez et al. 2020).

6.3.2 *Effects on Beneficial Fungi*

Beneficial soil fungi include symbionts such as arbuscular mycorrhizal fungi (AMF), biological control agents like *Trichoderma*, and decomposers known as saprophytic fungi (Fig. 6.3). Together, they enhance nutrient acquisition, suppress soil-borne pathogens, and decompose organic matter—functions essential to agroecosystem resilience. Many MNPs have broad-spectrum antimicrobial properties, meaning they often do not distinguish between pathogenic and beneficial fungi, disrupting essential soil functions (Chaudhary et al. 2025).

Arbuscular Mycorrhizal Fungi (AMF) colonise plant roots by forming highly branched arbuscules and vesicles, facilitating bidirectional nutrient exchange between the plant and fungal partner. However, recent studies have shown that exposure to MNPs can disrupt this finely tuned symbiosis and compromise AMF functionality (Berruti et al. 2016). ZnONPs and CuNPs have been shown to inhibit spore germination and mycelial extension, reducing colonisation potential. These effects are often mediated by oxidative stress, where excessive generation of ROS damages cellular organelles and DNA in fungal spores (Yang et al. 2023). MNPs disrupt the production or perception of signalling molecules, which are essential for root colonisation by AMF. CuNPs can downregulate strigolactone synthesis in roots, impairing AMF recruitment (Xia et al. 2025). Once inside the root cortex, AMF forms arbuscules, the sites of nutrient exchange. MNPs can accumulate in root tissues and disturb arbuscule integrity, leading to malformed structures or early degeneration. Also, fungal hyphae may take them up, causing internal toxicity and disrupting metal homeostasis within fungal cells (Yang et al. 2023). The impairment of AMF by MNPs translates into various adverse outcomes at the plant and ecosystem levels through decreased colonisation and limited phosphorus and micronutrient uptake, which are critical for photosynthesis and root development. AMF-impaired plants often exhibit stunted

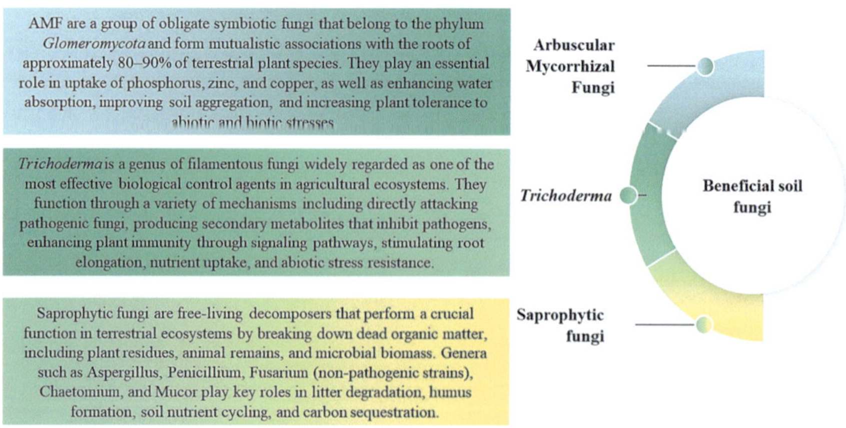

Fig. 6.3 Representatives of beneficial soil fungi

growth, lower chlorophyll content, and reduced drought tolerance. Reduced AMF activity leads to lower glomalin production and increased soil erosion, etc. (Wang et al. 2022).

Due to their widespread application as biopesticides and biofertilisers, *Trichoderma* is considered environmentally safe and beneficial. However, increasing exposure to MNPs raises concerns about unintended impacts on their beneficial roles (Guzmán-Guzmán et al. 2023). Exposure to AgNPs and CuNPs has been shown to inhibit *Trichoderma* spore germination and hyphal development significantly. The proposed mechanism involves ROS generation, leading to oxidative damage to cell membranes, DNA, and proteins. *Trichoderma*'s biocontrol activity relies heavily on the secretion of extracellular enzymes like chitinases, β-1,3-glucanases, and proteases. MNPs interfere with the transcription of genes encoding these enzymes, likely due to oxidative damage and metal ion stress (Zaki et al. 2021). Membrane blebbing, cytoplasmic condensation, and disrupted organelles, including the nucleus and mitochondria, result in reduced viability and functional impairment. AgNPs also impair mitochondrial respiration, limiting the energy available for growth and enzyme production (Takáč et al. 2023). Volatile organic compounds (VOCs) produced by *Trichoderma* contribute to pathogen suppression and plant growth stimulation. However, nanoparticle exposure downregulates the genes involved in VOC biosynthesis (Lee et al. 2016). The impairment of *Trichoderma* functionality by MNPs has significant downstream effects: increased disease pressure (reduction in mycoparasitic activity allows pathogens like *Rhizoctonia solani* and *Sclerotium rolfsii* to proliferate unchecked), decreased crop protection (without biocontrol activity, plants become more vulnerable to both root and foliar diseases), soil microbial imbalance (the suppression of *Trichoderma* can shift rhizosphere community dynamics in favour of opportunistic or less beneficial fungi), and loss of enzyme activity affects organic matter decomposition and nutrient mineralisation processes in the rhizosphere (Abbas et al. 2022).

Saprophytic fungi secrete an array of extracellular enzymes that catalyse the decomposition of complex polymers like cellulose, hemicellulose, and lignin, facilitating the mineralisation of organic matter into bioavailable nutrients. In agricultural soils, their activities are central to maintaining soil fertility, texture, and structure (Viret and Gindro 2024). Saprophytic fungi rely on their enzyme systems to degrade complex organic material. MNPs, particularly ZnONPs and CuNPs, bind to enzyme-active sites or alter their structure via metal ion interference, reducing enzyme activity (Kaur et al. 2024). The MNPs adhere to fungal membranes, increasing permeability and causing cytoplasmic leakage. This stress results in reduced growth, lower sporulation rates, and, in some cases, fungal death (Ogar et al. 2015). Chronic exposure to MNPs can selectively suppress sensitive fungal taxa while allowing resistant but often less functionally effective species to dominate (González-Fernández et al. 2025).

6.4 Main Mechanisms of Toxicity to Beneficial Soil Bacteria and Fungi

The destructive effect of MNPs on beneficial soil microbes may be due to one or more different mechanisms (Fig. 6.4), such as changes in cell surface morphology and growth behaviour, disruption of the cell membrane, oxidative stress, lipid peroxidation due to oxidative phosphorylation, destruction of enzyme and protein activity, DNA damage etc. (Ahmed et al. 2021).

6.4.1 Oxidative Stress and Reactive Oxygen Species Generation

Reactive oxygen species include molecules and reactive intermediates with strong positive redox potential. MNPs produce different types of ROS by reducing oxygen molecules (Šovljanski et al. 2025). Superoxide radicals (O^{-2}) and hydrogen peroxide (H_2O_2) have been shown to cause less acute stress reactions. Endogenous antioxidants can neutralise them, whereas hydroxyl radicals (·OH) and singlet oxygen (O_2) can lead to acute microbial death. With excessive production of ROS, the cell redox balance favours oxidation, which leads to oxidative stress (Wang et al. 2017). The ROS generation and induction of oxidative stress is described as one of the main nanotoxicity mechanisms responsible for damaging the cell membrane and cellular components such as lipids, proteins, and DNA, ultimately leading to microbial death (Sharma et al. 2023a). Studies suggest beneficial soil bacteria such as *Rhizobium* and *Pseudomonas* are particularly susceptible to oxidative stress induced by MNPs. Fan et al. (2014) investigated the effects of TiO_2 on *Rhizobium*-legume symbiosis and found an impact on *R. leguminosarum* bv. *viciae* 3841 strain, as well

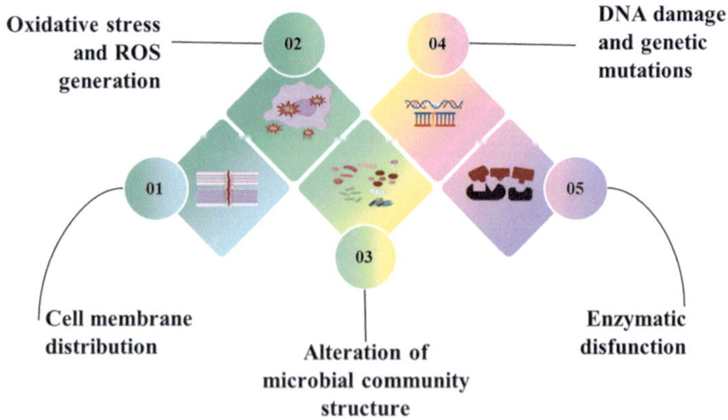

Fig. 6.4 Mechanisms of MNP toxicity in soil bacteria and fungi (ROS reactive oxygen species, DNA deoxyribonucleic acid)

as the decline in pea-*R. leguminosarum* symbiosis when exposed to this MNP, possibly due to stress caused by the generation of hydroxyl radicals. Lu et al. (2023) investigated the effect of MNPs, TiO_2, Al_2O_3, and ZnO on beneficial rhizosphere bacteria *Pseudomonas stutzeri* A1501, with ZnO having the most influence on the nitrogen fixation capacity of this bacteria. They concluded that these MNPs could cause the explosion of intracellular ROS that could further affect the membrane's permeability and inhibit the expression of nitrogenase synthesis-related genes and biofilm formation on the root surface, thus damaging the bacteria-rice system.

Similarly, CuNPs can produce all four previously mentioned ROS types, making them particularly dangerous to beneficial soil bacteria. *Bacillus subtilis* and *Pseudomonas fluorescens* showed higher ROS levels than untreated cells when exposed to CuNP, with higher concentrations leading to higher levels of ROS. Further, the ROS levels generated in Gram-negative *P. fluorescens* were higher than those measured in Gram-positive *B. subtilis*. Since there are differences in the cell membrane structures of these two bacteria, differential internalisation of CuNP is also present, as reported by Sharma et al. (2023a). Zhang et al. (2018) demonstrated that, among other mechanisms, AgNPs cause apoptosis of *Azotobacter vinelandii* cells by ROS generation and oxidative stress. AgNPs were internalised in bacterial cells, and their presence caused intracellular ROS production and generation of hydroxyl radicals.

6.4.2 Cell Membrane Disruption

MNPs attach to microbial cell membranes, which serve as a protective barrier from surrounding stressful environments, thus triggering structural changes in it and altering the permeability of the cell membrane, which ultimately leads to cell death either by immediate lysis or by disturbing cellular metabolism (Singh et al. 2016). Many studies reported that MNPs exhibit better activity against Gram-positive bacteria than against Gram-negative bacteria since the cell wall of Gram-negative bacteria consists of lipopolysaccharides and outer and inner cell membranes. However, the thickness of the membrane is less than that of Gram-positive bacteria (Šovljanski et al. 2024a, b). Ansari et al. (2014) images reveal the mechanism behind the action of AgNPs against Gram-negative bacterial cells. Namely, these NPs attach to the surface of the bacterial cell wall first, then permeate the cell membrane and enter the cell interior, which leads to significant gaps and pits in the cell membrane, leaving bacterial cells disorganised into several parts. Ahmed et al. (2020) concluded that when cultured in the presence of AgNPs and ZnONPs, the surface of beneficial soil bacteria *Bacillus thurigiensis*, *Pseudomonas mosselii*, *Sinorhizobium meliloti*, and *Azotobacter chroococcum* cells presented with visible damage and adsorbed NPs with small or bigger aggregates, while the surfaces of control cells were transparent, with standard cell shape. The NP-treated bacterial cells presented many gaps in cellular facets and fragmented and disorganised cell envelopes. Further, a fraction of NPs may have even reached the cytoplasm and interacted with the other cellular components, increasing membrane fluidity and causing leakage of cytoplasmic contents. Banik and Pérez-de-Luque (2017) examined the impact of copper NPs on two beneficial soil microorganisms: the fungus *Trichoderma harzianum* and bacteria *Rhizobium* spp. Their findings indicated that the mycelial growth of *T. harzianum* was entirely inhibited at a CuNP concentration of 1000 mg/L—the three *Rhizobium* spp. Strains exhibited varying responses to CuNP treatments, possibly due to physiological differences. Specifically, *R. meliloti* experienced complete inhibition at 200 mg/L CuNP, while *R. etli* and *R. tropici* showed substantial colony growth under the same conditions.

6.4.3 Enzymatic Dysfunction and Metabolic Inhibition

MNPs can also impair the metabolic functioning of cells, which may result in irreversible damage by interfering with enzyme activity essential for microbial respiration, nitrogen fixation, and organic matter decomposition. Microbial enzymes play an important role in transforming and mineralising unavailable forms of nutrients present as a part of soil organic matter to available forms. They are responsible for managing soil toxicity and biotransformation reactions of pollutants (Sharma et al. 2023a). Pulikova et al. (2024) observed a sharp decline in denitrification and dehydrogenase activity, the most significant representative of microbial activity when exposed to ZnO and CuO in floodplain soils. Huang et al. (2023) noticed that AgNPs also exhibit evident toxicity on soil enzymes, of which dehydrogenase presented with the most sensitivity and suppression of nitrogen transformation. Grün et al. (2018) tested the long-term effects of exposure to AgNPs in microbial soil communities over 1 year. They concluded that the tested AgNP concentrations significantly decreased the soil microbial biomass by affecting leucine aminopeptidase activity and caused severe disadvantages by influencing nitrogen cycling. Bouguerra et al. (2022) aimed to assess the impact of CoNPs on beneficial soil microorganisms by measuring the following soil enzyme activity: dehydrogenase, acid phosphatases, arylsulphatase, CM-cellulase, and urease, as well as the nitrogen mineralisation, soil microbial biomass carbon, and the soil basal respiration at a range of different concentrations. The results showed that the activities of arylsulphatase, acid phosphatase, and soil basal respiration were the most affected. Conversely, some stimulatory effects were observed in carbon, nitrogen cycling, and basal respiration activities. While microbial biomass carbon was not significantly affected, it showed a trend towards an increase with rising concentrations. The inhibition of enzyme activity by MNPs can occur when metal ions bind to the enzyme's active site or interact with its protein components. This leads to a conformational change that reduces the enzymes' ability to bind their substrate (Chhipa 2021). Jośko et al. (2014) observed that the surrounding clay or mineral in soil absorbed microbial extracellular enzymes, such as phosphatase and urease, thus protecting them from the interaction with MNPs and reducing the toxicity of the microbial enzyme activity.

6.4.4 DNA Damage and Genetic Mutations

The most important element regarding genotoxicity is the production of ROS, which, as previously mentioned, can be initiated by MNPs directly or indirectly. Oxidative stress could lead to various forms of DNA damage, such as lesions in bases and sugars, single or double-strand breaks, DNA-protein crosslinks, formations of abasic sites, and mutations in soil fungal and bacterial genomes (Sathyamoorthy et al. 2025). All of this could further impact microbial survival and adaptation mechanisms in soil (Wang et al. 2017). MNPs bind to sulfur and phosphorous-containing membranes, cellular proteins, or biomolecules, including DNA, which further adversely affects cells. For example, ZnONPs cause significant DNA damage in bacterial cells by ROS interaction with DNA and the production of chain breaks in nitrogenous bases and carbohydrates via nitration, methylation, deamination, or oxidation reaction (Singh et al. 2020). Sharan et al. (2023) concluded that when exposed to nickel, manganese, and cobalt oxide NPs, an environmentally relevant bacterial species, *Shewanella oneidensis*, presents with ROS-containing membrane vesicles, as well as an increased tendency to generate random DNA mutations and retain other DNA damage. When exposed to TiO_2, a reduction in the number of RNAs in the RNA/DNA ratio and compression, degeneration, and fragmentation of DNA in bacterial cells was noticeable, making them unable to perform biological functions and reducing the physiological activity of genes, which was the primary cause of cell death (Zhukova 2015). Additionally, whole-genome analysis can be utilised to explore the molecular mechanisms of bacterial apoptosis. Su et al. (2015) employed this method to investigate the mechanism of action of ZnONPs against *E. coli* DNA. Although *E. coli* is not a beneficial soil bacteria, the mechanisms of action could be similar. Namely, results revealed that mutant genes are concentrated in 10 regions of the bacterial genome, including those involved in gene expression, molecular structure-activity, ribosome composition, and RNA modification.

6.4.5 Alteration of Microbial Community Structure

Lastly, MNPs introduced in the environment accumulate in the soil and affect the characteristics of native inhabitants. Prolonged exposure to MNPs shifts microbial community composition, reducing beneficial populations while favouring more resistant species (Ameen et al. 2021). For example, the arbuscular mycorrhizal fungus *Glomus caledonium* exhibited higher ZnO toxicity tolerance than *G. versiforme*, resulting in root colonisation. This increased colonisation was attributed to *G. caledonium*'s enhanced resistance to metals like zinc, copper, lead, and cadmium (Wang et al. 2016). Also, a substantial shift in bacterial community composition was observed, where beneficial soil *Acidobacteria* and *Verrucomicrobiota* that have a role in the degradation of plant-derived organic matter (Rakitin et al. 2024) were decreased, and the *Proteobacteria phylum* was increased (Parada et al. 2019). Additionally, exposure to ZnONPs and CeO_2NPs affected the populations of nitrogen-fixing bacteria, such as *Azotobacter*, as well as phosphorus- and potassium-solubilising bacteria, leading to a reduction in enzymatic activities (Rajput et al. 2018). On the other hand, Afzal and Singh (2022) investigated the impact of ZnO and FeO on the microbial community structure within a rice-soil-microbial ecosystem. They noted that the most significant changes in microbial composition were observed in the abundance of three phyla: *Proteobacteria*, *Actinobacteria*, and *Planctomycetes*. While certain beneficial microbial groups, such as those involved in streptomycin production, carbon and nitrogen fixation, and lignin degradation, were increased in the rhizosphere, other key microbial functions related to chitin degradation, ammonia oxidation, and nitrite reduction were negatively affected.

6.5 Strategies for Mitigating MNP Toxicity

The immense production and uncontrolled discharge of MNPs from various industries threaten their accumulation in soil, negatively affecting the beneficial soil microbiota. Therefore, the development of efficient MNP mitigation strategies is a high priority. So far, the most popular approaches for regulating MNP-mediated toxicity towards the soil ecosystem involve coating nanoparticles, modifying their surface chemistry, and encapsulation (Abishagu et al. 2024).

6.5.1 Standard MNP Mitigation Strategies

Applying surface coating to MNPs reduces their toxicity, enhancing biocompatibility and facilitating cellular uptake. In contrast, surface chemistry alterations modify nanoparticles' charge density and hydrophobicity, reducing toxicity and increasing their efficacy in a particular application. Meanwhile, covalently attached functional groups to nanoparticle surfaces significantly influence their properties and toxicity levels. Alternatively, encapsulating metal nanomaterials within polymers, liposomes, or lipid-based structures enhances antimicrobial activity, lowers toxicity, and potentially improves delivery strategies (Mubeen et al. 2023). Some additional mitigation strategies involve forming metal-organic frameworks (MOF) that are considered innovative nanoparticle carriers with toxicity-reducing abilities. MOFs are composed of metal components and organic ligands, often aromatic acids. They possess a large surface area, extensive contact, high pore volume, and density while preserving non-toxicity, cohesiveness, and small size. The toxicity of metals such as zinc, zirconium, magnesium, and iron was significantly reduced when incorporated into MOFs (Manuja et al. 2021). Various physical, chemical, and biological remediation techniques have the potential to decontaminate an excess of accumulated MNPs in soil. They work by reducing the bioavailability and toxicity of metal nanoparticles in question. Besides bioremediation and phytoremediation, MNPs can be discarded using soil washing or electrokinetic (Padhye et al. 2023). Among various strategies investigated in the literature to mitigate MNP toxicity, several are summarised in Table 6.2.

6.5.2 Advanced Phytoremediation and Bioremediation Techniques

Since not many studies are reporting MNP accumulation in the soil as a burning issue for its microbiome, there is still not much research regarding their detoxification using innovative remediation techniques, especially in comparison with the vastly investigated remediation of heavy metals (Karnwal et al. 2024; Navina et al. 2024). This could also be connected with phytoremediation's compromised efficacy in the absorption of metal presented in nanovesicle form. Even though plant-to-metal interactions are well documented, there are still limitations on the exact plant interactions with MNPs (Fernandes et al. 2017). On the other hand, plenty of studies incorporate MNPs with bioremediation to reduce heavy metals from the soil ecosystem (Prakash and Smitha 2023; Ambaye et al. 2025; Gomez et al. 2025). Still, there are some examples of successful phytoremediation and bioremediation of MNP. Fernandes et al. (2017) investigated the potential of *Phragmites australis*, a common salt marsh plant, for the phytoremediation of AgNPs from damp soil and its impact on microbial communities in the sediment around the plant roots. The results concluded that *P. australis* can accumulate silver from the medium even in the form of nanoparticles, demonstrating its potential for phytoremediation of silver-contaminated environments. Still, accumulation was observed only efficiently without rhizosediment (Fernandes et al. 2017). The review article by Cota-Ruiz et al. (2018) extensively analyses nanomaterial phytoremediation from soil. Among the plants identified for their ability to remediate metal nanoparticles are *Arabidopsis thaliana*, *Lactuca sativa*, *Medicago sativa*, and *Prosopis juliflora-velutina*, which respectively accumulated TiO_2NPs, CuNPs, ZnONPs, and CeO_2NPs. Their phytotoxicity can cause a lack of success for phytoremediation efforts on nanomaterials. One example of overcoming this problem is utilising phosphorus supplementation to mitigate toxicity caused by ZnONPs in soil and decrease the sensitivity of plants *Triticum aestivum* and *Solanum lycopersicum*, as well as their root microorganisms (Yadav et al. 2024).

Fusarium solani was successfully utilised as a mycoremediation agent for the accumulated silver ions in the soil. The mechanism by which fungi reduce the toxicity of silver ions is connected to the increased production of oxalic acid. The variation in biosorption of silver ions depends on the specific biomass treatment, metal concentrations, temperature, pH value, and contact time (El Sayed and El-Sayed 2020). The synergistic relationship between arbuscular mycorrhizal fungi and the host plant and their beneficial interaction with different MNPs, like AgNPs,

Table 6.2 Mitigation strategies for reducing MNP impact on soil microbes

Mitigation strategy	Mechanism of action	Used materials	Metal NP	References
Surface coating	Allows controlled release of MNPs while limiting uptake into healthy tissues	Polymer micelles	AgNPs	Mohammadzadeh (2012)
	Functional groups on PEG significantly influence toxicity; PEG-NPs modified with thiol showed no cytotoxicity	Polyethylene glycol (PEG)	AuNPs	Reznickova et al. (2019)
	PEG coating provided colloidal stability and water dispersibility with reduced cytotoxicity while PA and HA helped with biocompatibility	Phosphonic acid (PA), hydroxamic acid (HA), PEG	Fe_3O_4NP	Patsula et al. (2019)
	PEG-coated MNPs demonstrated lower hepatotoxic outcomes compared to non-coated NPs	PEG	AuNP	Patlolla et al. (2019)
	The toxicity of MNPs could be modified based on the hydrophobicity and density of their surface coatings	Silane with different hydrophobicity	ZnONP	Lai et al. (2021)
	PVP-NP formed no aggregates, were more stable, and preserved their colloidal stability with lower toxicity	Polyvinyl-Pyrrolidone (PVP)	AgNP	Rónavári et al. (2021)
	Plant-based glucan was a great reducing agent with good capping ability, which helps reduce the toxicity of MNPs	Glucan extracted from *Pleurotus* spp.	AgNP	Irshad et al. (2024).
Surface coating and ozonation	PVP-NPs showed lower toxicity. Still, ozonation was used to remove PVP-INPs by degrading the PVP coating. The concentration of MNPs and pH had a greater effect on the stability than the ozonation time	PVP	Fe_3O_4NP	Amiri et al. (2025)
Surface modification	Reduces the formation of protein coronas on MNPs, lowering cytotoxicity and enhancing cellular compatibility	Polyethylene glycol (PEG)	ZnONPs	Luo et al. (2015)
	Blocks the release of toxic ions and prevents direct contact between bacteria/cells and the nanoparticle surface	SiO_2	AgNPs	Wei et al. (2015)
	Reduces toxicity to healthy cells and improves suitability for drug delivery (doxorubicin-loaded nanoparticles)	Chitosan	AuNPs	Laksee et al. (2018)
Polymer stabilisation	Provides biocompatibility through rubber particles that act as reducing agents and impart colloidal stability	Karaya gum/rubber particles	AuNPs	Zeng et al. (2014)
Encapsulation of MNPs	Facilitates co-encapsulation of antioxidants and MNPs and improves biocompatibility of the matrix	Chitosan; plant extract	AgNPs	Nayak et al. (2016)
MOF	The electron density of PtNPs was increased by creating a complex, functional, and stable environment	Bimetallic MOF with Ce and Zr atoms	PtNP	Luo et al. (2025)

TiO_2NPs, and ZnONPs, offers an interesting combined approach to metal detoxification. Since inoculation with AMF has been proven to increase the uptake of elements found in low concentrations in the soil, it helps the host plant facilitate the remediation of accumulated MNPs (Ameen et al. 2021). Similarly, the symbiosis between nitrogen-fixing bacteria and *Medicago truncatula* provided plants resilience against different treatments of silver nanoparticles added in a growing medium (Judy et al. 2016). This is still relatively uninvestigated, but initial studies offer promising potential for an innovative mitigation concept.

6.5.3 Sustainable Nanotechnology

Sustainable nanotechnology is regarded as a novel approach to mitigating the toxicity of metal nanoparticles to the ecosystem. It encompasses green synthesis as an alternative to traditional nanoparticle production by employing natural metal-reducing agents from yeast, fungi, bacteria, and plant extracts. Using biological materials ensures a non-toxic and non-pathogenic process safe for humans and the environment. Unlike conventional chemical and physical methods, often reliant on hazardous substances and harsh conditions, green nanotechnology promotes cleaner production, minimising environmental risks (Travičić et al. 2024; Travičić et al. 2025. Plant-based nanoparticles, in particular, are considered less likely to cause harmful effects in humans while offering broad practical applications. Natural plant extracts of numerous species possess a rich phytochemical content (Vučetić et al. 2025), which can serve as both reducing and stabilising agents in producing MNPs. When exposed to metal salts, plant extracts abundant in polysaccharides, vitamins, amino acids, proteins, saponins, alkaloids, terpenes, and phenolics cause the reduction of metal ions to their respective nanoparticles. Although many phytochemicals are known to contribute to the biosynthesis of MNPs, the precise mechanism of their interactions remains undefined. For example, alkaloids serve as reducing agents, while terpenoids and flavonoids mainly act as capping and stabilising agents.

Meanwhile, proteins and carbohydrates contribute to a reduction of metal salts, as well as the stabilisation of the formed metal nanoparticles. Different environmental factors also influence MNP biosynthesis, including pH value, temperature, concentration of metal ions, and reaction time (Edo et al. 2025). From what is currently known, the biosynthesis process is governed by two main mechanisms: redox reactions and ligand substitution. Phytochemicals in plant extracts function as redox mediators by donating electrons to metal ions. This electron transfer converts the ions into a neutral atomic form at the nanometre scale, initiating nucleation and forming the primary nanoparticles. Moreover, the transformation efficiency is affected by how strongly metal ions bind to intracellular reductive proteins and metabolites (El-Seedi et al. 2024).

Recent literature demonstrates the effectiveness of plant-based extracts for the green synthesis of nanoparticles. For example, both tea waste from *Camellia sinensis* and the aqueous flower extract of *Murraya paniculata* (L) Jack both served as potent reducing agents for the synthesis of silver nanoparticles from silver nitrate salt (Biswas et al. 2024; Mangamma et al. 2024). Additionally, eco-friendly methods have been developed for synthesising stannic dioxide nanoparticles from a 50% ethanolic extract of *Croton malabaricus* Bedd. (Sishu et al. 2025), Copper nanoparticles using the leaf extract of *I. chinensis* (Kirubakaran et al. 2024a), and zinc oxide nanoparticles using the leaf extract of *S. coccinea* (Kirubakaran et al. 2024b). On the other hand, when microbial cells encounter metals, they react by activating a protective response that inadvertently leads to the creation of metal nanoparticles. Their secondary metabolites, like intracellular proteins, enzymes, cofactors, and others, can reduce metal ions, converting them into less toxic forms. Also, the non-specific binding of metal cations to cell walls may further promote their reduction. Thus, various microorganisms can modify metals at the cellular level, forming MNPs in response to metal stress (El-Seedi et al. 2024). Other recent studies emphasise the potential of microbiologically synthesised metal nanoparticles for various biomedical and environmental applications. For example, silver nanoparticles were produced by a bacterial biosurfactant extracted from *Klebsiella* sp. strain. These MNPs exhibited non-cytotoxicity on red blood cells and earthworms (*Eudrilus eugeniae*) (Das et al. 2024). In another study, the bioproduction of silver nanoparticles was successfully achieved by utilising the fungus *Penicillium camemberti*, as confirmed by FTIR analysis, where amino acids, proteins, and heterocyclic compounds had key roles in reducing silver ions (Kaur et al. 2024). Also, a novel, simple, and long-lasting method was used to produce antifungal silver nanoparticles from *Brevibacillus* sp. supernatants (Abdulameer Kadhim AL-Ziadi et al. 2024). Meanwhile, zinc oxide nanoparticles were synthesised using *Bacillus paramycoides*, with prominent antioxidant, antimicrobial, and anticancer activities (El-Refai et al. 2024).

Even though biosynthesised MNPs are generally considered greener, safer alternatives to their chemically induced counterparts, nanoparticles' size, high surface area, and reactivity still present potential environmental and health risks. Therefore, it is imperative to consider several factors that directly influence the toxicity of biosynthesised nanoparticles, such as size, shape, surface chemistry, dosage, and exposure time, in designing safer MNPs (Edo et al. 2025).

6.5.4 Regulatory Frameworks

In order to create an integrated regulatory framework that addresses the technical and ethical aspects of applied nanotechnology, different aspects need to be considered and incorporated, including potential risks of MNP's utilisation and corresponding countermeasures, as seen in Fig. 6.5. The unique physicochemical properties of MNPs generate the need for industry-specific guidelines, demanding severe health and environmental protocols and standardised testing procedures to mitigate potential risks (Edo et al. 2025). Globally, regulatory agencies have set

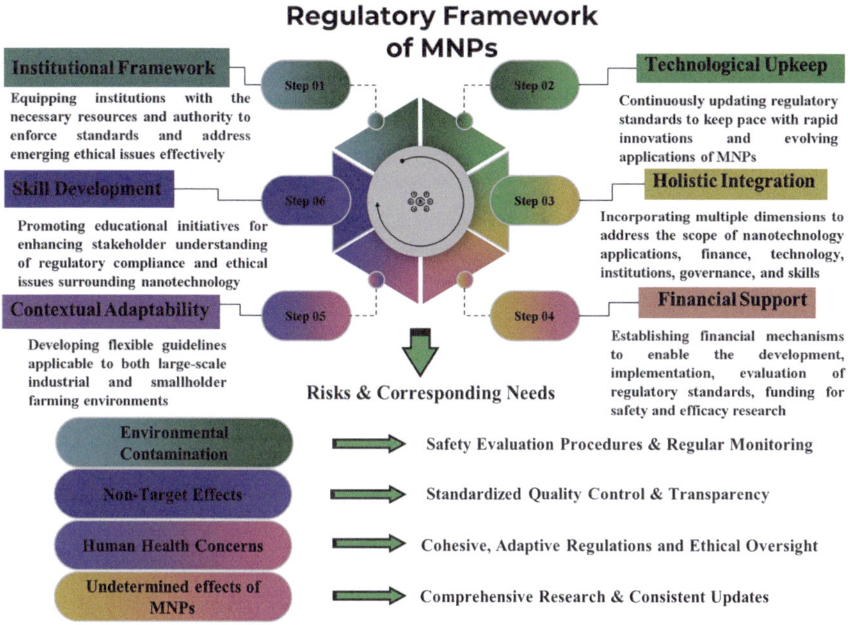

Fig. 6.5 Integrated regulatory framework and risk mitigation for metal-based nanoparticles (MNPs)

guidelines to warrant the safe use of nanotechnology and address these concerns. This is ensured by providing country-specific frameworks, mandatory risk assessments, labelling of nanotech-based products, and encouraging research and development in protecting human health and the environment.

Finally, safe regulatory framework implementation requires cooperation among researchers, manufacturers, regulators, and consumers (Kumari et al. 2023). Regulatory considerations and ethics are imperative in achieving acceptable and responsible practical MNP utilisation. These regulatory measures help maintain a balance between harnessing the benefits of nanotechnology and protecting human health and ecosystems. For example, under REACH, detailed data on a chemical's properties, potential hazards, and risk management measures must be submitted by manufacturers before being marketed in the EU. In the United States, however, EPA guidelines dictate that released nanomaterials must be monitored and managed through rigorous testing. FDA exercises its legal mandate by overseeing nanotechnology products and assessing the efficacy and safety of each particular nanomaterial. Unfortunately, gaps in the regulatory framework caused by a lack of standardisation present potential risks associated with the use of nanoparticles. Continuous updates in regulatory standards are necessary to keep up with constant technological progress and innovations. That is why institutions worldwide must be ready to enforce these rules while addressing new ethical challenges (Ahmad et al. 2024).

6.6 Future Perspectives

The impacts of MNPs lead to a cascade of ecological consequences: impaired nitrogen fixation, weakened biocontrol, disrupted nutrient cycling, and ultimately reduced crop productivity. Significant knowledge gaps remain despite growing awareness, especially concerning field-scale impacts, long-term persistence, and sub-lethal microbial responses to chronic MNP exposure. A sustainable path forward requires innovation in material design, policy interventions, and multidisciplinary collaboration to safeguard soil biodiversity. A promising direction in nanotechnology involves the green synthesis of nanoparticles using plant extracts, microbial metabolites, or biodegradable polymers. Adapting nanoparticle characteristics can help minimise adverse effects while retaining functional efficiency.

Additionally, incorporating biosafety screening protocols should become a standard practice in materials science and regulatory processes. Most current ecotoxicological data are based on short-term, laboratory-scale experiments, which fail to capture the complex interactions of MNPs in real-world soil environments. Therefore, long-term field studies are essential to assess the ageing, aggregation, dissolution, and transformation of MNPs under dynamic conditions. These studies should also monitor their cumulative effects on soil microbiota and crop health over multiple growing seasons to inform risk assessment models and agricultural best practices.

Additionally, emerging molecular tools offer unprecedented opportunities to unravel nanoparticles' subtle and complex effects on microbial systems. These tools can provide deep mechanistic insights into species-specific sensitivity, adaptive responses, and inter-microbial interactions in the presence of MNPs. Integrating these technologies into soil microbiome research will help bridge the gap between laboratory observations and ecological reality.

References

Abbas A, Mubeen M, Zheng H, Sohail MA, Moya-Elizondo EA, Zhou L (2022) *Trichoderma* spp. genes involved in the biocontrol activity against *Rhizoctonia solani*. Front Microbiol 13:884469

Abbas R, Luo J, Qi, X, Naz A, Khan I A, Liu H, Yu S, Wei J (2024) Silver nanoparticles: synthesis, structure, properties and applications. Nanomaterials 14(17):1425. https://doi.org/10.3390/nano14171425

Abdulameer Kadhim AL-Ziadi S, Yass Lahmood W, Abd Alabbas Muhammad D, Rashid Swadi R (2024) Bio-production of silver nanoparticles by a Brevibacillus sp. and testing of its inhibitory efficacy on pathogenic fungi. *Microbial*. Biosystems 9(2):58–66

Abishagu A, Kannan P, Sivakumar U, Manikanda Boopathi N, Senthilkumar M (2024) Metal nanoparticles and their toxicity impacts on microorganisms. Biologia 79(9):2843–2862

Abo-Zaid GA, Soliman NA-M, Abdullah AS, El-Sharouny EE, Matar SM, Sabry SA-F (2020) Maximization of Siderophores production from biocontrol agents, *Pseudomonas aeruginosa* F2 and *Pseudomonas fluorescens* JY3 using batch and exponential fed-batch fermentation. Processes 8(4):455

Afzal S, Singh NK (2022) Effect of zinc and iron oxide nanoparticles on plant physiology, seed quality and microbial community structure in a rice-soil-microbial ecosystem. Environ Pollut 314:120224

Ahmad Z, Niyazi S, Firdoos A, Wang C et al (2024) Enhancing plant resilience: nanotech solutions for sustainable agriculture. Heliyon 10(23):e40735

Ahmed B, Ameen F, Rizvi A, Ali K, Sonbol H, Zaidi A, Khan MS, Musarrat J (2020) Destruction of cell topography, morphology, membrane, inhibition of respiration, biofilm formation, and bioactive molecule production by nanoparticles of Ag, ZnO, CuO, TiO2, and Al2O3 toward beneficial soil bacteria. ACS Omega 5(14):7861–7876

Ahmed B, Rizvi A, Ali K, Lee J, Zaidi A, Khan MS, Musarrat J (2021) Nanoparticles in the soil–plant system: a review. Environ Chem Lett 19(2):1545–1609

Ajinkya N, Yu X, Kaithal P, Luo H, Somani P, Ramakrishna S (2020) Magnetic iron oxide nanoparticle (IONP) synthesis to applications: present and future. Materials 13(20):4644

Al Jabri H, Saleem MH, Rizwan M, Hussain I, Usman K, Alsafran M (2022) Zinc oxide nanoparticles and their biosynthesis: overview. Life 12(4):594. https://doi.org/10.3390/life12040594

Ambaye TG, Hassani A, Vaccari M, Franzetti A, Prasad S, Aminabhavi TM, Rtimi S (2025) Nanobioremediation for the removal of inorganic and organic pollutants from the soil. Environ Sci Pollut Res 32:1–19

Ameen F, Alsamhary K, Alabdullatif JA, ALNadhari S (2021) A review on metal-based nanoparticles and their toxicity to beneficial soil bacteria and fungi. Ecotoxicol Environ Saf 213:112027

Amiri M, Zinatloo-Ajabshir S, Ahmadi-Zeidabadi M, Sharifianjazi F (2025) Innovative ultrasound assisted synthesis of sponge like cerium dioxide nanostructure using Rosa Damascena extract and its efficient performance for cancer therapy. Sci Rep 15(1):933

Ansari MA, Khan HM, Khan AA, Ahmad MK, Mahdi AA, Pal R, Cameotra SS (2014) Interaction of silver nanoparticles with *Escherichia coli* and their cell envelope biomolecules. J Basic Microbiol 54(9):905–915

Ban Y, Xiao Z, Wu C, Lv Y, Meng F, Wang J, Xu Z (2021) The positive effects of inoculation using arbuscular mycorrhizal fungi and/or dark septate endophytes on the purification efficiency of CuO-nanoparticles-polluted wastewater in constructed wetland. J Hazard Mater 416:126095

Banik S, Pérez-de-Luque A (2017) In vitro effects of copper nanoparticles on plant pathogens, beneficial microbes and crop plants. Span J Agric Res 15(2):e1005–e1005

Baranowska-Wójcik E, Szwajgier D, Oleszczuk P, Winiarska-Mieczan A (2020) Effects of titanium dioxide nanoparticles exposure on human health—a review. Biol Trace Elem Res 193:118–129

Berruti A, Lumini E, Balestrini R, Bianciotto V (2016) Arbuscular mycorrhizal fungi as natural biofertilizers: let's benefit from past successes. Front Microbiol 6:1559

Biswas K, Ahamed Z, Dutta T, Mallick B, Khuda-Bukhsh AR, Biswas JK, Mandal SK (2024) Green synthesis of silver nanoparticles from waste leaves of tea (*Camellia sinensis*) and their catalytic potential for degradation of azo dyes. J Mol Struct 1318:139448

Boersma PJ, Lagugné-Labarthet F, McDowell T, Macfie SM (2023) Silver nanoparticles inhibit nitrogen fixation in soybean (Glycine max) root nodules. Environ Sci Pollut Res 30(11):32014–32031

Bouguerra S, Gavina A, Natal-da-Luz T, Sousa JP, Ksibi M, Pereira R (2022) The use of soil enzymes activity, microbial biomass, and basal respiration to assess the effects of cobalt oxide nanomaterial in soil microbiota. Appl Soil Ecol 169:104246

Chai H, Yao J, Sun J, Zhang C, Liu W, Zhu M, Ceccanti B (2015) The effect of metal oxide nanoparticles on functional bacteria and metabolic profiles in agricultural soil. Bull Environ Contam Toxicol 94(4):490–495

Chaudhary A, Poudyal S, Kaundal A (2025) Role of arbuscular mycorrhizal fungi in maintaining sustainable agroecosystems. Appl Microbiol 5(1):6

Chen L, Qiu T, Huang F, Zeng Y, Cui Y, Chen J, White JC, Fang L (2024a) Micro/nanoplastics pollution poses a potential threat to soil health. Glob Chang Biol 30(8):e17470

Chen Q, Song Y, An Y, Lu Y, Zhong G (2024b) Soil microorganisms: their role in enhancing crop nutrition and health. Diversity 16(12):734

Chhipa H (2021) Nano-toxicity to microbes: potential implications of nanomaterials on microbial activity. Nanotoxicol Nanoecotoxicol 1:99–123

Cota-Ruiz K, Delgado-Rios M, Martínez-Martínez A, Núñez-Gastelum JA, Peralta-Videa JR, Gardea-Torresdey JL (2018) Current findings on terrestrial plants–Engineered nanomaterial interactions: Are plants capable of phytoremediating nanomaterials from soil? Current Curr Opin Environ Sci Health 6:9–15

Cvanić T, Sulejmanović M, Perović M, Vulić J, Pezo L, Ćetković G, Travičić V (2024) Novel green strategy to recover bioactive compounds with different polarities from horned melon peel. Foods 13(18):2880

Das I, Bharali P, Gogoi P, Borah A, Borah D (2024) Green synthesis of silver nanoparticles using microbial biosurfactant produced by a newly isolated *Klebsiella* sp. strain RGUDBI03 through a single step method and exploring its role in enhancing the chickpea and rice seed germination. BioNanoScience:1–17

Din MI, Arshad F, Hussain Z, Mukhtar M (2017) Green adeptness in the synthesis and stabilization of copper nanoparticles: catalytic, antibacterial, cytotoxicity, and antioxidant activities. Nanoscale Res Lett 12(1):638

Dujković T, Pajčin I, Vlajkov V, Šovljanski O, Markov S, Loc M, Grahovac M, Grahovac J (2023) The effects of medium nutritional profile on *Bacillus* sp. Par 3 plant-growth promoting and biocontrol activity against *Botrytis cinerea*. Pesticides Phytomed 38(3):99–110

Edo GI, Mafe AN, Ali AB et al (2025) Green biosynthesis of nanoparticles using plant extracts: mechanisms, advances, challenges, and applications. BioNanoScience 15(2):267

El Sayed MT, El-Sayed AS (2020) Tolerance and mycoremediation of silver ions by *Fusarium solani*. Heliyon 6(5):e03866

El-Refai HA, Saleh AM, Mohamed SI, Aboul Naser AF, Zaki RA, Gomaa SK, Hamed MA (2024) Biosynthesis of zinc oxide nanoparticles using *Bacillus paramycoides* for *in vitro* biological activities and in vivo assessment against hepatorenal injury induced by CCl4 in rats. Appl Biochem Biotechnol:1–21

El-Seedi HR, Omara MS, Omar AH et al (2024) Updated review of metal nanoparticles fabricated by green chemistry using natural extracts: biosynthesis, mechanisms, and applications. Bioengineering 11(11):1095

Fahde S, Boughribil S, Sijilmassi B, Amri A (2023) Rhizobia: A promising source of plant growth-promoting molecules and their non-legume interactions: examining applications and mechanisms. Agriculture 13(7):1279

Fan R, Huang YC, Grusak MA, Huang CP, Sherrier DJ (2014) Effects of nano-TiO2 on the agronomically-relevant *Rhizobium*–legume symbiosis. Sci Total Environ 466:503–512

Fernandes JP, Mucha AP, Francisco T, Gomes CR, Almeida CMR (2017) Silver nanoparticles uptake by salt marsh plants–implications for phytoremediation processes and effects in microbial community dynamics. Mar Pollut Bull 119(1):176–183

Fernández-Triana I, Rubilar O, Parada J, Fincheira P, Benavides-Mendoza A, Durán P, ... Tortella GR (2024) Metal nanoparticles and pesticides under global climate change: assessing the combined effects of multiple abiotic stressors on soil microbial ecosystems. Sci Total Environ 942:173494

García PR, López ECA, Torres MAR, Frías MÁN, Delgadillo AT, Zumarán AM et al (2025) Strategies for overcoming bacterial resistance to nanoparticles: A systematic review. Cureus 17(1):e78064

Gomez LA, Bavlnka CQ, Nguyen PT, Alalq I, Sabisch JE, Boscoboinik JA, ... Crossley SP (2025) Evolution of Ni-Mo/MgO during catalytic methane pyrolysis to produce base-growth nanotubes. Cell Rep Phys Sci 6(4)

González-Fernández S, Blanco-Agudín N, Rodríguez D, Fernández-Vega I, Merayo-Lloves J, Quirós LM (2025) Silver nanoparticles: A versatile tool against infectious and non-infectious diseases. Antibiotics 14(3):289

Grün AL, Straskraba S, Schulz S, Schloter M, Emmerling C (2018) Long-term effects of environmentally relevant concentrations of silver nanoparticles on microbial biomass, enzyme activity, and functional genes involved in the nitrogen cycle of loamy soil. J Environ Sci 69:12–22

Gudkov SV, Burmistrov DE, Serov DA, Rebezov MB, Semenova AA, Lisitsyn AB (2021) Do iron oxide nanoparticles have significant antibacterial properties? Antibiotics 10(7):884

Guzmán-Guzmán P, Kumar A, de los Santos-Villalobos S, Parra-Cota FI, Orozco-Mosqueda MDC, Fadiji AE, Hyder S, Babalola OO, Santoyo G (2023) *Trichoderma* species: our best fungal allies in the biocontrol of plant diseases—a review. Plants 12(3):432

Haskett TL, Cooke L, Green P, Poole PS (2025) Regulation of Rhizobial nodulation genes by flavonoid-independent NodD supports nitrogen-fixing symbioses with legumes. Environ Microbiol 27(1):e70014

Huang J, Qian X, Li X, Hu Q, Cao C, Yan C (2023) A comparative study on responses of soil under three typical nanoparticles exposure: enzyme activities and microbial community structure. Water Air Soil Pollut 234(4):229

Hunter WJ, Kuykendall LD, Manter DK (2007) *Rhizobium selenireducens* sp. nov.: a selenite-reducing *α-Proteobacteria* isolated from a bioreactor. Curr Microbiol 55(5):455–460

Irshad A, Jawad R, Sharif S, Joly N, Ishtiaq U, Martin P, Mushtaq Q (2024) Bioengineering of glucan coated silver nanoparticles as dynamic biomedical compound; *in vitro* and *in vivo* studies. Microb Pathog 197:107005

Jośko I, Oleszczuk P, Futa B (2014) The effect of inorganic nanoparticles (ZnO, Cr_2O_3, CuO and Ni) and their bulk counterparts on enzyme activities in different soils. Geoderma 232:528–537

Judy JD, Kirby JK, McLaughlin MJ, McNear D Jr, Bertsch PM (2016) Symbiosis between nitrogen-fixing bacteria and *Medicago truncatula* is not significantly affected by silver and silver sulfide nanomaterials. Environ Pollut 214:731–736

Kamran M, Ali H, Saeed MF, Bakhat HF, Hassan Z, Tahir M, Abbas G, Naeem MA, Rashid MI, Shah GM (2020) Unraveling the toxic effects of iron oxide nanoparticles on nitrogen cycling through manure-soil-plant continuum. Ecotoxicol Environ Saf 205:111099

Karnwal A, Martolia S, Dohroo A, Al-Tawaha ARMS, Malik T (2024) Exploring bioremediation strategies for heavy metals and POPs pollution: the role of microbes, plants, and nanotechnology. Front Environ Sci 12:1397850

Kaur M, Singh K, Kumar V (2024) Green synthesis of silver nanoparticles using *Penicillium camemberti* and its biological applications. BioNanoScience 14(5):5179–5193

Kirubakaran D, Selvam K, Bagyalakshmi G, Manimegalai P, Adhavan R, Shivakumar MS (2024a) *Salvia coccinea* Buc'hoz ex Etl. Leaf extract mediated synthesis of zinc oxide nanoparticles: assessment of antimicrobial, antioxidant and anticancer activity. J Drug Del Sci Technol 96:105662

Kirubakaran D, Selvam K, Dhaneeshram M, Shivakumar MS, Rajkumar M, Shanmugarathinam A (2024b) Biogenic synthesis of copper nanoparticle using Impatiens chinensis L: insights into antimicrobial, antioxidant and anticancer activity. J Mol Struct 1317:138991

Kumari R, Suman K, Karmakar S, Mishra V, Lakra SG, Saurav GK, Mahto BK (2023) Regulation and safety measures for nanotechnology-based agri-products. Front Genome Editing 5:1200987

Lai RWS, Kang HM, Zhou GJ et al (2021) Hydrophobic surface coating can reduce toxicity of zinc oxide nanoparticles to the marine copepod *Tigriopus japonicus*. Environ Sci Technol 55(10):6917–6925

Laksee S, Puthong S, Kongkavitoon P, Palaga T, Muangsin N (2018) Facile and green synthesis of pullulan derivative-stabilized Au nanoparticles as drug carriers for enhancing anticancer activity. Carbohydr Polym 198:495–508

Lee S, Yap M, Behringer G, Hung R, Bennett JW (2016) Volatile organic compounds emitted by *Trichoderma* species mediate plant growth. Fungal Biol Biotechnol 3:1–14

Lu C, Hei R, Song X, Fan Z, Guo D, Luo J, Ma Y (2023) Metal oxide nanoparticles inhibit nitrogen fixation and rhizosphere colonization by inducing ROS in associative nitrogen-fixing bacteria *Pseudomonas stutzeri* A1501. Chemosphere 336:139223

Luo YH, Chang LW, Lin P (2015) Metal-based nanoparticles and the immune system: activation, inflammation, and potential applications. Biomed Res Int 2015(1):143720

Luo S, Fan S, Yuan J, Xiao J, Sun X, Wang L et al (2025) Regulation of electron density in Pt nanoparticles via bimetallic metal-organic frameworks for enhancing photothermal catalysis of toluene decomposition. J Hazard Mater 482:136561

Malhotra N, Ger TR, Uapipatanakul B, Huang JC, Chen KHC, Hsiao CD (2020) Review of copper and copper nanoparticle toxicity in fish. Nanomaterials 10(6):1126

Mangamma JL, Sagar PV, Kumar PK, Prameeladevi S, Basavaiah K (2024) Eco-conscious Silver nanoparticles through "flower extract" of Murraya paniculata (L) Jack for reduction of 4-nitro phenol. Mater Lett 377:137438

Manuja A, Kumar B, Kumar R, Chhabra D, Ghosh M, Manuja M et al (2021) Metal/metal oxide nanoparticles: toxicity concerns associated with their physical state and remediation for biomedical applications. Toxicol Rep 8:1970–1978

Markowicz A (2023) The significance of metallic nanoparticles in the emerging, development and spread of antibiotic resistance. Sci Total Environ 871:162029

Mgadi K, Ndaba B, Roopnarain A, Rama H, Adeleke R (2024) Nanoparticle applications in agriculture: overview and response of plant-associated microorganisms. Front Microbiol 15:1354440

Mohammadzadeh R (2012) Hypothesis: silver nanoparticles as an adjuvant for cancertherapy. Adv Pharm Bull 2(1):133

Moll J, Okupnik A, Gogos A, Knauer K, Bucheli TD, Van Der Heijden MG, Widmer F (2016) Effects of titanium dioxide nanoparticles on red clover and its rhizobial symbiont. PLoS One 11(5):e0155111

Montes de Oca-Vásquez G, Solano-Campos F, Vega-Baudrit JR, López-Mondéjar R, Odriozola I, Vera A et al (2020) Environmentally relevant concentrations of silver nanoparticles diminish soil microbial biomass but do not alter enzyme activities or microbial diversity. J Hazard Mater 391:122224

More PR, Pandit S, Filippis AD, Franci G, Mijakovic I, Galdiero M (2023) Silver nanoparticles: bactericidal and mechanistic approach against drug resistant pathogens. Microorganisms 11(2):369

Mubeen B, Hasnain A, Wang J, Zheng H, Naqvi SAH, Prasad R et al (2023) Current progress and open challenges for combined toxic effects of manufactured nano-sized objects (MNO's) on soil biota and microbial community. Coatings 13(1):212

Navina BK, Velmurugan NK, Kumar PS, Rangasamy G, Palanivelu J, Thamarai P et al (2024) Fungal bioremediation approaches for the removal of toxic pollutants: mechanistic understanding for biorefinery applications. Chemosphere 350:141123

Nayak D, Minz AP, Ashe S, Rauta PR, Kumari M, Chopra P, Nayak B (2016) Synergistic combination of antioxidants, silver nanoparticles and chitosan in a nanoparticle based formulation: characterization and cytotoxic effect on MCF-7 breast cancer cell lines. J Colloid Interface Sci 470:142–152

Noga M, Milan J, Frydrych A, Jurowski K (2023) Toxicological aspects, safety assessment, and green toxicology of silver nanoparticles (AgNPs)-Critical review: state of the art. Int J Mol Sci 24(6):5133

Ogar A, Tylko G, Turnau K (2015) Antifungal properties of silver nanoparticles against indoor mould growth. Sci Total Environ 521:305–314

Pacheco I, Buzea C (2020) Nanomaterials and nanocomposites: classification and toxicity. In: Handbook of nanomaterials and nanocomposites for energy and environmental applications, pp 1–38

Padhye LP, Jasemizad T, Bolan S, Tsyusko OV, Unrine JM, Biswal BK et al (2023) Silver contamination and its toxicity and risk management in terrestrial and aquatic ecosystems. Sci Total Environ 871:161926

Parada J, Rubilar O, Fernández-Baldo MA, Bertolino FA, Durán N, Seabra AB, Tortella GR (2019) The nanotechnology among US: are metal and metal oxides nanoparticles a nano or mega risk for soil microbial communities? Crit Rev Biotechnol 39(2):157–172

Parra-Ortiz E, Malmsten M (2022) Photocatalytic nanoparticles–from membrane interactions to antimicrobial and antiviral effects. Adv Colloid Interf Sci 299:102526

Patlolla AK, Kumari SA, Tchounwou PB (2019) A comparison of poly-ethylene-glycol-coated and uncoated gold nanoparticle-mediated hepatotoxicity and oxidative stress in Sprague Dawley rats. Int J Nanomedicine 14:639–647

Patsula V, Tulinska J, Trachtová Š, Kuricova M, Liskova A, Španová A et al (2019) Toxicity evaluation of monodisperse PEGylated magnetic nanoparticles for nanomedicine. Nanotoxicology 13(4):510–526

Peixoto S, Loureiro S, Henriques I (2022) The impact of silver sulfide nanoparticles and silver ions in soil microbiome. J Hazard Mater 422:126793

Prakash P, Smitha SC (2023) Nano-phytoremediation of heavy metals from soil: a critical review. Pollutants 3(3):360–380

Pulikova EP, Ivanov FD, Alliluev IA, Chernikova NP, Krepakova MR, Gorovtsov AV, Minkina TM, Bauer TV, Tsitsuashvili VS, Garg MC, Rajput VD (2024) Effects of bulk forms and nanoparticles of zinc and copper oxides on the abundance, nitrogen cycling and enzymatic activities of microbial communities, morphometric parameters and antioxidant status of *Hordeum vulgare* L. Environ Geochem Health 46(12):494

Rahman MS, Chakraborty A, Kibria A, Hossain MJ (2023) Effects of silver nanoparticles on seed germination and growth performance of pea (*Pisum sativum*). Plant Nano Biol 5:100042

Rajput VD, Minkina T, Sushkova S, Tsitsuashvili V, Mandzhieva S, Gorovtsov A, Nevidomskyaya D, Gromakova N (2018) Effect of nanoparticles on crops and soil microbial communities. J Soils Sediments 18:2179–2187

Rajput V, Minkina T, Ahmed B, Sushkova S, Singh R, Soldatov M et al (2020) Interaction of copper-based nanoparticles to soil, terrestrial, and aquatic systems: critical review of the state of the science and future perspectives. Rev Environ Contam Toxicol 252:51–96

Rakitin AL, Kulichevskaya IS, Beletsky AV, Mardanov AV, Dedysh SN, Ravin NV (2024) Verrucomicrobia of the family *Chthoniobacteraceae* participate in Xylan degradation in boreal peat soils. Microorganisms 12(11):2271

Rashid MI, Shahzad T, Shahid M, Imran M, Dhavamani J, Ismail IMI et al (2017a) Toxicity of iron oxide nanoparticles to grass litter decomposition in a sandy soil. Sci Rep 7(1):41965

Rashid MI, Shahzad T, Shahid M, Ismail IMI, Shah GM, Almeelbi T (2017b) Zinc oxide nanoparticles affect carbon and nitrogen mineralization of *Phoenix dactylifera* leaf litter in a sandy soil. J Hazard Mater 324(Part B):298–305

Reznickova A, Slavikova N, Kolska Z, Kolarova K, Belinova T, Kalbacova MH et al (2019) PEGylated gold nanoparticles: stability, cytotoxicity and antibacterial activity. Colloids Surf A Physicochem Eng Asp 560:26–34

Rodrigues AS, Batista JGS, Rodrigues MÁV, Thipe VC, Minarini LAR, Lopes PS, Lugão AB (2024) Advances in silver nanoparticles: a comprehensive review on their potential as antimicrobial agents and their mechanisms of action elucidated by proteomics. Front Microbiol 15:1440065

Rónavári A, Bélteky P, Boka E, Zakupszky D, Igaz N, Szerencsés B et al (2021) Polyvinylpyrrolidone-coated silver nanoparticles—the colloidal, chemical, and biological consequences of steric stabilization under biorelevant conditions. Int J Mol Sci 22(16):8673

Russo M, Oliva M, Hussain MI, Muscolo A (2023) The hidden impacts of micro/nanoplastics on soil, crop and human health. J Agric Food Res 14:100870

Sathyamoorthy S, Venugopal A, Lenin MJ, Tirkey S, Sevanan M (2025) Nanomaterials and microbial compatibility: synergistic and antagonistic mechanisms. In: Nano-bioremediation for wastewater treatment, pp 35–72

Shah GM, Shabbir Z, Rabbani F, Rashid MI, Bakhat HF, Naeem MA, Abbas G, Shah GA, Shahid N (2025a) Soil texture mediates the toxicity of ZnO and Fe_3O_4 nanoparticles to microbial activity. Toxics 13(2):84

Shah N, Shah M, Naz K, Muhammad J, Rehan T, Ali A, ... Humayun M (2025b) Silver immobilized magnetic iron nanoparticles fabricated via green method for detection of cobalt in water and their antibacterial activity. Environ Sci Pollut Res Int 1–14

Sharan D, Wolfson D, Green CM, Lemke P, Gavin AG, Hamers RJ, Feng ZV, Carlson EE (2023) Chronic exposure to complex metal oxide nanomaterials induces production of reactive oxygen species in bacteria. Environ Sci Nano 10(8):1978–1992

Sharma MMM, Kapoor D, Rohilla R, Sharma P (2023a) Nanomaterials and their toxicity to beneficial soil microbiota and fungi associated plants rhizosphere. In: Nanomaterials and nanocomposites exposures to plants: response, interaction, phytotoxicity and defense mechanisms. Springer Nature Singapore, Singapore, pp 353–380

Sharma S, Sharma S, Sharma N, Sharma V, Paul S (2023b) Synergistic promotion of Ag nanoparticles supported on magnetic Cu based metal organic framework for A 3-coupling and synthesis of benzopyranopyrimidines. New J Chem 47(10):4674–4686

Shen Z, Chen Z, Hou Z, Zhang P (2015) Ecotoxicological effect of zinc oxide nanoparticles on soil microorganisms. Front Environm Sci Eng 9:912–918

Shen K, Chen X, Chen J, Li Y (2016) Development of MOF-derived carbon-based nanomaterials for efficient catalysis. Acs Catalysis 6(9):5887–5903

Simonin M, Cantarel AA, Crouzet A, Gervaix J, Martins JM, Richaume A (2018) Negative effects of copper oxide nanoparticles on carbon and nitrogen cycle microbial activities in contrasting agricultural soils and in presence of plants. Front Microbiol 9:3102

Singh M, Mallick AK, Banerjee M, Kumar R (2016) Loss of outer membrane integrity in Gram-negative bacteria by silver nanoparticles loaded with *Camellia sinensis* leaf phytochemicals: plausible mechanism of bacterial cell disintegration. Bull Mater Sci 39:1871–1878

Singh A, Tyagi S, Mishra A, Verma PK, Kushwaha HR, Singh A (2024) Eco-compatible synthesis of metal nanoparticles using bacterial nanofactories and their applications in optoelectronics and magnetic technologies. In: materials for electronic, magnetic, and spintronic technologies: characterization and applications from energy storage to disease detection. Cham: Springer Nature Switzerland, pp 65–89

Singh R, Cheng S, Singh S (2020) Oxidative stress-mediated genotoxic effect of zinc oxide nanoparticles on *Deinococcus radiodurans*. 3 Biotech 10(2):66

Sishu NK, Karunakaran MKR, Hadkar VM, Mohanty C, Sharmila A, Selvaraj CI, Babu NG (2025) Phyto-mediated synthesis of SnO_2 nanoparticles using *Croton malabaricus* Bedd. for its antioxidant, antibacterial, hemocompatibility properties and photocatalytic activity. New J Chem 49(2):536–552

Slavin YN, Asnis J, Häfeli UO, Bach H (2017) Metal nanoparticles: understanding the mechanisms behind antibacterial activity. J Nanobiotechnology 15(1):65

Šovljanski O, Pezo L, Tomić A, Ranitović A, Cvetković D, Markov S (2021) Contribution of bacterial cells as nucleation centers in microbiologically induced $CaCO_3$ precipitation—A mathematical modeling approach. J Basic Microbiol 61(9):835–848

Šovljanski O, Pezo L, Grahovac J, Tomić A, Ranitović A, Cvetković D, Markov S (2022a) Best-performing *Bacillus* strains for microbiologically induced $CaCO_3$ precipitation: screening of relative influence of operational and environmental factors. J Biotechnol 350:31–41

Šovljanski O, Tomić A, Markov S (2022b) Relationship between bacterial contribution and self-healing effect of cement-based materials. Microorganisms 10(7):1399

Šovljanski O, Aćimović M, Tomić A, Lončar B, Miljković A, Čabarkapa I, Pezo L (2024a) Antibacterial and antifungal potential of *Helichrysum italicum* (Roth) G. Don Essential Oil. Antibiotics 13(8):722

Šovljanski O, Milović T, Bulatović V, Erceg T, Stanojev J, Bajac B, Tomić A (2024b) Insights into self-healing capacity of cement matrix containing high-efficiency bacteria under challenging conditions. J Build Eng 98:111094

Šovljanski O, Cvetanović Kljakić A, Saveljić A, Tomić A (2025) Bioactivity and bioavailability of phenols from plants. In: Ramawat KG, Mérillon JM (eds) Natural products. Springer, Berlin/Heidelberg

Su Y, Zheng X, Chen Y, Li M, Liu K (2015) Alteration of intracellular protein expressions as a key mechanism of the deterioration of bacterial denitrification caused by copper oxide nanoparticles. Sci Rep 5(1):15824

Sun W, Dou F, Li C, Ma X, Ma LQ (2021) Impacts of metallic nanoparticles and transformed products on soil health. Crit Rev Environ Sci Technol 51(10):973–1002

Sun L, Lv H, Feng J, Guselnikova O, Wang Y, Yamauchi Y, Liu B (2022) Noble-metal-based hollow mesoporous nanoparticles: synthesis strategies and applications. Adv Mater 34(31):2201954

Sungur Ş (2021) Titanium dioxide nanoparticles. In: Handbook of nanomaterials and nanocomposites for energy and environmental applications, pp 713–730

Takáč P, Michalková R, Čižmáriková M, Bedlovičová Z, Balážová Ľ, Takáčová G (2023) The role of silver nanoparticles in the diagnosis and treatment of cancer: are there any perspectives for the future? Life 13(2):466

Travičić V, Cvanić T, Šovljanski O, Erceg T, Perović M, Stupar A, Ćetković G (2024) Updating the status quo on the eco-friendly approach for antioxidants recovered from plant matrices using cloud point extraction. Antioxidants 13(3):280. https://doi.org/10.3390/antiox13030280

Travičić V, Cvanić T, Vučetić A, Kostić M, Perović M, Pezo L, Ćetković G (2025) Green extraction technologies for carotenoid recovery from citrus peel: comparative study and encapsulation for stability enhancement. Processes 13(7):1962

Tripathi GD, Javed Z, Dashora K (2024) Toxicity of copper oxide nanoparticles on agriculturally important soil rhizobacteria *Bacillus megaterium*. Emerging Contam 10(1):100280

Vallabani NS, Singh S (2018) Recent advances and future prospects of iron oxide nanoparticles in biomedicine and diagnostics. 3 Biotech 8(6):279

Verma Y, Singh SK, Jatav HS, Others (2022) Interaction of zinc oxide nanoparticles with soil: insights into the chemical and biological properties. Environ Geochem Health 44:221–234

Viret O, Gindro K (2024) Fungi and grapevine Mycobiota. In: Science of fungi in grapevine. Springer International Publishing, Cham, pp 109–195

Vučetić A, Rached RA, Manca ML, Šovljanski O, Cvetković D, Manconi M, Čanadanović-Brunet J (2025) Sustainable supramolecular extraction of phytocomplexes from microgreens and their eco-loading in nutriosomes: physicochemical characterization, stability, and In Vitro release behavior. Molecules 30(18):3774

Wang F, Liu X, Shi Z, Tong R, Adams CA, Shi X (2016) Arbuscular mycorrhizae alleviate negative effects of zinc oxide nanoparticle and zinc accumulation in maize plants–a soil microcosm experiment. Chemosphere 147:88–97

Wang L, Hu C, Shao L (2017) The antimicrobial activity of nanoparticles: present situation and prospects for the future. Int J Nanomedicine 12:1227–1249

Wang L, Yang D, Ma F, Wang G, You Y (2022) Recent advances in responses of arbuscular mycorrhizal fungi—plant symbiosis to engineered nanoparticles. Chemosphere 286:131644

Wang T, Xu J, Chen J, Liu P, Hou X, Yang L, Zhang L (2024) Progress in microbial fertilizer regulation of crop growth and soil remediation research. Plants 13(3):346

Wei L, Lu J, Xu H, Patel A, Chen ZS, Chen G (2015) Silver nanoparticles: synthesis, properties, and therapeutic applications. Drug Discov Today 20(5):595–601

Xia G, Wan T, Chen Z, Liu C, Li R (2025) Developmental toxicity of micro(Nano)plastics (MNPs) exposure in mammals: A mini-review. Toxics 13(3):224

Xie B, Wei X, Wan C, Zhao W, Song R, Xin S, Song K (2024) Exploring the biological pathways of Siderophores and their multidisciplinary applications: a comprehensive review. Molecules 29(10):2318

Yadav V, Arif N, Hussain I, Patel A, Tiwari S, Chauhan DK, Prasad SM (2024) Antioxidant machinery modulation by phosphorus mitigates zinc oxide nanoparticle toxicity in *Triticum aestivum* and *Solanum lycopersicum* seedlings. Nanotechnol Environm Eng 9(4):573–591

Yang Z, DeLoid GM, Zarbl H, Baw J, Demokritou P (2023) Micro-and nanoplastics (MNPs) and their potential toxicological outcomes: state of science, knowledge gaps and research needs. NanoImpact 32:100481

Yeremko L, Czopek K, Staniak M, Marenych M, Hanhur V (2025) Role of environmental factors in legume-rhizobium Symbiosis: a review. Biomolecules 15(1):118

Zaki SAEF, Kamal A, Ashmawy NA, Shoeib AA (2021) Nano-metals forming bacteria in Egypt I synthesis, characterization and effect on some phytopathogenic bacteria in vitro. Sci Rep 11(1):12876

Zeng Y, Zhang D, Wu M et al (2014) Lipid-AuNPs@ PDA nanohybrid for MRI/CT imaging and photothermal therapy of hepatocellular carcinoma. ACS Appl Mater Interfaces 6(16):14266–14277

Zhang L, Wu L, Si Y, Shu K (2018) Size-dependent cytotoxicity of silver nanoparticles to *Azotobacter vinelandii*: growth inhibition, cell injury, oxidative stress and internalization. PLoS One 13(12):e0209020

Zhu Y-G, Peng J, Chen C, Xiong C, Li S, Ge A, Wang E, Liesack W (2023) Harnessing biological nitrogen fixation in plant leaves. Trends Plant Sci 28(12):1391–1405

Zhukova LV (2015) Evidence for compression of *Escherichia coli* K12 cells under the effect of TiO_2 nanoparticles. ACS Appl Mater Interfaces 7(49):27197–27205

Chapter 7
Exploring Nanoscale Solutions for Enhancing Plant-Microbe Symbiosis

Ivana Danilov, Vanja Vlajkov, and Jovana Grahovac

Abstract The future of agricultural production, particularly of food products, depends on innovations that increase agricultural output within the constraints of limited available land. These innovations must simultaneously enhance crop quality and minimize the negative impacts of agricultural practices on the environment, biodiversity, and human and animal health. Nanotechnology offers a wide range of opportunities to develop novel agricultural tools, either through the use of nanoparticles alone or in combination with bio-based active components that enhance plant nutrition, growth, and quality. Nanobiofertilizers and nanobiopesticides—sourced from plants or microorganisms or formulated by integrating bio-based components with nanomaterials—have demonstrated significant influence on plant interactions with their environment, including both biotic and abiotic factors, resulting in improved agricultural productivity. This chapter aims to summarize the main mechanisms by which nanoscale solutions interact with plants and surrounding microbial communities, particularly soil microbiota, which collectively contribute to strengthening beneficial (i.e., symbiotic) plant–microbe relationships, supporting improved plant growth, yield, and soil health. While many potential interactions between nanomaterials and living systems, especially long-term effects, remain poorly understood, this chapter provides an overview of the documented benefits of nanomaterial applications in plant production with an emphasis on microbially mediated mechanisms of plant growth promotion.

Keywords Nanomaterials · Nanoparticles · Nanobiofertilizers · Nanobiopesticides · Microbial bionanofactories · Plant growth promotion

I. Danilov (✉) · V. Vlajkov · J. Grahovac
Faculty of Technology Novi Sad, University of Novi Sad,
Bulevar cara Lazara 1, Novi Sad, Serbia
e-mail: ivana.pajcin@uns.ac.rs

K. A. Abd-Elsalam (ed.), *Plant-Microbiome Nanotechnology*, Sustainability Sciences in Asia and Africa, https://doi.org/10.1007/978-981-95-5056-2_7

7.1 Introduction

The European Commission declares nanomaterials as “natural, incidental, or manufactured materials containing particles, in an unbound state or as aggregates or agglomerates, and where, for 50% or more of the particles in the size distribution, one or more external dimensions are in the size range of 1 nm–100 nm” (Recommendation on the Definition of a Nanomaterial 2011/696/EU). In other words, nanotechnologies define the manipulation of materials at various scales, including macromolecular, molecular, and atomic levels, providing the development of a new generation of products that differ from their larger-scale origin materials, significantly contributing to the generation of novel functional properties. Among a wide spectrum of possible applications, some nano-based products are relevant to agriculture as promising solutions in the context of more sustainable farming and crop management. The principles of the Green Revolution, followed by recent challenges including rapid population growth, urbanization, and escalating environmental pressures, have been defined as the global imperative to optimize agricultural yields and grow healthier crops. Nano(bio)technology could offer multiple benefits related to agricultural sustainability, starting with the targeted and controlled delivery of active components, resulting in a beneficial action that could be manipulated on a stimuli basis, thus minimizing nonselective effects on nontarget organisms (Ahmed et al. 2023a). Nanosystems enable the transport, reduction, and capping of agents that work as active components themselves or could enhance the activity of other active components. This is especially important considering the low reach of chemical pesticides to target pathogens, as well as the low utilization efficiency of inorganic fertilizers, resulting in the environmental accumulation of agrochemicals and unavoidable negative consequences for the biosphere.

Beneficial actions of nanoscale technologies for plant growth and the environment could be directed towards pest/pathogen suppression or control, plant growth promotion (PGP), or soil quality enhancement/bioremediation. Furthermore, nano-enabled immunomodulation in plants has become one of the major research topics in plant nanobiotechnology, considering that the application of nanomaterials could increase plants’ resilience and tolerance to various biotic and abiotic stresses (Zhang et al. 2024). The novel toolkit for sustainable agricultural development now includes a nanotechnological approach to the manufacturing of next-generation agrochemicals, in addition to the continuous discovery of novel beneficial PGP/endophytic strains and their application as single active components or synthetic microbial communities, supported by microbiome engineering, gene editing tools such as CRISPR, and artificial intelligence. This chapter aims to provide background for further research on the potential benefits offered by a nanotechnological approach to symbiotic relationships between plants and the surrounding microbes, diving deeper into the potential mechanisms and actions to be pursued in order to achieve improved agricultural output, food safety, and future food sustainability.

7.2 The Basic Considerations and Classification of Nanoscale Solutions for Agricultural Applications

Specific size range of nanoscale particles (1–100 nm) offers novel benefits in comparison to individual larger particles and bulk materials, mostly in terms of more efficient cell entry due to lower biological membrane resistance to the particles in this size range (Xu et al. 2022a). This is also reflected in particle-specific surface and charge, increasing their target reach efficiency and selectivity, as well as in their improved biodegradability (Ahmed et al. 2023a). Other beneficial properties of nanoscale particles include advanced and modifiable physical (e.g., surface/volume ratio) and chemical traits (surface potential, higher reactivity, quantum properties), subject to (bio)chemical manipulations that tune the properties towards the target application requirements (Khan et al. 2017). Furthermore, nano-based formulations provide protection for active components, thus prolonging their beneficial in situ action and reducing the deteriorative effects of radiation (photodecomposition), drought, temperature variations, and other unfriendly environmental conditions (Zong et al. 2022).

Nano-based agrochemicals could be applied via the same application routes as conventional agrochemicals, including foliar, soil/root application, irrigation, hydroponics, and seed treatment, providing both downward and upward transport and nutrient delivery inside the plant depending on the selected application route (Liao et al. 2023). Modifiable traits of nanoscale materials provide a vast array of possibilities to adjust surface chemistry (e.g., hydrophobicity/hydrophilicity), morphology, and interactions at the plant-nanoparticle (NP) interface in order to enhance adhesion to the plant surface, plant tissue entry (uptake), and transport of nutrients/bioactive compounds inside the plant organism (Grillo et al. 2021), even their localization inside the plant cells (e.g., in chloroplasts—Wong et al. 2016). One of the major advantages of nanoscale particles is their size, providing the possibility of internalization and translocation of NPs across the plant cell walls via cuticular and stomatal pathways for hydrophobic and hydrophilic NPs, respectively (Xia et al. 2021). Although there are still many uncertainties regarding NP movement in the inner plant tissues, there are several methods to track their distribution inside the plant, such as ICP-MS coupled with several other techniques and isotope-tracing methods (Liao et al. 2023).

Antimicrobial activity of nanoscale particles is based on different cell death-inducing mechanisms, including selective adhesion to target microbial cells, dissolution and degradation of the cell wall and membrane, production of reactive oxygen species, etc. (Albalghiti et al. 2020), presenting the foundation for their pesticidal applications. Nanopesticides can be classified into two major classes (Wang et al. 2022):

- Type 1—metal-based nanopesticides used as active components (Cu, Ti, Ag, Fe, Zn, Al, etc.)

- Type 2—polymers, clays, biochar, and zein used as nanomaterial carriers for active components.

While class I nanopesticides provide selectivity and targeted action, class II offers more customization options in terms of the possibility to develop and deliver synchronized and controlled stimuli-responsive action at the target site, simultaneously providing biocompatibility and high loading/encapsulation capacity (Ahmed et al. 2023a). Furthermore, nanoscale-based immobilization of (bio)pesticidal active components via encapsulation, binding, doping, and adsorption (Dutta et al. 2022) could provide adaptation to surface-related requirements of the target plant tissue, contributing to their more efficient adhesion and preventing their premature loss before the target action delivery (Kumar et al. 2018), while simultaneously providing stimuli-based release/delivery that is usually based on osmotic pressure, passive diffusion, desorption, erosion/physical dissolution of nanocarrier formulations, capsule swelling, or biodegradation (physical or chemical—Liao et al. 2023). General mechanisms of action of nanopesticides rely on improved pathogen cell adhesion/entry, followed by cell wall dissolution, reactive oxygen species production, cytotoxicity, and cell death induction (Albalghiti et al. 2020). On the other hand, nanofertilizers could be classified as:

- Macronutrient-based active components (C, N, P).
- Micronutrient-based active components (Cu, Fe, Zn, Mn, etc.)
- Carriers for macro- and micronutrients aimed at improving plant nutrition (Kah et al. 2018).

Nanoregulators could be considered an additional category of plant-beneficial nanoscale products, which could also incorporate bioactive compounds or act solely as NP-based systems in the advancement of plant biosynthetic activity or abiotic/biotic stress tolerance (Hu and Xianyu 2021). The main classification of nanoagrochemicals, including nanofertilizers and nanopesticides, is provided in Fig. 7.1. Nanofertilizers contribute to the improvement of plant nutrient uptake and further utilization, providing their sustainable, controlled, and targeted delivery and preventing their loss throughout the plant-soil-air ecosystem (Tarafder et al. 2020), mostly via two main axes:

1. Temporal—via a slow-release function prolonging nutrient delivery from 4 to 10 days in the case of conventional fertilizers to 40–50 days in the case of nanofertilizers, thus minimizing nutrient losses due to their fast release.
2. Spatial—via controlled nutrient release adjusted to target humidity levels, pH values, temperature, and enzyme activity (Xu et al. 2022b), thus delivering nutrients to the target plant absorption points (roots or leaves) (Seleiman et al. 2020; Liao et al. 2023), providing a basis for precision fertilization in the concept of smart/precision agriculture aimed at minimizing resource requirements and utilization in terms of time, agrochemicals, and labor.

The nanopesticides market is foreseen to double its value in the eight-year period between 2019 and 2027—from US$410 million to US$940 million (https://www.

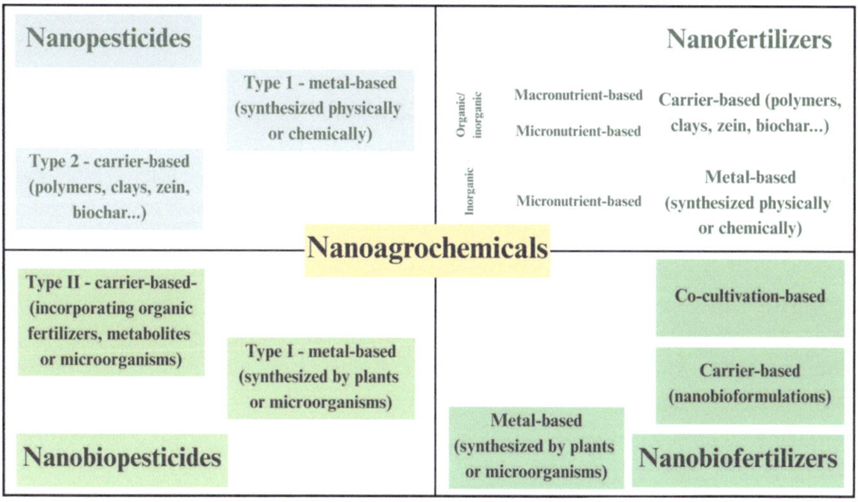

Fig. 7.1 Classification of nanoagrochemicals

credenceresearch.com/report/nanopesticide-market). It is also important to take into account the remark showing that the cost of nanopesticides is 50% lower compared to conventional chemical pesticides, contributing to their cost-effectiveness (Su et al. 2022), while class II nanopesticides are up to 35% more efficient compared to their non-nanoscale analogues, opening a new chapter of possibilities for techno-economic savings and improved environmental footprint (Kah et al. 2018; Wang et al. 2022; Liao et al. 2023). When it comes to nanofertilizers, they show 29% higher efficacy compared to their conventional mineral/inorganic counterparts (Ahmed et al. 2023a), mostly due to the significantly higher loss rate of conventional fertilizers, ranging from 40% to 90% (Ditta and Arshad 2015; Zulfiqar et al. 2019).

7.3 Nanobiopesticides and Nanobiofertilizers

Metal NPs could also be considered class II nanobiopesticides and micronutrient-based nanobiofertilizers exhibiting noticeable antimicrobial and plant growth regulation traits, taking into account the possibility of their nature-based sourcing from plant and microbial biosynthesis routes (Bahrulolum et al. 2021; Sharma et al. 2022). As mentioned before, NPs could be applied solely as active components contributing to plant growth and pathogen protection (e.g., Zn and Fe NPs—Sahai et al. 2019), or in synergistic combination with organic fertilizers, plant/microbial metabolites (Dikbaş and Cinisli 2020), or live microbial active components themselves (e.g., *Bacillus*-based bionanoformulations—Kumar et al. 2021), offering a spectrum of novel green nanobiochemicals. Microbial nanobiofertilizers could be classified as:

(a) microbially synthesized NPs and (b) nanobioformulations enhancing the synergistic activity of NPs and biological active components, including PGP microorganisms (Liao et al. 2023).

Nanotechnology opens a new avenue of possibilities to overcome the current drawbacks of microbial PGP biofertilizers, including temperature sensitivity, dehydration, short storage stability, and production/quality reproducibility (De Moraes et al. 2021). The critical limitations of PGP microorganism application for plant-beneficial purposes are their minimal threshold number and viability in the open field, as well as viability concerns following the downstream procedure and storage period to retain maximal viability before field application (Kour et al. 2019). Hence, a specific carrier/formulation of PGP microorganisms is required to support their viability during transport, storage, and field applications, where nanoformulations could offer a viable alternative for UV radiation, heat, desiccation, and other environmental stresses protection, together with possible synergistic effects in terms of PGP. The lack of literature data on the development of both small-scale and large-scale testing of nano-based bioformulations of biopesticides/biofertilizers leads to the conclusion that this field is still in its infancy, but on the other hand, this fact opens new perspectives towards the development of next-generation nanobiochemicals (Bhattacharjee et al. 2023).

An additional avenue of nanotechnology application in sustainable agriculture is related to genetic engineering pathways, providing the possibility for NPs to deliver carefully selected genetic elements and modify plant traits, as well as possibly silencing/enhancing the expression of certain genes in insects and microbial plant pathogens (Hofmann et al. 2020; Zhang et al. 2021). Another research direction is aimed at the development of smart seed priming nanobiocomposites, providing "scheduled" seed germination in the field depending on temperature, moisture content, or pH value (El-Ramady et al. 2018). The sensing properties of nanomaterials could be employed in the development of next-generation biosensors monitoring plant health status and growth-related parameters, including symbiotic and antagonistic relationships with the surrounding microbial communities (Giraldo et al. 2019).

7.4 Microbial Bionanofactories as a Tool to Promote Plant-Microbe Symbiosis

Besides the physical (top-down) and chemical (bottom-up) approaches in NP synthesis, green or biological synthesis of NPs is receiving rising attention in light of environmental sustainability (Zahra et al. 2022). Microbial biosynthesis of NPs can be considered a sustainable nanomaterials sourcing route for their subsequent agricultural applications, serving as a more sustainable and eco-friendlier alternative compared to conventional agrochemicals in terms of production equipment complexity and cost, chemical utilization, energy consumption, and time (taking approximately 24–48 hours in the case of bacteria) (Xu et al. 2021; Balusamy et al. 2023). It also contributes to improved biocompatibility, less or no toxicity of NPs, along with techno-economic advantages, including easier scale-up and long-term production stability (Kaur et al. 2022). A significant number of research studies have dealt with biopesticidal applications of microbially synthesized NPs, which are summarized by Ahmed et al. (2023b).

On the other hand, microbial synthesis of NPs can also be viewed as a phenomenon taking place at the plant-microbiome-nanomaterials interface. The bottom-up approach assumes the biosynthesis of metal NPs starting from metal ionic solutions, using the metabolic machinery of bacteria, fungi, yeasts, or microalgae, followed by their subsequent downstream processing to obtain metal NPs applicable as nanobiopesticides or nanobiofertilizers (Singh et al. 2016). In contrast, the direct approach considers on-site synthesis of metal NPs by the plant microbiome and their introduction into the plant-microbiome interplay. This approach can be considered highly significant in the case of NPs-synthesizing plant endophytes (Fadiji et al. 2022), providing synergism in the endophyte-NPs correlation and contributing to improved plant physiological states (Adeleke et al. 2022) and responses to biotic and abiotic stresses (Kumar and Nautiyal 2022).

Microbe-nanomaterial interaction can be observed from two different points of view:

(a) Considering the effects of nanomaterials on microbial cell structure and physiological functions, which are mostly influenced by nanomaterials' size and surface characteristics, providing the possibility to overcome cell wall and membrane barriers, but also to remain adhered to the microbial cell surface (Stauber et al. 2018)
(b) Taking into account the effects of microbes on nanomaterials, where it can be observed that a significant share of microorganisms exerts tolerance/resistance to the presence of metal ions, which can be converted to NPs/nanoscale forms by employing diverse and yet underexplored intracellular and extracellular mechanisms of microbial metabolism using the major cell-building and functional macromolecule blocks, including enzymes and other proteins, carbohydrates, and lipids (Huang et al. 2015; Kumari and Singh 2019).

The main currently known mechanisms involved in the microbial biosynthesis of metal NPs are related to the protective responses of microbial cells to the presence of metal ions, including oxido-reduction reactions (with a significant role of enzymes (e.g., reductase) and secondary metabolites in the reduction of metal ions to stable nanoscale forms (Singh et al. 2016)). Microorganisms are considered better NPs sourcing agents than plants due to their modifiable, stimuli-dependent, and possibly unlimited production of the reducing metabolic products (Adeleke et al. 2022), with a special emphasis on plant endophytes (Rahman et al. 2019). Intracellular synthesis of NPs is inevitably followed by biomass recovery (e.g., using centrifugation, microfiltration (Jokić et al. 2019, 2021; 2022), or flocculation (Dmitrović et al. 2022, 2023, 2025) and cell disruption downstream steps to successfully recover NPs, creating significant production costs. On the other hand, extracellular synthesis of NPs is mediated via microbial extracts/microbial biomass-free culture filtrates or supernatants containing cofactor-released reductases (Jadoun et al. 2022). The main factors contributing to the properties of the microbially synthesized NPs include metal salt concentration, temperature, incubation time, and pH value (Ahmed et al. 2023b).

Furthermore, the presence and close proximity of intra- and extracellular biomolecules eliminate the necessity for additional stabilizing agents, considering that the aforementioned biomolecules usually form a capping layer or biocorona around the biosynthesized NPs, supporting their protection, stability, and biocompatibility (Bahrulolum et al. 2021; Balusamy et al. 2023), as well as their specific binding to plant cell surfaces, internalization, and subsequent transport in plant tissues and biodistribution in plant cells, forming a smart delivery system (Doorley and Payne 2012; Chugh et al. 2021). The corona forming on the NP surface can be observed as a reversible and dynamically changeable process dependent on the environmental conditions, where hard coronas last longer due to stronger attachment of biomolecules, while soft coronas can be quickly exchanged depending on the local macromolecular milieu in the rhizosphere/phyllosphere/endosphere, thus affecting short-term and long-term effects of NPs on plant growth (Miclăuş et al. 2016).

However, significant knowledge gaps can be identified in the area of microbial NPs biosynthesis, starting from microbial genetic/phenotypic backgrounds highly dependent on the surrounding conditions, which are further related to biosynthesis/cultivation parameters in terms of nutritional content, environmental/technological parameters, the presence of other microorganisms/contaminants, and time (Liao et al. 2023). The large amount of uncertainty in understanding biosynthesis mechanisms and the resulting by-products generation/interactions with target products result in still underdeveloped processes to obtain NPs of consistent quality and size over time (i.e., on demand) and to provide reproducibility whether it comes to industrial scale-up or on-site biosynthesis of microbial NPs in the plant-microbiome ecosystem.

7.5 Plant-Microbe-Nanomaterial Interactions from the Perspective of Microbial Nano-Based Formulations Development and Application

Microorganisms in the soil-plant ecosystem play a tremendous role in agricultural sustainability, primarily as plant nutrition facilitators via organic matter decomposition and nutrient cycling (CNPS—carbon, nitrogen, phosphorus, and sulfur; Fe, Zn—Sokol et al. 2022), followed by direct PGP and biocontrol mechanisms (Vlajkov et al. 2022), soil remediation ability, and as an overall soil health indicator. The summary of basic direct and indirect PGP mechanisms mediated by microbial influence is provided in Fig. 7.2.

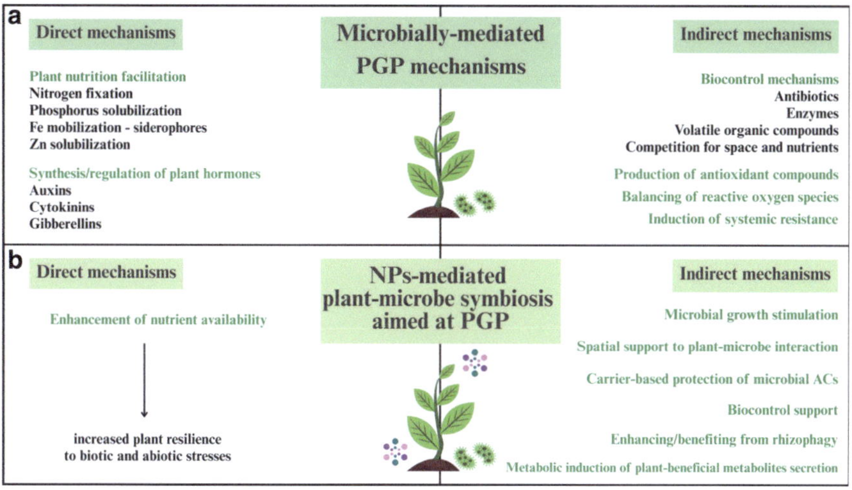

Fig. 7.2 Plant growth promotion mechanisms mediated by: (**a**) microbial active components and (**b**) NPs-induced plant-microbe symbiosis

Introduction of nanoagrochemicals aims to improve plant nutrition directly, via the direct utilization of NPs as nutrient sources, or their microbially modified forms, thus enhancing nutrient availability and increasing plant resilience to various biotic and abiotic stresses (Kalia and Kaur 2019). The indirect mechanisms improving nano-mediated PGP traits of microbial-based bioproducts for sustainable agriculture are exhibited by enhancing microbial growth (cell viability and count, as well as microbial growth rate), establishment of beneficial microbial populations via plant epiphytic/rhizospheric colonization (improving plant surface adhesion due to increased surface area and other mechanisms), as well as via plant penetration by endophytic microorganisms, together with providing long-term positive changes in plant physiological functions (Liao et al. 2023). Furthermore, NPs mediate the metabolic induction of plant-beneficial compounds secretion by microbial cells (such as plant growth hormones, siderophores, enzymes involved in nitrogen fixation, and volatile organic compounds, etc. (Grahovac et al. 2023)), biocontrol support (via competition and antibiosis mechanisms), simultaneously providing protection of microbial active components in the nanocarrier-based formulations (De Moraes et al. 2021). Another important mechanism of nitrogen supplementation in plants is rhizophagy, i.e., the microbiovory process that includes the extraction of nitrogen from microbial cells (cellular membranes, cell walls, proteins, and other nitrogen-containing molecules) which are symbiotic endophytes via the production of reactive oxygen species by plants. The rhizophagy cycle includes nitrogen fixation from the air, endophytic colonization of plants by nitrogen-fixing bacteria, oxidative processes aimed at nutrient extraction, and the exit of the surviving bacterial cells into the rhizosphere to start the novel rhizophagy cycle (White et al. 2018). In the aforementioned cycle, the role of NPs could be predicted in facilitating bacterial plant penetration, as well as in the induction of plant oxidative metabolism, with the possibility for further metabolic processing of the NPs internalized together with symbiotic endophytes (De Moraes et al. 2021). The summary of NPs-mediated plant-microbe symbiotic interactions resulting in PGP is provided in Fig. 7.2.

The synergism between nanoscale materials and plant-beneficial microbes could be achieved through three main pathways:

(a) Co-cultivation-based bioproducts (simultaneous production of joint bio-based products via microbial cultivation processes involving NPs)
(b) Co-inoculation (simultaneous application of separate microbial and nano-based products) and
(c) Protective/carrier-based microbial nanobioformulations (the use of nanomaterials as formulation constituents for microbial-based bioproducts).

Examples of previous investigations of these complex microbes-NPs systems and their resulting effects on plant growth, soil quality, and microbial communities are presented in Table 7.1.

7.5.1 Surface Interactions and Enhancement of Microbe-Plant Attachment

NPs-microbe axis interaction could be observed at the extracellular and intracellular levels, considering the effects of nanomaterials on microbial growth and metabolism in relation to plant-beneficial activities. At the extracellular level, the presence of NPs can induce the synthesis of extracellular polymeric substances facilitating the adhesion of microbial cells to plant tissue, biofilm or clump formation, thus forming cohesive microenvironment units, especially in the root zone, providing better survival and establishment of the dominant target microbial population, thus offering the possibility for the rational design of holobionts (Timmusk et al. 2018). The interaction between nanomaterials and bacterial cells mostly depends on the surface charge and functional groups of NPs, considering the predominantly negative surface charge of bacterial cells (Kurdish 2019). Furthermore, binding patterns of NPs to bacterial cell surfaces usually include the induction of hydrophobic regions or occur at existing hydrophobic points, resulting in bacterial cell aggregation (Hayden et al. 2012). Besides electrostatic interactions, different surface chemical reactions take place, including functional groups both at bacterial cells and NPs' surfaces, mostly targeting phospholipids, proteins, lipoteichoic acid, and lipopolysaccharides (Jiang et al. 2010). Palmqvist et al. (2015) have shown beneficial effects of Ti NPs on clustering and adhesion of plant-beneficial bacteria to oilseed rape, thus contributing to the enhancement of oilseed rape and stress management. Similarly, a higher extent of bacterial attachment and increased metabolic activity were observed on titanium nanophase compared to conventional titania in the case of *Pseudomonas fluorescens* 5RL and *Pseudomonas putida* TVA8 (Park et al. 2008).

Table 7.1 The interactions between nanoscale materials and plant beneficial microbes regarding plant growth modulation under different experimental and environmental conditions: a review of literature data

Nanoscale material(s)	Microorganism(s)	Plant	Experimental conditions	Plant-beneficial or antagonistic outcome	Reference
SiO_2 NPs synthesized from rice husk—foliar spraying (10 mg/L)	*Bacillus megaterium*, *Bacillus brevis*, *Pseudomonas fluorescens*, and *Azotobacter vinelandii*	Maize (*Zea mays*)	Co-inoculation—pot experiments	Improved viable count of PGP bacteria in the soil	Karunakaran et al. (2013)
Fe, Zn, and Mn NPs	*Azotobacter* sp. and *Pseudomonas* sp.	Wheat (*Triticum aestivum*)	Nanobiofertilizer Biozar—field experiments varying sowing date—Foliar spraying (4 L/ha)	Increased spike length and number, seed number, and weight	Mardalipour et al. (2014)
Combined Fe- and K-chelated NPs—soil treatment (0.002375 g/kg and 0.00475 g/kg, respectively)	*Azotobacter* sp.—soil irrigation (10^7–10^8 CFU/g)	Sorghum (*Speedfeed hybrid*)	Co-inoculation—pot experiments	Increased content of chlorophyll a and b, carotenoids, and carbohydrates	Mir et al. (2015)
TiO_2 NPs—seed treatment (50 μg/mL)	*Bacillus amyloliquefaciens* subsp. *plantarum* UCMB5113—seed treatment (10^7 CFU/mL)	Oilseed rape (*Brassica napus*)	Co-inoculation—In vitro plate experiments	Improved adhesion of bacteria to plant root, improved protection against *Alternaria brassicicola* MUCL20297	Palmqvist et al. (2015)
Ag NPs—seed soaking (20 ppm)	*Pseudomonas* sp., *Pseudomonas fluorescens*, and *Bacillus cereus*—seed treatment (10^5 CFU/seed)	Maize (*Zea mays*)	Co-inoculation—open field pot experiments	Increase in root length, modulated levels of abscisic acid, indole acetic acid, and gibberellic acid, increase in proline content, augmented bioremediation potential for Ni, Cd, and Pb	Khan and Bano (2015)

(continued)

Table 7.1 (continued)

Nanoscale material(s)	Microorganism(s)	Plant	Experimental conditions	Plant-beneficial or antagonistic outcome	Reference
ZnO NPs—foliar treatment (0.9 g/L)	*Bradyrhizobium japonicum*, *Azosprillum lipoferum* OF, and *Pseudomonas putida*—seed treatment	Soybean (*Glycine max*)	Co-inoculation—field experiments	Increased dry matter, crop growth rate, plant height, number of nodules and pods per plant, grain yield, and mass	Seyed Sharifi and Khoramdel (2016)
Nano-zeolite-loaded nitrogen nanofertilizer—Soil treatment (100 kg/fed) + foliar treatments	*Bacillus megaterium* and *Azotobacter chroococcum*—Root dipping + soil treatment (1:1 *v/v* suspension, 10^8 CFU/mL)	Caraway (*Carum carvi*)	Co-inoculation—Field experiments	Improved growth characteristics and chemical composition, resulting in maximized gross income per hectare	Mahmoud et al. (2017)
Ag NPs synthesized by *Stenotrophomonas* sp. BHU-S7 (MTCC 5978)—foliar spraying (2–10 μg/mL)	–	Chickpea (*Cicer arietinum*)	Microbial NPs biosynthesis—Greenhouse experiments	Successful management of collar rot caused by *Sclerotium rolfsii* by H_2O_2 production, altered lignification, induction of phenolic acids, and reduced germination of sclerotia	Mishra et al. (2017)
ZnO NPs—Foliar treatment (50 mg Zn/L)	*Rhizobium leguminosarum* biovar *phaseoli*—Soil treatment	Snap beans (*Phaseolus vulgaris*)	Co-inoculation—Field experiments	Increased plant height, leaf area, and pod yield	Morsy et al. (2017)
Gelatin-coated Cu NPs—Root inoculation (10 μg/mL)	*Piriformospora indica*—Root inoculation (10 μg/mL)	Pigeon pea (*Cajanus cajan*)	Co-inoculation—Pot experiments	Increased chlorophyll a content	Rajak et al. (2017)

TiO_2 NPs—Naturally present in sandy soil	*Bacillus thuringiensis* AZP233 and *Paenibacillus polymyxa* A26—Via irrigation (1 mL of 10^5 CFU/mL cultivation broth) + foliar spraying (1 mL of cultivation broth filtrate)	Wheat (*Triticum aestivum*)	Co-inoculation—Pot experiments including drought, salt, and pathogen stress	Improved root attachment of bacteria, increase in seedling biomass	Timmusk et al. (2018)(
Silica NPs synthesized from the stems of *Equisetum telmateia*—Soil treatment (0.05 and 0.07 ppm)	*Pseudomonas stutzeri* and *Mesorhizobium* sp.—Soil treatment	Land cress (*Barbarea verna*)	Co-inoculation—Pot experiments	Increased root and shoot dry weight, as well as N and P soil content	Boroumand et al. (2019)
Fe_2O_3 NPs—Via irrigation (0.5–1.5 g/L)	*Pseudomonas putida* 146—Seed treatment (10^7 CFU/g + 15% w/v gum arabic) Mycorrhizal fungi (*Glomus imoseaea*)—Soil treatment (20 g/m^2)	Maize (*Zea mays*)	Co-inoculation—Field experiments including drought stress	Increased grain yield, 1000 grains mass, and chlorophyll content by the microbial inoculants, without measurable benefits of NPs application	Eliaspour et al. (2020)
Ag NPs synthesized from *Tagetes erecta* (marigold) leaf and flower extracts—Seed treatment (25 ppm) and foliar spraying (50 ppm)	*Bacillus cereus* LPR2—Seed treatment with CMC (carboxymethyl cellulose) solution and bacterial biomass slurry	Maize (*Zea mays*)	Co-inoculation—Pot experiments	Improved seed germination, root and shoot length	Kumar et al. (2020)

(continued)

Table 7.1 (continued)

Nanoscale material(s)	Microorganism(s)	Plant	Experimental conditions	Plant-beneficial or antagonistic outcome	Reference
Ag NPs—Foliar spraying (5 ppm)	*Pseudomonas putida* KX574857 and *Pseudomonas stutzeri* KX574858—Seed treatment (10^6 CFU/mL)	Cucumber (*Cucumis sativus*)	Co-inoculation—Pot experiments	Suppressed root and shoot length, but improved antoxidant and defense enzyme activity	Nawaz and Bano (2019)
ZnO-FeO NPs—Foliar treatment (1.5 g/L)	*Azotobacter chroococcum* 5—Seed inoculation (10^8 CFU/g)	Wheat (*Triticum aestivum* L.)	Co-inoculation—Field experiments including drought stress	Increase in grain yield under severe water limitation	Sharifi et al. (2020)(
ZnO NPs—Soil treatment (40–400 mg Zn/kg)	*Rhizobium japonicum*—Seed treatment	Soybean (*Glycine max* cv. Kowsar)	Co-inoculation—Pot experiments	Concentration-, particle size-, and morphology-dependent influence of NPs on yield, lipid peroxidation, and antioxidant biomarkers	Yusefi-Tanha et al. (2020)
TiO_2 NPs—Soil treatment (500 mg/kg)	*Pseudomonas fluorescens*—Soil treatment (10^7 CFU/kg)	White clover (*Trifolium repens*)	Co-inoculation—Pot experiments	Increased plant biomass, but observed dose-dependent-inhibitory activity of NPs at 1000 mg/kg, increased uptake and accumulation capacity of Cd for phytoremediation purposes	Zand et al. (2020)
BioSe NPs synthesized by *Lactobacillus acidophilus* ML14—Seed treatment (100 µg/mL)	–	Wheat (*Triticum aestivum*)	Microbial NPs biosynthesis—Greenhouse experiments including heat and drought stress	Reduced incidence and severity of crown and root rot diseases (CRDs) induced by *Fusarium culmorum* and *Fusarium graminearum*, enhanced tolerance to heat and drought stress, enhanced plant growth and productivity	El-Saadony et al. (2021)

Silica NPs—Foliar spraying (500 mg/L)	*Azospirillum lipoferum* SP2 and *Bacillus circulance* NCAIM B.02324—Seed treatment (15 mL of 10^8 CFU/mL suspension +30 g of peat)	Maize (*Zea mays*)	Co-inoculation—Pot experiments including drought stress	Increased productivity and nutrient uptake (N, P, K)	Hafez et al. (2021)
Ag NPs synthesized by *Rosmarinus officinalis*—Foliar spraying	*Comamonas testosteroni*—Soil spraying + foliar spraying (10^7 CFU/mL)	Linseed (*Linum usitatissimum*)	Co-inoculation—Pot experiments including salinity (NaCl) stress	Improved content of photosynthetic pigments, soluble sugars, proteins, and proline, increases in enzymatic and nonenzymatic antioxidants content	Khalofah et al. (2021)
ZnO NPs synthesized by *Azospirillum* spp.—Seed treatment (250 μg/mL)	*Azospirillum* sp. 1 and 2	Mung bean (*Vigna radiata*)	Microbial NPs biosynthesis—Pot experiments	Increase of germination rate, leaf area, number of roots, root length, shoot length, fresh weight, and dry weight, increase in chlorophyll a, b, and ab, as well as in carotenoid content	Manivannan et al. (2021)
Magnetite NPs—Soil treatment	*Bacillus* sp. MR-1/2—Seedlings dipping (10^8 CFU/mL)	Rice (*Oryza sativa*)	Co-inoculation—Pot experiments including water stress	Increased N content, plant dry matter content, and kernel weight, reduced content of ethylene, catalase, malondialdehyde, and ascorbate peroxidase	Tahir et al. (2021)

(continued)

Table 7.1 (continued)

Nanoscale material(s)	Microorganism(s)	Plant	Experimental conditions	Plant-beneficial or antagonistic outcome	Reference
Nanophosphozinc—Foliar spraying (200 ppm)	*Acinetobacter baumannii*—Soil drenching	Barley (*Hordeum vulgare*)	Co-inoculation—Field experiments including salinity stress	Increase in dry weight, straw, grain, and plant height	Ibrahim and Hegab (2022)
Silica NPs—Foliar spraying (500 mg/L)	*Azotobacter chroococcum* SARS 10 and *Pseudomonas koreensis* MG209738—Seed treatment (950 g/ha)	Barley (*Hordeum vulgare*)	Co-inoculation—Field experiments including salinity stress	Improved soil microbial count (*Bacillus* and *Azotobacter*), as well as soil enzymatic activity (dehydrogenase and urease), increase in chlorophyll and water content, K/N ratio, N, P, K uptake, stomatal conductivity, and plant enzyme activity	Alharbi et al. (2022)
SiO_2 NPs—Foliar spraying (100 ppm)	*Bacillus cereus*-Amazcala	Chilli pepper (*Capsicum annuum*)	Co-inoculation—Greenhouse pot experiments	Increase in seed germination percentage, plant height, leaves number, and fruit yield	Ferrusquía-Jiménez et al. (2022)
ZnO NPs—Foliar spraying (100 mg/L)	Biozote commercial biofertilizer—Seed treatment (10^7 CFU/mL)	Maize (*Zea mays*)	Co-inoculation—Greenhouse pot experiments including As contamination	Increased water content, fresh and dry mass, soluble sugars, protein and proline content, improved enzymatic antioxidant defense	Khan et al. (2022)
Fe NPs—Via irrigation (1% w/v)	*Bacillus aryabhattai* RSO25—Via irrigation (0.5% inoculum solution in 1 L of water was added per tray containing 12 plants)	Wheat (*Triticum aestivum*)	Co-inoculation—Greenhouse pot experiments	Antagonistic effect between the NPs and bacterial strain was observed, resulting in intermediate Fe accumulation in spikes	Merinero et al. (2022)

ZnO NPs—Via irrigation (150 mg/L)	*Azospirillum brasilense*—Seed treatment	Wheat (*Triticum aestivum*)	Co-inoculation—Field experiments including drought stress	Improved drought tolerance via sustained nutrient uptake and increase in production of photosynthetic pigments and antioxidants	Muhammad et al. (2022)
TiO_2 NPs—Foliar spraying (100 mg/L)	Arbuscular mycorrhizal fungi (AMF—*Funneliformis mosseae*)—Soil treatment (80 g of AMF-enriched soil per planting hole)	Sage (*Salvia officinalis*)	Co-inoculation—Field experiments including drought stress	Increased dry matter yield, water use efficiency, essential oil content, and the net income index	Ostadi et al. (2022a, b)
Chitosan NPs—1 mg/mL	*Pseudomonas monteilii*	Cowpea (*Vigna unguiculata*)	Co-cultivation—Cultivation broth filtrate obtained after cultivation of microbial strain with chitosan NPs—Pot experiments	Increase in shoot length, leaf number, and fresh weight	Panichikkal et al. (2022)
Chitosan-coated mesoporous silica NPs	*Azotobacter vinelandii* DSM 2289 and *Bacillus megaterium* DSM 32	Tomato (*Solanum lycopersicum*) and watermelon (*Citrullus lanatus*)	Microbial nanobiofertilizer (0.1 g of SiNPs +2 mL of bacterial suspensions (10^8 CFU/mL in 50 mL of sterile distilled water)—Greenhouse pot experiments (250 mg/L)	Improvement in number of N-fixing and phosphate-solubilizing bacteria in the soil, increase in chlorophyll, N and P content, and antioxidative properties	Pavlicevic et al. (2022)

(continued)

Table 7.1 (continued)

Nanoscale material(s)	Microorganism(s)	Plant	Experimental conditions	Plant-beneficial or antagonistic outcome	Reference
Carbon nanotubes and SiO_2 NPs	*Bacillus velezensis*	Pistachio (*Pistacia vera*)	Microbial nanobiofertilizer encapsulated in sodium alginate-gelatine microcapsules—Greenhouse experiments—Soil treatment (10 g/kg)	Effective biocontrol of pistachio gummosis (*Phytophthora drechsleri*) + growth-promoting effects on pistachios	Pour et al. (2022)
Fe NPs—Soil treatment (0.5–1%)	*Azotobacter chroococcum* and *Pseudomonas putida*—Seed treatment (10^7 CFU/g, 7 g/ha)	Triticale (*Triticosecale*)	Co-inoculation—Field experiments	Increase in chlorophyll a and b, as well as carotenoid content, increase in N, P, and Fe content	Sepehrzadegan et al. (2021)
CeO_2 NPs nanocomposite with soya chunks as a carrier—Seed treatment (75 ppm/seed)	*Pseudomonas gessardi* and *Pseudomonas azotoformans*—Seed treatment (0.1 mL/seed)	Fenugreek (*Trigonella foenum-graecum*)	Co-inoculation—Pot experiments	Increased seed germination percentage and seedlings vigor index	Sonali et al. (2021)
Se NPs—Seed priming (25 ppm)	*Nocardiopsis* sp.—Seed priming ($2.5 \cdot 10^7$ CFU/mL suspension + vermiculite)	Broccoli (*Brassica oleracea*)	Co-inoculation—In vitro experiments	Increase in fresh and dry weight, accumulation of glucosinolate, improved antioxidant and antibacterial activity	AbdElgawad et al. (2023)
Chitosan NPs—Foliar spraying (0.5% w/v)	Arbuscular mycorrhizal fungi—Soil treatment	Thyme (*Thymus vulgaris*)	Co-inoculation—Field experiments including drought stress	Increased dry yield, essential oil content and composition	Machiani et al. (2023)

TiO_2 NPs—Foliar spraying (100 mg/L)	Arbuscular mycorrhizal fungi (Myco-root—*Funneliformis mosseae*)—70 g of soil containing 1000 spores/10 g per one planting hole	Peppermint (*Mentha piperita*)	Co-inoculation—Field experiments including drought stress	Increase in dry yield and essential oil content	Ostadi et al. (2022a, b)
Fe_2O_3 NPs synthesized from the leaf extract of *Syzigium cumini*—Soil treatment (500 mg/kg)	*Rhizobium pusense*—Seed treatment (10^8 CFU/seed)	Greengram—Mung bean (*Vigna radiata*)	Co-inoculation—Pot experiments	Increased seed germination, plant growth parameters, seed protein content, and pigment content	Saleem and Khan (2022)
ZnS NPs synthesized from *Jacaranda mimosifolia* flower extracts—Seed treatment (2.5 mg/mL)	*Acinetobacter pittii* and *Bacillus velezensis*—Seed treatment (10^6 CFU/mL)	Tomato (*Solanum lycopersicum*)	Co-inoculation—Pot experiments	Increased content of photosynthetic pigments, soluble proteins, sugars, and phenolics, increase in plant fresh and dry biomass, reduced root rot infection incidence caused by *Rhizoctonia solani*, improved activity of antioxidant defense enzymes	Shah et al. (2023)

TiO_2 NPs were also shown to promote wheat root attachment of *Bacillus thuringiensis* AZP233 and *Paenibacillus polymyxa* A26 (Timmusk et al. 2018). The surplus in negative charge of gram-negative bacteria was observed during co-cultivation with SiO_2 NPs, explained by the porin conformation changes due to the adsorption of NPs on the bacterial cell surface (Gordienko and Kurdish 2007), which was facilitated by the hydration of the silica NPs' surface, attracting bacterial cells and providing their acid resistance (Hirota et al. 2009). On the other hand, NPs could improve oxygen transfer, as well as ionic exchange, in the aforementioned microenvironment, supporting microbial growth and beneficial microbes' responses to abiotic stresses (Liao et al. 2023). Additional benefits in terms of upregulation of motility-related genes, thus resulting in improved plant-microbe interaction, were observed, e.g., in the case of Se NPs, which contributed to increased upregulation of *Sphingomonas* flagellar genes and enhanced interaction with rice roots (Jiao et al. 2023).

7.5.2 Induction of Microbial Metabolism Aimed at Plant-Beneficial Compounds Secretion

From the intracellular perspective, the modulating effect of NPs on the secretive metabolism of microbial enzymes and plant growth regulatory molecules was observed (Hussain et al. 2023). Some research results hypothesize that the presence of metal oxide NPs is essential for the microbial synthesis of specific secondary metabolites (Haris and Ahmad 2017). The ATPase activity of *Azotobacter vinelandii* and *Bacillus subtilis* was positively affected by bentonite NPs, while silica NPs contributed to its decrease (Chobotarov et al. 2017). The same study revealed the stimulatory effect of vermiculite NPs on the dehydrogenase activity exhibited by *Azotobacter vinelandii* and the peroxidase activity exhibited by *Bacillus subtilis*, as well as the increase in abscisic acid content in both investigated strains following NP exposure, as it is the main regulator of plant resilience to various abiotic and biotic stresses (Chobotarov et al. 2017). Furthermore, the effects of NPs on microbial synthesis of IAA as the PGP hormone could be modulated depending on the type of NPs (e.g., CuO NPs stimulate IAA production, while the presence of ZnO NPs contributes to lower levels in bacteria—Dimkpa et al. 2012a) and concentration—with higher concentrations of NPs contributing to decreased IAA productivity (Haris and Ahmad 2017). The stimulatory effect of silica NPs was recorded in terms of cytokinin production by *Bacillus subtilis* (Kurdish et al. 2015), similarly to the case of vermiculite NPs and IAA production (Chobotarov et al. 2017). The increased gibberellin productivity was also recorded as induced by silica NPs in the case of *Bacillus cereus*-Amazcala, together with improved phosphate solubilization activity (Ferrusquía-Jiménez et al. 2022). The NPs' dose-dependent phosphate solubilization activity was also observed in the case of *Bacillus amyloliquefaciens*, *Pseudomonas aeruginosa*, and *Pseudomonas fluorescens* (Haris and Ahmad 2017), while the immobilization of *Pseudomonas kilonensis* in a carrier based on alginate-nanoclay maximized phosphatase activity (Safari et al. 2020). Silver NPs contributed to an increase in the nitrogen-fixing capacity of *Nitrosomonas europaea* by modulating the expression of the nitrification-involved genes *amoC2* and *amoA1* (Yang et al. 2013). Similar results were observed in the case of increased expression of nitrogen-fixing related genes by the soil microbial communities in a wheat field because of increased nitrite concentration in the rhizosphere following the presence of CuO NPs (Guan et al. 2020). The stimulatory effects could also be observed in the case of ZnO NPs' effects on siderophore production by the rhizospheric root colonizer *Pseudomonas chlororaphis* O6, where the binding of NPs to siderophores contributes to the release of Zn^{2+} ions into the rhizosphere and enhanced plant nutrition (Dimkpa et al. 2012b). These effects are not only limited to NPs-microbe interactions, but they could also be stimulatory for the plants themselves, affecting the secretion and content of root exudates, thus offering the ability to attract beneficial surrounding motile microorganisms exhibiting both PGP and plant nutrition facilitation traits (Jiao et al. 2023).

7.5.3 Enhanced Plant Nutrition Via Dissolution of NPs

Microbes-NPs interaction axis is usually characterized by the dissolution of NPs in the rhizosphere or phyllosphere by the secreted microbial extracellular substances, thus enabling enhanced plant nutrition (Lv et al. 2018). The role of the plant rhizosphere, epiphytic, and endophytic microbiome is still insufficiently understood, but some studies show the potential of plant symbionts, such as *Bacillus subtilis*, and their extracellular metabolites to have a significant role in the biotransformation of NPs, such as Ag, CdSe, ZnS, CuO, etc., through sulfidation, biomethylation, and dissolution (Peng et al. 2017; Eymard-Vernain et al. 2018; Wei et al. 2023). The beneficial adjustable properties of NPs play a tremendous role, especially when it comes to interaction with other biological systems other than plants (i.e., plant beneficial microbes), offering the possibility to fine-tune NPs' traits to correspond to the target-specific responsive cell gatekeepers, recognition biomolecules, and ligands, resulting in direct PGP, on-demand, and in-time nutrient delivery in a target-specific, time- and stimuli-controlled manner, as well as in improved plant tolerance to adverse environmental conditions, including both biotic and abiotic stresses (Liao et al. 2023).

7.5.4 Induction of Root Exudate Secretion

The special aspect to be looked at more deeply includes the plant metabolic shifts affecting root exudate production, which could be modulated by nano-based and microbially mediated approaches. The plant-synthesized root exudate biomolecules, including malic acid, flavonoids, camalexin, coumarin, and benzoxazinoids, significantly contribute to the establishment of beneficial microbial consortia in the rhizosphere, providing insect and other pest protection (Meel and Saharan 2024). The organic compounds in root exudates are usually classified as low molecular weight compounds, including sugars, amino acids, organic acids, phenolic compounds, etc., and high molecular weight compounds, such as proteins and polysaccharides, in addition to water, ions, free oxygen, and other primary and secondary plant metabolites, whose secretion could be modulated by both microbiome and nanomaterial influences (Das et al. 2022). Increased nitrate accumulation could be observed in the wheat rhizosphere under the influence of CuO NPs, while a similar effect was not recorded in the surrounding soil, emphasizing the necessity to include both temporal and spatial dimensions in the investigation of NPs' effects on soil quality and the associated microbial communities (Guan et al. 2020). Se NPs were found to modulate root exudation of rice, resulting in increased secretion of malic and citric acid, contributing to the increased number of plant-beneficial bacteria from the genera *Streptomyces*, *Sphingomonas*, *Bacillus*, and *Pseudomonas* in the rice rhizosphere (Jiao et al. 2023). Soil microbiome engineering aimed at maximizing these positive effects on soil microbial communities and plants could consequently be achieved by the targeted application of nanobiochemicals, thus enlarging the basis of microbiome- and nano-based agricultural services, resulting in improved nutrient cycling, soil quality enhancement, and advanced plant productivity.

7.5.5 Interaction with Soil Microbial Communities and Soil Quality Considerations

Another important dimension in the plant-microbe-NPs interaction matrix includes soil and its affiliated microbial communities. The impact of NPs application in agriculture, with an emphasis on the consequential effects on the plant-associated microbiome—including the rhizoplane and rhizosphere, phyllosphere, and endosphere—remains largely unexplored, especially considering the feedback loop of phytobiome responses to plant growth and productivity, as well as immune responses, overall fitness, and health, together with insufficiently documented soil quality changes due to the absence of long-term field studies including nanomaterials. For example, the application of ZnO NPs resulted in improved synthesis of exopolysaccharides by *Bacillus* spp., contributing to moisture retention in the soil as well as to improvement in its aggregation properties (You et al. 2017). The contrasting direction also includes the influence of nano-based objects on soil microbial communities, depending on the soil physico-chemical properties, including organic matter and moisture content, buffering and aggregation capacity, the formation of coronas on nanoscale objects, their soil depositions, etc. (Zhang et al. 2024). Furthermore, soil pollutant degradation activity mediated by the soil microbial communities could be enhanced by the introduction of Ag NPs (Chavan and

Nadanathangam 2019). Additionally, the bioremediation potential of rhizospheric plant-beneficial bacteria could be stimulated regarding heavy metals bioremediation, as mediated by Ag NPs too (Khan and Bano 2015). Similar results were reported by Duan et al. (2023) in terms of the co-application of biochar and *Bacillus cereus* in Cu-contaminated soil, resulting in improved lettuce growth. When it comes to bioremediation purposes, nanobiochar represents one of the promising materials (Bhandari et al. 2023), especially considering its porosity and surface-active properties, which provide the possibility of binding plant-beneficial microorganisms and supporting their PGP traits (Vlajkov et al. 2023). Chavan and Nadanathangam (2019) reported bactericidal effects of silver NPs against nitrogen fixators, phosphate mobilizers, and biofilm-forming bacteria; however, the abundance of *Pseudomonas* spp. and *Stenotrophomonas* spp. as plant growth promoters significantly increased. Silver NPs at a concentration of 75 ppm inhibited the population growth of nitrogen fixators and siderophore-producing bacteria in wheat and cowpea; however, a lower concentration of 50 ppm stimulated the growth of phosphosolubilizers (Pallavi et al. 2016), indicating the necessity to enlarge the current knowledge base with novel research results, taking into account variables affecting plant-related, soil-related, microbes-related, and NPs-related parameters. The summary of interactions in the plant-microbe-nanoscale materials-soil matrix is provided in Fig. 7.3.

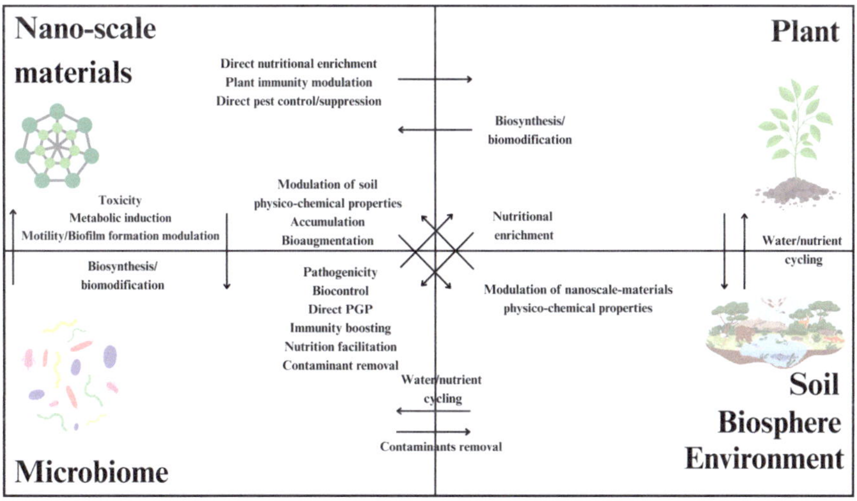

Fig. 7.3 Plant-microbe-nanoscale materials-soil matrix: the summary of main interactions

7.6 Future Research and Application Considerations Regarding Nanoscale Solutions in Agriculture

Continuous efforts are still necessary to provide an understanding of the complex mechanisms of plant-microbe-nanomaterials interaction in these novel systems enriched with additional nanoscale components, especially considering the scarcity of studies investigating the nonselective effects of nano(bio)pesticides and nano(bio) fertilizers, and the resulting effects of their application on plant health, soil composition, quality, and microbial community structure, biodiversity, as well as the consequential food chain effects on the human population. Nanotoxicology explores the ecotoxicological implications of nanoscale particle application in agriculture and other applicable areas, taking into account their size and other physicochemical properties, aggregation tendency, dose/concentration, and exposure time, as well as their interaction with plant hosts and other (micro)organisms in the target ecosystem/soil niche (Mishra et al. 2019). Some general conclusions can be drawn regarding NP toxicity depending on their size, shape, charge, ionic state, and concentration: downsizing NP dimensions contributes to their increased reactivity, uptake, and toxicity, with higher toxicity exhibited in the case of positively charged NPs that are rod-shaped and anisotropic (Singh et al. 2019). Another toxicity variable includes the surface characteristics of bacteria, where it was observed that TiO_2 NPs were more toxic to gram-positive bacteria (Chavan et al. 2020), while on the plant side, the main variables include plant species, growth phase, and exposure time (Jha and Pudake 2016). Due to the lack of documented experimental and field data, it is still uncertain what the fate of NPs introduced into soil/plant ecosystems is, as well as their adsorption, absorption, translocation, and biotransformation in plants (Paramo et al. 2020; Murali et al. 2021). The novel insights in this area are usually obtained through studies investigating plant vasculature loading, symplastic, and apoplastic pathways (Lv et al. 2018), but further research is required regarding the interactions of plant-introduced NPs with plant macromolecules and the endophytic microbiome. Hence, there is a necessity to contribute to the state of the art in the area of long-term effects of NP application in agriculture to prevent nanotoxicity and toxic effects on soil (microbial) communities and the surrounding agricultural ecosystem. One of the major areas to explore is the possible toxic (bio)transformation of the applied NPs, potentially resulting in nonselective transfer across the food chain, including both animals and humans, and biomagnification (Wei et al. 2023). Besides the biosafety and environmental footprint of nanoscale particles, knowledge gaps are also evident in the areas of organic and microbial-based nanobiochemicals, as well as in the techno-economic viability of production and adoption of nanoscale solutions in agriculture, paving novel research directions to be pursued in the future to provide real-life contributions of nanobiotechnology to agricultural and food sustainability.

References

AbdElgawad H, Korany SM, Reyad AM, Zahid I, Akhter N, Alsherif E, Sheteiwy MS, Shah AA, Selim S, Hassan AHA, Khanghahi MY, Beemster GTS, Crecchio C (2023) Synergistic impacts of plant-growth-promoting bacteria and selenium nanoparticles on improving the nutritional value and biological activities of three cultivars of *Brassica* sprouts. ACS Omega 8(29):26414–26424. https://doi.org/10.1021/acsomega.3c02957

Adeleke BS, Akinola SA, Adedayo AA, Glick BR, Babalola OO (2022) Synergistic relationship of endophyte-nanomaterials to alleviate abiotic stress in plants. Front Environ Sci 10. https://doi.org/10.3389/fenvs.2022.1015897

Ahmed A, He P, He P, Wu Y, He Y, Munir S (2023a) Environmental effect of agriculture-related manufactured nano-objects on soil microbial communities. Environ Int 173:107819. https://doi.org/10.1016/j.envint.2023.107819

Ahmed T, Luo J, Noman M, Ijaz M, Wang X, Masood HA, Manzoor N, Wang Y, Li B (2023b) Microbe-mediated nanoparticle intervention for the management of plant diseases. Crop Health 1(1):3. https://doi.org/10.1007/s44297-023-00006-9

Albalghiti E, Stabryla LM, Gilbertson LM, Zimmerman JB (2020) Towards resolution of antibacterial mechanisms in metal and metal oxide nanomaterials: a meta-analysis of the influence of study design on mechanistic conclusions. Environ Sci Nano 8(1):37–66. https://doi.org/10.1039/d0en00949k

Alharbi K, Rashwan E, Mohamed HH, Awadalla A, Omara AE, Hafez EM, Alshaal T (2022) Application of silica nanoparticles in combination with two bacterial strains improves the growth, antioxidant capacity and production of barley irrigated with saline water in salt-affected soil. Plants 11(15):2026. https://doi.org/10.3390/plants11152026

Bahrulolum H, Nooraei S, Javanshir N, Tarrahimofrad H, Mirbagheri VS, Easton AJ, Ahmadian G (2021) Green synthesis of metal nanoparticles using microorganisms and their application in the agrifood sector. J Nanobiotechnol 19(1). https://doi.org/10.1186/s12951-021-00834-3

Balusamy SR, Joshi AS, Perumalsamy H, Mijakovic I, Singh P (2023) Advancing sustainable agriculture: a critical review of smart and eco-friendly nanomaterial applications. J Nanobiotechnol 21(1). https://doi.org/10.1186/s12951-023-02135-3

Bhandari G, Gangola S, Dhasmana A, Rajput V, Gupta S, Malik S, Slama P (2023) Nano-biochar: recent progress, challenges, and opportunities for sustainable environmental remediation. Front Microbiol 14. https://doi.org/10.3389/fmicb.2023.1214870

Bhattacharjee A, Dubey S, Sharma S (2023) "Next-generation bioformulations" for plant growth promotion and stress mitigation: a promising approach for sustainable agriculture. J Plant Growth Regul 42(10):6741–6759. https://doi.org/10.1007/s00344-023-10996-z

Boroumand N, Behbahani M, Dini G (2019) Combined effects of phosphate solubilizing bacteria and nanosilica on the growth of land cress plant. J Soil Sci Plant Nutr 20(1):232–243. https://doi.org/10.1007/s42729-019-00126-8

Chavan S, Nadanathangam V (2019) Effects of nanoparticles on plant growth-promoting bacteria in Indian agricultural soil. Agronomy 9(3):140. https://doi.org/10.3390/agronomy9030140

Chavan S, Sarangdhar V, Nadanathangam V (2020) Toxicological effects of TiO_2 nanoparticles on plant growth promoting soil bacteria. Emerg Contamin 6:87–92. https://doi.org/10.1016/j.emcon.2020.01.003

Chobotarov A, Volkogon M, Voytenko L, Kurdish I (2017) Accumulation of phytohormones by soil bacteria *Azotobacter vinelandii* and *Bacillus subtilis* under the influence of nanomaterials. J Microbiol Biotechnol Food Sci 7(3):271–274. https://doi.org/10.15414/jmbfs.2017/18.7.3.271-274

Chugh G, Siddique KHM, Solaiman ZM (2021) Nanobiotechnology for agriculture: smart technology for combating nutrient deficiencies with nanotoxicity challenges. Sustainability 13(4):1781. https://doi.org/10.3390/su13041781

Das PP, Singh KR, Nagpure G, Mansoori A, Singh RP, Ghazi IA, Kumar A, Singh J (2022) Plant-soil-microbes: a tripartite interaction for nutrient acquisition and better plant growth for sustainable agricultural practices. Environ Res 214:113821. https://doi.org/10.1016/j.envres.2022.113821

De Moraes ACP, Da Silva Ribeiro L, De Camargo ER, Lacava PT (2021) The potential of nanomaterials associated with plant growth-promoting bacteria in agriculture. 3 Biotech 11(7):318. https://doi.org/10.1007/s13205-021-02870-0

Dikbaş N, Cinisli KT (2020) Microbial metabolites powered by nanoparticles could be used as pesticides in future? (NanoBioPesticides). Biotechnol J Int:1–4. https://doi.org/10.9734/bji/2019/v23i430088

Dimkpa CO, Zeng J, McLean JE, Britt DW, Zhan J, Anderson AJ (2012a) Production of indole-3-acetic acid via the indole-3-acetamide pathway in the plant-beneficial bacterium *Pseudomonas chlororaphis* O6 is inhibited by ZnO nanoparticles but enhanced by CuO nanoparticles. Appl Environ Microbiol 78(5):1404–1410. https://doi.org/10.1128/aem.07424-11

Dimkpa CO, Mclean JE, Britt DW, Anderson AJ (2012b) CuO and ZnO nanoparticles differently affect the secretion of fluorescent siderophores in the beneficial root colonizer, *Pseudomonas chlororaphis* O6. Nanotoxicology 6(6):635–642. https://doi.org/10.3109/17435390.2011.598246

Ditta A, Arshad M (2015) Applications and perspectives of using nanomaterials for sustainable plant nutrition. Nanotechnol Rev 5(2). https://doi.org/10.1515/ntrev-2015-0060

Dmitrović S, Pajčin I, Lukić N, Vlajkov V, Grahovac M, Grahovac J, Jokić A (2022) Taguchi grey relational analysis for multi-response optimization of *bacillus* bacteria flocculation recovery from fermented broth by chitosan to enhance biocontrol efficiency. Polymers 14(16):3282. https://doi.org/10.3390/polym14163282

Dmitrović S, Lukić N, Pajčin I, Vlajkov V, Grahovac J, Jokić A (2023) The use of chitosan for flocculation recovery of *bacillus* biomass grown on dairy and wine industry effluents. Processes 11(4):1099. https://doi.org/10.3390/pr11041099

Dmitrović S, Lukić N, Danilov I, Vlajkov V, Grahovac J, Jokić A (2025) Synergistic effect of bacillus and chitosan: from flocculation to enhanced antimicrobial activity. Antibiotics 14(4):412. https://doi.org/10.3390/antibiotics14040412

Doorley GW, Payne CK (2012) Nanoparticles act as protein carriers during cellular internalization. Chem Commun 48(24):2961. https://doi.org/10.1039/c2cc16937a

Duan M, Li Z, Yan R, Zhou B, Su L, Li M, Xu H, Zhang Z (2023) Mechanism for combined application of biochar and *Bacillus cereus* to reduce antibiotic resistance genes in copper contaminated soil and lettuce. Sci Total Environ 884:163422. https://doi.org/10.1016/j.scitotenv.2023.163422

Dutta S, Pal S, Panwar P, Sharma RK, Bhutia PL (2022) Biopolymeric nanocarriers for nutrient delivery and crop biofortification. ACS Omega 7(30):25909–25920. https://doi.org/10.1021/acsomega.2c02494

Eliaspour S, Sharifi RS, Shirkhani A, Farzaneh S (2020) Effects of biofertilizers and iron nano-oxide on maize yield and physiological properties under optimal irrigation and drought stress conditions. Food Sci Nutr 8(11):5985–5998. https://doi.org/10.1002/fsn3.1884

El-Ramady H, El-Ghamry A, Mosa A, Alshaal T (2018) Nanofertilizers vs. biofertilizers: new insights. Environment, Biodiversity and Soil Security 2(1):40–50. https://doi.org/10.21608/jenvbs.2018.3880.1029

El-Saadony MT, Saad AM, Najjar AA, Alzahrani SO, Alkhatib FM, Shafi ME, Selem E, Desoky EM, Fouda SE, El-Tahan AM, Hassan MA (2021) The use of biological selenium nanoparticles to suppress *Triticum aestivum* L. crown and root rot diseases induced by *Fusarium* species and improve yield under drought and heat stress. Saudi J Biol Sci 28(8):4461–4471. https://doi.org/10.1016/j.sjbs.2021.04.043

Eymard-Vernain E, Lelong C, Del Real AP, Soulas R, Bureau S, Suarez VT, Gallet B, Proux O, Castillo-Michel H, Sarret G (2018) Impact of a model soil microorganism and of its secretome on the fate of silver nanoparticles. Environ Sci Technol 52(1):71–78. https://doi.org/10.1021/acs.est.7b04071

Fadiji AE, Mortimer PE, Xu J, Ebenso EE, Babalola OO (2022) Biosynthesis of nanoparticles using endophytes: a novel approach for enhancing plant growth and sustainable agriculture. Sustainability 14(17):10839. https://doi.org/10.3390/su141710839

Ferrusquía-Jiménez NI, González-Arias B, Rosales A, Esquivel K, Escamilla-Silva EM, Ortega-Torres AE, Guevara-González RG (2022) Elicitation of *Bacillus cereus*-Amazcala (B.C-A) with SiO_2 nanoparticles improves its role as a plant growth-promoting bacteria (PGPB) in chili pepper plants. Plants 11(24):3445. https://doi.org/10.3390/plants11243445

Giraldo JP, Wu H, Newkirk GM, Kruss S (2019) Nanobiotechnology approaches for engineering smart plant sensors. Nat Nanotechnol 14(6):541–553. https://doi.org/10.1038/s41565-019-0470-6

Gordienko AS, Kurdish IK (2007) Surface electrical properties of *Bacillus subtilis* cells and the effect of interaction with silicon dioxide particles. Biophysics 52(2):217–219. https://doi.org/10.1134/s0006350907020121

Grahovac J, Pajčin I, Vlajkov V (2023) *Bacillus* VOCs in the context of biological control. Antibiotics 12(3):581. https://doi.org/10.3390/antibiotics12030581

Grillo R, Mattos BD, Antunes DR, Forini MM, Monikh FA, Rojas OJ (2021) Foliage adhesion and interactions with particulate delivery systems for plant nanobionics and intelligent agriculture. Nano Today 37:101078. https://doi.org/10.1016/j.nantod.2021.101078

Guan X, Gao X, Avellan A, Spielman-Sun E, Xu J, Laughton S, Yun J, Zhang Y, Bland GD, Zhang Y, Zhang R, Wang X, Casman EA, Lowry GV (2020) CuO nanoparticles alter the rhizospheric bacterial community and local nitrogen cycling for wheat grown in a calcareous soil. Environ Sci Technol 54(14):8699–8709. https://doi.org/10.1021/acs.est.0c00036

Hafez EM, Osman HS, Gowayed SM, Okasha SA, Omara AE, Sami R, El-Monem AMA, El-Razek UA (2021) Minimizing the adversely impacts of water deficit and soil salinity on maize growth and productivity in response to the application of plant growth-promoting rhizobacteria and silica nanoparticles. Agronomy 11(4):676. https://doi.org/10.3390/agronomy11040676

Haris Z, Ahmad I (2017) Impact of metal oxide nanoparticles on beneficial soil microorganisms and their secondary metabolites. SSR Inst Int J Life Sci 3(3). https://doi.org/10.21276/ijlssr.2017.3.3.10

Hayden SC, Zhao G, Saha K, Phillips RL, Li X, Miranda OR, Rotello VM, El-Sayed MA, Schmidt-Krey I, Bunz UHF (2012) Aggregation and interaction of cationic nanoparticles on bacterial surfaces. J Am Chem Soc 134(16):6920–6923. https://doi.org/10.1021/ja301167y

Hirota R, Hata Y, Ikeda T, Ishida T, Kuroda A (2009) The silicon layer supports acid resistance of *Bacillus cereus* spores. J Bacteriol 192(1):111–116. https://doi.org/10.1128/jb.00954-09

Hofmann T, Lowry GV, Ghoshal S, Tufenkji N, Brambilla D, Dutcher JR, Gilbertson LM, Giraldo JP, Kinsella JM, Landry MP, Lovell W, Naccache R, Paret M, Pedersen JA, Unrine JM, White JC, Wilkinson KJ (2020) Technology readiness and overcoming barriers to sustainably implement nanotechnology-enabled plant agriculture. Nat Food 1(7):416–425. https://doi.org/10.1038/s43016-020-0110-1

Hu J, Xianyu Y (2021) When nano meets plants: a review on the interplay between nanoparticles and plants. Nano Today 38:101143. https://doi.org/10.1016/j.nantod.2021.101143

Huang J, Lin L, Sun D, Chen H, Yang D, Li Q (2015) Bio-inspired synthesis of metal nanomaterials and applications. Chem Soc Rev 44(17):6330–6374. https://doi.org/10.1039/c5cs00133a

Hussain M, Shakoor N, Adeel M, Ahmad MA, Zhou H, Zhang Z, Xu M, Rui Y, White JC (2023) Nano-enabled plant microbiome engineering for disease resistance. Nano Today 48:101752. https://doi.org/10.1016/j.nantod.2023.101752

Ibrahim G, Hegab R (2022) Improving yield of barley using bio and nano fertilizers under saline conditions. Egypt J Soil Sci 62:41–53. https://doi.org/10.21608/ejss.2022.124377.1496

Jadoun S, Chauhan NPS, Zarrintaj P, Barani M, Varma RS, Chinnam S, Rahdar A (2022) Synthesis of nanoparticles using microorganisms and their applications: a review. Environ Chem Lett 20(5):3153–3197. https://doi.org/10.1007/s10311-022-01444-7

Jha S, Pudake RN (2016) Molecular mechanism of plant–nanoparticle interactions. In: Kole C, Kumar DS, Khodakovskaya MV (eds) Plant nanotechnology: principles and practices. Springer, Germany, pp 155–181. https://doi.org/10.1007/978-3-319-42154-4_7

Jiang W, Yang K, Vachet RW, Xing B (2010) Interaction between oxide nanoparticles and biomolecules of the bacterial cell envelope as examined by infrared spectroscopy. Langmuir 26(23):18071–18077. https://doi.org/10.1021/la103738e

Jiao L, Cao X, Wang C, Chen F, Zou H, Yue L, Wang Z (2023) Crosstalk between in situ root exudates and rhizobacteria to promote rice growth by selenium nanomaterials. Sci Total Environ 878:163175. https://doi.org/10.1016/j.scitotenv.2023.163175

Jokić A, Pajčin I, Grahovac J, Lukić N, Dodić J, Rončević Z, Šereš Z (2019) Energy efficient turbulence promoter flux-enhanced microfiltration for the harvesting of rod-shaped bacteria using tubular ceramic membrane. Process Saf Environ Prot 150:359–368. https://doi.org/10.1016/j.cherd.2019.08.014

Jokić A, Pajčin I, Lukić N, Vlajkov V, Kiralj A, Dmitrović S, Grahovac J (2021) Modeling and optimization of gas sparging-assisted bacterial cultivation broth microfiltration by response surface methodology and genetic algorithm. Membranes 11(9):681. https://doi.org/10.3390/membranes11090681

Jokić A, Lukić N, Pajčin I, Vlajkov V, Dmitrović S, Grahovac J (2022) Kenics static mixer combined with gas sparging for the improvement of cross-flow microfiltration: modeling and optimization. Membranes 12(7):690. https://doi.org/10.3390/membranes12070690

Kah M, Kookana RS, Gogos A, Bucheli TD (2018) A critical evaluation of nanopesticides and nanofertilizers against their conventional analogues. Nat Nanotechnol 13(8):677–684. https://doi.org/10.1038/s41565-018-0131-1

Kalia A, Kaur H (2019) Nano-biofertilizers: harnessing dual benefits of nano-nutrient and biofertilizers for enhanced nutrient use efficiency and sustainable productivity. In: Pudake R, Chauhan N, Kole C (eds) Nanoscience for sustainable agriculture. Springer, Germany, pp 51–73. https://doi.org/10.1007/978-3-319-97852-9_3

Karunakaran G, Suriyaprabha R, Manivasakan P, Yuvakkumar R, Rajendran V, Prabu P, Kannan N (2013) Effect of nanosilica and silicon sources on plant growth promoting rhizobacteria, soil nutrients and maize seed germination. IET Nanobiotechnol 7(3):70–77. https://doi.org/10.1049/iet-nbt.2012.0048

Kaur M, Gautam A, Guleria P, Singh K, Kumar V (2022) Green synthesis of metal nanoparticles and their environmental applications. Curr Opin Environ Sci Health 29:100390. https://doi.org/10.1016/j.coesh.2022.100390

Khalofah A, Kilany M, Migdadi H (2021) Phytostimulatory influence of *Comamonas testosteroni* and silver nanoparticles on *Linum usitatissimum* L. under salinity stress. Plants 10(4):790. https://doi.org/10.3390/plants10040790

Khan N, Bano A (2015) Role of plant growth promoting rhizobacteria and Ag-nano particle in the bioremediation of heavy metals and maize growth under municipal wastewater irrigation. Int J Phytoremediation 18(3):211–221. https://doi.org/10.1080/15226514.2015.1064352

Khan I, Saeed K, Khan I (2017) Nanoparticles: properties, applications and toxicities. Arab J Chem 12(7):908–931. https://doi.org/10.1016/j.arabjc.2017.05.011

Khan MA, Yasmin H, Shah ZA, Rinklebe J, Alyemeni MN, Ahmad P (2022) Co application of biofertilizer and zinc oxide nanoparticles upregulate protective mechanism culminating improved arsenic resistance in maize. Chemosphere 294:133796. https://doi.org/10.1016/j.chemosphere.2022.133796

Kour D, Rana KL, Yadav AN, Yadav N, Kumar M, Kumar V, Vyas P, Dhaliwal HS, Saxena AK (2019) Microbial biofertilizers: bioresources and eco-friendly technologies for agricultural and environmental sustainability. Biocatal Agric Biotechnol 23:101487. https://doi.org/10.1016/j.bcab.2019.101487

Kumar V, Nautiyal CS (2022) Plant abiotic and biotic stress alleviation: from an endophytic microbial perspective. Curr Microbiol 79(10):311. https://doi.org/10.1007/s00284-022-03012-2

Kumar S, Nehra M, Dilbaghi N, Marrazza G, Hassan AA, Kim K (2018) Nano-based smart pesticide formulations: emerging opportunities for agriculture. J Control Release 294:131–153. https://doi.org/10.1016/j.jconrel.2018.12.012
Kumar P, Pahal V, Gupta A, Vadhan R, Chandra H, Dubey RC (2020) Effect of silver nanoparticles and *Bacillus cereus* LPR2 on the growth of *Zea mays*. Sci Rep 10(1). https://doi.org/10.1038/s41598-020-77460-w
Kumar P, Pandhi S, Mahato DK, Kamle M, Mishra A (2021) *Bacillus*-based nano-bioformulations for phytopathogens and insect–pest management. Egypt J Biol Pest Control 31(1). https://doi.org/10.1186/s41938-021-00475-6
Kumari R, Singh DP (2019) Nano-biofertilizer: an emerging eco-friendly approach for sustainable agriculture. Proc Natl Acad Sci India, Sect B 90(4):733–741. https://doi.org/10.1007/s40011-019-01133-6
Kurdish IK (2019) Interaction of microorganisms with nanomaterials as a basis for creation of high-efficiency biotechnological preparations. In: Prasad R, Kumar V, Kumar M, Choudhary D (eds) Nanotechnology in the life sciences. Springer, Germany, pp 259–287. https://doi.org/10.1007/978-3-030-17061-5_11
Kurdish I, Roy A, Skorochod I, Chobotarov A, Herasimenko I, Plotnikov V, Gylchuk V, Korniychuk A (2015) Free-flowing complex bacterial preparation for crop and efficiency of its use in agroecosystems. J Microbiol Biotechnol Food Sci 4(6):527–531. https://doi.org/10.15414/jmbfs.2015.4.6.527-531
Liao Y, Xu D, Cao Y, Zhu Y (2023) Advancing sustainable agriculture: enhancing crop nutrition with next-generation nanotech-based fertilizers. Nano Res 16(12):13205–13225. https://doi.org/10.1007/s12274-023-6284-8
Lv J, Christie P, Zhang S (2018) Uptake, translocation, and transformation of metal-based nanoparticles in plants: recent advances and methodological challenges. Environ Sci Nano 6(1):41–59. https://doi.org/10.1039/c8en00645h
Machiani MA, Javanmard A, Ostadi A, Alizadeh K (2023) Improvement in essential oil quantity and quality of thyme (*Thymus vulgaris* L.) by integrative application of chitosan nanoparticles and arbuscular mycorrhizal fungi under water stress conditions. Plants 12(7):1422. https://doi.org/10.3390/plants12071422
Mahmoud AWM, El-Attar ABE, Mahmoud AA (2017) Economic evaluation of nano and organic fertilisers as an alternative source to chemical fertilisers on *Carum carvi* L. plant yield and components. Agriculture (Pol Nohospodárstvo) 63(1):35–51. https://doi.org/10.1515/agri-2017-0004
Manivannan N, Aswathy S, Malaikozhundan B, Boopathi T (2021) Nano-zinc oxide synthesized using diazotrophic *Azospirillum* improves the growth of mung bean, *Vigna radiata*. Int Nano Lett 11(4):405–415. https://doi.org/10.1007/s40089-021-00351-z
Mardalipour M, Zahedi H, Sharghi Y (2014) Evaluation of nano biofertilizer efficiency on agronomic traits of spring wheat at different sowing date. Biol Forum 6:349
Meel S, Saharan BS (2024) Enhancing crop resilience towards drought: by integrating nanotechnology, microbiomes, and growth-promoting rhizobacteria. Discov Agric 2(1). https://doi.org/10.1007/s44279-024-00131-1
Merinero M, Alcudia A, Begines B, Martínez G, Martín-Valero MJ, Pérez-Romero JA, Mateos-Naranjo E, Redondo-Gómez S, Navarro-Torre S, Torres Y, Merchán F, Rodríguez-Llorente ID, Pajuelo E (2022) Assessing the biofortification of wheat plants by combining a plant growth-promoting rhizobacterium (PGPR) and polymeric Fe-nanoparticles: allies or enemies? Agronomy 12(1):228. https://doi.org/10.3390/agronomy12010228
Miclăuş T, Beer C, Chevallier J, Scavenius C, Bochenkov VE, Enghild JJ, Sutherland DS (2016) Dynamic protein coronas revealed as a modulator of silver nanoparticle sulphidation *in vitro*. Nat Commun 7(1). https://doi.org/10.1038/ncomms11770
Mir S, Sirousmehr A, Shirmohammadi E (2015) Effect of nano and biological fertilizers on carbohydrate and chlorophyll content of forage sorghum (Speedfeed hybrid). Int J Biosci 6(4):157–164. https://doi.org/10.12692/ijb/6.4.157-164

Mishra S, Singh BR, Naqvi AH, Singh HB (2017) Potential of biosynthesized silver nanoparticles using *Stenotrophomonas* sp. BHU-S7 (MTCC 5978) for management of soil-borne and foliar phytopathogens. Sci Rep 7(1). https://doi.org/10.1038/srep45154

Mishra M, Dashora K, Srivastava A, Fasake VD, Nag RH (2019) Prospects, challenges and need for regulation of nanotechnology with special reference to India. Ecotoxicol Environ Saf 171:677–682. https://doi.org/10.1016/j.ecoenv.2018.12.085

Morsy NM, Shams AS, Abdel-Salam MA (2017) Zinc foliar spray on snap beans using nano-Zn with N-soil application using mineral, organic and biofertilizer. Middle East J Agric Res 6:1301–1312

Muhammad F, Raza M a S, Iqbal R, Zulfiqar F, Aslam MU, Yong JWH, Altaf MA, Zulfiqar B, Amin J, Ibrahim MA (2022) Ameliorating drought effects in wheat using an exclusive or co-applied rhizobacteria and ZnO nanoparticles. Biology 11(11):1564. https://doi.org/10.3390/biology11111564

Murali M, Gowtham H, Singh SB, Shilpa N, Aiyaz M, Alomary MN, Alshamrani M, Salawi A, Almoshari Y, Ansari MA, Amruthesh K (2021) Fate, bioaccumulation and toxicity of engineered nanomaterials in plants: current challenges and future prospects. Sci Total Environ 811:152249. https://doi.org/10.1016/j.scitotenv.2021.152249

Nawaz S, Bano A (2019) Effects of PGPR (*pseudomonas* sp.) and ag-nanoparticles on enzymatic activity and physiology of cucumber. Recent Pat Food Nutr Agric 11(2):124–136. https://doi.org/10.2174/2212798410666190716162340

Ostadi A, Javanmard A, Machiani MA, Kakaei K (2022a) Optimizing antioxidant activity and phytochemical properties of peppermint (*Mentha piperita* L.) by integrative application of biofertilizer and stress-modulating nanoparticles under drought stress conditions. Plants 12(1):151. https://doi.org/10.3390/plants12010151

Ostadi A, Javanmard A, Machiani MA, Sadeghpour A, Maggi F, Nouraein M, Morshedloo MR, Hano C, Lorenzo JM (2022b) Co-application of TiO_2 nanoparticles and arbuscular mycorrhizal fungi improves essential oil quantity and quality of sage (*Salvia officinalis* L.) in drought stress conditions. Plants 11(13):1659. https://doi.org/10.3390/plants11131659

Pallavi N, Mehta CM, Srivastava R, Arora S, Sharma AK (2016) Impact assessment of silver nanoparticles on plant growth and soil bacterial diversity. 3 Biotech 6(2):254. https://doi.org/10.1007/s13205-016-0567-7

Palmqvist NGM, Bejai S, Meijer J, Seisenbaeva GA, Kessler VG (2015) Nano titania aided clustering and adhesion of beneficial bacteria to plant roots to enhance crop growth and stress management. Sci Rep 5:10146. https://doi.org/10.1038/srep10146

Panichikkal J, Mohanan DP, Koramkulam S, Krishnankutty RE (2022) Chitosan nanoparticles augmented indole-3-acetic acid production by rhizospheric *pseudomonas monteilii*. J Basic Microbiol 62(12):1467–1474. https://doi.org/10.1002/jobm.202100358

Paramo LA, Feregrino-Pérez AA, Guevara R, Mendoza S, Esquivel K (2020) Nanoparticles in agroindustry: applications, toxicity, challenges, and trends. Nanomaterials 10(9):1654. https://doi.org/10.3390/nano10091654

Park MR, Banks MK, Applegate B, Webster TJ (2008) Influence of nanophase titania topography on bacterial attachment and metabolism. Int J Nanomedicine 3(4):497–504. https://doi.org/10.2147/ijn.s4399

Pavlicevic M, Abdelraheem W, Zuverza-Mena N, O'Keefe T, Mukhtar S, Ridge G, Ranciato J, Haynes C, Elmer W, Pignatello J, Pagano L, Caldara M, Marmiroli M, Maestri E, Marmiroli N, White JC (2022) Engineered nanoparticles, natural nanoclay and biochar, as carriers of plant-growth promoting bacteria. Nanomaterials 12(24):4474. https://doi.org/10.3390/nano12244474

Peng C, Xu C, Liu Q, Sun L, Luo Y, Shi J (2017) Fate and transformation of CuO nanoparticles in the soil–rice system during the life cycle of rice plants. Environ Sci Technol 51(9):4907–4917. https://doi.org/10.1021/acs.est.6b05882

Pour MM, Riseh RS, Skorik YA (2022) Sodium alginate–gelatin nanoformulations for encapsulation of *bacillus velezensis* and their use for biological control of pistachio gummosis. Materials 15(6):2114. https://doi.org/10.3390/ma15062114

Rahman S, Rahman L, Khalil AT, Ali N, Zia D, Ali M, Shinwari ZK (2019) Endophyte-mediated synthesis of silver nanoparticles and their biological applications. Appl Microbiol Biotechnol 103(6):2551–2569. https://doi.org/10.1007/s00253-019-09661-x

Rajak J, Bawaskar M, Rathod D, Agarkar G, Nagaonkar D, Gade A, Rai M (2017) Interaction of copper nanoparticles and an endophytic growth promoter *Piriformospora indica* with *Cajanus cajan*. J Sci Food Agric 97(13):4562–4570. https://doi.org/10.1002/jsfa.8324

Safari M, Motamedi E, Dolatabad HK, Sanavy S a MM (2020) Nano-carriers effects on the viability and efficiency of *pseudomonas* strains as phosphate solubilizing bacteria. Heliyon 6(10):e05076. https://doi.org/10.1016/j.heliyon.2020.e05076

Sahai P, Sinha VB, Dutta R (2019) Bioformulation and nanotechnology in pesticide and fertilizer delivery system for eco-friendly agriculture: a review. Acta Sci Agric 3(11):02–10. https://doi.org/10.31080/asag.2019.03.0675

Saleem S, Khan MS (2022) Phyto-interactive impact of green synthesized iron oxide nanoparticles and *Rhizobium pusense* on morpho-physiological and yield components of greengram. Plant Physiol Biochem 194:146–160. https://doi.org/10.1016/j.plaphy.2022.11.013

Seleiman MF, Almutairi KF, Alotaibi M, Shami A, Alhammad BA, Battaglia ML (2020) Nano-fertilization as an emerging fertilization technique: why can modern agriculture benefit from its use? Plants 10(1):2. https://doi.org/10.3390/plants10010002

Sepehrzadegan Z, Alizadeh O (2021) Investigation of the growth bacteria and nano iron on the chlorophyll and some nutrients triticale. Rev Agrogeoambiental 13(1). https://doi.org/10.18406/2316-1817v13n120211572

Seyed Sharifi R, Khoramdel S (2016) Effects of nano-zinc oxide and seed inoculation by plant growth promoting rhizobacteria (PGPR) on yield, yield components and grain filling period of soybean (*Glycine max* L.). Iran J Field Crops Res 13:738–753

Shah ZM, Naz R, Naz S, Zahoor S, Nosheen A, Shahid M, Anwar Z, Keyani R (2023) Incorporation of zinc sulfide nanoparticles, *Acinetobacter pittii* and *Bacillus velezensis* to improve tomato plant growth, biochemical attributes and resistance against *Rhizoctonia solani*. Plant Physiol Biochem 202:107909. https://doi.org/10.1016/j.plaphy.2023.107909

Sharifi RS, Khalilzadeh R, Pirzad A, Anwar S (2020) Effects of biofertilizers and nano zinc-iron oxide on yield and physicochemical properties of wheat under water deficit conditions. Commun Soil Sci Plant Anal 51(19):2511–2524. https://doi.org/10.1080/00103624.2020.1845350

Sharma B, Tiwari S, Kumawat KC, Cardinale M (2022) Nano-biofertilizers as bio-emerging strategies for sustainable agriculture development: potentiality and their limitations. Sci Total Environ 860:160476. https://doi.org/10.1016/j.scitotenv.2022.160476

Singh P, Kim Y, Zhang D, Yang D (2016) Biological synthesis of nanoparticles from plants and microorganisms. Trends Biotechnol 34(7):588–599. https://doi.org/10.1016/j.tibtech.2016.02.006

Singh AV, Laux P, Luch A, Sudrik C, Wiehr S, Wild A, Santomauro G, Bill J, Sitti M (2019) Review of emerging concepts in nanotoxicology: opportunities and challenges for safer nanomaterial design. Toxicol Mech Methods 29(5):378–387. https://doi.org/10.1080/15376516.2019.1566425

Sokol NW, Slessarev E, Marschmann GL, Nicolas A, Blazewicz SJ, Brodie EL, Firestone MK, Foley MM, Hestrin R, Hungate BA, Koch BJ, Stone BW, Sullivan MB, Zablocki O, LLNL Soil Microbiome Consortium, Pett-Ridge J (2022) Life and death in the soil microbiome: how ecological processes influence biogeochemistry. Nat Rev Microbiol 20(7):415–430. https://doi.org/10.1038/s41579-022-00695-z

Sonali JMI, Kavitha R, Kumar PS, Rajagopal R, Gayathri KV, Ghfar AA, Govindaraju S (2021) Application of a novel nanocomposite containing micro-nutrient solubilizing bacterial strains and CeO_2 nanocomposite as bio-fertilizer. Chemosphere 286:131800. https://doi.org/10.1016/j.chemosphere.2021.131800

Stauber RH, Siemer S, Becker S, Ding G, Strieth S, Knauer SK (2018) Small meets smaller: effects of nanomaterials on microbial biology, pathology, and ecology. ACS Nano 12(7):6351–6359. https://doi.org/10.1021/acsnano.8b03241

Su Y, Zhou X, Meng H, Xia T, Liu H, Rolshausen P, Roper C, McLean JE, Zhang Y, Keller AA, Jassby D (2022) Cost–benefit analysis of nanofertilizers and nanopesticides emphasizes the need to improve the efficiency of nanoformulations for widescale adoption. Nat Food 3(12):1020–1030. https://doi.org/10.1038/s43016-022-00647-z

Tahir M, Imran M, Nawaz F, Shahid M, Naeem M, Ahmad I, Akram M, Khalid U, Farooq A, Bakhat H, Kamran M, Shah Z (2021) Effects of *Bacillus* sp. MR-1/2 and magnetite nanoparticles on yield improvement of rice by urea fertilizer under different watering regimes. J Appl Microbiol 131(5):2433–2447. https://doi.org/10.1111/jam.15110

Tarafder C, Daizy M, Alam MM, Ali MR, Islam MJ, Islam R, Ahommed MS, Aly M a S, Khan MZH (2020) Formulation of a hybrid nanofertilizer for slow and sustainable release of micronutrients. ACS Omega 5(37):23960–23966. https://doi.org/10.1021/acsomega.0c03233

Timmusk S, Seisenbaeva G, Behers L (2018) Titania (TiO_2) nanoparticles enhance the performance of growth-promoting rhizobacteria. Sci Rep 8(1). https://doi.org/10.1038/s41598-017-18939-x

Vlajkov V, Pajčin I, Loc M, Budakov D, Dodić J, Grahovac M, Grahovac J (2022) The effect of cultivation conditions on antifungal and maize seed germination activity of *Bacillus*-based biocontrol agent. Bioengineering 9(12):797. https://doi.org/10.3390/bioengineering9120797

Vlajkov V, Pajčin I, Vučetić S, Anđelić S, Loc M, Grahovac M, Grahovac J (2023) *Bacillus*-loaded biochar as soil amendment for improved germination of maize seeds. Plants 12(5):1024. https://doi.org/10.3390/plants12051024

Wang D, Saleh NB, Byro A, Zepp R, Sahle-Demessie E, Luxton TP, Ho KT, Burgess RM, Flury M, White JC, Su C (2022) Nano-enabled pesticides for sustainable agriculture and global food security. Nat Nanotechnol 17(4):347–360. https://doi.org/10.1038/s41565-022-01082-8

Wei L, Liu J, Hou X, Chen W, Feng Y, Kong W, Tang Y, Zhong C, Zhang S, Wang T, Zhao G, Jiao S, Jiang G (2023) Rice seedlings and microorganisms mediate biotransformation of Se in CdSe/ZnS quantum dots to volatile alkyl selenides. Environ Sci Technol 57(48):20261–20271. https://doi.org/10.1021/acs.est.3c07094

White JF, Kingsley KL, Verma SK, Kowalski KP (2018) Rhizophagy cycle: an oxidative process in plants for nutrient extraction from symbiotic microbes. Microorganisms 6(3):95. https://doi.org/10.3390/microorganisms6030095

Wong MH, Misra RP, Giraldo JP, Kwak S, Son Y, Landry MP, Swan JW, Blankschtein D, Strano MS (2016) Lipid Exchange envelope penetration (LEEP) of nanoparticles for plant engineering: a universal localization mechanism. Nano Lett 16(2):1161–1172. https://doi.org/10.1021/acs.nanolett.5b04467

Xia X, Shi B, Wang L, Liu Y, Zou Y, Zhou Y, Chen Y, Zheng M, Zhu Y, Duan J, Guo S, Jang HW, Miao Y, Fan K, Bai F, Tao W, Zhao Y, Yan Q, Cheng G, Liu H, Jiao Y, Liu S, Huang Y, Ling D, Kang W, Xue X, Cui D, Huang Y, Cui Z, Sun X, Qian Z, Gu Z, Han G, Yang Z, Leong DT, Wu A, Liu G, Qu X, Shen Y, Wang Q, Lowry GV, Wang E, Liang X–J, Gardea-Torresdey J, Chen G, Parak WJ, Weiss PS, Zhang L, Stenzel MM, Fan C, Bush AI, Zhang G, Grof CPL, Wang X, Galbraith GW, Tang BZ, Offler CE, Patrick JW, Song C-P (2021) From mouse to mouse-ear cress: Nanomaterials as vehicles in plant biotechnology. Exploration 1(1):9–20. https://doi.org/10.1002/exp.20210002

Xu L, Zhu Z, Sun D (2021) Bioinspired nanomodification strategies: moving from chemical-based agrosystems to sustainable agriculture. ACS Nano 15(8):12655–12686. https://doi.org/10.1021/acsnano.1c03948

Xu T, Wang Y, Aytac Z, Zuverza-Mena N, Zhao Z, Hu X, Ng KW, White JC, Demokritou P (2022a) Enhancing agrichemical delivery and plant development with biopolymer-based stimuli responsive core–shell nanostructures. ACS Nano 16(4):6034–6048. https://doi.org/10.1021/acsnano.1c11490

Xu L, Wang X, Shi H, Hua B, Burken JG, Ma X, Yang H, Yang JJ (2022b) Uptake of engineered metallic nanoparticles in soil by lettuce in single and binary nanoparticle systems. ACS Sustain Chem Eng 10(50):16692–16700. https://doi.org/10.1021/acssuschemeng.2c04748

Yang Y, Wang J, Xiu Z, Alvarez PJJ (2013) Impacts of silver nanoparticles on cellular and transcriptional activity of nitrogen-cycling bacteria. Environ Toxicol Chem 32(7):1488–1494. https://doi.org/10.1002/etc.2230

You T, Liu D, Chen J, Yang Z, Dou R, Gao X, Wang L (2017) Effects of metal oxide nanoparticles on soil enzyme activities and bacterial communities in two different soil types. J Soils Sediments 18(1):211–221. https://doi.org/10.1007/s11368-017-1716-2

Yusefi-Tanha E, Fallah S, Rostamnejadi A, Pokhrel LR (2020) Zinc oxide nanoparticles (ZnONPs) as a novel nanofertilizer: Influence on seed yield and antioxidant defense system in soil grown soybean (*Glycine max* cv. Kowsar). Sci Total Environ 738:140240. https://doi.org/10.1016/j.scitotenv.2020.140240

Zahra Z, Habib Z, Hyun H, Shahzad HMA (2022) Overview on recent developments in the design, application, and impacts of nanofertilizers in agriculture. Sustainability 14(15):9397. https://doi.org/10.3390/su14159397

Zand AD, Tabrizi AM, Heir AV (2020) Application of titanium dioxide nanoparticles to promote phytoremediation of Cd-polluted soil: contribution of PGPR inoculation. Bioremed J 24(2–3):171–189. https://doi.org/10.1080/10889868.2020.1799929

Zhang H, Goh NS, Wang JW, Pinals RL, González-Grandío E, Demirer GS, Butrus S, Fakra SC, Del Rio Flores A, Zhai R, Zhao B, Park S, Landry MP (2021) Nanoparticle cellular internalization is not required for RNA delivery to mature plant leaves. Nat Nanotechnol 17(2):197–205. https://doi.org/10.1038/s41565-021-01018-8

Zhang P, Jiang Y, Schwab F, Monikh FA, Grillo R, White JC, Guo Z, Lynch I (2024) Strategies for enhancing plant immunity and resilience using nanomaterials for sustainable agriculture. Environ Sci Technol 58(21):9051–9060. https://doi.org/10.1021/acs.est.4c03522

Zong X, Wu D, Zhang J, Tong X, Yin Y, Sun Y, Guo H (2022) Size-dependent biological effect of copper oxide nanoparticles exposure on cucumber (*Cucumis sativus*). Environ Sci Pollut Res Int 29(46):69517–69526. https://doi.org/10.1007/s11356-022-20662-8

Zulfiqar F, Navarro M, Ashraf M, Akram NA, Munné-Bosch S (2019) Nanofertilizer use for sustainable agriculture: advantages and limitations. Plant Sci 289:110270. https://doi.org/10.1016/j.plantsci.2019.110270

Chapter 8
The Combined Use of Nanoparticles and Beneficial Microbes in Remediation and Crop Production in Heavy Metal-Contaminated Soils

Heriberto Fortino Ramírez-Cariño, Daniel Alejandro Ramírez-Villanueva, Paula Cecilia Guadarrama-Mendoza, Mary Carmen Pacheco-Esteva, Elizabeth Gonzalez-Terreros, and Rogelio Valadez-Blanco

Abstract The increased accumulation of heavy metals (HMs) in the environment, primarily caused by human activities, reduces crop productivity, exacerbates environmental contamination, and poses risks to human health. Due to their unique physicochemical properties, nanomaterials (NMs) hold great potential for enhancing crop production and remediating soils contaminated with heavy metals. NM can mitigate HM toxicity in plants by reducing oxidative damage in plants through the activation of plant defense systems, which include the stimulation of antioxidant enzymes and secondary metabolite production. In addition, nanoparticles facilitate the immobilization of heavy metals through various physicochemical mechanisms that reduce their bioavailability and toxicity. Nanoparticles (NPs) have been successfully employed as nanofertilizers in food crops under HM stress, with most studies focusing on cereals, particularly rice, wheat, and corn. Beneficial microorganisms, such as plant-growth-promoting rhizobacteria (PGPR), have also been applied together with nanoparticles to ameliorate HM-induced stress and improve crop productivity. Moreover, nanoparticles have been employed in the remediation of heavy-metal-contaminated soils. Remarkably, remediation strategies that combined all three components (nanoparticles, hyperaccumulating plants, and PGPR), nPB-remediation, were reported more frequently than individual or dual nanoremediation approaches (NP, NP-plant, or NP-PGPR). This trend reflects the enhanced efficiency of nPB remediation due to the synergic effects of the combined treatments.

H. F. Ramírez-Cariño · D. A. Ramírez-Villanueva · E. Gonzalez-Terreros (✉)
Universidad de la Sierra Juárez, Oaxaca, Mexico
e-mail: gterrero@unsij.edu.mx

P. C. Guadarrama-Mendoza · R. Valadez-Blanco (✉)
Instituto de Agroindustrias, Universidad Tecnológica de la Mixteca, Oaxaca, Mexico
e-mail: rvaladez@mixteco.utm.mx

M. C. Pacheco-Esteva
Instituto Politécnico Nacional, CIIDIR Unidad Oaxaca, Oaxaca, Mexico

K. A. Abd-Elsalam (ed.), *Plant-Microbiome Nanotechnology*, Sustainability Sciences in Asia and Africa, https://doi.org/10.1007/978-981-95-5056-2_8

The most commonly studied contaminants were Cd, Cr, Pb, As, and Cu, whereas Zn- and Fe-based nanoparticles were predominantly employed in fertilization and remediation applications. In conclusion, NP-based strategies represent a highly efficient and sustainable approach for both crop cultivation and remediation in heavy-metal-contaminated soils.

Keywords Agronanotechnology · Heavy metal nanoremediation · Nanofertilizers · Plant-growth-promoting rhizobacteria · Abiotic plant stress · Sustainable agriculture

8.1 Introduction

Agricultural productivity is increasingly threatened by fluctuating environmental conditions and declining biological resources. Abiotic stressors, including drought, salinity, and extreme temperatures, impair soil fertility, reduce plant vigor, and hinder nutrient uptake throughout all plant life stages (Mangena 2025). For over five decades, human activities like mining, industry, and wastewater disposal have intensified soil contamination with heavy metals (Zaynab et al. 2022). Despite occurring naturally, heavy metals can become harmful even at low levels when biogeochemical cycles are disrupted (Rajput et al. 2022). Despite the fact that the physical and chemical characteristics of heavy metals are not clearly defined, this group of elements includes highly toxic metals such as Cd, As, Hg, and Pb, which are deleterious for living cells and human health. The toxicity of these heavy metals has been related to their persistence in the environment and their capacity to bioaccumulate in plants, fauna, and human cells through the entire food chain, which is caused by diverse biochemical mechanisms (Balali-Mood et al. 2021; Kiran et al. 2021). These traits cause detrimental effects on the environment and on human health. Particularly, this chapter focuses on the reduction of heavy metal toxicity in agricultural or contaminated soils using nanotechnology-based treatments.

Great efforts have been made to mitigate stress caused by both biotic and abiotic factors in various crops, and in recent years, nanotechnology has been used as a valuable tool to mitigate oxidative stress in plants. In this regard, engineering-synthesized nanoparticles (NPs) have been developed as a modern approach to aid in the sustainable production of agricultural crops as well as in the treatment of toxic pollutants. The use of nanoparticles formulated with different materials has emerged as a strategy to mitigate heavy metal toxicity in plants, with a focus on reducing heavy metal accumulation and enhancing the plant's growth and health. Among the characteristics of nanoparticles, which make them excellent components in the formulation of fertilization enhancers and heavy metal removal agents, are their extremely reduced size (equal to or less than 100 nm in one of its dimensions) and large surface area per volume. These properties allow formulated nanomaterials (NMs) to enter the plant cells directly as nutrients or as growth-promoter elicitors.

It has also been well documented in the literature that NM formulations can help to alleviate heavy metal toxicity in crop plants cultivation (Singh et al. 2023).

In addition, nanomaterials have been used as an aid in the remediation treatment of soils contaminated with heavy metals (nanoremediation). The combination of nanoparticles with both microorganisms and plants has led to the development of new and promising remediation strategies, respectively: nanophytoremediation (Milano et al. 2024; Rajput et al. 2022) and nanobioremediation (Milano et al. 2024). In both strategies, treatments are aimed to limit the exposure of the host cells to heavy metals or to reduce the toxicity of heavy metal compounds by physicochemical or biochemical mechanisms (Prakash and S 2023). According to this, nanobioremediation provides a successful and cost-effective approach for large-scale remediation (Chaudhary et al. 2023). Among nanobioremediation strategies, there is increasing interest in the use of plant-associated microbes, such as plant-growth-promoting rhizobacteria (PGPR), which have been demonstrated to be a sustainable alternative, in which not only heavy metal remediation can be achieved, but also biological and chemical restoration of soils (Milano et al. 2024). At the front core of nanoremediation research, the use of treatments combining nanomaterials, microbes, and plants has been successfully employed (nanophytobioremediation). This advanced treatment aims to exploit the heavy metal remediation capacities of nanomaterials, plants, and microbial components.

In this chapter, we critically examine the use of engineering-synthesized nanomaterials in combination with microorganisms and plants for their application as crop fertilizers and remediation agents in heavy metal-contaminated soils. Additionally, mechanisms involved in nano-, bio-, and phyto-remediation are addressed. Finally, recent reports on the advanced combined treatment of nanophytobioremediation are discussed.

8.2 Nanofertilization as a Strategy to Mitigate Heavy Metal Stress in Crop Plants

Agricultural soils can be contaminated by heavy metals derived from natural or anthropological sources. Crop plants cultivated in soils polluted by heavy metals are subject to oxidative stress. Heavy metal stress cannot only hamper plant growth and development but also deteriorate the plant's health and survival. High concentrations of heavy metals compromise plant development by damaging membranes, denaturing proteins, and disrupting cellular processes, ultimately reducing crop yield (Saleem et al. 2023). In addition, plants can accumulate these toxic metals in their edible parts, which in turn may pose a health risk to fauna and humans who consume them, thus extending heavy metal toxicity throughout the whole food chain. In this regard, nanoparticles or nanomaterials have been successfully used to alleviate oxidative stress in crop plants exposed to heavy metal-contaminated soils (Maryam et al. 2025).

Due to their physicochemical properties, metal nanoparticles have demonstrated the capacity to ameliorate oxidative stress caused by several abiotic factors, such as extreme temperatures, high salinity, drought, and heavy metal toxicity. The main general mechanism for the alleviation of abiotic stress factors is based on the reduction of the plant's oxidative damage by priming the defense system of the plant (Singh et al. 2023). Nanomaterials act as stress-signaling molecules which in turn induce the synthesis and secretion of antioxidant enzymes by the host plant, such as superoxide dismutase (SOD), catalase (CAT), ascorbate peroxidase (APX), glutathione reductase (GR), and peroxidase (POD), and reduce the production of malonaldehyde (MDA) and the cell's content of reactive oxygen species (ROS). Besides the antioxidant mechanism, the plant prompts root morphological changes and phytohormone synthesis, thus preserving and promoting the plant's growth (Singh et al. 2023). In addition, in the case of heavy metal contamination, nanomaterials aid in the immobilization of toxic metals through mechanisms such as chemical complexation and chelation, depending on the nature of the nanomaterial and its functional groups. These processes prevent the translocation of heavy metals to different plant organs, thus reducing their toxicity (Pande et al. 2022). An additional mechanism of heavy metal stress alleviation is related to the increase in the soil's pH by some types of nanomaterials, such as potassium silicate NPs. The change in soil's acidity makes heavy metals less soluble and available for plant uptake, thus causing a reduction in their toxicity to the plant (Rachappanavar et al. 2024). In this section, we examine the application of nanoparticles in crop cultivation subjected to heavy metal stress.

8.2.1 The Use of Nanoparticles to Alleviate Heavy Metal Stress in Plant Crops

Table 8.1 summarizes recent studies on the use of nanoparticles and microorganisms for crop cultivation in heavy metal-polluted soils, for both plant fertilization and stress alleviation in crop plants. Anwar et al. (2025) evaluated the effect of NiO nanoparticles in tomato seedlings subjected to Cd stress. The nanoparticles decreased Cd accumulation in roots and shoots, thus reducing Cd toxicity and leading to enhanced plant growth. In addition, it was observed that the improved plant's resistance to Cd stress corresponded to an increase in the synthesis of antioxidant enzymes, and a reduction of MDA and ROS in the roots and leaves. Furthermore, the authors investigated the stress-resistance mechanism involved, concluding that the nanoparticles induced the expression of stress-defense related genes SlERF4 and SlWRI1. In a similar work, Sun et al. (2023) studied the effect of ZnO nanoparticles on the resistance of tomato seedlings to Cd toxic concentrations. The treatment reduced Cd-induced stress and mitigated cellular oxidative damage by activating the plant's defense system, thus enhancing its antioxidant capacity. The nanoparticles decreased the accumulation of Cd in roots and leaves by 30.2 and

Table 8.1 Application of nanoparticles and co-application of plant-growth-promoting rhizobacteria (PGPR) on food crop cultivation under heavy metal stress

Crop	Heavy metal	NP metal	Effect	Reference
Tomato (*Solanum lycopersicum*)	Cd	Ni	Decreased Cd toxicity in roots and shoots, increased growth of seedlings	Anwar et al. (2025)
Tomato (*Solanum lycopersicum*)	Cd	Zn	Reduced Cd accumulation in roots and leaves of seedlings. Enhanced photosynthetic activity	Sun et al. (2023)
Peppers (*Capsicum chinense*)	Cd	Zn	Decreased accumulation of Cd in shoots and roots. Improved plant growth and photosynthetic activity	Tahira et al. (2025)
Eggplant (*Solanum melongena*)	Cd	Se	Reduced toxicity of Cd. Increased chlorophyll and photosynthetic activity, improving growth of plant	Ahmed et al. (2024)
Shallot (*Allium schoenoprasum*)	Cd	Se	Biochar and Se-NPs reduced the availability of Cd by 38.6%	Lai et al. (2025)
Rice *(Oryza sativa)*, wheat *(Triticum aestivum)*, and sunflower (*Helianthus annuus)*	Cd	Fe/Si	Decreased accumulation of Cd in plant tissue. Photosynthesis and growth were improved	Maryam et al. (2025)
Rice (*Oryza sativa*)	Cd	Si	Inhibited heavy metal transporting genes, reduced Cd bioaccumulation in plant tissues. Activated phytohormone synthesis	Muhammad et al. (2025)
Barley (*Hordeum vulgare* L.)	Ni	C/Si	Carbon and Si-NPs alleviated Ni toxicity in plant reducing metal uptake	Luo et al. (2024)
Soybean (*Glycine max*)	Cd	Zn	Improved Cd tolerance through epigenetic modulation	Wang and Fang (2024)
Coriander (*Coriandrum sativum*)	Cd	Ti	Increased biomass, photosynthetic activity, and seedling vigor	Sardar et al. (2022)
Tobacco (*Nicotiana tabacum*)	Cd	Mg	NPs mitigated Cd toxicity, increased the fresh weight of shoots and roots, reduced Cd accumulation in roots and shoots, and increased lignin content	Qin et al. (2025)

Key. *NP* Nanoparticle

34.5%, respectively, in comparison with the control without nanoparticle treatment. In addition, the NP treatment improved the absorption of micronutrients such as Zn, Cu, and Mn. A metabolomics analysis revealed that the secondary metabolites, flavonoids and alkaloids, may be related to the increase in the plant's resistance to Cd in the presence of Zn nanoparticles.

Tahira et al. (2025) applied ZnO nanoparticles in habanero pepper cultivation (*Capsicum chinense*) subjected to Cd stress. The authors reported that the nanoparticles were effective not only to reduce Cd toxicity in the plant, observing Cd reductions of 30% and 50% in leaf and root, respectively, but also to promote plant growth and synthesis of phytochemical compounds. These results were attributed to the activation of the plant's defense mechanisms by the presence of nanoparticles, through the induction of the synthesis of antioxidant enzymes (SOD, POD, CAT, GR, and APX) and compounds (phenols and flavonoids). In addition, the NP treatment caused a reduction in MDA and reactive oxygen species (ROS, H_2O_2, and O_2^-) in the plant cells. In another work, Ahmed et al. (2024) applied Se nanoparticles to eggplants (*Solanum melongena*) cultivated under Cd stress during both the germination and the plant-development stages. Foliar applications of the NP treatment improved chlorophyll, photosynthetic activity, plant height, fresh weight, ascorbic acid concentration, and antioxidant activity, and reduced heavy metal toxicity. Furthermore, the nanoparticles applied to the seeds increased the contents of proline, proteins, phenols, and essential minerals (Mg, K, Zn, Na), thus enhancing stress tolerance to heavy metals and yield. Lai et al. (2025) demonstrated that biochar enriched with Se nanoparticles reduced the availability of Cd by 38.6% in contaminated soils, limiting its uptake in shallot (*Allium schoenoprasum*) tissues. The treatment also raised Se concentrations in plants by 21–30% and reduced stomatal deformation by 44.2% compared to the untreated controls.

In a similar study, Muhammad et al. (2025) evaluated the individual and combined application of silicon dioxide (SiO_2) nanoparticles and salicylic acid (SA) to mitigate Cd stress in rice seedlings. The nanoparticles reduced the accumulation of Cd by 42–56%, boosted photosynthetic activity and biomass, and consequently decreased overall toxicity. The combined treatment of Si nanoparticles and SA activated the synthesis of antioxidant enzymes and phytohormones, inhibited heavy metal transporting genes, and reduced Cd bioaccumulation by 41–56% in plant tissues. Similarly, Luo et al. (2024) reported that the individual and combined application of carbon and silica nanoparticles alleviated Ni toxicity in barley plants by enhancing growth and biomass while reducing metal uptake. The NP treatments decreased proline metabolism, a stress-related pathway, and improved the ascorbate-glutathione cycle responsible for hydrogen peroxide detoxification and redox homeostasis in the cell. Wang and Fang (2024) explored the effect of ZnO nanoparticles on the expression of stress-related genes in soybean plants exposed to Cd toxicity. The authors concluded that the nanoparticles induced the activation of the stress-response genes NRT1 and GM35E, thus improving the plant's resistance to Cd.

Qin et al. (2025) reported that Mg nanoparticles were effective to mitigate Cd toxicity in tobacco plants by increasing their shoot and root fresh weight by 35.1% and 45.7%, respectively, in comparison with the non-treated control. Additionally, the nanoparticles reduced Cd accumulation in plant tissues by increasing the lignin content of tobacco cell walls by 50.6% compared to the control treatment without NP application. Similarly, Sardar et al. (2022) reported that TiO_2 nanoparticles improved the growth performance and health of coriander plants (*Coriandrum sativum*) exposed to Cd toxicity. Treated plants showed increased biomass, seed development, and seedling vigor, likely due to enhanced nutrient absorption and Rubisco activity. The nanofertilization treatment also improved photosynthetic efficiency, antioxidant response, germination rates, and overall yield.

8.2.2 *Application of Nanoparticles and Plant-Growth-Promoting Rhizobacteria (PGPR) for Alleviating Heavy Metal Stress in Crop Cultivation*

The beneficial role of plant-growth-promoting rhizobacteria (PGPR) on food crop cultivation has been widely documented in the literature (Ramírez-Cariño et al. 2023). Different biofertilization mechanisms have been disclosed among which the most important are: induction of phytohormone synthesis, nutrient mineralization and solubilization, and stimulation of the plant's immune system. The latter is achieved primarily by priming the plant's antioxidant capacity through the synthesis of secondary metabolites and antioxidant enzymes. Plant defense mechanisms play a critical role not only in combating pests and phytopathogens but also in enabling the plant to withstand abiotic stress factors, such as the presence of heavy metals in agricultural soils. In addition, recently, some studies have demonstrated the synergy between PGPR strains and nanomaterial-based fertilization (nanofertilization) toward plant growth and nutrition, and crop yield (Verma et al. 2024). In addition to this, these combined treatments have also been demonstrated to be effective in the alleviation of plant's oxidative damage due to abiotic factors, such as heavy metal stress. Table 8.2 enlists the latest studies reported in the literature regarding the application of nanoparticles combined with beneficial microorganisms, particularly PGPR, for crop cultivation in heavy metal-polluted soils, for both plant fertilization and stress alleviation in crop plants.

Al-Huqail et al. (2024) applied the combination of CaO nanoparticles and *Bacillus mycoides* to corn plants grown in Cr-contaminated soil. As expected, without treatment, the exposure of the plants to Cr decreased plant biomass and altered antioxidant responses (SOD, POD, CAT, APX). The treatment with combined PGPR and Ca nanoparticles attenuated Cr-induced oxidative stress, reduced Cr accumulation in roots/shoots, and increased plant growth. Khan et al. (2022) evaluated the combined treatment of ZnO nanoparticles and commercial PGPR strains in corn cultivated under toxic concentrations of As (50 and 100 ppm). The combined treatment improved plant growth and increased the phytochemical parameters: biomass, protein (117–241%) and soluble sugars (114–170%), water content (43–50%), and antioxidant enzyme activity. On the other hand, the content of malondialdehyde (MDA) in the plant was reduced 46–57% by the NP-bacterial treatment. In another study related to corn cultivation, El Sharkawy et al. (2024) investigated the combination of Mg nanoparticles with the bacterial strains *Rhizobium nepotum* and *Glutamicibacter halophytocola* to reduce stress in the plants grown under As contamination. The combination of NPs with bacterial strains decreased arsenic uptake, increased corn plant growth parameters, and enhanced soil biological activity.

Ahmed et al. (2022) evaluated the combination of a bacterial strain of *Bacillus* sp. with Mo nanoparticles in wheat plants cultivated in As-contaminated soil. The authors reported that the dual treatment improved nutrient levels and ion homeostasis and reduced As translocation by 30.3% in comparison to the non-treated control. According to the authors, the NPs activated bacterial growth-promoting

Table 8.2 Combined application of nanoparticles and beneficial microbes on food crop cultivation under heavy metal stress

Crop	Heavy metal	NP element	PGPR	Effect	Reference
Corn (*Zea mays*)	Cr	Ca	*Bacillus mycoides*	Reduced Cr accumulation in roots and shoots. Improved plant growth	AL-Huqail et al. (2024)
Corn (*Zea mays*)	As	Zn	Commercial biofertilizer	Improved plant growth, biomass, protein, soluble sugars, and water content.	Khan et al. (2022)
Corn (*Zea mays*)	As	Mg	*Rhizobium nepotum* and *Glutamicibacter halophytocola*	Increased chlorophyll content and growth of maize plants. Reduced uptake of As	El Sharkawy et al. (2024)
Wheat (*Triticum aestivum*)	As	Mo	*Bacillus* sp	Reduced As translocation. Improved nutrient content	Ahmed et al. (2022)
Wheat (*Triticum aestivum*)	Cd	Zn	Commercial biofertilizer	Reduced Cd content in roots, shoots, and grains. Increased antioxidant enzyme activity	Sehrish et al. (2024)
Wheat (*Triticum aestivum*)	Cd	Ti	*Staphylococcus aureus*	Improved growth, chlorophyll, and carotenoid content. Reduction in Cd accumulation in roots, shoots, and grains	Younas et al. (2025)
Rice (*Oryza sativa*)	Cd	Ca/Mg	*B. pumilus*	Reduced Cd stress and increased chlorophyll and carotenoids. Reduced Cd accumulation in roots, shoots, and grains	Ali et al. (2024)
Rice (*Oryza sativa*)	Pb, Cu	Zn	*B. cereus* and *Lysinibacillus macrolides*	Co-application decreased the uptake of Pb and Cu in roots, shoots, and leaves, improving plant growth	Akhtar et al. (2021)
Rice (*Oryza sativa*)	Cr	Fe	*S. aureus*	Increased chlorophyll, plant growth, biomass, and micronutrient bioavailability. Decreased Cr uptake and accumulation in rice plants. Increased reduction of Cr^{6+} to Cr^{2+}	Alharby and Ali (2022)
Barley (*Hordeum vulgare*)	Cd	Ti	*B. mycoides*	Reduced the Cd content in roots and shoots	Ma et al. (2023)
Indian mustard (*Brassica juncea*)	Cr	Zn	*Klebsiella* sp.	Cr adsorption and toxicity were reduced in seedlings	Shah et al. (2024b)

(continued)

Table 8.2 (continued)

Crop	Heavy metal	NP element	PGPR	Effect	Reference
Sunflower (*Helianthus annus*)	Cr	Ce	*S. aureus*	Reduced Cr toxicity and stress, enhanced plant growth and biomass	Ma et al. (2022)
Rapeseed (*Brassica napus* L.)	Pb	C	Arbuscular mycorrhizal fungi	Increased the activity of antioxidant enzymes and improved nutrient uptake	Shah et al. (2024a)

Key. *NP* Nanoparticle

mechanisms that were previously inhibited by the presence of the heavy metal. In a similar study, Sehrish et al. (2024) investigated the combined effect of ZnO nanoparticles and a PGPR *Rhizobium* strain on wheat plants cultivated under Cd stress conditions. The exposure of the seeds to Cd impaired plant growth and altered the plant's physiological and biochemical properties, as well as the expression of the gene TaEIL1. The treatments with the nanoparticles (100 mg/L) reversed these effects, by increasing the antioxidant enzyme activity of CAT (52.4%), POD (57.4%), SOD (60.1%), and APX (47.4%), while decreasing MDA (47.4%), hydrogen peroxide (38.2%), and electrolyte leakage (47.3%) in plants subjected to 20 mg/kg of Cd. The combined NP-bacteria treatment reduced Cd concentration in roots (56.3%), shoots (49.4%), and grains (59.4%) in comparison to the control plants. In addition, the expression of the gene TaEIL 1 was reduced, thus confirming the role of the nanoparticles in Cd detoxification. In a study on wheat cultivation subjected to Cd-induced stress, Younas et al. (2025) reported that the combined treatment of TiO_2 nanoparticles and *Staphylococcus aureus* improved growth of wheat plants compared to the control treatment. Agronomic parameters such chlorophyll (12%) and carotenoids (13%) were increased in the plants in which both NPs and bacteria were added. In addition, the combined treatment reduced Cd accumulation in roots, shoots, and grains (34%, 23%, and 27%, respectively). Furthermore, the parameters indicating oxidative damage, MDA, H_2O_2, and electrolyte leakage, were reduced (4, 3, and 5%, respectively), whereas the activities of the antioxidant enzymes increased: CAT (17%), SOD (7%), POD (8%), and APX (29%).

Several NP-PGPR applications have been reported in rice cultivation. Ali et al. (2024) performed the application of the combined treatment of Ca-Mg nanocomposites with *B. pumilus* in rice plants grown in Cd-contaminated soil. Compared to the single treatments, the combined NP-PGPR treatment enhanced the defense mechanism of the plant against Cd toxicity, increasing chlorophyll and carotenoids by 18% and 21%, respectively. In addition, the combined treatment reduced oxidative markers MDA, H_2O_2, and other ROS concentrations by 15%, 13%, and 24%, respectively. In addition, the treatment also reduced Cd accumulation in roots (22%), shoots (13%), and leaves (20%) in comparison with the non-treated control. Furthermore, the combined treatment improved the overall growth and yield of rice with respect to the control. Similarly, Akhtar et al. (2021) studied the synergism between *B. cereus* and *Lysinibacillus macroides* bacteria and Zn nanoparticles in rice plants grown in contaminated water with Pb and Cu. The combined treatment removed a greater amount of heavy metals from contaminated water while decreasing their uptake by the plant's roots, shoots, and leaves, thus improving plant growth with respect to the individual treatments. Furthermore, the authors reported that the combined treatment also reduced the heavy metal bioaccumulation rate and the content of metallothionein. In a related study, Alharby and Ali (2022) reported the effect of the combined treatment of Fe nanoparticles and the bacterium *Staphylococcus aureus* in rice plants cultivated in a Cr-contaminated environment. Chromium addition without NP-bacteria application affected the plants' growth, biomass, and chlorophyll content, causing severe oxidative damage. The joint application of Fe nanoparticles and the microorganism mitigated this damage by reducing Cr absorption and content in the plants. Additionally, the combined treatment increased the transformation of Cr^{6+} to the less harmful form Cr^{3+}.

According to Ma et al. (2023), the combination of TiO_2 nanoparticles and *B. mycoides* enhanced the scavenging of reactive oxygen species in barley plants under Cd stress and reduced the content of this metal in the roots and shoots, thus boosting plant growth. In another work, Shah et al. (2024a, b) used ZnO nanoparticles together with *Klebsiella* sp. to reduce Cr adsorption and toxicity in Indian mustard (*Brassica juncea*) seedlings by activation of the plant's antioxidant enzymes. Ma et al. (2022) combined Ce nanoparticles and *Staphylococcus aureus* to reduce chromium (Cr) toxicity in sunflower plants (*Helianthus annuus*) grown in Cr-contaminated soil. The combined treatment increased growth, biomass generation, and enzyme activity, consequently reducing oxidative stress in the plants. Furthermore, the treatment increased chlorophyll a and b and Mg content, and reduced electrolyte losses. In a study on rapeseed cultivation (*Brassica napus*), Shah et al. (2024a, b) investigated the effect of carbon-based fullerenol nanoparticles in combination with arbuscular mycorrhizal fungi (AMF) on the plant's resistance to toxic concentrations of Pb. The foliar application of carbon NPs together with the soil's inoculation with AMF increased the activity of antioxidant enzymes, thus reducing Pb-induced oxidative stress. The activities of the antioxidant enzymes (CAT, POD, SOD, GR, and APX) increased by 37%, 19%, 96%, 47%, and 200%, respectively. The NP-AMF combination also improved nutrient uptake and regulated photosynthetic capacity and redox balance in the plants.

8.2.3 Current Status of Nanofertilization in Heavy Metal-Contaminated Soils

In this review, we critically assessed the latest works related to the effect of nanofertilizers on the alleviation of the toxicity caused by plant's exposure to heavy metal toxicity. A simple bibliometric analysis of the relevant literature reports was performed. In terms of the types of crop products ($n = 26$) employing nanofertilization, with or without microorganisms, most reports were related to cereals (54%), from which rice was the most studied crop (19%), followed by wheat (15%) and corn (12%). The other crop groups were non-cereal seeds (19%) and fruit vegetables (12%). Interestingly, from the works surveyed ($n = 24$), a larger number of studies were reported regarding the use of nanoparticles coupled with plant-growth-promoting rhizobacteria (PGPR), with 54% of the articles, in comparison to the studies using nanofertilizers alone (46%). A possible reason for this may be related to the enhanced efficiency of the combined use of nanoparticles and beneficial microbes in comparison with the individual treatments.

In terms of the types of heavy metals reported ($n = 25$), most studies focused on the treatment of Cd-contaminated agricultural soils (56%), which were followed by Cr and As, with 16% and 12% of the nanofertilizing works, respectively. This is mainly because cadmium is highly toxic and easily accumulates in plant tissues. Additionally, this metal is a critical contaminant in the food chain due to its high mobility and bioaccumulation. In addition, according to the bibliometric analysis on the type of nanomaterial employed for crop fertilization on heavy metal-contaminated soils ($n = 27$), Zn nanoparticles have been the most widely reported materials in the literature (26%) followed by Ti, Mg, and Si nanoparticles with an equal number of publications (11%). Zinc nanoparticles are essential for plants because they are vital micronutrients in plant metabolism, and they also have a high capacity to increase plant's growth and its tolerance to adverse abiotic factors, such as heavy metal toxicity.

8.3 Use of Nanomaterials for Remediation of Soils Contaminated with Heavy Metals

In order to counteract heavy metal contamination in soils, diverse remediation strategies have been developed, including physical, chemical, and biological methods. Physical methods can remove large volumes of contaminated soil but are often costly, whereas chemical treatments may neutralize pollutants but are prone to generate secondary toxicity. In contrast, biological methods are in principle environmentally friendly, though they are also typically slower and less efficient than the other remediation techniques (Raffa et al. 2021). Biological remediation relies on living organisms to remove pollutants and can be applied either in situ, using native species directly at the site, or ex situ, where contaminated material is treated elsewhere. Microbial and plant-based in situ approaches, involving bacteria, fungi, and algae, have shown encouraging remediation results (Abo-Alkasem et al. 2023; Ayilara and Babalola 2023). For clarity, in this review, we distinguish between bioremediation, driven by microorganisms, and phytoremediation, which is based on plants.

The combination of different remediation techniques can improve the overall effectiveness of the decontamination process. In the last decade, nanotechnology has gained attention for its role in heavy metal remediation. In this regard, engineered nanoparticles (NPs) are increasingly used to treat contaminated soils and water (nanoremediation), providing efficient and sustainable alternatives to conventional treatments (Rajput et al. 2022). Moreover, the combined application of nanoparticles, plants, and plant-growth-promoting rhizobacteria (PGPR) to restore heavy metal-contaminated soils has been demonstrated to be a very promising strategy in terms of efficacy and sustainability. In this section, we discuss the latest research on nanoremediation of heavy metal-contaminated soils.

8.3.1 Remediation of Heavy Metal-Contaminated Soils Using Nanoparticles

The use of nanoparticles and nanomaterials can participate in the remediation of heavy metal soils through different mechanisms like adsorption, redox reactions, and precipitation. These capabilities derive from their high surface reactivity and molecular-scale interactions (Raffa et al. 2021). After application, nanoparticles penetrate soil micropores and participate in catalytic and absorptive processes thanks to their large surface area (Ali et al. 2023).

Among the metals that have been employed for heavy metal nanoremediation, nanoparticles made of Zn and Fe have gained great relevance due to their removal efficacy. The effectiveness of these nanoparticles derives from their high reactivity, large surface area, and strong reducing capacity, which enable the removal of heavy metals and metalloids. Other remediation mechanisms include redox reactions and the formation of secondary products, such as iron oxides and oxyhydroxides, which also play a role in immobilizing heavy metals through adsorption, changes in speciation, surface precipitation, and co-precipitation (Rajput et al. 2022). Table 8.3 summarizes the studies on the remediation of heavy metals using nanoparticles (nanoremediation).

Deng et al. (2020) synthesized chlorapatite nanoparticles for Pb remediation in contaminated sediments. Due to its internal stratification and enhanced dispersibility, the nanoparticles improved remediation performance, achieving greater Pb immobilization efficiency (80.3%) in comparison with bulk chlorapatite (45.6%). The authors reported that the nanoparticles facilitated the release of soluble P, which in turn generates stable complexes with Pb through precipitation. Similarly, Lu et al. (2023) synthesized magnetic hydroxyapatite nanoparticles for remediation of Cd-contaminated soil. The authors reported that the NP remediation treatment removed 16.5% of Cd from the contaminated soil, concluding that the application of nanoparticles decreased the exchangeable forms of Cd and promoted their transformation to a more stable form. Bolaños-Guerrón et al. (2021) evaluated the remediation of soils contaminated with mining tailings containing Cu and Cd using iron-rich technosols and composite Fe/FeS nanoparticles. The combined technosol-NM treatment retained up to 70% of metals in fixed beds and 90% in batch tests. In another study, Caballero-Mejía et al. (2025) utilized magnetite nanoparticles using

Table 8.3 Remediation of heavy metal-contaminated soils through the use of nanoparticles: nanoremediation

Contaminant source	Heavy metal	NP element	Effects	Reference
Contaminated sediment	Pb	Chlorapatite	Nano-chlorapatite immobilizes Pb and releases P, which combines with Pb through precipitation, reducing both P and Pb in the sediment	Deng et al. (2020)
Contaminated soil	Cd	Hydroxyapatite (HAP)	Nano-hydroxyapatite was synthesized by coprecipitation. Nano-HAP increased Cd removal rate in contaminated soil by the magnetic method to separate NP and adsorb Cd	Lu et al. (2023)
Mining tailings	Cu, Cd	Fe/FeS	Iron nanoparticles retain heavy metals from mining tailings and form stable complexes	Bolaños-Guerrón et al. (2021)
Contaminated mine soils	As, Cd, Cu, Fe, Mn, Pb, Zn	Fe	Removal and immobilization facilitated by the magnetic properties of NPs, their adsorption capacity, and their ability to form aggregates	Caballero-Mejía et al. (2025)
Contaminated soil from a sawmill and contaminated mining soil	As, Cu, Cr, Zn	Fe	Immobilization of heavy metals with high retention of As and Cu. Long-term retention with ZVI was more effective	Danila et al. (2020)

Key: *NP* Nanoparticle

magnetic-separation mechanisms to remove heavy metals from contaminated mine soils. The application of nanoparticles removed 44% of As, 65% of Cd, 60% of Cu, 47% of Fe, 40% of Mn, 65% of Pb, and 62% of Zn. The optimized removal was attributed to the specific characteristics of the nanoparticles, including their ability to adsorb and form aggregates.

8.3.2 Combined Use of Nanoparticles and Plants for Remediation of Heavy Metal-Contaminated Soils: Nanophytoremediation

As stated above, nanomaterials can be combined with hyperaccumulator plants to enhance the efficiency of heavy metal remediation of soils in comparison with the individual treatments. Plants mitigate heavy metal stress through several mechanisms, such as restricting metal uptake at the root level, enhancing the expression of detoxification proteins, increasing the synthesis of organic acids and flavonoids, and activating defense-related genes. These responses contribute directly to the plant's heavy metal tolerance and overall stress adaptation (Huibo et al. 2023). In this regard, the combination of nanoparticles and hyperaccumulating plants (nanophytoremediation) provides a promising strategy for restoring soils contaminated with heavy metals.

In nanophytoremediation treatments, nanoparticles must not exert toxicity on plants or rhizospheric microbiota. Additionally, they should possess the capacity to bind to metal pollutants, improving their capture and facilitating their extraction. The interaction of nanoparticles with plants stimulates root development and biomass, largely by inducing phytohormone synthesis, thus enhancing contaminant uptake (Khan et al. 2022). According to current studies, nanoparticles also increase the adaptive capacity of plants subjected to biotic and abiotic stress, thus strengthening their phytoremediation potential. Aseri et al. (2024) reported that the effectiveness of nano-phytoremediation depends on several factors, among which the most relevant are the following: (i) nanoparticle traits (structure, size, shape, composition, physicochemical properties, and surface activity); (ii) soil characteristics (temperature, pH, moisture, toxicity of mineral salts, light, oxygen availability and microbial diversity); (iii) metal-contaminant nature (atomic mass, bonding type, mobility, toxicity, interaction behavior, and reactivity); and (iv) plant attributes (branching pattern, root system architecture, tolerance capacity, accumulation potential, growth rate and biomass).

However, the use of phytoremediation combined with nanoparticles is still incipient. In Table 8.4, we report recent studies reported in the literature regarding the remediation of heavy metals using nanoparticles coupled to plants and PGPR strains, both separately and in combination. Zhang et al. (2025) reported the combined application of TiO_2 nanoparticles with humic acids to remediate soil contaminated with Cd and the polycyclic aromatic hydrocarbon pyrene. This was achieved by enhancing Cd extraction by the sedum plant (*Hylotelephium spectabile*). The composite NP-plant treatment and humic acid improved Cd availability, leading to increased heavy metal assimilation and accumulation in plant tissues. The contents of Cd in the plant's roots and leaves increased by 57.7% and 81.4%, respectively, and decreased by 45% in the shoots. The combined treatment improved root growth, increased the concentration of the antioxidant enzymes superoxide dismutase (SOD) and peroxidase (POD), decreased Cd stress, and enhanced remediation efficiency.

Souri et al. (2017) utilized salicylic acid (SA)-loaded chitosan nanoparticles for As phytoremediation in *Isatis cappadocica* plants. According to the results, the authors reported that the nanohybrid treatment enhanced plant's growth by increasing its root height, length, and biomass production with respect to the control treatment without nanomaterial's addition. In addition, the combined NP-plant remediation treatment increased As concentrations in the roots and shoots of *I. cappadocica* in comparison with the treatment without NP application. The authors concluded that the application of the SA composite nanoparticles mitigated As toxicity, thereby improving the growth and remediation efficiency of *I. cappadocica*. In a phytoremediation study in in-vitro plant tissue cultivation (Emamverdian et al. 2022), the combined use of TiO_2 nanoparticles and the phytohormones

Table 8.4 Studies on the remediation of heavy metal-contaminated soils through the combined use of nanoparticles with hyperaccumulating plants and plant-growth-promoting rhizobacteria (PGPR)

Contaminant source	Heavy metal	NP element	Plant	Microorganism	Effects	Reference
Nanophytoremediation						
Artificial	Cu, Cd	Ti	Bamboo (*Pleioblastus pygmaeus*)	–	Enhanced plant Cd tolerance, reduced translocation from roots to shoots, and decreased accumulation in plant	Emamverdian et al. (2022)
Artificial	Cd	Ti	Sedum (*Hylotelephium spectabile*)	–	Increased availability, assimilation, and accumulation of Cd in plant tissues	Zhang et al. (2025)
Artificial	As	Salicylic acid- chitosan	*Isatis cappadocica*	–	Increased As content in roots and shoots	Souri et al. (2017)
Nanobioremediation						
Artificial	Pb, Cd	Fe	–	*Halomonas* sp.	Increased Pb and Cd removal primarily through exopolysaccharides secretion. NPs enhanced bacterial performance	Cao et al. (2020)
Nanophytobioremediation						
Contaminated area	Cd, Pb, Zn	C	Nightshade (*Solanum nigrum*)	*Bacillus* sp.	Increased Cd, Pb, and Zn removal in plant. Increased growth and bacterial survival	Ding et al. (2024)
Artificial	As	Mg	Corn (*Zea mays*)	*Rhizobium nepotum* and *Glutamicibacter halophytocola*	Reduced As uptake	El Sharkawy et al. (2024)
Artificial	Pb	Ti	Sorghum (*Sorghum bicolor*)	*Pseudomonas fluorescens*	Increased Pb accumulation in roots and plant	Zand et al. (2022)
Contaminated soil	Cd	Ti	White clover (*Trifolium repens*)	*Pseudomonas fluorescens*	Increased accumulation of Cd in plant	Zand et al. (2020a)

(continued)

Table 8.4 (continued)

Contaminant source	Heavy metal	NP element	Plant	Microorganism	Effects	Reference
Contaminated soil	Sb	Fe	White clover (*Trifolium repens*)	*Pseudomonas fluorescens*	Increased Sb accumulation in the plant shoots and enhanced plant growth	Zand et al. (2020b)
Contaminated soil	Cd	Hydroxyapatite	Rapeseed (*Brassica campestris*)	*Bacillus subtitlis*	Elevated Cd content in plant and enhanced microbial diversity	Liu et al. (2018)
Artificial	Pb, Cu, Cd	Fe	White willow (*Salix alba*)	*Rhizophagus irregularis* and *Pseudomonas fluorescens*	Improved the bioconcentration factor of HMs	Mokarram-Kashtiban et al. (2019)

Key: *NP* nanoparticle

24-epibrassinolide effectively reduced Cu and Cd toxicity in bamboo (*Pleioblastus pygmaeus*) plants in a synergic manner. The nanophytoremediation treatment enhanced the plant's tolerance to Cd stress by reducing heavy metal accumulation and restricting metal translocation from roots to shoots.

8.3.3 Combined Use of Nanoparticles and Microorganisms for Heavy Metal Removal: Nanobioremediation

The diverse and extensive soil microbial community inhabiting the plant's rhizosphere holds great potential for heavy metal detoxification of contaminated soils. Microbiological remediation of soils contaminated with heavy metals relies on the metabolic activity of microorganisms, since heavy metals are not easily degradable. However, microorganisms possess several biochemical properties related to adaptation mechanisms that confer them resistance to heavy metals. As regards heavy metal removal from soils, potential bioremediation microorganisms require, apart from resistance to heavy metals, the capability to detoxify and immobilize these toxic compounds in contaminated soils.

Microorganisms have developed diverse biochemical strategies to survive in contaminated environments, particularly those with high concentrations of heavy metals. These strategies enable them to detoxify and immobilize toxic metals through different processes such as biosorption, bioaccumulation, biotransformation, and biomineralization (Kushwaha and Kashyap 2021). At the cellular level, metal ions can be sequestered by functional groups present on microbial surfaces, including the cell membrane and extracellular structures (Mogazy 2024), thus promoting metal immobilization and exclusion. Additionally, microorganisms can also transform metals through redox reactions, promote the formation of extracellular complexes, and facilitate intracellular accumulation, volatilization, and precipitation (Olaya-Abril et al. 2024; Raklami et al. 2022).

Certain bacteria and cyanobacteria produce extracellular polysaccharides capable of sequestering heavy metals, thus preventing their entry into the cell. Other microorganisms, including bacteria, fungi, and algae, permit metal ions to cross the cell membrane, where they are deposited in the cytoplasm and subsequently detoxified, excreted, or transformed through metabolic pathways (Jeyakumar et al. 2023). Among remediation mechanisms for heavy metals, biosorption plays a central role in microbial detoxification. Reduction, chelation, and biotransformation of heavy metals often occur extracellularly as processes preceding cellular biosorption (Milano et al. 2024). Biosorption is a physicochemical process that involves the adherence of metal ions to microbial surfaces through various mechanisms: physical binding to polysaccharides, chemical attraction to negatively charged functional groups, and complex formation involving multiple ions and surface ligands (Jeyakumar et al. 2023). In addition to biosorption, bacteria employ other resistance mechanisms to counteract heavy metal toxicity, such as metal exclusion mediated by the cell wall, plasma membrane barrier, and active efflux systems that expel metals from the cytoplasm. Other strategies include extracellular and intracellular sequestration, precipitation, and biotransformation (Olaya-Abril et al. 2024).

Among bioremediation bacteria, *Bacillus* species are frequently used for heavy metal removal due to their ability to perform biosorption, bioaccumulation, bioprecipitation, and metal-binding mediated by biopolymers. In addition to their detoxification roles, *Bacillus* spp. contribute to plant growth in phytoremediation systems (Wróbel et al. 2023). Other genera, including *Pseudomonas*, *Flavobacterium*, *Enterobacter*, and *Micrococcus*, have also demonstrated biosorption capacity through active chemisorption sites located in their cell walls (Chaudhary et al. 2023). Mondal et al. (2025) reported that in a medium supplemented with heavy metals, the strain *Enterobacter* sp. reduced the concentrations of As, Pb, and Cd by 92%, 98%, and 97%, respectively. Additionally, the PGPR bacterial strain increased the production of exopolysaccharides and exhibited biofertilizer characteristics. The authors proposed a remediation mechanism based on heavy metal biosorption by the bacterial cells, in which metals are bound to the functional groups present on the cell surface. The groups identified to participate in heavy metal binding are amino, hydroxyl, carboxyl, alkene, and phosphate. Furthermore, the rhizobacteria diminished the contents of As, Cd, and Pb by 37.2%, 42.8%, and 21.5% respectively, in polluted soil in comparison with the non-treated control, thus demonstrating their bioremediation potential.

The use of nanoparticles, combined either simultaneously or sequentially with certain bioremediation microorganisms, enhances heavy metal removal, a process in which microbial agents act as NP carriers. Nanoparticles can improve pollutant removal by stimulating microbial communities and modulating nutrient availability, which in turn supports microbial metabolism in bioremediation systems (Patle et al. 2024). Interestingly, only a few applications have been reported regarding the combined use of nanoparticles and microorganisms toward remediation of heavy metal-contaminated soils. Figure 8.1 represents the different mechanisms involved in the remediation of heavy metal-contaminated soils by using microorganisms and plants.

Cao et al. (2020) investigated the effect of a nanobioremediation treatment on an artificially contaminated aqueous media with Pb and Cd. In the study it was tested the use of polyvinylpyrrolidone-coated Fe nanoparticles alone or in combination with a *Halomonas* sp. strain isolated from oil-polluted soil. The authors reported that complete removal of lead and cadmium was attained at 24 and 48 h, respectively, by the synergic action of the NPs and the bacteria. The nanoparticles reduced metal toxicity by enhancing microbial activity. The proposed mechanism involved secretion of exopolysaccharides, with *Halomonas* sp. forming a protective and absorptive barrier that promoted cell's resistance to both heavy metals. In particular, the results demonstrated that the reduction of Pb toxicity to the cells was achieved through the decrease of metal transportation into the cell by the action of exopolysaccharides.

A possible reason for the limited number of nanobioremediation applications is that potential bioremediation bacteria provide better detoxification efficiency as part of the rhizospheric microenvironment of plants. In this regard, the most recent

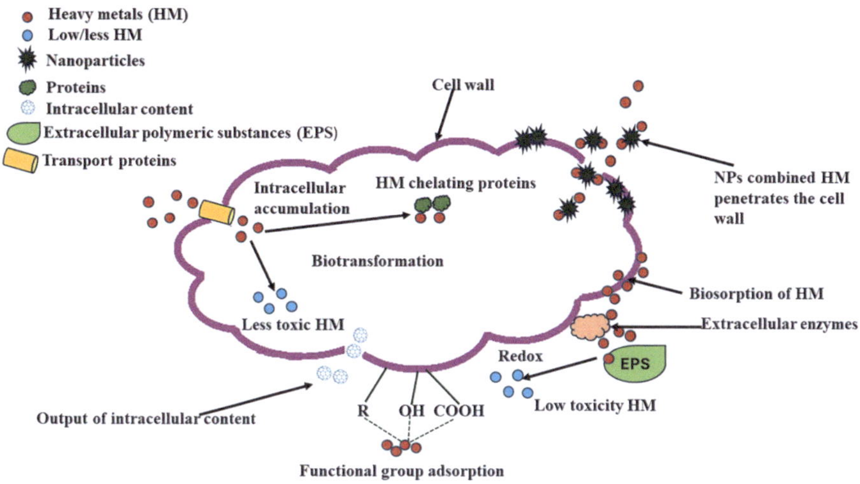

Fig. 8.1 Synergies and mechanisms of plant-growth-promoting rhizobacteria combined with nanoparticles for remediation of heavy metal-contaminated soils (Mai et al. 2025)

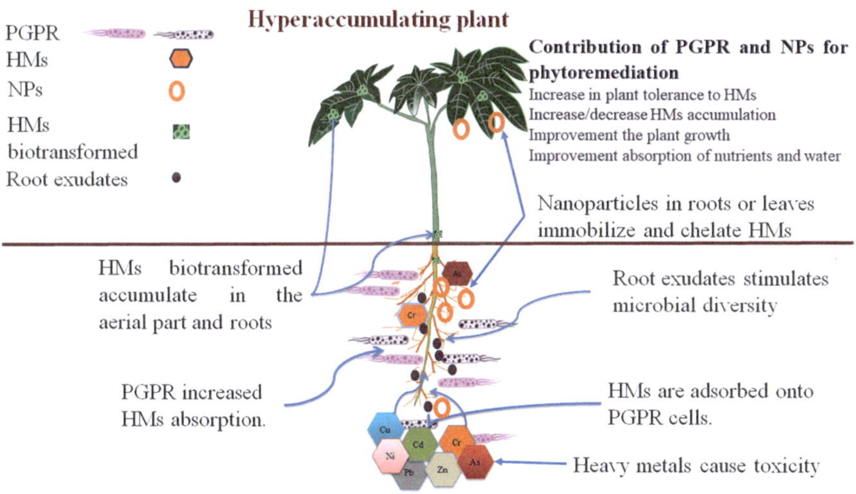

Fig. 8.2 Combined use of nanoparticles, plants, and plant-growth-promoting rhizobacteria (PGPR) for heavy metal nanoremediation

trends in remediation of heavy metal-contaminated soils include treatments incorporating nanoparticles, bacteria (particularly PGPR), and hyperaccumulating plants (Fig. 8.2). The use of nanophytobioremediation treatments (nPB-remediation) has been demonstrated to enhance the detoxification efficacy of single-factor remediation strategies (Verma et al. 2024; Akhtar et al. 2021).

8.3.4 *Application of Nanoparticles, Bacteria, and Plants for Heavy Metal Remediation: Nanophytobioremediation*

As previously mentioned, nanophytobioremediation integrates the use of nanomaterials, plants, and beneficial microbes, particularly PGPR, in order to provide a more efficient remediation treatment for soils contaminated with heavy metals. Approaches combining plant species such as *Hibiscus* and *Helianthus* with microorganisms of the genera *Bacillus*, *Streptomyces*, and *Aspergillus* have shown a good potential in NP-assisted remediation of contaminated soils (Karnwal et al. 2024). However, the development of effective plant–microbe–nanoparticle systems for heavy metal remediation of soils requires a deep understanding of the physiological and biochemical mechanisms that confer detoxification capacity and biological stress tolerance in highly toxic contaminated soils.

Ding et al. (2024) combined multi-walled carbon nanotubes with *Bacillus* sp. to enhance phytoremediation of heavy metal-contaminated soils (Cd, Pb, and Zn) in *Solanum nigrum* plants. This nPB-remediation treatment boosted plant growth and heavy metal accumulation in the plant. The authors reported that the application of C nanotubes promoted bacterial survival and improved soil microbial communities in comparison with single bacterial or plant treatments. The combined approach increased Cd, Pb, and Zn removal in *S. nigrum* by 62.0%, 69.4%, and 61.3%, respectively, compared to the single treatments. Additionally, in the aforementioned study by El Sharkawy et al. (2024) on corn cultivation, the strains *Rhizobium nepotum* and *Glutamicibacter halophytocola*, which were resistant to arsenic, were used alone or together with MgO nanoparticles as a nanobioremediation treatment in corn cultivation under As-induced stress. The combined NP-bacterium treatment reduced As uptake by 31.2% and 36.6% for *R. nepotum* and *G. halophytacola*, respectively, in comparison with the non-treated control. According to the authors, the main remediation mechanism taking place is the synthesis of exopolysaccharides by the bacteria, which bind metal cations due to their negative charges and functional groups. The combined PGPR-NP-plant system reduced As toxicity to the plant and limited its translocation to aerial parts.

In another report, the combination of *Sorghum bicolor* seedlings, TiO_2 nanoparticles, and *Pseudomonas fluorescens* was evaluated for Pb remediation (Zand et al. 2022). The application of the PGPR strain and NPs alone or in combination enhanced Pb accumulation in the roots. In addition, the nanoparticles at concentrations from 100 to 500 mg/kg increased the Pb uptake capacity of *S. bicolor* seedlings. The combined use of NPs and *P. fluorescens* proved to be the best remediation and fertilization treatment, resulting in improved biomass, photosynthesis, chlorophyll production, and overall plant growth. Moreover, the nPB-remediation treatment mitigated Pb toxicity in the soil without causing plant-growth inhibition, while simultaneously increasing Pb accumulation in *S. bicolor*.

Zand et al. (2020a) evaluated the co-application of different concentrations of TiO_2 nanoparticles and a PGPR strain of *Pseudomonas fluorescens* in seedlings of *Trifolium repens* subjected to Cd stress. The combination of the bacterial strain with 500 mg/kg of TiO_2 NPs yielded the highest cadmium accumulation (~1200 mg/pot) and improved the plant's chlorophyll content. However, concentrations of 1000 mg/kg negatively affected seedling development. In a study of the same research group on Sb remediation (Zand et al. 2020b), *Trifolium repens* seedlings were inoculated with *P. fluorescens* together with Fe (nZVI) nanoparticles. The nPB-remediation treatment increased Sb accumulation in the plant shoots, with the highest accumulation attained at an Fe NPs concentration of 500 mg/kg. Additionally, the authors reported that the combined treatment enhanced plant's photosynthesis and growth. Liu et al. (2018) examined the effect of *Bacillus subtilis* and hydroxyapatite nanoparticles, applied individually and in combination, on rapeseed plants (*Brassica campestris*) grown in Cd-contaminated soil. The individual treatments resulted in moderate changes in Cd distribution, with an increase in the content of Cd in the soil and a reduction in the plant's metal content. On the other hand, the combined treatment not only boosted the heavy metal removal efficiency of the individual treatments but also enhanced microbial diversity, plant growth, and overall bioremediation efficacy in comparison with the control. In another study, Mokarram-Kashtiban et al. (2019) evaluated the application of a consortium of rhizospheric microorganisms, comprising the arbuscular mycorrhizal fungi, *Rhizophagus irregularis*, and the PGPR *Pseudomonas fluorescens*, combined with different concentrations of Fe (nZVI) nanoparticles on white willow plants (*Salix alba*) grown in soil contaminated with Pb, Cu, and Cd. The combined treatment improved the physiological, biochemical, and growth parameters of white willow, while also increasing the bioconcentration factor of the heavy metals in the plant. Low nanoparticle doses (150 mg/kg) were effective in increasing the plant's leaf area and root length in comparison to the control plants, whereas the use of high concentrations (300 mg/kg) of NPs was detrimental for plant growth and heavy metal accumulation.

8.3.5 Current Status on the Nanoremediation of Heavy Metals

A critical assessment was performed on the surveyed literature using a simple bibliometric analysis. According to the overall literature analysis on nanoremediation of heavy metals ($n = 16$), it is remarkable that the treatments that combined all three remediation factors—nanomaterials, hyperaccumulating plants, and beneficial microbes (nPB-remediation) with 44% of the works—were more studied than the individual treatments with the simple nanoremediation having the second larger number of publications (31%). As it was discussed above, the reason for this may be that the use of nPB treatments significantly improves heavy metal removal from contaminated soils in comparison with single-factor remediation strategies. Furthermore, as it was discussed in the previous sections, combined remediation treatments often have synergic effects on heavy metal detoxification of contaminated soils since in these systems the individual remediation capabilities are greatly potentiated. In terms of the type of heavy metal treated (n = 32), similarly to the nanofertilization results, cadmium is the most widely reported (31%), followed by Pb (19%), and As and Cu, with the rate of publications (13%). On the other hand, with regard to the nanoparticles used on remediation treatments, Fe nanoparticles are the most widely reported (35%), followed by Ti and P, with 29 and 18%, respectively. It was noticeable that the use of Zn nanoparticles was not prominent for remediation treatments, as it was in the case of nanofertilization. Finally, in the analysis of the microorganisms which are coupled to nanoremediation systems (n = 10), *Pseudomonas fluorescence* strains have been the most widely used (40%) followed by *Bacillus* spp. with 20% of the sites.

8.4 Nanoparticle Toxicity on Biologically Based Nanoremediation Systems

Despite the positive effects of nanoparticles on crop cultivation and remediation in heavy metal-contaminated soils, some reports indicate that, depending on the plant species and nanoparticle concentration, harmful outcomes for the plant may occur. In an aforementioned study, Zand et al. (2022) applied TiO_2 nanoparticles and a PGPR strain on *Trifolium repens* plants for Cd remediation. The authors reported that the maximum NP concentration tested (1000 mg/kg) was toxic for the plant's growth. Notably, a 500 mg/kg dose was effective for Cd remediation. Similarly, Mokarram-Kashtiban et al. (2019) detected TiO_2 toxicity at a concentration of 300 mg/kg on the plants in a nanophytobioremediation system for heavy metal removal (Cd, Cu, and Pb), whereas a concentration of 150 mg/kg yielded considerable plant growth and remediation efficiency. Wang et al. (2021) evaluated the effect of TiO_2 nanoparticles with two heavy metal:nanoparticle concentration ratio on *Scenedesmus obliquus* algae grown in a Cd-contaminated medium. From the results of the study, the authors postulated that higher concentrations of nanoparticles led to increased absorption and introduction of Cd into algal cells, resulting in the destruction of cell walls and membranes. This destruction was attributed to either mechanical damage caused by the nanoparticles or Cd-induced oxidative damage, thereby increasing Cd toxicity in the presence of Ti nanoparticles. Zhang et al. (2025) studied the remediation of Cd and pyrene in contaminated soils employing the plant *Hylotelephium spectabile* and TiO_2 nanoparticles combined with humic acids. The authors reported that Ti nanoparticles can inhibit plant nutrient uptake by competing with Fe, P, and Zn in the plant's rhizosphere. In addition, the roots exposed to the nanoparticles presented changes indicating NP toxicity: decrease in tip diameter, increase in Ti accumulation, and overall damage of root hairs. When Cd interacted with high concentrations of TiNPs and humic acids, it increased toxicity by synergism, causing a decrease in plant growth of *H. spectabile*. In a similar study, Qian et al. (2024) evaluated the toxicity of different concentration ratios of Ti nanoparticles and As(V) on the algae *Phaeodactylum tricornutum*. The toxicity of the NP treatment depended on the ratio of NP-heavy metal concentration, with the 1:4 ratio yielding synergic toxic effects on the algal cells, whereas the ratios 1:1 and 4:1 resulted in neutral or antagonistic dependencies, respectively. The authors reported that the synergic toxicity of the 1:4 ratio favored cell internalization of Ti nanoparticles, which in turn caused As accumulation and toxicity. In addition, this toxicity-inducer treatment upregulated the pst gene, related to As transport and As accumulation, whereas it downregulated the ArsM gene, related to As biotransformations, which in turn caused greater As bioabsorption and toxicity.

Xiao et al. (2022) observed that when Cu nanoparticles were applied to soybean (*Glycine max*) plants, growth was inhibited above 200 mg Cu/kg. Copper uptake by soybeans decreased with increasing Cu-NP concentrations, possibly due to the efficient Cu uptake at low concentrations. TEM images showed that at high concentrations of Cu nanoparticles, soybean root structures exhibited an irregular morphology and the presence of particles or aggregates in the cells. In a similar work, Yusefi-Tanha et al. (2020) assessed the application of Cu nanoparticles and conventional $CuCl_2$ micronutrient solution in soybean cultivation (*Glycine max*) for the Cu enrichment of the seeds. The authors evaluated the effect of different Cu concentrations and nanoparticle sizes on Cu accumulation and toxicity in different parts of the plant. Copper toxicity was directly proportional to Cu concentration without regard to the particle size or nanoparticle-conventional treatment. On the other hand, Cu toxicity increased with reduced NP sizes, observing the highest toxicity on the plant for the 25-nm nanoparticles. The bioaccumulation of Cu was greater in the root, which was followed by the leaf, the stem, and the seed, with the lowest Cu concentration. In the study, imaging analysis through transmission electron microscopy (TEM) revealed that 25-nm Cu nanoparticles were stored in the cell walls, cell membranes, and cytoplasmic vacuoles of the seed embryo (Fig. 8.3).

In conclusion, nanoparticle toxicity is caused by a variety of factors, among which, NP concentration and elemental composition, as well as particle size, are the main factors. However, other factors include the characteristics of the heavy metal, since according to the literature, toxic synergies can take place in nanoremediation systems, which comprise biological components such as plants and microorganisms. In these cases, other important factors regarding NP toxicity are the characteristics of the plant or microorganism used in the remediation treatment. Finally, it should be noted that high NP concentrations have been employed in the literature, which can be close to the toxic limits for plants and microbes. Therefore, the relatively toxic high concentrations reported may be too high for practical applications. However, to make an overall assessment of the viability of the nanoremediation system, it is important to consider both the effective and the toxic NP concentrations.

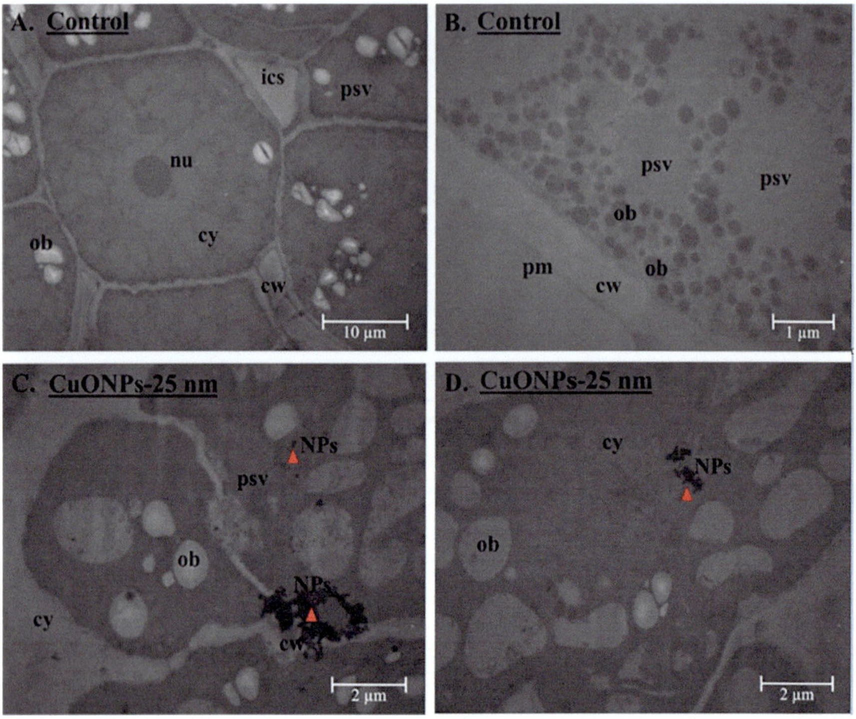

Fig. 8.3 The structure of the soybean (*Glycine max*) seed embryo was observed using TEM with Cu NPs-25 nm (500 mg kg^{-1} of soil). A and B represent control without nanoparticles, while C and D represent treatments with Cu NPs. Metal aggregates are observed within the cell wall (cw) and/or plasma membrane (pm), as well as within the cytoplasm (cy) and/or protein storage vacuoles (psv). nu nucleus, ics intracellular space, ob oil bodies, NPs nanoparticles (red triangle). (Source: Yusefi-Tanha et al. (2020)

8.5 Conclusions and Future Prospects

Recent research efforts aimed at the application of nanoparticles (NPs) in agriculture have acquired great prominence due to the proven capacity of these materials to act as nanofertilizers by promoting plant growth and improving the quality of the harvested products. Furthermore, according to the surveyed literature, these nanomaterials are not only effective as fertilizers but also contribute to ameliorate the toxicity caused to crop plants that are cultivated in heavy metal-contaminated soils. The main mechanism by which nanofertilizers mitigate this toxicity is the reduction of the plant's oxidative stress through the activation of the plant defense system of the plant, which in turn prompts the plant's synthesis of antioxidant enzymes and secondary metabolites related to stress relief.

According to the literature review, most studies on heavy metal-related nanofertilization have focused on cereals, with rice being the most extensively investigated crop, followed by wheat and corn. Notably, a greater number of reports addressed nanofertilization coupled with beneficial microorganisms in comparison with those relying solely on NP treatments. In this context, plant-growth-promoting rhizobacteria (PGPR), mainly *Bacillus* spp. and *Staphylococcus aureus*, have been applied, since their co-application with NPs has been shown to foster crop fertilization and alleviate heavy metal stress in plants. Regarding the contaminants studied, Cd was the most frequently reported heavy metal, followed by Cr, As, and Pb. With respect to nanomaterial formulations, most studies employed Zn-based nanoparticles for the nanofertilization of heavy-metal-contaminated soils.

In addition, several recent studies have explored the use of nanoparticles for the remediation of heavy metal-contaminated soils. Among these, Fe- and Ti-based nanoparticles were the most extensively employed formulations. Remarkably, according to our review, remediation strategies that combined all three factors (nanomaterials, hyperaccumulating plants, and beneficial microbes), nPB-remediation, were reported more frequently than individual nanoremediation approaches (NP, NP-plant, or NP-PGPR). This trend probably reflects the enhanced efficiency of the nPB strategy, which potentiates the remediation capacity of single and dual treatments. In this context, PGPR strains, particularly *Pseudomonas fluorescens* and *Bacillus* spp., have been successfully applied as nanobioremediation agents in heavy metal-contaminated soils.

For the further consolidation of the nanofertilization and nanoremediation applications, safety studies are still required, particularly those addressing the toxicity of nanoparticles in microbial cells and host plants (including both food crops and hyperaccumulating species). Overall, the reviewed literature indicates that nanoparticle-based treatments for crop fertilization and remediation represent a highly efficient and sustainable strategy for managing heavy metal-contaminated soils. Future research should prioritize the development of nanomaterial synthesis and formulation methods that ensure low toxicity and maximize fertilization/remediation efficiency. Such advances will provide a stronger foundation for the widespread application of nanoparticles to ameliorate the serious agricultural and environmental challenges posed by heavy metal contamination.

Acknowledgments Ramírez-Cariño thanks Secretaría de Ciencia, Humanidades, Tecnología e Innovación (SECIHTI), Mexico, for the financial support of his postdoctoral fellowship (No. 644716).

Declaration of Competing Interests The authors declare that they have no known competing financial interests or personal relationships that could have appeared to influence the work reported in this paper.

References

Abo-Alkasem MI, Hassan NH, Abo Elsoud MM (2023) Microbial bioremediation as a tool for the removal of heavy metals. Bull Natl Res Cent 47(1):31. https://doi.org/10.1186/s42269-023-01006-z

Ahmed T, Noman M, Rizwan M, Ali S, Ijaz U, Nazir MM, ALHaithloul HAS, Alghanem SM, Abdulmajeed AM, Bin Li (2022) Green molybdenum nanoparticles-mediated bio-stimulation of *Bacillus* sp. strain ZH16 improved the wheat growth by managing in planta nutrients supply, ionic homeostasis and arsenic accumulation. J Hazard Mater 423:127024. https://doi.org/10.1016/j.jhazmat.2021.127024

Ahmed S, Fatima M, Sardar R, Yasin NA (2024) Application of nano selenium alleviates cd-induced growth inhibition and enhances biochemical responses and the yield of Solanum melongena L. J Soil Sci Plant Nutr 24:8099–8120. https://doi.org/10.1007/s42729-024-02101-4

Akhtar N, Khan S, Rehman SU, Rehman ZU, Khatoon A, Rha ES, Jamil M (2021) Synergistic effects of zinc oxide nanoparticles and bacteria reduce heavy metals toxicity in rice (*Oryza sativa* L.) plant. Toxics 9(5):113. https://doi.org/10.3390/toxics9050113

Alharby HF, Ali S (2022) Combined role of Fe nanoparticles (Fe NPs) and Staphylococcus aureus L. in the alleviation of chromium stress in rice plants. Life 12(3):338. https://doi.org/10.3390/life12030338

AL-Huqail AA, Alghanem SMS, Alghamdi SA, Alhaithloul HAS, Al-Robai SA, Alalawy AI, Alzuaibr FM, Alasmari A, Ali B, Saleem MH, Fahad S, Ali S, Abeed AHA (2024) Coactive application of bacillus mycoides PM35 and calcium oxide nanoparticles stimulate gene expression responses in maize (*Zea mays* L.) under chromium stress. J Soil Sci Plant Nutr:4224–4242. https://doi.org/10.1007/s42729-024-01826-6

Ali Q, Zia MA, Kamran M, Shabaan M, Zulfiqar U, Ahmad M, Iqbal R, Maqsood MF (2023) Nanoremediation for heavy metal contamination: a review. Hybrid Adv 4:100091. https://doi.org/10.1016/j.hybadv.2023.100091

Ali MA, Nafees M, Waseem M, Alomrani SO, Al-Ghanim KA, Alshehri MA, Zheng H, Ali S, Li F (2024) Modulation of cd carriers by innovative nanocomposite (ca+Mg) and cd-resistance microbes (*Bacillus pumilus*): a mechanistic approach to enhance growth and yield of rice (Oryza sativa L.). Front Plant Sci 15:1387187. https://doi.org/10.3389/fpls.2024.1387187

Anwar A, Chen M, Chen C, Hu C, Su W, Chen R, Song S (2025) Nickel oxide nanoparticles (NiO-NPs) alleviate cadmium (Cd) stress in tomato by activating the expression of SlERF4 and SlWRI1. Ind Crop Prod 230:121051. https://doi.org/10.1016/j.indcrop.2025.121051

Aseri V, Chopade LR, Godara V, Lohar S, Kumari P, Mavry B, Nagar V, Sharma V, Awasthi G, Awasthi KK, Sankhla MS (2024) Advancing phytoremediation of contaminated soil through nanoparticle-based approaches: current progress and future perspectives. Lett Appl Nano Biol Sci 13(2):1–15. https://doi.org/10.33263/LIANBS132.057

Ayilara MS, Babalola OO (2023) Bioremediation of environmental wastes: the role of microorganisms. Front Agron 5:1183691. https://doi.org/10.3389/fagro.2023.1183691

Balali-Mood M, Naseri K, Tahergorabi Z, Khazdair MR, Sadeghi M (2021) Toxic mechanisms of five heavy metals: mercury, lead, chromium, cadmium, and arsenic. Front Pharmacol 12:1–19. https://doi.org/10.3389/fphar.2021.643972

Bolaños-Guerrón D, Capa J, Flores LC (2021) Retention of heavy metals from mine tailings using Technosols prepared with native soils and nanoparticles. Heliyon 7(7):e07631. https://doi.org/10.1016/j.heliyon.2021.e07631

Caballero-Mejía B, Moliner A, Escolástico C, Hontoria C, Mariscal-Sancho I, Pérez-Esteban J (2025) Use of magnetite nanoparticles and magnetic separation for the removal of metal(loid) s from contaminated mine soils. J Hazard Mater 486:137081. https://doi.org/10.1016/j.jhazmat.2024.137081

Cao X, Alabresm A, Chen YP, Decho AW, Lead J (2020) Improved metal remediation using a combined bacterial and nanoscience approach. Sci Total Environ 704:135378. https://doi.org/10.1016/j.scitotenv.2019.135378

Chaudhary P, Ahamad L, Chaudhary A, Kumar G, Chen WJ, Chen S (2023) Nanoparticle-mediated bioremediation as a powerful weapon in the removal of environmental pollutants. J Environ Chem Eng 11(2):109591. https://doi.org/10.1016/j.jece.2023.109591

Danila V, Kumpiene J, Kasiuliene A, Vasarevičius S (2020) Immobilisation of metal(loid)s in two contaminated soils using micro and nano zerovalent iron particles: evaluating the long-term stability. Chemosphere 248:126054. https://doi.org/10.1016/j.chemosphere.2020.126054

Deng R, Huang D, Xue W, Lei L, Chen S, Zhou C, Liu X, Wen X, Li B (2020) Eco-friendly remediation for lead-contaminated riverine sediment by sodium lignin sulfonate stabilized nano-chlorapatite. Chem Eng J 397:125396. https://doi.org/10.1016/j.cej.2020.125396

Ding S, Liang Y, Wang M, Hu R, Song Z, Xu X, Zheng L, Shen Z, Chen C (2024) Less is more: A new strategy combining nanomaterials and PGPB to promote plant growth and phytoremediation in contaminated soil. J Hazard Mater 469:134110. https://doi.org/10.1016/j.jhazmat.2024.134110

El Sharkawy M, AL-Huqail AA, Aljuaid AM, Kamal N, Mahmoud E, Omara AED, El-Kader NA, Li J, Mahmoud NN, El Baroudy AA, Ghoneim AM, Ismail SM (2024) Nano-bioremediation of arsenic and its effect on the biological activity and growth of maize plants grown in highly arsenic-contaminated soil. Nanomaterials 14(13):1164. https://doi.org/10.3390/nano14131164

Emamverdian A, Ding Y, Barker J, Liu G, Hasanuzzaman M, Li Y, Ramakrishnan M, Mokhberdoran F (2022) Co-application of 24-Epibrassinolide and titanium oxide nanoparticles promotes Pleioblastus pygmaeus plant tolerance to cu and cd toxicity by increasing antioxidant activity and photosynthetic capacity and reducing heavy metal accumulation and Translocati. Antioxidants 11(3):1–26. https://doi.org/10.3390/antiox11030451

Huibo Z, Yong Z, Rui L, Guorui L, Jianjun D, Qi W, Xiaotian L, Mingda Y, Yanpeng W, Zhiyan W, Fenglan H (2023) Analysis of the mechanism of Ricinus communis L. tolerance to Cd metal based on proteomics and metabolomics. PLoS One 18(3):1–24. https://doi.org/10.1371/journal.pone.0272750

Jeyakumar P, Debnath C, Vijayaraghavan R, Muthuraj M (2023) Trends in bioremediation of heavy metal contaminations. Environ Eng Res 28(4):220631. https://doi.org/10.4491/eer.2021.631

Karnwal A, Martolia S, Dohroo A (2024) Exploring bioremediation strategies for heavy metals and POPs pollution: the role of microbes, plants, and nanotechnology. Front Environ Sci 12:1397850. https://doi.org/10.3389/fenvs.2024.1397850

Khan MA, Yasmin H, Shah ZA, Rinklebe J, Alyemeni MN, Ahmad P (2022) Co application of biofertilizer and zinc oxide nanoparticles upregulate protective mechanism culminating improved arsenic resistance in maize. Chemosphere 294:133796. https://doi.org/10.1016/j.chemosphere.2022.133796

Kiran, Bharti R, Sharma R (2021) Effect of heavy metals: an overview. Mater Today Proc 51:880–885. https://doi.org/10.1016/j.matpr.2021.06.278

Kushwaha P, Kashyap PL (2021) A review of advances in bioremediation of heavy metals by microbes and plants. J Nat Resour Conserv Manag 2(1):65–80. https://doi.org/10.51396/anrcm.2.1.2021.65-80

Lai C, Yang H, Guo Z, Yi H, He T, Chen M, He G (2025) Nano-selenium modified green eggshell biochar reduces cadmium accumulation in shallots (*Allium schoenoprasum* L.). Environ Res 277:121635. https://doi.org/10.1016/j.envres.2025.121635

Liu W, Zuo Q, Zhao C, Wang S, Shi Y, Liang S, Zhao C, Shen S (2018) Effects of Bacillus subtilis and nanohydroxyapatite on the metal accumulation and microbial diversity of rapeseed (Brassica campestris L.) for the remediation of cadmium-contaminated soil. Environ Sci Pollut Res 25:25217–25226. https://doi.org/10.1007/s11356-018-2616-8

Lu J, Li Y, Yang Z (2023) Adsorption and removal effect of magnetic nanohydroxyapatite on cadmium in contaminated soil. J Ecol Rural Environ 39(6):803–809. https://doi.org/10.19741/j.issn.1673-4831.2021.0822

Luo Z, Mfarrej MFB, Saleem MH, Ma J, Saleh IA, Abdel-Maksoud MA, Zakri AM, Chen F, Oliván LMG (2024) Individual and combinatorial application of nanosilica and carbon nanoparticles alleviate nickel stress in barley (*Hordeum vulgare* L.): impacts on gene expression, AsA – GSH cycle, cellular fractionation, and proline metabolism. Sci Total Environ 954:176304. https://doi.org/10.1016/j.scitotenv.2024.176304

Ma J, Alshaya H, Okla MK, Alwasel YA, Chen F, Adrees M, Hussain A, Hameed S, Shahid MJ (2022) Application of cerium dioxide nanoparticles and chromium-resistant bacteria reduced chromium toxicity in sunflower plants. Front Plant Sci 13:876119. https://doi.org/10.3389/fpls.2022.876119

Ma J, Li Y, Chen F, Sun Y, Zhu Y, Wang L (2023) Bacillus mycoides PM35 in combination with titanium dioxide (TiO2)-nanoparticles enhanced morpho-physio-biochemical attributes in Barley (Hordeum vulgare L.) under cadmium stress. Chemosphere 323:138224. https://doi.org/10.1016/j.chemosphere.2023.138224

Mai X, Tang J, Tang J, Zhu X, Yang Z, Liu X, Zhuang X, Feng G, Tang L (2025) Research progress on the environmental risk assessment and remediation technologies of heavy metal pollution in agricultural soil. J Environ Sci 149:1–20. https://doi.org/10.1016/j.jes.2024.01.045

Mangena P (2025) A review on the role of algae-nanoparticles as a sustainable agricultural strategy to improve resilience against abiotic stresses in leguminous crops. Plant Nano Biol 11:100125. https://doi.org/10.1016/j.plana.2024.100125

Maryam, El-Mogy MM, Jan MF, Naz I, Ahmad I, Ahmad R, Altaf MT (2025) Nanoparticle Innovations for Mitigating Metal Toxicity in Plants. Phyton-Int J Exp Bot *94*(3):623–640. https://doi.org/10.32604/phyton.2025.063763

Milano F, Giotta L, Lambreva MD (2024) Perspectives on nanomaterial-empowered bioremediation of heavy metals by photosynthetic microorganisms. Plant Physiol Biochem 216:109090. https://doi.org/10.1016/j.plaphy.2024.109090

Mogazy AM (2024) Bacterial exopolysaccharides for heavy metals and toxic dye removal: biosynthesis, mechanism, and remediation strategies. In: Bacterial secondary metabolites. Elsevier, pp 15–27. https://doi.org/10.1016/B978-0-323-95251-4.00006-5

Mokarram-Kashtiban S, Hosseini SM, Tabari Kouchaksaraei M, Younesi H (2019) The impact of nanoparticles zero-valent iron (nZVI) and rhizosphere microorganisms on the phytoremediation ability of white willow and its response. Environ Sci Pollut Res 26(11):10776–10789. https://doi.org/10.1007/s11356-019-04411-y

Mondal S, Mondal T, Pal P, Mitra S, Ghosh SK, Soren T, Maiti TK (2025) Bioprotective mechanisms of Enterobacter sp. against arsenic, cadmium, and lead toxicity and its potential role in soil bioremediation. *J Environ Chem Eng* 13(2):115432. https://doi.org/10.1016/j.jece.2025.115432

Muhammad S, Ulhassan Z, Munir R, Yasin MU, Islam F, Zhang K, Chen W, Jan M, Afzal M, Muhammad A, Hannan F, Zhou W (2025) Nanosilica and salicylic acid synergistically regulate cadmium toxicity in rice. Environ Pollut 364:125331. https://doi.org/10.1016/j.envpol.2024.125331

Olaya-Abril A, Biełło K, Rodríguez-Caballero G, Cabello P, Sáez LP, Moreno-Vivián C, Luque-Almagro VM, Roldán MD (2024) Bacterial tolerance and detoxification of cyanide, arsenic and heavy metals: holistic approaches applied to bioremediation of industrial complex wastes. Microb Biotechnol 17(1):1–22. https://doi.org/10.1111/1751-7915.14399

Pande A, Mun BG, Methela NJ, Rahim W, Lee DS, Lee GM, Hong JK, Hussain A, Loake G, Yun BW (2022) Heavy metal toxicity in plants and the potential NO-releasing novel techniques as the impending mitigation alternatives. Front Plant Sci 13:1–13. https://doi.org/10.3389/fpls.2022.1019647

Patle T, Tomar B, Parihar SS, Tomar SS, Singh P (2024) Nanotechnology-based soil improvement and conservation for enhancement of crop production. In: Harnessing nano omics and nanozymes for sustainable agriculture. IGI Global Scientific Publishing. https://doi.org/10.4018/979-8-3693-1890-4.ch011

Prakash P, S SC (2023) Nano-phytoremediation of heavy metals from soil: a critical review. Pollutants 3(3):360–380. https://doi.org/10.3390/pollutants3030025

Qian W, Chen CC, Huang Y, Zhu X (2024) Exposure concentration ratios and biological responses play a critical role in determining the joint toxicity of TiO2 nanoparticles and As(V) to the organism: the case study in marine algae Phaeodactylum tricornutum. Sci Total Environ 909:168508. https://doi.org/10.1016/j.scitotenv.2023.168508

Qin M, Yan J, Li R, Jia T, Sun X, Liu Z, El-Sheikh MA, Ahmad P, Liu P (2025) Integrated physiological, transcriptomic, and metabolomic investigation reveals that MgO NPs mediate the alleviation of cadmium stress in tobacco seedlings through ABA-regulated lignin synthesis. J Hazard Mater 483(61):136693. https://doi.org/10.1016/j.jhazmat.2024.136693

Rachappanavar V, Gupta SK, Jayaprakash GK, Abbas M (2024) Silicon mediated heavy metal stress amelioration in fruit crops. Heliyon 10(18):e37425. https://doi.org/10.1016/j.heliyon.2024.e37425
Raffa CM, Chiampo F, Shanthakumar S (2021) Applied sciences remediation of metal / metalloid-polluted soils : A short review. Appl Sci 11(9):4134
Rajput VD, Minkina T, Upadhyay SK, Kumari A, Ranjan A, Mandzhieva S, Sushkova S, Singh RK, Verma KK (2022) Nanotechnology in the restoration of polluted soil. Nano 12(5):1–16. https://doi.org/10.3390/nano12050769
Raklami A, Meddich A, Oufdou K (2022) Plants-microorganisms-based bioremediation for heavy metal cleanup: recent developments phytoremediation techniques, regulation mechanisms and molecular responses. Int J Mol Sci 23:5031. https://doi.org/10.3390/ijms23095031
Ramírez-Cariño HF, Morales I, Guadarrama-Mendoza PC, González-Terreros E, Martínez-Gutiérrez GA, Dunlap CA, Valadez-Blanco R (2023) Biofertilizing effect of putative plant growth promoting rhizobacteria in vitro and in tomatillo seedlings (*Physalis ixocarp*a Brot.). Sci Hortic 308:111567. https://doi.org/10.1016/j.scienta.2022.111567
Saleem S, Rizvi A, Khan MS (2023) Microbiome-mediated nano-bioremediation of heavy metals: a prospective approach of soil metal detoxification. Int J Environ Sci Technol 20(11):12823–12846. https://doi.org/10.1007/s13762-022-04684-w
Sardar R, Ahmed S, Yasin NA (2022) Titanium dioxide nanoparticles mitigate cadmium toxicity in Coriandrum sativum L. through modulating antioxidant system, stress markers and reducing cadmium uptake. Environ Pollut 292:118373. https://doi.org/10.1016/j.envpol.2021.118373
Sehrish AK, Ahmad S, Nafees M, Mahmood Z, Ali S, Du W, Kashif Naeem M, Guo H (2024) Alleviated cadmium toxicity in wheat (Triticum aestivum L.) by the coactive role of zinc oxide nanoparticles and plant growth promoting rhizobacteria on TaEIL1 gene expression, biochemical and physiological changes. Chemosphere 364:143113. https://doi.org/10.1016/j.chemosphere.2024.143113
Shah AA, Usman S, Noreen Z, Kaleem M, Raja V, El-Sheikh MA, Ibrahim Z, Sehar S (2024a) Fullerenol nanoparticles and AMF application for optimization of Brassica napus L. resilience to lead toxicity through physio-biochemical and antioxidative modulations. Sci Rep 14(1):1–17. https://doi.org/10.1038/s41598-024-82086-3
Shah AA, Zafar S, Usman S, Javad S, Zaib-un-Nisa, Aslam M, Noreen Z, Elansary HO, Almutairi KF, Ahmad A (2024b) Zinc oxide nanoparticles and Klebsiella sp. SBP-8 alleviates chromium toxicity in *Brassica juncea* by regulation of antioxidant capacity, osmolyte production, nutritional content and reduction in chromium adsorption. Plant Physiol Biochem 210:108624. https://doi.org/10.1016/j.plaphy.2024.108624
Singh A, Rajput VD, Al Tawaha ARM, Al Zoubi OM, Habeeb T, Rawat S, Ghazaryan K, Minkina T (2023) A review on crop responses to nanofertilizers for mitigation of multiple environmental stresses. Ecol Eng Environ Technol 24(7):280–296. https://doi.org/10.12912/27197050/169313
Souri Z, Karimi N, Sarmadi M, Rostami E (2017) Salicylic acid nanoparticles (SANPs) improve growth and phytoremediation efficiency of Isatis cappadocica Desv., under As stress. IET Nanobiotechnol 11(6):650–655. https://doi.org/10.1049/iet-nbt.2016.0202
Sun L, Wang R, Ju Q, Xing M, Li R, Li W, Li W, Wang W, Deng Y, Xu J (2023) Mitigation mechanism of zinc oxide nanoparticles on cadmium toxicity in tomato. Front Plant Sci 14:1–14. https://doi.org/10.3389/fpls.2023.1162372
Tahira S, Bahadur S, Lu X, Liu J, Wang Z (2025) ZnONPs alleviate cadmium toxicity in pepper by reducing oxidative damage. J Environ Manag 373:123796. https://doi.org/10.1016/j.jenvman.2024.123796
Verma KK, Joshi A, Song XP, Singh S, Kumari A, Arora J, Singh SK, Solanki MK, Seth CS, Li YR (2024) Synergistic interactions of nanoparticles and plant growth promoting rhizobacteria enhancing soil-plant systems: a multigenerational perspective. Front Plant Sci 15:1–14. https://doi.org/10.3389/fpls.2024.1376214

Wang B, Fang R (2024) ZnO nanoparticles enhances cadmium tolerance by modulating N6-methyladenosine (m6A) level of stress-responsive genes NRT1 and GM35E in vegetable soybean. Plant Physiol Biochem 217:109303. https://doi.org/10.1016/j.plaphy.2024.109303

Wang P, Zhao L, Huang Y, Qian W, Zhu X, Wang Z, Cai Z (2021) Combined toxicity of nano-TiO_2 and Cd^{2+} to *Scenedesmus obliquus*: effects at different concentration ratios. J Hazard Mater 418:126354. https://doi.org/10.1016/j.jhazmat.2021.126354

Wróbel M, Śliwakowski W, Kowalczyk P, Kramkowski K, Dobrzyński J (2023) Bioremediation of heavy metals by the genus Bacillus. Int J Environ Res Public Health 20(6):4964. https://doi.org/10.3390/ijerph20064964

Xiao Y, Ma J, Xian J, Peijnenburg WJGM, Du Y, Tian D, Xiao H, He Y, Luo L, Deng O, Tu L (2022) Copper accumulation and physiological markers of soybean (*Glycine max*) grown in agricultural soil amended with copper nanoparticles. Ecotoxicol Environ Saf 229:113088. https://doi.org/10.1016/j.ecoenv.2021.113088

Younas M, Nafees M, Munir M, Alomrani SO, Waseem M, Alshehri MA, Sarker PK, Ali S (2025) Cadmium resistance microbes and TiO 2 nanoparticles alleviate cadmium toxicity in wheat. Sci Rep 15(1):5557. https://doi.org/10.1038/s41598-025-88371-z

Yusefi-Tanha E, Fallah S, Rostamnejadi A, Pokhrel LR (2020) Root system architecture, copper uptake and tissue distribution in soybean (*Glycine max* (L.) Merr.) grown in copper oxide nanoparticle (CuONP)-amended soil and implications for human nutrition. Plants 9(10):1326. https://doi.org/10.3390/plants9101326

Zand AD, Tabrizi AM, Heir AV (2020a) Application of titanium dioxide nanoparticles to promote phytoremediation of Cd-polluted soil: contribution of PGPR inoculation. Biorem J 24(2–3):171–189. https://doi.org/10.1080/10889868.2020.1799929

Zand AD, Tabrizi AM, Heir AV (2020b) The influence of association of plant growth-promoting rhizobacteria and zero-valent iron nanoparticles on removal of antimony from soil by Trifolium repens. Environ Sci Pollut Res 27(34):42815–42829. https://doi.org/10.1007/s11356-020-10252-x

Zand AD, Heir AV, Khodaei H (2022) Integrated remediation approach for metal polluted soils using plants, nanomaterials and root-associated bacteria. J Dispers Sci Technol 43(11):1674–1688. https://doi.org/10.1080/01932691.2021.1878900

Zaynab M, Al-Yahyai R, Ameen A, Sharif Y, Ali L, Fatima M, Khan KA, Li S (2022) Health and environmental effects of heavy metals. J King Saud Univ 34(1):101653. https://doi.org/10.1016/j.jksus.2021.101653

Zhang Y, Wang J, Liu X, Zhang L, Wang J, Xu C, Qian T, Guo J (2025) Nano-TiO2 with humic acid to promote remediation of cadmium/pyrene composite contaminated soil by Hylotelephium spectabile. J Environ Chem Eng 13(3). https://doi.org/10.1016/j.jece.2025.116044

Chapter 9
Microbial Nanoscale Advances in Crop Health Management

Sandeep Jain, Jyoti Jain, Jayesh Singh, Simarjot Kaur, Pooja Salaria, and Ritu Rani

Abstract The convergence of microbiology and nanoscience is transforming sustainable agriculture through targeted and eco-friendly management of major biotic and abiotic stresses. This chapter explores the conceptual and practical integration of nanoscale microbial formulations to boost crop resilience, nutrient uptake, and physiological performance. Emphasizing green chemistry principles, the chapter outlines the concepts and application of microbial nanomaterials tailored for improved water-use efficiency, salinity tolerance, and pest and disease management. Case studies and mechanistic insights demonstrate how microbial nanoparticles modulate stress response pathways, ion homeostasis, and photosynthetic efficiency. The chapter also highlights the regulatory, environmental, and translational challenges associated with nano-biotechnology in agriculture, advocating for interdisciplinary research to bridge fundamental knowledge with field-level applications. Further studies are needed to evaluate long-term field efficacy, biosafety, ecological impacts, and species-specific responses of microbial nanotechnology across diverse agroecosystems.

Keywords Nanotechnology · Microbes · Plants · Disease · Stress · Abiotic · Biotic · Drought · Soil · Salinity · IPM

9.1 Introduction

Maintaining crop health and reducing environmental pollution from chemical inputs are essential to achieve sustainable agriculture and food security. Because nanoparticles (NPs <100 nm) have special qualities like increased surface area and

S. Jain · J. Jain · S. Kaur · P. Salaria · R. Rani
Department of Plant Pathology, Punjab Agricultural University, Ludhiana, India

J. Singh (✉)
Department of Plant Breeding and Genetics, College of Agriculture, Punjab Agricultural University, Ludhiana, Punjab, India
e-mail: jayesh_singh@pau.edu

K. A. Abd-Elsalam (ed.), *Plant-Microbiome Nanotechnology*, Sustainability Sciences in Asia and Africa, https://doi.org/10.1007/978-981-95-5056-2_9

reactivity, nanotechnology offers creative alternatives. Traditional NPs, however, give rise to toxicity issues. This has increased interest in biogenic nanoparticles (bio-NPs), which are made from plants, algae, and microbes. According to Noman et al. (2023) and Hoballah et al. (2019), bio-NPs are more eco-friendly, biocompatible, and have the ability to detoxify metals, increase microbial activity, and improve plant stress tolerance. Microbiology and nanotechnology work together because of their compatible sizes. Gold and silver NPs are examples of functionalized NPs that have antimicrobial properties and help detect pathogens. By neutralizing contaminants, nanomaterials such as carbon nanotubes and iron oxide nanoparticles (NPs) aid bioremediation in environmental microbiology (Ashraf et al. 2024). Nanofertilizers improve nitrogen fixation in agriculture by boosting nodulation, encouraging nitrogen-fixing microbes (like Rhizobium), and increasing nutrient use efficiency (Kots et al. 2024; Ghalamboran 2011). Important examples include cobalt/iron NPs which promote soybean nodulation; magnetite NPs increase nodule size; and nano-molybdenum increases microbial diversity. Agro-input targeted nano-delivery systems are currently being developed (Cherian et al. 2023).

Biotic agents (bacteria, viruses, and fungi) cause variation in plant structure, impact on nutrient uptake, and reduction in yield (Haggag 2024). Drought, salinity, temperature fluctuations, and pollution are examples of abiotic stressors that interfere with metabolism and root growth (Zhang et al. 2023). Due to their immobility, plants depend on complex defense mechanisms. Plant immunity is frequently weakened by abiotic stress, making them more susceptible to biotic agents and vice versa. Complex signaling networks control these interactions (Dixit et al. 2024), and among the remedies are PGRs, silicon supplementation, and advantageous microorganisms (biofilms, molecular defenses). Nanotechnological treatments like nanoprimers, nanopesticides, and nanoherbicides activate antioxidant enzymes, and nanosilica from algae (*Oscillatoria agardhii*) increases wheat resistance (Haggag 2018). Ag, TiO_2, ZnO, CNTs, or graphene nanoprimers improve seed vigor and defense gene expression (Shelar et al. 2024).

Green chemistry, which promotes less waste and safer materials, has gained popularity due to growing concerns about ecological degradation. It is founded on 12 principles that were introduced by Anastas and Warner (Idoko et al. 2024). This is in line with the green synthesis of NPs, which uses plants and microbes rather than hazardous chemicals. These techniques guarantee enhanced stability, decreased toxicity, and compatibility with the environment. Applications for biogenic nanoparticles in agriculture are promising and include: Nanoagrosomes for targeted delivery of pesticides and nutrients. According to Dong et al. (2013a, b), probiotic bacteria (*Lactobacillus casei*, *L. fermentum*) efficiently absorb heavy metals like cadmium, with the help of the selenium interaction. To remove heavy metals, fungi (like *Aspergillus tubingensis*) biosynthesize iron oxide nanoparticles (NPs) (Mahanty et al. 2020).

It is still difficult to find information about the interactions between microbiology and nanoscience in agricultural practices. This chapter was therefore written to provide a thorough and simple understanding that will aid in bridging these two crucial fields.

9.2 Biotic Stress Management with Microbial Nanotechnology

Biotic stress factors are a serious menace to sustainable farming and global food security. The demand for food production is rising due to a growing population, which makes it crucial to protect crop yields by managing plant diseases caused by biotic agents. These biotic agents cause an estimated loss of 20–40% in agricultural output worldwide (Cabral-Pinto et al. 2020). The farmers have mostly relied on chemical pesticides and fungicides for managing these issues, and the overuse of these chemicals causes various problems, such as the emergence of resistant pests and pathogens, disruption of nitrogen fixation, and the buildup of toxic substances in the ecosystem. These problems harm wildlife as well as pose risks to human health, which makes it urgent to find safer and more sustainable applications. This is where nanotechnology comes into the picture. Compared to traditional methods, the nano-based approaches offer new possibilities for managing diseases in crops. Researchers are actively involved in exploring nanotechnology as a replacement to conventional agrochemicals. This field holds great potential even in its early stages (Khan et al. 2021), and the studies show that nanomaterials are significant in their performance in managing plant diseases. The ultra-small size and unique properties highlight the potential of nanomaterials in suppressing the pests and pathogens for protecting crops and as a powerful tool for eco-friendly practices (Umar et al. 2021).

9.2.1 Microbial Nanostructures: A New Frontier Against Plant Pathogens

Scientists are molding their way towards more sustainable and eco-friendly solutions like microbial-based nanoparticles and biofertilizers for reducing the harmful effects of excessive chemical use in farming. The nanoparticles show high reactivity due to their nanoscale size and large surface area, which improves the plant's ability to manage diseases. The nanoparticles were mostly produced through physical and chemical methods, but nowadays, green-synthesized ones using natural organisms like bacteria, fungi, algae, plants, yeast, and actinomycetes are becoming popular. Another promising development is the use of plant probiotics as bioinoculants, and these microbes are beneficial in helping plants to tolerate stress by increasing photosynthesis, raising the production of osmolytes and phenolics, and activating antioxidant enzymes and stress-responsive genes. They reduce the signs of oxidative damage, like malondialdehyde (MDA) levels and electrolyte leakage, which indicates the improvement of plant health under stress (Upadhayay et al. 2023b) (Fig. 9.1).

The microbial nanoparticles are a cost-effective, eco-friendly option for disease management in crops (Meena et al. 2021). They combat the major plant pathogens, like *Fusarium oxysporum*, *Alternaria solani*, *Aspergillus niger*, *Ralstonia*

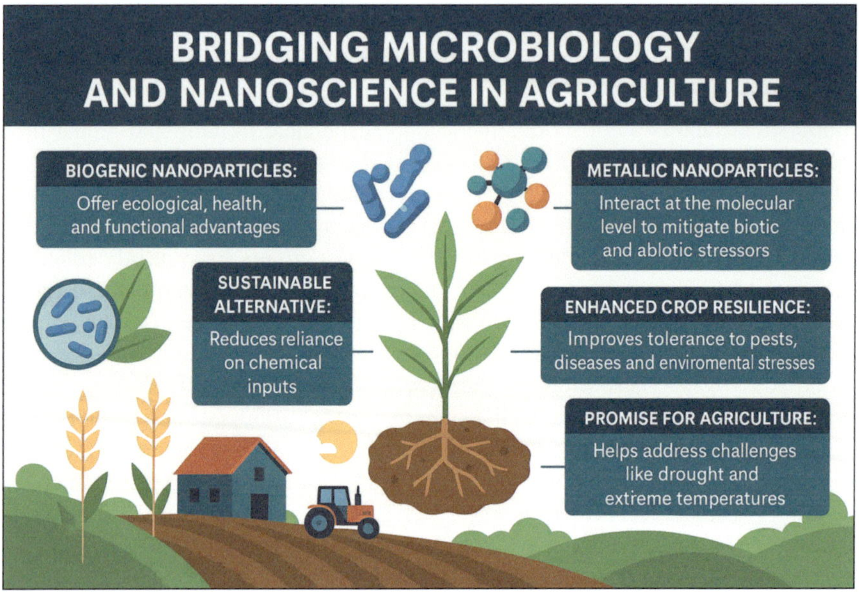

Fig. 9.1 The convergence of nanotechnology and microbiology in agriculture

solanacearum, and *Erwinia amylovora* (El-Batal et al. 2020). Microbial bioherbicides are synthesized using fungi, bacteria, or viruses or their natural metabolites, and they show efficacy in the suppression or killing of weeds. Various beneficial microbes produce phytotoxins that act as natural herbicides. The stability and ability to penetrate the weed tissues improve when these bioherbicides are encapsulated in nanoparticles, which controls weed with lower doses. Titanium dioxide nanoparticles (TiO_2-NPs) show antimicrobial activity which work by generating reactive oxygen species (ROS), which damage cell structures and ultimately kill harmful bacteria. These nanoparticles can disrupt the bacterial communication and alter gene expression by interfering with transcription factors (Meena et al. 2021). The research continues by highlighting the potential of encapsulated bioagents to protect the crops. Mousa et al. (2024) demonstrated that *Bacillus megaterium* after encapsulated in alginate microbeads and enriched with humic acid showed effective control of root-knot nematode (*Meloidogyne javanica*) infestations in tomato plants. This approach reduces pest pressure and boosts plant growth by improving photosynthetic pigment levels and antioxidant enzyme activity.

Based on expanding research on nanoscale microbial formulations, Table 9.1 compiles the different studies from both plant and non-plant models and illustrates the broader biological mechanisms and potential cross-domain applications (Fig. 9.2).

Table 9.1 Examples of role of microbe-mediated nanoparticles in plant defense against biotic stressors

Nanoparticle	Source organism	Nature of Action	Target organism	References
Silicon NPs	*Fusarium oxysporum SM5*	Nematicidal	*Meloidogyne incognita*	El-Ashry et al. (2022)
	Aspergillus niger	Antifungal	*Alternaria solani*	Albalawi et al. (2022)
	Penicillium sp	Pesticide	*Myzus persicae*	Hersanti et al. (2018)
	Penicillium sp	Pesticide		Hersanti et al. (2018)
MgO-NPs	*Aspergillus niger*	Antibacterial	*Staphylococcus aureus, P. aeruginosa*	Ibrahem et al. (2017)
TiO_2-NPs	*Aspergillu flavus*	Antibacterial	*E. coli*	Rajakumar et al. (2012)
Silver/silicon dioxide (Ag/SiO2) nanocomposite	*Escherichia coli D8*	Antifungal	*Botrytis cinerea*	Baka and El-Zahed (2022)
TiO_2-NPs	*Aeromonas hydrophila*	Antibacterial	*E. coli, P. aeruginosa, Staphylococcus aureus, Streptococcus pyogenes, and E. faecalis*	Jayaseelan et al. (2013)
Ag_2O NPs	*Bacillus paramycoides*	Antibacterial	*Vibrio parahaemolyticus, Salmonella* sp., *Enterobacter* sp., *and Micrococcus* sp.	Dharmaraj et al. (2021)
AgNPs	*Lactobacillus plantarum TA4*	Antibacterial	Gram-positive and Gram-negative bacteria	Yusof et al. (2020)
AgNPs	*Escherichia coli, Exiguobacterium aur antiacumm, and Brevundimonas diminuta*	Antibacterial	Pseudomonas aeruginosa, Staphylococcus aureus, and Salmonella typhi	Saeed et al. (2020)
Iron oxide NPs	*Proteus vulgaris*	Antibacterial, antioxidant, and cytocompatible	*Staphylococcus aureus*	Majeed et al. (2021)

(continued)

Table 9.1 (continued)

Nanoparticle	Source organism	Nature of Action	Target organism	References
ZnO NPs	*Bacillus subtilis ZBP4*	Antibacterial	Gram-positive and Gram-negative foodborne pathogens	Hamk et al. (2023)
Ag-NPs	*Marinobacter lipolyticus (EMBL\LN835275.1)*	Antifungal	*C. albicans*	Oves et al. (2016)

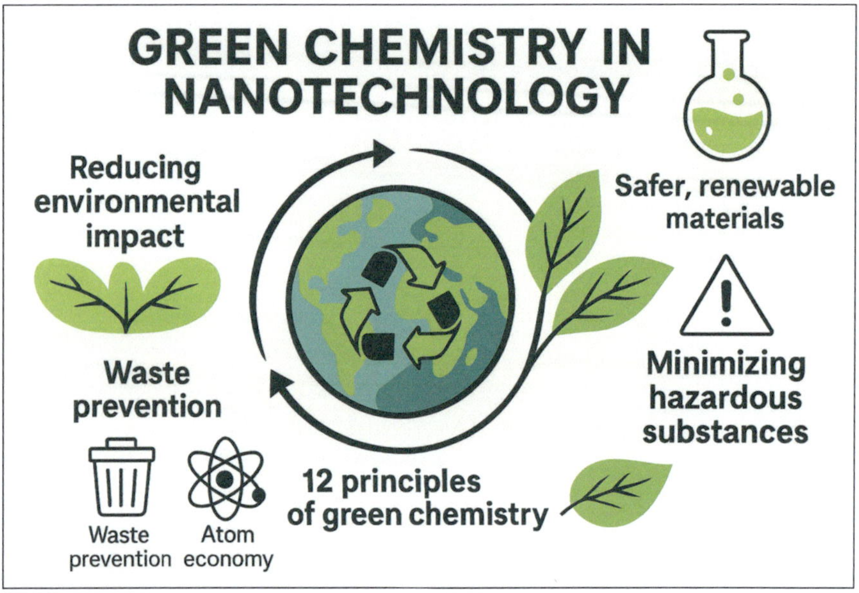

Fig. 9.2 Green chemistry principles driving sustainable nanotechnology

9.3 Abiotic Stress Management with Microbial Nanotechnology

Numerous abiotic stresses, such as drought, salinity, high temperatures, waterlogging, and pollution, can affect plants. Globally, the prevalence of these stressors has grown, resulting in significant financial losses. By controlling nutritional deficiencies, increasing enzymatic activity, and improving the adherence of plant growth-promoting bacteria (PGPB) to roots, nanoparticles can help manage abiotic stresses. In numerous ways, these interactions increase crop resilience to environmental stressors (Iqbal et al. 2020). Additionally, some endophytic and rhizospheric beneficial microorganisms with plant growth-promoting (PGP) characteristics show promise in promoting plant growth under stress. Through processes including the synthesis of phytohormones, exopolysaccharides, ACC deaminase, volatile organic compounds, osmolytes, antioxidant enzymes, and enzymes that break down pathogen cell walls, they reduce stress. They also control genes that respond to stress and encourage systemic resistance, which enhances plant resilience overall. The efficiency of nano-bioinoculants in controlling abiotic stressors has been demonstrated by recent research. In addition to improving root architecture and activating innate defense mechanisms, these formulations increase stress tolerance by controlling physiological and molecular responses. This provides both biotic and abiotic stress protection. A viable and sustainable way to address climate-related issues in crop production is through the synergistic interaction of microbial inoculants and nanomaterials (Upadhayay et al. 2023a).

9.3.1 Drought Stress: Enhancing Water Use Efficiency with Microbial Nanotools

One of the most harmful stresses to the environment is drought, which affects crops everywhere. Drought has become more frequent and more intense because of climate change, and this has been seen as a serious threat to plant growth and yields. From seed germination to flowering, drought affects every phase of a plant's life cycle and interferes with critical physiological functions like photosynthesis, nutrient uptake, and hormone regulation. One early indicator of drought stress is a change in the root structure. In order to find water, plants in arid soils often grow longer, deeper roots; however, as the plant grows, the total surface area of the roots may decrease, making it more difficult for the plant to absorb nutrients. Low seed yield results from the membranes becoming more permeable and reactive oxygen species (ROS) building up (Parrish and Fike 2005). ROS damages compounds that affect proteins, lipids, and DNA, and releases electrolytes. Researchers have been looking into a number of methods to protect crops from these detrimental effects, such as creating crop cultivars that can withstand drought, using exogenous stress-protectant substances (hormones, osmolytes), adjusting nutrient management, and using beneficial microbes. By enhancing their internal defenses, nanoparticles (NPs) can help plants manage drought stress. They help in stabilizing the photosynthetic apparatus, increase the activity of antioxidant enzymes, and reduce the accumulation of harmful substances like hydrogen peroxide and malondialdehyde (MDA). Because of their extremely small size, NPs can penetrate plant tissues and chloroplasts, increasing electron transport (ET), PSII activity, and light absorption. This increases photosynthesis efficiency even in stressful conditions.

Furthermore, nanoparticles help plants become more stable and structurally sound, which facilitates drought adaptation. Water absorption and retention are enhanced when plant cell membranes are strengthened. According to Zahedi et al. (2020), ZnO nanoparticles can enhance water absorption by altering root architecture through increased lateral root and biomass development. Additionally, these particles increase drought resilience by upregulating hormone signaling and genes linked to stress. These nanoparticles help maintain nutrient homeostasis by improving the uptake and distribution of nitrogen, potassium, manganese, and zinc, which are frequently deficient during drought conditions because of decreased transpiration and root activity (Fig. 9.3).

These nutrient levels can be restored by foliar application or soil soaking. For instance, Hafez et al. (2021) found that a treatment that combined Si NPs and plant growth-promoting rhizobacteria (PGPR) improved photosynthesis, antioxidant capacity, and nutrient uptake in drought-stressed maize. By increasing chlorophyll synthesis, PSII function, and RuBisCO activity, certain nanoparticles, such as ZnO and TiO_2, reduce the effects of drought. Additionally, they aid in fixing CO_2, shield chloroplasts and mitochondrial structures, and boost the synthesis of carotenoid pigments that shield chlorophyll from oxidative damage (Rasheed et al. 2022). In addition, NPs regulate electrolyte osmosis and hormone secretion, which contribute to greater drought tolerance. For example, TiO_2 NPs stimulate the production of proline and the expression of genes that produce glycine betaine, certain soluble sugars, and phenolic compounds that aid in cellular and osmotic defense. NPs suppress ABA, which promotes growth in conditions of water deficit, and raise levels of some growth hormones, especially gibberellins and indole-3-acetic acid (IAA). Additionally, NPs can enhance melatonin biosynthesis, which increases plants' resistance to drought (Zahedi et al. 2021). Similarly, TiO_2 NPs have been shown to enhance the defense response of plants by upregulating the expression of genes that produce ferulic, sinapic, and syringic acids (Shoarian et al. 2020).

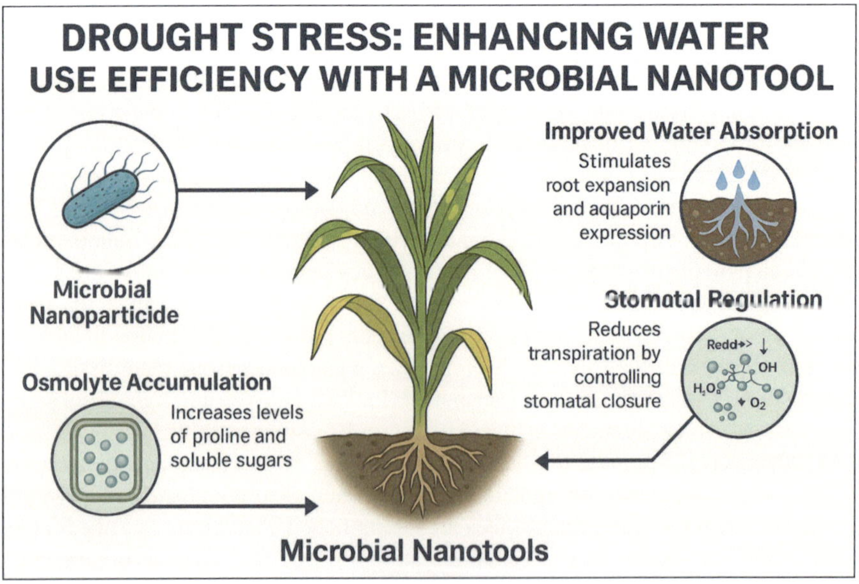

Fig. 9.3 Microbial nanotools for water-efficient agriculture

In terms of anatomy, NPs such as zinc at 100 ppm improve vascular development, stem thickness, and cortex diameter, which help in the movement of nutrients and water during drought (Semida et al. 2021; Al-Dhalimi and Al-Ajeel 2020). By enhancing root function, antioxidant capacity, and hormonal signaling, NPs support plant growth even when stomatal closure causes CO_2 limitation. Applying chitosan nanoparticles as a foliar spray or soil amendment during drought conditions has also shown promise in increasing yield metrics such as biomass weight and grain weight (Abdel-Aziz et al. 2016). Zn NPs raise protein and starch content while enhancing nutrient mobilization and amylase activity. While ZnO NPs have been shown to support rice biofortification through seed priming, Fe and Cu NPs help wheat grow under drought stress by boosting enzyme activity and nutrient uptake (Khanra et al. 2018). NPs like ZnO, Si, and Se counteract these effects by increasing non-enzymatic antioxidants like phenolics and ascorbic acid and enzymatic antioxidants like ascorbate peroxidase (APX), catalase (CAT), and superoxide dismutase (SOD). SiO_2 NPs have been demonstrated to raise total phenolic content (TPC), DPPH scavenging ability, and FRAP values in addition to promoting the expression of antioxidant genes and the accumulation of amino acids and osmoprotectants. Furthermore, by promoting nitrate reductase activity, which in turn encourages the synthesis of proline and glycine betaine, NPs raise the production of nitric oxide (NO). They provide extra protection against drought-induced stress by encouraging the buildup of sugars such as sucrose, fructose, trehalose, and glucose. Also, the use of NPs impacts the expression of drought-responsive genes, such as GmRD20A, GmDREB2, GmERD1, GmFDL19, GmNAC11, GmWRKY27, GmMYB118, and GmMYB174, which are implicated in hormone signaling, secondary metabolite biosynthesis, lignin formation, seed germination, and general stress adaptation (Kandhol et al. 2022).

9.3.2 Salinity and Soil Health: Nanoscale Microbes for Ionic Balance

Salinity is one of the most harmful environmental stresses that impacts over 20% of agricultural soils and over one-third of all arable land worldwide. The primary cause of soil salinity is the increasing use of saline irrigation water, which is exacerbated by global water scarcity and climate change. This problem is becoming worse due to a combination of natural and man-made factors, including soil texture, overfertilization, poor drainage, saline groundwater intrusion, insufficient rainfall, and inefficient wastewater management systems. High salinity levels greatly inhibit plant growth and survival, particularly when sodium chloride (NaCl) concentrations exceed 200 mM. This stress affects all the main stages of development, which includes seed germination, seedling establishment, vegetative growth, and flowering. This stress disrupts ionic equilibrium and interferes with critical biochemical, physiological, and metabolic pathways required for plant growth and productivity (Singh et al. 2024).

Osmotic stress, which is brought on by elevated osmotic pressure from high salt levels, limits water absorption and disrupts internal water balance while affecting plant growth, metabolism, and nutrient uptake. Moreover, salinity induces oxidative stress, which harms proteins, DNA, and lipids inside cells, by generating an excess of reactive oxygen species (ROS) and reactive nitrogen species (RNS). These effects lead to decreased crop yields, reduced biomass, and hindered development. Reproductive processes also show notable declines in pollen viability, flower formation, and seed set. As a result, improving plant resistance to salinity is crucial for the global sustainability of agriculture and has led to the development of numerous mitigation strategies (Singh et al. 2024). Among these, nanotechnology presents a novel and intriguing approach to increasing agricultural output and resilience to stress.

Salt stress causes osmotic and ionic problems that hinder plant development. Plants can absorb salts using both apoplastic and symplastic pathways, in which monocots mainly use the apoplastic pathway, while dicots use the symplastic pathway. One of the key channels and transporters that mediate Na^+ transport via the symplastic route is the Na^+/H^+ antiporter (SOS1). Calcium (Ca^{2+}) inhibits non-selective cation channels (NSCCs), which are modulated by glutamate receptor-like channels (GLRs) and cyclic nucleotide-gated channels (CNGCs) (Blumwald 2000). Long-distance Na^+ transport via xylem and phloem is also made possible by low-affinity cation transporters (LCT1), cation/H^+ exchangers (CHXs), and high-affinity potassium transporters (HKTs) (Isayenkov and Maathuis 2019). The movement and absorption of chloride (Cl^-) depend on nitrate transporters (NRTs), H^+/Cl^- symporters, and Cl^-/H^+ co-transporters. Long-distance Cl^- transport is facilitated by members of the nitrate/peptide transporter family (NPF) and anion channel-associated homolog 1 (SLASH). The passive transport of Cl^- is mediated by cation chloride cotransporters (CCCs), while the intracellular distribution of Cl^- is mediated by chloride channels (CLCs) and aluminum-activated malate transporters (ALMTs) on the tonoplast. Together, these systems regulate the uptake and distribution of Na+ and Cl+, which regulates the level of ion toxicity caused by salinity (Assaha et al. 2017; Horie et al. 2001). Nanoparticles (NPs) must first enter the symplastic pathway and pass through the plasma membrane in order to start having an impact. There are several ways in which this internalization occurs:

- Disruption of the membrane, where NPs create transient openings to permit direct entry.
- Endocytosis, in which NPs are transported to target organelles by vesicles.
- Protein-mediated transport, in which NPs bind to transporter membrane proteins.

Aquaporins, which have pore diameters between 2.8 and 3.4 Å, may allow NP movement if their structure is changed (Sohail et al. 2022). NPs must first enter the symplast, but plasmodesmata facilitate their passage through the phloem between cells. Ion channels, which typically have a diameter of 1 nm, are typically too small for NPs unless significant changes are made (Paramo et al. 2020).

9.3.3 Oxidative Stress in Plants: Nanomaterials/Nanomicrobials as Antioxidant Mediators

Environmental stressors interfere with the metabolism of reactive oxygen species (ROS) and reactive nitrogen species (RNS), which frequently leads to their buildup beyond the antioxidant defense systems' ability to scavenge them. Redox homeostasis is disrupted by such imbalances, which can result in nitro-oxidative stress and damage vital macromolecules such as proteins, nucleic acids, and membrane lipids, ultimately leading to cell death (Chaki et al. 2020). Depending on their characteristics and application context, nanomaterials (NMs) can either exacerbate or mitigate oxidative stress in plant systems. By either stimulating the plant's own antioxidant defense mechanisms or imitating the action of antioxidant molecules, NMs can alter the levels of ROS in stressed cells. Because of these characteristics, NMs are promising instruments for reducing the negative consequences of abiotic stress (Sachdev and Ahmad 2021) (Fig. 9.4).

Hosseinpour et al. (2020) investigated the protective effects of zinc oxide nanoparticles (ZnO-NPs) at concentrations of 0, 20, and 40 mg L^{-1} in conjunction with plant growth-promoting bacteria (PGPB) such as *Lactobacillus casei*, *Bacillus subtilis*, and *Bacillus pumilus* on cytosine methylation and DNA damage in tomato (*Solanum lycopersicum* L. "Linda") seedlings under salinity stress caused by 250 mM NaCl. Cytosine methylation and genomic stability under stress were evaluated using two molecular methods: CRED-RA and RAPD. The results demonstrated that while the combined application of ZnO-NPs and PGPB significantly

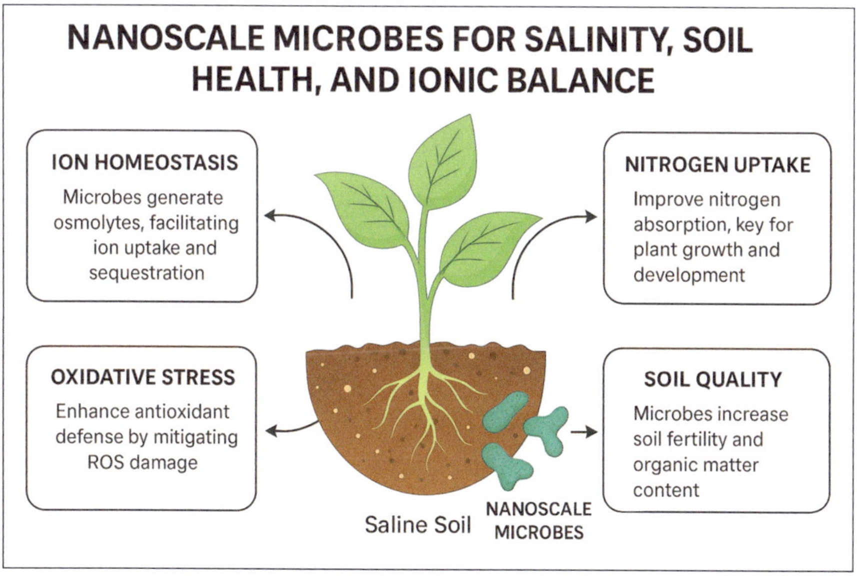

Fig. 9.4 Nanoscale microbial interventions for salinity stress and soil health restoration

reduced these effects, salinity stress increased DNA polymorphism, indicating genomic instability. The combination of 40 mg L^{-1} ZnO-NPs and Lactobacillus casei resulted in the highest genomic stability, and all treatments demonstrated decreased DNA methylation, suggesting a stress-reduction effect. All things considered, the combination of PGPBs and ZnO-NPs worked well to reduce the genetic harm caused by salinity stress, increasing tomato seedlings' ability to withstand stress.

9.3.4 Temperature Extremes: Neutralizing Thermal Stress with Nanomicrobes

Temperature stress, whether due to extreme heat or cold, disrupts key biochemical, physiological, and molecular processes in plants. These disruptions ultimately impair development and yield. Recent studies have shown that nanoparticles like silver (Ag), silicon (Si), titanium dioxide (TiO_2), and selenium (Se) can influence plant physiological and biochemical responses under temperature stress. For instance, Qi et al. (2013) demonstrated that nano-TiO_2 application on tomato plants under heat stress improved net photosynthesis, water conductance, and transpiration, while reducing electron transport and chlorophyll fluorescence, thereby enhancing photosynthetic efficiency. Selenium nanoparticles (SeNPs) have been widely studied under both heat and cold stress conditions. El-Saadony et al. (2021) reported that Bio-SeNP treatment (100 g/mL) in heat-stressed wheat increased chlorophyll and carotenoid content by 12–32% and improved gas exchange parameters, including transpiration rate, stomatal conductance, and net photosynthesis. Further, Seliem et al. (2020) used biogenically synthesized SeNPs (50–100 nm) derived from *Lactobacillus casei* to treat heat-stressed chrysanthemum cultivars, which enhanced growth, physiological traits, and flower quality.

Djanaguiraman et al. (2018) found that the application of 10 mg/L SeNPs at the boot stage in sorghum enhanced the antioxidant defense system, improved pollen germination, seed set, yield, and reduced oxidative stress under high temperatures. In a comparative study, Elsheery et al. (2020) evaluated the effects of various nanoparticles ($nSiO_2$, nZnO, nSe, and graphene nanoribbons) on sugarcane under chilling stress. These treatments improved photosynthetic efficiency (Fv/Fm, Pm), maintained gas exchange, and increased pigment levels. Notably, $nSiO_2$ showed the most significant ameliorative effect. In mung bean, Kareem et al. (2022) observed that temperatures above 28–30 °C during flowering reduced plant performance. Foliar application of ZnO NPs (15–60 mg/L) enhanced chlorophyll content, gas exchange, and enzyme activity, resulting in improved pod development and grain yield. Silver nanoparticles (AgNPs) have also shown benefits in cold stress tolerance. Prazak et al. (2020) reported that low concentrations of AgNPs (0.25, 1.25 mg/dm^3) improved germination, plant height, biomass, and photosynthetic activity in green beans, both in lab and field conditions under high temperature stress.

9.3.5 *Heavy Metal Toxicity: Microbial Nanoparticles as Soil Detoxifiers*

Bio-organic nanoparticles have shown promising potential in removing heavy metals from soil. For instance, silver nanoparticles biosynthesized by *Morganella psychrotolerans* were effectively used for heavy metal remediation (Arif et al. 2016; Enez et al. 2018). Polyvinyl pyrrolidone (PVP)-coated iron oxide nanoparticles in combination with *Halomonas* sp. (a Gram-negative bacterium) have been employed for the remediation of lead and cadmium contamination (Alabresm et al. 2018). Similarly, phthalic acid-coated magnetic Fe_3O_4 nanoparticles, treated with *Staphylococcus aureus*, demonstrated high removal efficiencies 83%–89% for Cu^{2+}, 99.4%–100% for Pb^{2+}, and 92.6%–97.5% for Ni^{2+}, highlighting the importance of microbial surface functional groups and nanoparticle cores in metal removal (Mahmoud et al. 2016). In a different strategy, zinc oxide nanoparticles combined with heavy metal-resistant bacteria such as *Bacillus cereus* and *Lysinibacillus macroides* aided in the remediation of Pb, Cd, Cr, and Cu, with *B. cereus* contributing significantly to the process (Baragaño et al. 2020). A strain of *B. cereus* (XMCr-6) was found to reduce Cr^{6+} to Cr^{3+}, which subsequently reacted with oxygen to form chromium oxide nanoparticles as a by-product (Laslo et al. 2022). Dong et al. (2013a, b) reported that the combination of selenium nanoparticles with the probiotic bacterium *Lactobacillus casei* significantly improved cadmium remediation in soil and water, achieving 65% absorption, substantially higher than the 43–78% range observed with *L. casei* alone. The bio-organic synthesis of nanoparticles has dual advantage because, selective microbes can detoxify heavy metal pollutants and production of valuable nanomaterials. For example, *Enterococcus faecalis* was used to synthesize lead nanoparticles, with sizes around 10 nm which showed strong catalytic properties, reducing 5.0 μmol of Cr^{6+} within 12 hours (Cao et al. 2013).

Ansari et al. (2009) demonstrated that magnetic nanoparticles in combination with *Rhodococcus erythropolis* facilitated the desulfurization of dibenzothiophene (DBT). Bokare et al. (2012) demonstrated that the bimetallic Pd/nFe nanoparticles when paired with *Sphingomonas wittichii* effectively aided in the remediation of NBR and 2,3,7,8-tetrachlorodibenzo-p-dioxin. Silica nanoparticles functionalized with lipid bilayers from *Pseudomonas aeruginosa* effectively removed polycyclic aromatic hydrocarbons (PAHs) like benzo[a]pyrene, aided by the bacterium's membrane lipids (Wang et al. 2015). Graphene oxide nanoparticles combined with the laccase enzyme from *Trametes versicolor* also proved efficient in PAH degradation (Patil et al. 2016). In crude oil-contaminated soils, *Alcaligenes faecalis* with iron oxide nanoparticles enhanced hydrocarbon breakdown (Oyewole et al. 2019). Similarly, *Sphingomonas* strain NM05's ability to degrade hexachlorocyclohexane (HCH) was nearly doubled when treated with Pd/FeO bimetallic nanoparticles (Singh et al. 2013). Perovskite nanoparticles ($LaFeO_3$), used in conjunction with proteobacteria, contributed to the degradation of organic pollutants in marine sediments (Hung et al. 2021). Beyond remediation, nanoparticles have also been noted for their positive impact on soil health. Silicon nanoparticles, for instance, have been shown to enhance rhizospheric microbial growth and biomass, thereby improving overall soil microflora (Rajput et al. 2022).

9.4 Nanoscale Nutrient Management: Microbial Nanoparticles in Precision Fertilization

Precision agriculture uses nanofertilizers to maintain soil microbial diversity, fauna, and water retention capacity while supplying vital nutrients in smaller, more effective doses that support plant growth and productivity (Liu and Lal 2015; Singh 2017). As "smart fertilizers," nanofertilizers can serve as pesticides, plant growth regulators, and nutrient carriers (Sodhi et al. 2025). Encasing nutrients in nanomaterials is the most popular technique for creating nanofertilizers. Fertilizers can be delivered in three main ways using nanoparticles: (i) encapsulation in non-porous materials, (ii) formulations coated with polymers, and (iii) nutrients in nanosized form (Naderi and Danesh-Shahraki 2013). In these formulations, chitosan and oleylamine are frequently employed as nanocarriers in conjunction with surfactants. According to Dhoke et al. (2013), applying nanofertilizers topically improves the amount of dry matter and chlorophyll, which in turn supports the general health and growth of plants. Natural biofertilizers like *Bacillus subtilis*, *Paenibacillus elgii*, and *Pseudomonas fluorescens* embedded in nanoparticles have demonstrated remarkable potential in fostering plant growth in controlled settings. When compared to traditional fertilizers, nano-biofertilizer is an economical and effective solution because a single liter can treat multiple hectares. Furthermore, the fertilizing potential of carbon-based nanomaterials, zeolites, and nano-clay minerals has been investigated (Guo et al. 2018; Bisinoti et al. 2019). Silver nanoparticles that were biosynthesized from *Pseudomonas* and *Achromobacter* species demonstrated antifungal activity against *Fusarium oxysporum* in chickpeas (Kaur et al. 2018). In both in vitro and in vivo tests, their findings demonstrated potent antifungal activity. Similarly, Amna et al. (2021) reported that biologically synthesized AgNPs were more effective against *Fusarium moniliforme* than *Colletotrichum falcatum* in sugarcane. According to Kant et al. (2022), at a concentration of just 0.125%, zinc oxide nanoparticles (ZnONPs) extracted from *Bacillus subtilis* supernatant inhibited the growth of major plant pathogens, such as *Sclerotium rolfsii*, *Fusarium oxysporum*, *Colletotrichum capsici*, and *Alternaria solani*, by up to 99%. Furthermore, Gaba et al. (2022) used *Trichoderma harzianum* to synthesize copper oxide nanoparticles (CuONPs) and evaluated their antifungal efficacy against *Alternaria brassicicola*. Higher CuONP concentrations (0.1 and 0.15 mg/mL) were found to be especially successful in suppressing foliar fungal infections.

9.4.1 Improving Plant Microbe Symbiosis Through Nano-Bioengineering

Investigating novel strategies for the mutually beneficial interactions between microbes, nanomaterials, and plants may help to reduce various stresses and promote plant growth. According to Campos et al. (2023), this three-way interaction enhances plant health and presents a viable plan for guaranteeing sustainability in farming methods. Nanotechnology's ability to precisely and effectively deliver biomolecules across a variety of plant tissues is its main contribution to plant genetic engineering. Because some nanoparticles are naturally fluorescent, cargo delivery and release can be imaged in real time. Research has also demonstrated that NPs give cargoes resistance to deterioration while in transit (Chiu and Choi 2020). According to Landry and Mitter (2019), nanoparticles improve disease resistance by carrying siRNA for gene silencing and facilitate genetic transformation by delivering plasmid DNA. sRNA for gene silencing has been effectively delivered by carbon nanotubes and DNA nanostructures (Zhang et al. 2020; Demirer et al. 2019a, b). Wang et al. (2018) emphasized that proteins can be delivered into plants by nanoparticles, facilitating DNA-free genome editing and advancing plant genetic engineering. Both suspension cells and mature plants are the targets of gene transformation in plants mediated by nanoparticles. Suspended plant cells have reportedly shown more promise for biomedical applications, despite the fact that mature plants undergo transformation more frequently. While transformation is more uniform and efficient in suspension cells, it is restricted to small areas in mature plants. Furthermore, suspension cells' improved accessibility and uptake efficiency make them more favorable for nanoparticle entry (Chiu and Choi 2020). Organelle engineering allows for maternal inheritance and circumvents transgene constraints, whereas conventional transformation targets the nucleus to produce transgenic plants (Lv et al. 2020). Effective gene transfer to the nucleus and chloroplast is made possible by biomolecule-functionalized nanoparticles (Kwak et al. 2019; Newkirk et al. 2021). Transgene-free applications in both model and non-model plants are made possible by such techniques (Demirer et al. 2018). There is still much to learn about how nanoparticles (NPs) are transported through the plasma membrane into subcellular organelles (Lew et al. 2018). Protoplasts have been used as alternate targets for passing the obstacle that the plant cell wall presents when delivering genes. However, many plant species still face difficulties with effective protoplast culture and regeneration (Jat et al. 2020). Because NPs can encapsulate and deliver plasmid DNA, mRNA, and ribonucleoproteins (RNPs) with high specificity, they have emerged as promising carriers despite the cell wall's inhibitory role in delivering genome-editing cargos into mature plants. Improved gene-editing efficiency has been demonstrated by direct cytoplasmic delivery of Cas9 protein complexed with sgRNA (Mout et al. 2017).

For effective cargo delivery in CRISPR/Cas systems, a variety of nanocarriers have been used, such as polymeric nanoparticles, DNA nanoclews, and lipid-based nanoparticles (e.g., CRISPR-Gold) (Duan et al. 2021). The creation of multiplexed

CRISPR platforms has made it possible to edit multiple genomic loci at once, even though single-sgRNA-based editing still has limited multiplexing capabilities (McCarty et al. 2020). Notably, it has been reported that NP-mediated nano-biomimetic transformation systems allow for stable, residue-free transformation in crop species (Lew et al. 2018; Izuegbunam et al. 2021). Combining gene editing and nanotechnology is a potent way to speed precision breeding and crop improvement.

9.5 Integrated Pest Management Through Nano-Biocontrol

The main goal of integrated pest management (IPM) is to reduce risks to the economy, human health, and the environment by combining commonsense methods. The effectiveness, stability, and targeted delivery of biological control agents are greatly enhanced by nano-biocontrol, which enables IPM to incorporate them into sustainable agricultural practices and makes them more competitive with traditional chemical pesticides. The protective function of biogenic silver nanoparticles (AgNPs) in treating *Alternaria solani* in tomatoes was examined by Kumari et al. (2017a). Using biosynthesized AgNPs at a concentration of 5 μg/mL, they successfully decreased the pathogenic population of *A. solani* in a concentration-dependent manner without negatively impacting the native soil microflora, demonstrating an environment-friendly disease control strategy. Animal cell lines were used for toxicity and cell migration tests, which verified the non-toxic nature of these nanoparticles and validated their biocompatibility (Giri et al. 2019; Kumari et al. 2017b). It has been demonstrated that microbially produced silver nanoparticles (AgNPs) inhibit phytopathogenic fungi, such as *Alternaria brassicicola* and *Fusarium oxysporum*, by increasing reactive oxygen species (ROS) levels and weakening the fungal antioxidant defense system (Kumari et al. 2019). These nanoparticles are created using extracellular metabolites from *Trichoderma viride,* which were optimized under different physicochemical parameters, including pH, temperature, and reaction time. This showed superior antimicrobial activity when compared to their chemically synthesized counterparts (Kumari et al. 2017a). Silver/silver chloride (Ag/AgCl) nanoparticles synthesized using extracellular exudates of *Macrophomina phaseolina* showed strong antibacterial activity against both Gram-positive and Gram-negative bacterial strains (Spagnoletti et al. 2019). In a different method, Sathiyabama and Charles (2015) used a cell wall polymer derived from *Fusarium oxysporum* f.sp. *lycopersici* to create nanoparticles intended to suppress Fusarium wilt in greenhouse tomatoes. Treated plants displayed reduced disease severity, more flowers, and increased fruit yield. Furthermore, chickpea seeds were treated with chitosan nanoparticles made from extracellular proteins from the endophytic fungus *Penicillium oxalicum.* In comparison to untreated controls, this treatment improved seed germination, the seedling vigor index, and the length of the shoots and roots. The ability of microbially produced nanoparticles to combat important plant pathogens was further confirmed by in vitro investigations by Bartolucci et al (2020) and Sathiyabama and Parthasarathy (2016).

Zinc oxide nanoparticles (ZnO NPs) coated with *Bacillus thuringiensis* biomolecules were created by Nandhini et al. in 2019. In addition to encouraging plant growth, these ZnO NPs demonstrated antimicrobial qualities that strengthened plant defenses against infections. The mycelia of *Pyricularia oryzae* underwent morphological deformations as a result of the combined action of ZnO NPs and ecalbasaponins that were isolated from *Eclipta alba* (Zhao et al. 2001). By causing plasmolysis of zoospores, ZnO NPs at a 50 ppm concentration dramatically reduced the sporangia germination of *Sclerospora graminicola* on leaf discs, which is how downy mildew is managed in pearl millet. Similar results were also reported by Wagner et al. (2016), who found that Zn and ZnO NPs inhibited *Peronospora tabacina* spore germination on tobacco. Mishra et al. (2017) showed the significance of transferring biosynthesized nanoparticles from lab to field conditions for efficient field application. In both greenhouse and field settings, ZnO nanoparticles' capacity to induce resistance against downy mildew was shown. The incidence of disease was considerably decreased by combination of seed treatment and foliar application at 100 ppm. In comparison to untreated seedlings, disease incidence was decreased by 36% at 200 ppm. Chitosan (Siddaiah et al. 2018) and selenium (Nandini et al. 2017) nanoparticles have shown comparable disease suppression outcomes, but ZnO NPs have the added advantage of being a pesticide and fertilizer. Iron and zinc oxide nanoparticles having irregular shapes, measuring 16.8 and 18.53 nm, respectively, were synthesized by Kamal et al. (2023) from Daedalea mushrooms and it was noticed that the *Aspergillus niger* was susceptible to the antifungal effects of these nanoparticles. Hermosilla et al. (2023) used *Stereum hirsutum* to create 25 nm-sized chitosan-encapsulated silver nanoparticles (Cs-Ag NPs). The antimicrobial activity of chitosan-encapsulated silver nanoparticles (Cs-Ag NPs) was 2–40-fold higher than that of AgNPs made from non-encapsulated fungi against a variety of phytopathogenic and human pathogens, such as *Pseudomonas syringae*, *Escherichia coli*, *Staphylococcus aureus*, and *Candida albicans* (Fig 9.5).

The cost of development and production, regulatory barriers for new nanoproducts, and the requirement for thorough research on their environmental fate and long-term effects on nontarget organisms and ecosystems are some of the obstacles that stand in the way of the potential. Essentially, the biological control element of IPM is greatly enhanced by nano-biocontrol. By using the accuracy and defense provided by nanotechnology, it turns naturally occurring pest control methods into extremely efficient, focused, and eco-friendly instruments, bringing agriculture one step closer to real sustainability.

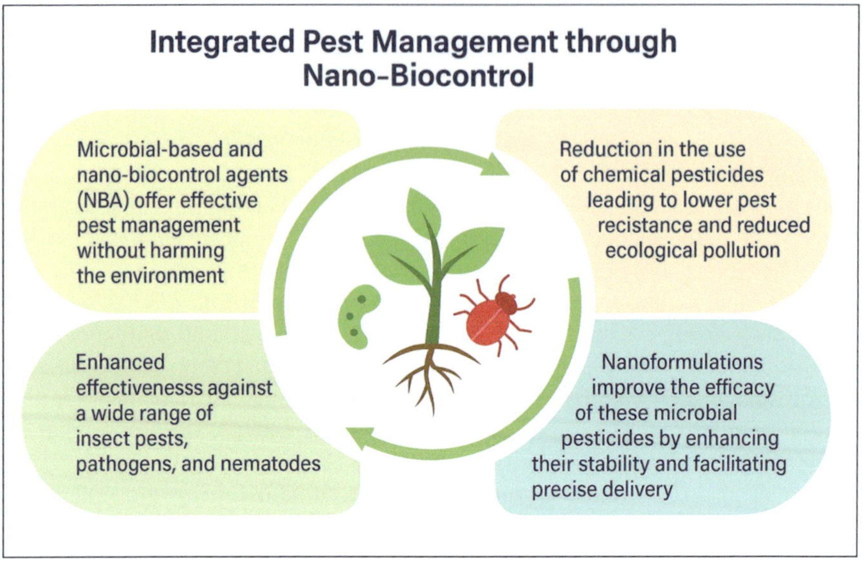

Fig. 9.5 Nanotechnology for eco-friendly pest control in IPM frameworks

9.6 Microbial Nanoformulations for Improved Seed Germination

The basic idea behind a microbial nanoformulation for seeds is to use nanoparticles to encapsulate or associate advantageous microorganisms or their active metabolites. These nanoparticles, which are usually between 1 and 100 nanometers in size, act as facilitators, carriers, and shields for the microbial agents. The final product is directly applied to the soil surrounding seeds or used as a seed coating or seed priming agent. Furthermore, since microbial nanoformulations greatly increase seed germination and general plant vitality, carbon nanotubes (CNTs) can be integrated with microbial applications. Transmission electron microscopy was used by Khodakovskaya et al. (2009) to verify that carbon nanotubes (CNTs) had penetrated the tomato seed coat, improving seed germination and early plant growth significantly. The uptake of nanoparticles by seeds has been associated with enhanced seed quality and vigor. Similarly, multi-walled carbon nanotubes (MWCNTs) were found to stimulate the growth of roots and shoots in *Brassica juncea* (mustard) by Mondal et al. (2011). According to Tarafdar et al. (2014), pearl millet seedlings treated with zinc nanofertilizers showed a notable increase in shoot length, root length, enzymatic activity, and grain yield.

The germination rate and seedling vigor of pearl millet were enhanced in the current study by the combination of zinc nanoparticles and saponin, which increased germination from 77% in the control to 95% at a concentration of 200 ppm. Yang et al. (2018) observed a similar pattern and showed that seed priming with saponin improved quinoa germination under salt stress. Furthermore, when compared to untreated controls, priming pearl millet seeds with ZnO nanoparticles increased plant height and dry biomass in a greenhouse setting.

9.7 Mitigating Post-Harvest Losses Through Nanomicrobes

After harvest, improper crop handling results in significant yield and quality losses, which raises the market prices of agricultural products. Harvesting, handling, storing, processing, distributing, and consuming are some of the phases that lead to the majority of postharvest losses in fruits and vegetables. To reduce such losses, it is crucial to take care of suitable postharvest procedures, such as better handling, packaging, transportation, and storage. The demand for fruits and vegetables has grown along with living standards, expanding supply chains from local, national, and international levels. Better fresh produce preservation techniques are required as a result of this expansion (Elik et al. 2019). Conventionally produced nanoparticles (NPs), particularly those that employ chemical reducing agents, are cytotoxic and inappropriate for use in food or biomedical applications. However, a sustainable and environment-friendly substitute is provided by green nanotechnology which uses biological systems like plants, bacteria, algae, and fungi to synthesize NPs (Babu et al. 2010, 2012; Sharma et al. 2012; Panpatte et al. 2016; Park et al. 2016; Dahoumane et al. 2017; Babu and Doble 2018; Babu et al. 2018, 2020). For fruits like citrus, grapes, bananas, apples, mangoes, papayas, and guavas, a variety of nanomaterials have been created and used as postharvest disease management tools (Ruffo Roberto et al. 2019). Shakeel et al. (2025) used dried fruit and vegetable peels to create silver nanoparticles in an environmentlly friendly manner. These nanoparticles demonstrated their potential for preserving fresh produce by aiding in the preservation of important quality attributes like pH. Because of their potent antimicrobial qualities, silver nanoparticles (AgNPs) are one of the most widely used nanomaterials in packaging. Chitosan nanoparticles (CNPs) and chitosan/silica nanocomposites have been successfully used to stop bunch loss in "Italia" grapes without compromising fruit quality (Rubentheren et al. 2015). Grapes coated with AgNPs (25–45 nm) and chitosan/gelatin maintained their fresh look, experienced no juice leakage, and had an increase in shelf life by two weeks. CNPs have demonstrated several advantages in fruit preservation. Particularly when refrigerated, they slow down ripening, preserve moisture and acidity, and lessen general deterioration (Rubentheren et al. 2015). For instance, silver nanoparticles (AgNPs) made from extracts of banana peels demonstrated potent antibacterial activity against Shigella, Klebsiella, and *Escherichia coli* (Siddiqi et al. 2018). At a 0.2% concentration, banana fruits coated with chitosan nanoparticles (CNPs; 102.4–370 nm) showed improved control of postharvest infections, decreased moisture and aroma loss, and limited oxygen permeability (Esyanti et al. 2019).

Furthermore, postharvest treatment with CNPs (121.2 nm) and chitosan (1.25%) preserved an ideal pulp-to-peel ratio and enhanced sensory attributes, prolonging shelf life by delaying ripening (Lustriane et al. 2018). An inorganic nanocomposite film made of ZnO nanoparticles, cinnamaldehyde, and soybean protein isolate is another promising strategy. According to Ruffo Roberto et al. (2019), this film has excellent mechanical and barrier qualities as well as potent antimicrobial activity.

9.8 Carbon Sequestration and Climate Resilience via Microbial Nanoscale Innovations

Climate change poses a significant threat to agriculture, primarily through the increased frequency and intensity of extreme weather events such as floods, droughts, and heatwaves. In the context of climate change, one of the main concerns is the increased buildup of greenhouse gases (GHGs), especially carbon dioxide (CO_2), in the atmosphere. The most prevalent GHG is CO_2, which is growing at a rate of about 2 parts per million per year, causing a serious environmental concern on a worldwide scale (Lal 2009). Improving terrestrial carbon storage is one of the most important ways to address this issue. In order to ascertain whether soils serve as a net source or sink of atmospheric CO_2, agricultural management techniques are essential. It has been demonstrated that nanomaterials help stabilize and sequester carbon in soils because of their unique physicochemical characteristics at the nanoscale (Monreal et al. 2010; Calabi-Floody et al. 2011). Numerous elements, such as soil characteristics, environmental circumstances, and management techniques, all have an impact on soil carbon sequestration; however, soil aggregation and structure are especially important (Singh et al. 2017). Finding appropriate nanomaterials that can increase carbon retention through enhanced aggregation, thus, requires an understanding of how soil aggregation contributes to carbon stabilization. It is commonly acknowledged that soil carbon sequestration is a "win–win" strategy that improves soil functionality and stabilizes atmospheric CO_2 levels. Stabilizing soil organic carbon (SOC) entails using a variety of defense mechanisms to stop its release into the atmosphere. According to Barré et al. (2014), these include:

(i) Making it more resistant to degradation by increasing its chemical recalcitrance.
(ii) Limiting accessibility by fostering strong organo-mineral associations; and
(iii) Reducing decomposition by isolating organic matter from microbial activity.

Natural nanoparticles (oxyhydroxides, hydrous iron oxides, and nanoclays) demonstrated promising results in stabilizing carbon in soils (Calabi-Floody et al. 2011, 2015; Filimonova et al. 2016). This function is aided by their unique kinetic, optical, magnetic, and electronic characteristics (Monreal et al. 2010). For example, in Andisols, allophane, a naturally occurring nanoclay and non-crystalline aluminosilicate, makes up roughly 22–28% of the soil's total clay fraction (Calabi-Floody et al. 2015). The effectiveness of allophane in carbon stabilization has been highlighted by its notable ability to retain approximately 11.8% of organic carbon even after intensive peroxide treatment (Calabi-Floody et al. 2011).

9.9 Challenges and Future Prospectives

A survey of the literature demonstrated that the microbial agents and nanoparticles (NPs) work together to reduce plant stress. Nevertheless, most of the research is concentrated on reducing the toxicity of heavy metals and metalloids and few studies have addressed other stressors like drought and salinity. This emphasizes the necessity of conducting more extensive research under a greater variety of stressors. According to recent research, using NPs and microorganisms in concert is still a developing area of nanobiology. Our knowledge of the molecular reactions of plant systems to these nanocarriers is also conspicuously lacking. In-depth research is necessary for crucial elements like the processes of nanocarrier uptake, transport, and distribution within plant tissues, especially in edible sections. This is necessary to make sure that future generations won't be exposed to unanticipated toxicological risks from such interventions. All agri-food industry stakeholders, including farmers, researchers, consumers, and policymakers, must work together on a single platform considering these worries. To create smart crops that support the objectives of safe, resilient, and sustainable agriculture in the face of current and upcoming environmental challenges, such cooperation will be essential for the responsible and strategic use of nanocarriers (Shaikh et al. 2024).

References

Abdel-Aziz HM, Hasaneen MN, Omer AM (2016) Nano chitosan-NPK fertilizer enhances the growth and productivity of wheat plants grown in sandy soil. Span J Agric Res 14:e0902

Alabresm A, Chen YP, Decho AW, Lead J (2018) A novel method for the synergistic remediation of oil-water mixtures using nanoparticles and oil-degrading bacteria. Sci Total Environ 630:1292–1297

Albalawi MA, Abdelaziz AM, Attia MS, Saied E, Elganzory HH, Hashem AH (2022) Mycosynthesis of silica nanoparticles using *Aspergillus Niger*: control of *Alternaria solani* causing early blight disease, induction of innate immunity and reducing oxidative stress in eggplant. Antioxidants 11:2323

Al-Dhalimi AMA, Al-Ajeel SAH (2020) Anatomical study on effect of plant growth regulators and zinc nanoparticles in peduncle anatomy of sunflower (*Helianthus annuus* L.) plant under water stress. Indian J Ecol 47:7–16

Amna T, Mahmood UN, Khan B, Amin MT, Javed S, Mehmood S, Farooq MA, Sultan T, Munis MFH, Chaudhary HJ (2021) Characterization of bio-fabricated silver nanoparticles for distinct anti-fungal activity against sugarcane phytopathogens. Microsc Res Tech 84:1522–1530

Ansari F, Grigoriev P, Libor S, Tothill IE, Ramsden JJ (2009) DBT degradation enhancement by decorating *Rhodococcus erythropolis* IGST8 with magnetic Fe_3O_4 nanoparticles. Biotechnol Bioeng 102:1505–1512

Arif N, Yadav V, Singh S, Singh S, Ahmad P, Mishra RK (2016) Influence of high and low levels of plant-beneficial heavy metal ions on plant growth and development. Front Environ Sci 4:69

Ashraf S, Nazir M, Khan MH, Fazil I, Ahmed A, Fatima M et al (2024) Nanotechnology in microbiology: bridging tiny worlds for big impact. In: Current studies in health and life sciences. ISRES Publishing, Istanbul, p 74

Assaha DVM, Ueda A, Saneoka H, Al-Yahyai R, Yaish MW (2017) The role of Na^+ and K^+ transporters in salt stress adaptation in glycophytes. Front Physiol 8:509

Babu PJ, Doble M (2018) Albumin capped carbon-gold (C-Au) nanocomposite as an optical sensor for the detection of Arsenic (III). Opt Mater 84:339–344

Babu PJ, Sharma P, Borthakur BB, Das RK, Nahar P, Bora U (2010) Synthesis of gold nanoparticles using *Mentha arvensis* leaf extract. Int J Green Nanotechnol Phys Chem 2:P62–P68

Babu PJ, Saranya S, Sharma P, Tamuli R, Bora U (2012) Gold nanoparticles: sonocatalytic synthesis using ethanolic extract of *Andrographis paniculata* and functionalization with polycaprolactone-gelatin composites. Front Mater Sci 6:236–249

Babu PJ, Raichur AM, Doble M (2018) Synthesis and characterization of biocompatible carbon-gold (C-Au) nanocomposites and their biomedical applications as an optical sensor for creatinine detection and cellular imaging. Sens Actuat B-Chem 258:1267–1278

Babu PJ, Saranya S, Raichur AM, Doble M (2020) Design of photoluminescence point-of-care membrane strip for the detection of dopamine. Mater Lett 277:128316

Baka ZA, El-Zahed MM (2022) Antifungal activity of silver/silicon dioxide nanocomposite on the response of faba bean plants (*Vicia faba* L.) infected by *Botrytis cinerea*. Bioresour Bioprocess 9:102

Baragaño D, Forján R, Welte L, Gallego JLR (2020) Nanoremediation of As and metals polluted soils by means of graphene oxide nanoparticles. Sci Rep 10:1–10

Barré P, Fernandez-Ugalde O, Virto I, Velde B, Chenu C (2014) Impact of phyllosilicate mineralogy on organic carbon stabilization in soils: incomplete knowledge and exciting prospects. Geoderma 235–236:382–395

Bisinoti MC, Moreira AB, Melo CA, Fregolente LG, Bento LR, dos Santos JV, Ferreira OP (2019) Application of carbon-based nanomaterials as fertilizers in soils. In: Nanomaterials applications for environmental matrices: water, soil and air. Elsevier Science, Amsterdam

Blumwald E (2000) Sodium transport and salt tolerance in plants. Curr Opin Cell Biol 12:431–434

Bokare V, Murugesan K, Kim JH, Kim EJ, Chang YS (2012) Integrated hybrid treatment for the remediation of 2,3,7,8-tetrachlorodibenzo-p-dioxin. Sci Total Environ 435–436:563–566

Cabral-Pinto MMS, Inácio M, Neves O, Almeida AA, Pinto E, Oliveiros B, Ferreira da Silva EA (2020) Human health risk assessment due to agricultural activities and crop consumption in the surroundings of an industrial area. Expo Health 12:629–640

Calabi-Floody M, Bendall JS, Jara AA, Welland ME, Theng BKG, Rumpel C, Mora ML (2011) Nanoclays from an Andisol: extraction, properties and carbon stabilization. Geoderma 161:159–167

Calabi-Floody M, Rumpel C, Velásquez G, Violante A, Bol R, Condron LM, Mora ML (2015) Role of nanoclays in carbon stabilization in Andisols and Cambisols. J Soil Sci Plant Nutr 15:587–604

Campos EVR, Pereira A, Aleksieienko I, do Carmo GC, Gohari G, Santaella C et al (2023) Encapsulated plant growth regulators and associative microorganisms: nature-based solutions to mitigate the effects of climate change on plants. Plant Sci 331:111688

Cao M, Wang L, Wang L, Chen J, Lu X (2013) Remediation of DDTs contaminated soil in a novel Fenton-like system with zero-valent iron. Chemosphere 90:2303–2308

Chaki M, Begara-Morales JC, Barroso JB (2020) Oxidative stress in plants. Antioxidants 9:481

Cherian T, Alghamdi S, Ansari MA, Ali K, Shesha NT, Shukla AK, Al-Dossary HA (2023) Microbial nanotechnology. In: Nanoremediation. Elsevier, Amsterdam, Netherlands, pp 313–340

Chiu YTE, Choi CHJ (2020) Enabling transgenic plant cell–derived biomedicines with nanotechnology. Adv NanoBiomed Res 1:2000028

Dahoumane SA, Jeffryes C, Mechouet M, Agathos SN (2017) Biosynthesis of inorganic nanoparticles: a fresh look at the control of shape, size and composition. Bioengineering 4:14

Demirer GS, Zhang H, Goh NS, Chang R, Landry MP (2019a) Carbon nanocarriers deliver siRNA to intact plant cells for efficient gene knockdown. Sci Adv 6:eaaz0495

Demirer GS, Zhang H, Goh NS, Matos JL, Cunningham FJ, Sung Y, Chang R, Aditham AJ, Chio L, Cho MJ, et al (2018) Nanoparticle-mediated delivery towards advancing plant genetic engineering. Trends Biotechnol 36:882–897

Demirer GS, Zhang H, Goh NS, Gonzalez-Grandio E, Landry MP (2019b) Carbon nanotube-mediated DNA delivery without transgene integration in intact plants. Nat Protoc 14:2954–2971
Dharmaraj D, Krishnamoorthy M, Rajendran K, Karuppiah K, Annamalai J, Durairaj KR, Ethiraj K (2021) Antibacterial and cytotoxicity activities of biosynthesized silver oxide (Ag_2O) nanoparticles using *Bacillus paramycoides*. J Drug Deliv Sci Technol 61:102–111
Dhoke S, Mahajan P, Kamble R, Khanna A (2013) Effect of nanoparticle suspension on the growth of mung (*Vigna radiata*) seedlings by foliar spray method. Nanotechnol Dev 3:10.4081/nd.2013.e1
Dixit S, Sivalingam PN, Baskaran RM, Senthil-Kumar M, Ghosh PK (2024) Plant responses to concurrent abiotic and biotic stress: unravelling physiological and morphological mechanisms. Plant Physiol Rep 29(1):6–17
Djanaguiraman M, Belliraj N, Bossmann SH, Prasad PVV (2018) High-temperature stress alleviation by selenium nanoparticle treatment in grain sorghum. ACS Omega 3:2479–2491
Dong G, Wang Y, Gong L, Wang M, Wang H, He N, Zheng Y, Li Q (2013a) Formation of soluble Cr (III) end-products and nanoparticles during Cr (VI) reduction by *Bacillus cereus* strain XMCr-6. Biochem Eng J 70:166–172
Dong H, Rowland I, Thomas LV, Yaqoob P (2013b) Immunomodulatory effects of a probiotic drink containing *Lactobacillus casei* Shirota in healthy older volunteers. Eur J Nutr 52:1853–1863
Duan L, Ouyang K, Xu X, Xu L, Wen C, Zhou X, Qin Z, Xu Z, Sun W, Liang Y (2021) Nanoparticle delivery of CRISPR/Cas9 for genome editing. Front Genet 12:673286
El-Ashry RM, El-Saadony MT, El-Sobki AEA, El-Tahan AM, Al-Otaibi S, El-Shehawi AM, Saad AM, Elshaer N (2022) Biological silicon nanoparticles maximize the efficiency of nematicides against biotic stress induced by *Meloidogyne incognita* in eggplant. Saudi J Biol Sci 29:920–932
El-Batal AI, Balabel NM, Attia MS, El-Sayyad GS (2020) Antibacterial and antibiofilm potential of mono-dispersed stable copper oxide nanoparticles-streptomycin nano-drug: implications for some potato plant bacterial pathogen treatment. J Clust Sci 31:1021–1040. https://doi.org/10.1007/s10876-019-01707-4
Elik A, Yanik DK, Istanbullu Y, Guzelsoy NA, Yavuz A, Gogus F (2019) Strategies to reduce post-harvest losses for fruits and vegetables. Strategies 5:29–39
El-Saadony MT, Saad AM, Najjar AA, Alzahrani SO, Alkhatib FM, Shaf ME, Hassan MAA (2021) The use of biological selenium nanoparticles to suppress *Triticum aestivum* L. crown and root rot diseases induced by *Fusarium* species and improve yield under drought and heat stress. Saudi J Biol Sci 28:4461
Elsheery NI, Sunoj VSJ, Wen Y, Zhu JJ, Muralidharan G, Cao KF (2020) Foliar application of nanoparticles mitigates the chilling effect on photosynthesis and photoprotection in sugarcane. Plant Physiol Biochem 149:50–60
Enez A, Hudek L, Bräu L (2018) Reduction in trace element mediated oxidative stress towards cropped plants via beneficial microbes in irrigated cropping systems: a review. Appl Sci 8:1953
Esyanti RR, Zaskia H, Amalia A (2019) Chitosan nanoparticle-based coating as post-harvest technology in banana. J Phys Conf Ser 1204:012109
Filimonova S, Kaufhold S, Wagner FE, Häusler W, Kögel-Knabner I (2016) The role of allophane nanostructure and Fe oxide speciation for hosting soil organic matter in an allophanic Andosol. Geochim Cosmochim Acta 180:1–15
Gaba S, Varma A, Prasad R, Goel A (2022) Exploring the impact of bioformulated copper oxide nanoparticles on cytomorphology of *Alternaria brassicicola*. Curr Microbiol 79:244
Ghalamboran MR (2011) Symbiotic nitrogen fixation enhancement due to magnetite nanoparticles. PhD thesis, Cranfield University
Giri VP, Katiyar R, Nautiyal CS, Mishra A (2019) Biogenic silver nanoparticles as a more efficient contrivance for wound healing acceleration than common antiseptic medicine. FEMS Microbiol Lett 366:fnz201
Guo F, Wu J, Wu W, Huang D, Yan Q, Yang Q, Gao Y, Yang G (2018) PEGylated self-assembled enzyme-responsive nanoparticles for effective targeted therapy against lung tumors. J Nanobiotechnol 16:57

Hafez EM, Osman HS, Gowayed SM, Okasha SA, Omara AE-D, Sami R et al (2021) Minimizing the adverse impacts of water deficit and soil salinity on maize growth and productivity in response to the application of plant growth-promoting rhizobacteria and silica nanoparticles. Agronomy 11:676

Haggag WM (2018) Development and application of biotechnological products for sustainable corn and soybean production under stresses condition. Int J Agric Technol 14:1209–1224

Haggag WM (2024) Plant biotic and abiotic stresses combinations and their management: review article. Int J Agric Technol 20(1):105–122

Hamk M, Akçay FA, Avcı A (2023) Zinc oxide nanoparticles using *Bacillus subtilis* ZBP4 and their antibacterial potential against foodborne pathogens. Prep Biochem Biotechnol 53(3):255–264

Hermosilla E, Díaz M, Vera J, Contreras MJ, Leal K, Salazar R, Rubilar O (2023) Synthesis of antimicrobial chitosan-silver nanoparticles mediated by reusable chitosan fungal beads. Int J Mol Sci 24(3):2318

Hersanti H, Susanto A, Virgiawan R, Joni M (2018) The effectiveness of *Penicillium* sp. mixed with silica nanoparticles in controlling *Myzus persicae*. AIP Conf Proc 1927:030029

Hoballah E, Saber M, Zaghloul A (2019) Nanotechnology applications in agriculture. Int J Environ Pollut Environ Model 2(4):196–211

Horie T, Yoshida K, Nakayama H, Yamada K, Oiki S, Shinmyo A (2001) Two types of HKT transporters with different properties of Na^+ and K^+ transport in *Oryza sativa*. Plant J 27:129–138

Hosseinpour A, Haliloglu K, Cinisli KT, Ozkan G, Ozturk HI, Pour-Aboughadare A, Poczai P (2020) Application of zinc oxide nanoparticles and plant growth-promoting bacteria reduces genetic impairment under salt stress in tomato (*Solanum lycopersicum* L. 'Linda'). Agriculture 10(11):521

Hung CM, Huang CP, Chen CW, Dong CD (2021) Degradation of organic contaminants in marine sediments by peroxymonosulfate over $LaFeO_3$ nanoparticles supported on water caltrop shell-derived biochar and the associated microbial community responses. J Hazard Mater 420:126553

Ibrahem EJ, Thalij KM, Badawy AS (2017) Antibacterial potential of magnesium oxide nanoparticles synthesized by *Aspergillus Niger*. Biotechnol J Int 18(1):1–7

Idoko FA, Ezeamii GC, Ojochogwu OJ (2024) Green chemistry in manufacturing: innovations in reducing environmental impact. World J Adv Res Rev 23(3):2826–2841

Iqbal S, Waheed Z, Naseem A (2020) Nanotechnology and abiotic stresses. In: Nanoagronomy. Springer, Cham, pp 37–52

Isayenkov SV, Maathuis FJM (2019) Plant salinity stress: many unanswered questions remain. Front Plant Sci 10:80

Izuegbunam CL, Wijewantha N, Wone B, Ariyarathne MA, Sereda G, Wone BWM (2021) A nano-biomimetic transformation system enables *in planta* expression of a reporter gene in mature plants and seeds. Nanoscale Adv 3:3240–3250

Jat SK, Bhattacharya J, Sharma MK (2020) Nanomaterial-based gene delivery: a promising method for plant genome engineering. J Mater Chem B 8:4165–4175

Jayaseelan C, Rahuman AA, Roopan SM, Kirthi AV, Venkatesan J, Kim SK et al (2013) Biological approach to synthesize TiO_2 nanoparticles using *Aeromonas hydrophila* and its antibacterial activity. Spectrochim Acta A Mol Biomol Spectrosc 107:82–89

Kareem HA, Saleem MF, Saleem S, Rather SA, Wani SH, Siddiqui MH, Alamri S, Kumar R, Gaikwad NB, Guo Z, Niu J, Wang Q (2022) Zinc oxide nanoparticles interplay with physiological and biochemical attributes in terminal heat stress alleviation in mungbean (Vigna radiata L.). Front Plant Sci 13:842349

Kamal A, Saba M, Kamal A, Batool M, Asif M, Al-Mohaimeed AM, Ahmad S (2023) Bioinspired green synthesis of bimetallic iron and zinc oxide nanoparticles using mushroom extract and use against *Aspergillus Niger*; the most devastating fungi of the green world. Catalysts 13(2):400

Kandhol N, Jain M, Tripathi DK (2022) Nanoparticles as potential hallmarks of drought stress tolerance in plants. Physiol Plant 174(2):e13665

Kant V, Kumar V, Joshi VG, Chhabra R, Kamothi DJ (2022) Evaluation of antibacterial potential of bilirubin nanoparticles against Staphylococcus aureus and Escherichia coli. Nanosyst Phys Chem Math 13(3):342–348

Kaur P, Thakur R, Duhan JS, Chaudhury A (2018) Management of wilt disease of chickpea *in vivo* by silver nanoparticles biosynthesized by rhizospheric microflora of chickpea (*Cicer arietinum*). J Chem Technol Biotechnol 93:3233–3243

Khan M, Khan AU, Hasan MA, Yadav KK, Pinto MM, Malik N et al (2021) Agro-nanotechnology as an emerging field: a novel sustainable approach for improving plant growth by reducing biotic stress. Appl Sci 11(5):2282

Khanra S, Abdullah-Al Mamun M, Ferreira FF, Ghosh K, Guha S (2018) Functionalized self-assembled peptide nanotubes with cobalt ferrite nanoparticles for applications in organic electronics. ACS Appl Nano Mater 1:1175–1187

Khodakovskaya M, Dervishi E, Mahmood M, Xu Y, Li Z, Watanabe F, Biris AS (2009) Carbon nanotubes are able to penetrate plant seed coat and dramatically affect seed germination and plant growth. ACS Nano 3:3221–3227

Kots S, Khrapova A, Rybachenko LI, Pukhtaievych P (2024) The influence of metal nanocarboxylates on the nitrogen-fixing activity of symbiotic soybean systems grown under field conditions. Stud Biol. https://doi.org/10.30970/sbi.1801.751

Kumari M, Pandey S, Giri VP, Bhattacharya A, Shukla R, Mishra A, Nautiyal CS (2017a) Tailoring shape and size of biogenic silver nanoparticles to enhance antimicrobial efficacy against MDR bacteria. Microb Pathog 105:346–355

Kumari M, Pandey S, Bhattacharya A, Mishra A, Nautiyal CS (2017b) Protective role of biosynthesized silver nanoparticles against early blight disease in Solanum lycopersicum. Plant Physiol Biochem 121:216–225

Kumari M, Giri VP, Pandey S, Kumar M, Katiyar R, Nautiyal CS, Mishra A (2019) An insight into the mechanism of antifungal activity of biogenic nanoparticles than their chemical counterparts. Pestic Biochem Physiol 157:45–52

Kwak SY, Lew TTS, Sweeney CJ, Koman VB, Wong MH, Bohmert-Tatarev K, Snell KD, Seo JS, Chua NH, Strano MS (2019) Chloroplast-selective gene delivery and expression *in planta* using chitosan-complexed single-walled carbon nanotube carriers. Nat Nanotechnol 14:447–455

Lal R (2009) Challenges and opportunities in soil organic matter research. Eur J Soil Sci 60:158–169

Landry MP, Mitter N (2019) How nanocarriers delivering cargos in plants can change the GMO landscape. Nat Nanotechnol 14:512–514

Laslo V, Pinzaru SC, Zaguła G, Kluz M, Vicas SI, Cavalu S (2022) Synergic effect of selenium nanoparticles and lactic acid bacteria in reduction of cadmium toxicity. J Mol Struct 1247:131325

Lew TTS, Wong MH, Kwak SY, Sinclair R, Koman VB, Strano MS (2018) Rational design principles for the transport and subcellular distribution of nanomaterials into plant protoplasts. Small 14:e1802086

Liu R, Lal R (2015) Potentials of engineered nanoparticles as fertilizers for increasing agronomic productions. Sci Total Environ 514:131–139

Lustriane C, Dwivany FM, Suendo V, Reza M (2018) Effect of chitosan and chitosan nanoparticles on fruit quality of banana fruits. J Plant Biotechnol 45:36–44

Lv Z, Jiang R, Chen J, Chen W (2020) Nanoparticle-mediated gene transformation strategies for plant genetic engineering. Plant J 104:880–891

Mahanty S, Chatterjee S, Ghosh S, Tudu P, Gainc T, Bakshi M, Das S, Das P, Bhattacharyya S, Bandyopadhyay S (2020) Synergistic approach towards the sustainable management of heavy metals in wastewater using mycosynthesized iron oxide nanoparticles: biofabrication, adsorptive dynamics and chemometric modeling study. J Water Process Eng 37:101426

Mahmoud ME, Abdou AE, Mohamed SM, Osman MM (2016) Engineered *Staphylococcus aureus* via immobilization on magnetic Fe_3O_4-phthalate nanoparticles for biosorption of divalent ions from aqueous solutions. J Environ Chem Eng 4:3810–3824

Majeed S, Danish M, Mohamad Ibrahim MN, Sekeri SH, Ansari MT, Nanda A, Ahmad G (2021) Bacteria-mediated synthesis of iron oxide nanoparticles and their antibacterial, antioxidant, cytocompatibility properties. J Clust Sci 32:1083–1094

McCarty NS, Graham AE, Studena L, Ledesma-Amaro R (2020) Multiplexed CRISPR technologies for gene editing and transcriptional regulation. Nat Commun 11:1281
Meena M, Zehra A, Swapnil P, Harish, Marwal, Yadav G, Sonigra P (2021) Endophytic nanotechnology: an approach to study scope and potential applications. Front Chem 9:613343
Mishra S, Keswani C, Abhilash PC, Fraceto LF, Singh HB (2017) Integrated approach of agri-nanotechnology: challenges and future trends. Front Plant Sci 8:471
Mondal A, Basu R, Das S, Nandy P (2011) Beneficial role of carbon nanotubes on mustard plant growth: an agricultural prospect. J Nano Res 13:4519
Monreal CM, Sultan Y, Schnitzer M (2010) Soil organic matter in nano-scale structures of a cultivated black Chernozem. Geoderma 159:237–242
Mousa ESM, Elbagory M, Mahdy ME, Abo-Koura HA, Omara AED (2024) Microencapsulation of *Bacillus megaterium* in humic acid-supplied alginate beads enhances tomato growth and suppresses the root-knot nematode *Meloidogyne javanica* under greenhouse conditions. Horticulturae 10(12):1284
Mout R, Ray M, Yesilbag Tonga G, Lee YW, Tay T, Sasaki K, Rotello VM (2017) Direct cytosolic delivery of CRISPR/Cas9-Ribonucleoprotein for efficient gene editing. ACS Nano 11:2452–2458
Naderi M, Danesh-Shahraki A (2013) Nanofertilizers and their roles in sustainable agriculture. Int J Agric Crop Sci 5:2229
Nandini B, Hariprasad P, Prakash HS, Shetty HS, Geetha N (2017) Trichogenic-selenium nanoparticles enhance disease suppressive ability of *Trichoderma* against downy mildew disease caused by *Sclerospora graminicola* in pearl millet. Sci Rep 7:2612
Newkirk GM, de Allende P, Jinkerson RE, Giraldo JP (2021) Nanotechnology approaches for chloroplast biotechnology advancements. Front Plant Sci 12:691295
Noman M, Ahmed T, Ijaz U, Hameed A, Shahid M, Azizullah et al (2023) Microbe-oriented nanoparticles as phytomedicines for plant health management: an emerging paradigm to achieve global food security. Crit Rev Food Sci Nutr 63(25):7489–7509
Oves M, Qari HA, Felemban NM, Khan MZ, Rehan ZA, Ismail IM (2016) *Marinobacter lipolyticus* from Red Sea for lipase production and modulation of silver nanomaterials for anticandidal activities. IET Nanobiotechnol 11(4):403–410
Oyewole OA, Raji RO, Musa IO, Enemanna CE, Abdulsalam ON, Yakubu JG (2019) Enhanced degradation of crude oil with *Alcaligenes faecalis* ADY25 and iron oxide nanoparticle. Int J Biol Res 10:62–72
Panpatte DG, Jhala YK, Shelat HN, Vyas RV (2016) Nanoparticles: the next generation technology for sustainable agriculture. In: Microbial inoculants in sustainable agricultural productivity. Springer, New Delhi, pp 289–300
Paramo LA, Feregrino-Pérez AA, Guevara R, Mendoza S, Esquivel K (2020) Nanoparticles in agroindustry: applications, toxicity, challenges, and trends. Nanomaterials. 10:1654
Park TJ, Lee KG, Lee SY (2016) Advances in microbial biosynthesis of metal nanoparticles. Appl Microbiol Biotechnol 100:521–534
Parrish DJ, Fike JH (2005) The biology and agronomy of switchgrass for biofuels. Crit Rev Plant Sci 24:423–459
Patil S, Deshmukh A, Mane R, Kulkarni S (2016) Green synthesis of silver nanoparticles using Ocimum tenuiflorum leaf extract and their antibacterial activity. J Mater Environ Sci 7:3256–3263
Prazak R, Swiecilo A, Krzepilko A, Michalek S, Arczewska M (2020) Impact of Ag nanoparticles on seed germination and seedling growth of green beans in normal and chill temperatures. Agriculture 10:312
Qi M, Liu Y, Li T (2013) Nano-TiO_2 improves the photosynthesis of tomato leaves under mild heat stress. Biol Trace Elem Res 156:323–328
Rajakumar G, Rahuman AA, Roopan SM, Khanna VG, Elango G, Kamaraj C et al (2012) Fungus-mediated biosynthesis and characterization of TiO_2 nanoparticles and their activity against pathogenic bacteria. Spectrochim Acta A Mol Biomol Spectrosc 91:23–29

Rajput VD, Minkina T, Kumari A, Shende SS, Ranjan A, Faizan M et al (2022) A review on nano-bioremediation approaches for restoration of contaminated soil. Eurasian J Soil Sci 11:43–60

Rasheed A, Li H, Tahir MM, Mahmood A, Nawaz M, Shah AN et al (2022) The role of nanoparticles in plant biochemical, physiological, and molecular responses under drought stress: a review. Front Plant Sci 13:976179

Rubentheren V, Ward TA, Chee CY, Tang CK (2015) Processing and analysis of chitosan nanocomposites reinforced with chitin whiskers and tannic acid as a crosslinker. Carbohydr Polym 115:379–387

Ruffo Roberto S, Youssef K, Hashim A F, Ippolito A (2019) Nanomaterials as alternative control means against post-harvest diseases in fruit crops. Nanomaterials 9:1752

Sachdev S, Ahmad S (2021) Role of nanomaterials in regulating oxidative stress in plants. In: Nanobiotechnology: mitigation of abiotic stress in plants. Springer, Cham, Switzerland, pp 305–326

Saeed S, Iqbal A, Ashraf MA (2020) Bacterial-mediated synthesis of silver nanoparticles and their significant effect against pathogens. Environ Sci Pollut Res 27:37347–37356

Sathiyabama M, Charles RE (2015) Fungal cell wall polymer based nanoparticles in protection of tomato plants from wilt disease caused by Fusarium oxysporum f.sp. lycopersici. Carbohydr Polym 133:400–407

Sathiyabama M, Parthasarathy R (2016) Biological preparation of chitosan nanoparticles and its in vitro antifungal efficacy against some phytopathogenic fungi. Carbohydr Polym 151:321–325

Siddiqi KS, Husen A, Rao RAK (2018) A review on biosynthesis of silver nanoparticles and their biocidal properties. J Nanobiotechnol 16:14

Seliem MK, Hafez Y, El-Ramady H (2020) Using nano-selenium in reducing the negative effects of high temperature stress on Chrysanthemum morifolium Ramat. J Sustain Agric Sci 46:47–60

Semida WM, Abdelkhalik A, Mohamed GF, Abd El-Mageed TA, Abd ElMageed SA, Rady MM et al (2021) Foliar application of zinc oxide nanoparticles promotes drought stress tolerance in eggplant (*Solanum melongena* L.). Plants 10:421

Shaikh II, Shafi S, Kazi A, Sante R, Jadhav A, Mohini G, Ritthe P (2024) Silver nanoparticle as a modern era in antimicrobial activity. Asian J Pharm Res Dev 12:144–149

Shakeel A, Rohi M, Batool R, Tehseen S, Aziz M, Malik K, Abdul Sattar M, Raza A, Diantom A (2025) Green synthesis of nanoparticles in mitigating postharvest losses of fruits and vegetables. Food Sci Nutr 13:e70017

Sharma P, Babu PJ, Bora U (2012) *Sapindus mukorossi* aqueous fruit extract as reducing, capping and dispersing agents in synthesis of gold nanoparticles. Micro Nano Lett 7:1296–1299

Shelar A, Singh AV, Chaure N, Jagtap P, Chaudhari P, Shinde M et al (2024) Nanoprimers in sustainable seed treatment: molecular insights into abiotic-biotic stress tolerance mechanisms for enhancing germination and improved crop productivity. Sci Total Environ 951:175118

Shoarian N, Jamei R, Pasban Eslam B, Salehi Lisar SY (2020) Titanium dioxide nanoparticles increase resistance of *L. iberica* to drought stress due to increased accumulation of protective antioxidants. Iran J Plant Physiol 10:3343–3354

Siddaiah CN, Prasanth KVH, Satyanarayana NR, Mudili V, Gupta VK, Kalagatur NK, Singh BP (2018) Chitosan nanoparticles having higher degree of acetylation induce resistance against pearl millet downy mildew through nitric oxide generation. Sci Rep 8:2485

Singh MD (2017) Nano-fertilizers: a new way to increase nutrient use efficiency in crop production. Int J Agric Sci 09:75–3710

Singh R, Manickam N, Mudiam MKR, Murthy RC, Misra V (2013) An integrated (nano-bio) technique for degradation of γ-HCH contaminated soil. J Hazard Mater 258–259:35–41

Singh M, Sarkar B, Sarkar S, Churchman J, Bolan N, Mandal S, Menon M, Purakayastha TJ, Beerling DJ (2017) Stabilization of soil organic carbon as influenced by clay mineralogy. Adv Agron 143:1–97

Singh A, Rajput VD, Varshney A, Sharma R, Ghazaryan K, Minkina T et al (2024) Revolutionizing crop production: nanoscale wonders—current applications, advances, and future frontiers. Egypt J Soil Sci 64(1):0

Sohail A, Sawati L, Ferrari E, Stierhof YD, Kemmerling B, Mashwani ZUR (2022) Molecular effects of biogenic zinc nanoparticles on the growth and development of *Brassica napus* L., revealed by proteomics and transcriptomics. Front Plant Sci 13:875

Sodhi GK, Wijesekara T, Kumawat KC, Singh A, Mehta P, Kumar V, et al (2025) Nanomaterials–plants–microbes interaction: plant growth promotion and stress mitigation. Front Microbiol 15:1516794

Spagnoletti FN, Spedalieri C, Kronberg F, Giacometti R (2019) Extracellular biosynthesis of bactericidal Ag/AgCl nanoparticles for crop protection using the fungus Macrophomina phaseolina. J Environ Manag 231:457–466

Tarafdar JC, Raliya R, Mahawar H, Rathore I (2014) Development of zinc nanofertilizer to enhance crop production in pearl millet (*Pennisetum americanum*). Agric Res 3:257–262

Umar OB, Ranti LA, Abdulbaki AS, Bola AL, Abdulhamid AK, Biola MR, Victor KO (2021) Stresses in plants: biotic and abiotic. In: Current trends in wheat research. IntechOpen, London, UK

Upadhayay VK, Kumari G, Jaiswal A, Chaudhary P (2023a) Role of nano-bioinoculants in alleviation of abiotic stresses for sustainable agriculture. In: Challenges in the green world: abiotic stress and plants. Elite Publishing House, New Delhi, p 49.

Wagner G, Korenkov V, Judy JD, Bertsch PM (2016) Nanoparticles composed of Zn and ZnO inhibit *Peronospora tabacina* spore germination *in vitro* and *P. tabacina* infectivity on tobacco leaves. Nanomaterials 6:50

Wang H, Kim B, Wunder SL (2015) Nanoparticle-supported lipid bilayers as an *in situ* remediation strategy for hydrophobic organic contaminants in soils. Environ Sci Technol 49:529–536

Wang J, Jia S, Cao Y, et al (2018) Design principles for nanoparticle plasmon-enhanced organic solar cells. Nanoscale Res Lett 13:211

Yang A, Akhtar SS, Iqbal S, Qi Z, Alandia G, Saddiq MS, Jacobsen SE (2018) Saponin seed priming improves salt tolerance in quinoa. J Agron Crop Sci 204:31–39

Yusof MH, Abdul Rahman NA, Mohamad R, Zaidan UH (2020) Microbial mediated synthesis of silver nanoparticles by *Lactobacillus plantarum* TA4 and its antibacterial and antioxidant activity. Appl Sci 10(19):6973

Zahedi SM, Karimi M, Teixeira Da Silva JA (2020) The use of nanotechnology to increase quality and yield of fruit crops. J Sci Food Agric 100:25–31

Zahedi SM, Hosseini MS, Daneshvar Hakimi Meybodi N, Peijnenburg W (2021) Mitigation of the effect of drought on growth and yield of pomegranates by foliar spraying of different sizes of selenium nanoparticles. J Sci Food Agric 101:5202–5213

Zhang H, Zhang H, Demirer GS, González-Grandio E, Fan C, Landry MP (2020) Engineering DNA nanostructures for siRNA delivery in plants. Nat Protoc 15:3064–3087

Zhang Y, Jing X, Ruofan L, Yanrui G, Yufei L, Ruili L (2023) Plant's response to abiotic stress: mechanisms and strategies. Int J Mol Sci 30;24(13):10915. https://doi.org/10.3390/ijms243110915

Zhao YP, Tang HF, Jiang YP, Wang ZZ, Yi YH, Lei QY (2001) Triterpenoid saponins from *Eclipta prostrata* L. Acta Pharm Sin B 36:660–663

Chapter 10
Plant Growth-Promoting Rhizobacteria: A Strategy for Alleviating Nanoparticle-Induced Stress in Plants

Anu Kalia, Jayesh Singh, Megha Handa, and Swetha Muduku

Abstract The plant growth-promoting rhizobacteria (PGPR) encourage development of plant by several processes such as siderophore synthesis, biological nitrogen fixation, phosphate solubilization, quorum sensing (QS) and inhibit the formation of biofilms, produce phytohormones, antifungal activity, generation of volatile organic compounds (VOCs), induce systemic resistance, encourage beneficial plant-microbe symbioses, interfere with the production of pathogen toxins, etc. Also, to enhance the production and growth of crops, modern techniques that incorporate nanotechnology are widely used in agricultural practices. Despite its scientific name, nanotechnology has revolutionized a variety of global sectors. Our everyday lives already frequently involve metal nanoparticles and metal oxide nanoparticles (MPs/MONPs), which are widely utilized in a variety of products and industrial processes. Accordingly, once MNPs/MONPs are purposefully or inadvertently released into the environment, soil has been proposed as the primary sink of these pollutants. Environmental concerns related to disturbances in the soil microbial populations and related ecological processes compounded due to rise in episodes of drastic climate-change related abiotic stresses have been witnessed in the last two decades. The crop plants are now being exposed to two most predominant stresses including the oxidative and heavy metal stress due to an emerging environmental contaminant i.e. nanoparticles of metal/ metal oxide/ non-metal oxides and other forms of dual or multiple element nanomaterials (as most efficient NPs used are of metal origin). PGPRs present in soil and rhizosphere of plants have various

A. Kalia
Electron Microscopy and Nanoscience Laboratory, Punjab Agricultural University, Ludhiana, Punjab, India

J. Singh (✉)
Department of Plant Breeding and Genetics, College of Agriculture, Punjab Agricultural University, Ludhiana, Punjab, India
e-mail: jayesh_singh@pau.edu

M. Handa · S. Muduku
Department of Microbiology, College of Basic Sciences and Humanities, Punjab Agricultural University, Ludhiana, Punjab, India

K. A. Abd-Elsalam (ed.), *Plant-Microbiome Nanotechnology*, Sustainability Sciences in Asia and Africa, https://doi.org/10.1007/978-981-95-5056-2_10

interactions with nanoparticles, which exert an influence on the plant's overall health and development. Through a variety of methods, either by themselves or in conjunction with nanoparticles, PGPRs assist in reducing various stresses that plants encounter. This manuscript summarizes the overall mechanism and effect of PGPRs and nanoparticles on plants.

Keywords Plant growth · Plant probiotic microbes · Nanomaterial · Nanotoxicity · Oxidative stress

10.1 Introduction

Nanotechnology has ushered in a new era of applications across diverse domains including medicine, electronics, and agriculture. The synthesis and utilization of nanoparticles have engendered apprehensions regarding their possible deleterious impacts on the environment, particularly detrimental impacts on living organisms, including crop plants, microbes, animals, and human beings. Therefore, elucidation of the potential phytotoxic effects of metal-based nanoparticles (NPs) has emerged as a significant area of research worldwide. Nanomaterials are ultra-fine structures with size dimensions ranging from 1 to 100 nm (Nowack and Bucheli 2007). The nanoscale dimensions render novel chemical and physical attributes of the bulk materials from which the nanomaterials are derived (Jefferson 2000). Furthermore, materials at the nanoscale are highly reactive and could perform improved catalytic activities due to their high surface area-to-volume ratios (Mekuye and Abera 2023). On application, the nanomaterials especially the metal/metal oxide nanoparticles can exhibit multitude of interactions with the soil, and living cells of varied origins leading to substantial environmental implications (Gakis et al. 2023; Soares and Soares 2021).

Heavy metal or transition metal-based nanomaterials have been generated and utilized most extensively by different industries. As a result, soil and water pollution with heavy metal contaminants would become a significant environmental issue. Since soil and water are the ultimate sinks, these nanoparticles exhibit potential interactions with the living organisms, particularly the plants and microflora, through absorption, transformation, and accumulation processes. These mechanisms affect how nanoparticles behave, travel, and change in the environment (Handa and Kalia 2024). Nanoparticles can move inside the plant tissues through multiple pathways, encompassing root absorption, foliar application, and soil exposure. The consequences of their movement may include the initiation of oxidative stress, disruption of nutrient uptake mechanisms, hindrance of photosynthetic processes, and occurrence of genotoxic responses in plants. Hence, it becomes imperative to devise effective strategies aimed at ameliorating nanoparticle-induced stress in plants, with a view to foster sustainable agricultural practices (Gulzar and Mazumder 2022).

Beneficial bacteria called PGPRs reside in the soil around plant roots. On inoculation, these bacteria exhibit positive effects on plants such as improvement in plant growth traits, and enhanced acquisition of essential nutrients, besides serving as probiotic microflora that protects the plants from attack of harmful phytopathogenic microbes (Singh et al. 2018). Through various processes, such as nitrogen fixation, production of growth-promoting chemicals, and disease control, these bacteria have been demonstrated to enhance plant development and growth. Furthermore, they can aid plants in surviving various abiotic stressors, such as heavy metals, salt, and drought. This manuscript aims to highlight the role of PGPRs to help in dealing with the stress caused due to advertent or inadvertent supplementation or application of nanomaterials, particularly the metal and metal oxide particles. Furthermore, the fundamental mechanisms that are engaged in the process, including the negative impacts of the leaching or release of ions of heavy metals (HMs) from the M/MO nanoparticles on the environment and living organisms, are also discussed.

10.2 Understanding Nanoparticles (NPs) and Their Interaction with Plants

A know-how of the complex interactions existing between the plants and nanoparticles includes examining the NP uptake mechanisms of the host plant and their process of transportation across the above- and below-ground tissues. Studies showed that both the roots and the leaves of plants may allow NPs to penetrate through the tissues. The internalization of NPs elicits a variety of chemical or physical mechanisms through which NPs interact with plants. Reactive oxygen species (ROS) are produced through the former methods, can cause damage due to oxidation, impairment of the ion transport function of cell membranes, and lipid peroxidation events (Table 10.1). After entering the plant cells, single, dual, or multiple MNPs/MONPs release the metal ions from their surface, which react with carboxyl and sulfhydryl functional groups of a variety of proteins, thereby altering their function. Furthermore, the nanoscale dimensions of the nanomaterials (NMs) potentially cause cellular toxicity, thereby impacting plant development and growth. However, it is challenging to precisely quantify the changes that NMs experience throughout the exposure time, such as aggregation, adsorption, dissolution, precipitation, sedimentation, and many other processes. Interestingly, the larger surface area, surface charge, and other characteristics of the NMs allow for easy adsorption/absorption of organic molecules and inorganic ions from the nutritional media, which may lead to nutritional deprivation and indirect toxicity symptoms, including wilting and chlorosis (Begum et al. 2012).

Table 10.1 Negative impacts on different tissues of the test plants as induced by nanomaterials and manifested as phyto-nanotoxicity

Type of nanoparticle	Conc. of nanoparticle evaluated	Plant type	Toxic effects	References
Silver nanoparticle (AgNPs)	1, 5, 20, 50, and 100 mg/L	Wheat (*Triticum aestivum*) and Cucumber (*Cucumis sativus*)	Exposure to high concentration proved toxic to test plants A discernible drop in cucumber roots' and shoots' dry biomass (at 50 and 100 mg/L) and wheat (20–100 mg/L), respectively.	Cui et al. (2014)
	Broad range of conc. (0.01 to 100 mg/L)	*Arabidopsis thaliana* and *Populus*	Nanotoxicity attributed to hormesis above a specific concentration Root length inhibited at high concentration.	Wang et al. (2013)
	AgNPs solution and leaf extracts	Lettuce (*Lactuca sativa*)	AgNPs (all different sizes) decreased seedling vegetative parameters Leaf length and width decreased by large-sized (70 nm) particles.	Wang et al. (2018a)
	10, 15, and 20 mg/L	Tomato (*Lycopersicum esculentum*) and Corn (*Zea mays*)	Reduced root length and progressive inhibition in germination with the increase in concentration Exhibited more toxicity than its ion concentration.	Ravindran et al. (2012)
	0,0.5,1,2,3, and 4 mg/L	*Camelina sativa*	Decrease in plant height, total dry weight, and photosynthetic pigments Increase in MDA, H_2O_2(oxidative stress)	Mirmoeini et al. (2021)
AgNPs and Titanium dioxide (TiO_2)	0.001, 1, 100, 10,000 mg/L	Onion (*Allium cepa*) and Radish (*Raphanus sativus*)	Inhibition of root growth TiO_2 were more toxic than AgNPs suspensions for both the test plants	Pittol et al. (2017)
TiO_2	0, 50, 100, 200 or 400 mg/L for 1 week	Lettuce (*Lactuca sativa*)	Plant experienced oxidative stress on NP's exposure. Dose-dependent (50–400 mg/L) negative effect on produce nutritional quality Decreased Ca, Mg, Mn, Fe, Zn, and B content in edible tissues of lettuce on comparing with control	Hu et al. (2020)

ZnO and MWCNTs	0–750 mg/Kg	Cucumber (*Cucumis sativus*)	Increased catalase activity in cucumber leaves at 500 mg/Kg, suggesting stress caused by NP	Servin et al. (2013)
	(ZnO: 40 and 50 mg/L MWCNTs: 100 and 500 mg/L)	*Arabidopsis thaliana*	ZnO had greater inhibitory effects compared to MWCNTs Showed reduced length of root, chlorophyll content and higher ROS concentration On exposure of both NPs (ZnO, MWCNTs) showed even worse effects on plant growth indices and antioxidant enzymes (SD, CAT, GR)	Yang et al. (2023)
ZnO and CeO_2	(ZnO: 100 and 200 mg/L CeO_2: 100 and 200 mg/L)	*Pisum sativum*	Inhibited root and stem elongation, restricted stomata and leaf transpiration Affected the mineral elements balance	Skiba et al. (2021)

10.3 Mechanism of Nanoparticle Toxicity

10.3.1 Nanoparticle Uptake and Transport in Plants

Plant surfaces, including the above-ground leaf and stem morpho-phenology, allow deposition of various nanomaterials. The epicuticular structures on shoot surfaces usually reduce wettability (lotus effect), favoring deposition of various nanoparticles (Barthlott and Neinhuis 1997). These nanoparticles then move across the barriers (cuticle, exudates, mucilage, waxy layer, pectin) through various natural openings. Nanomaterials further reach inside the cell cytoplasm through ion channels, aquaporins, and other carrier proteins. The cuticle is the initial hydrophobic layer that protects plants and prevents many nanomaterials from entering the plant. Positively charged lipid-based liquid crystal nanoparticles (150–300 nm) (Nadiminti et al. 2013) and negatively charged mesoporous silica nanoparticles (MSNs, 20 nm) (Hussain et al. 2013) tend to accumulate within negatively charged anticlinal cell wall structure. Larger nanoparticles pass through cuticle-free regions, whereas tiny particles enter the cuticle after making holes in it (Larue et al. 2014). Plants preferentially absorb metal-based NPs via certain ion transporters (Hall 2003).

The interphase between the plant tissue and the soil environment comprises single or multi-layer complex compounds of fatty acid, i.e., suberin in plant roots and tuber surface (Schreiber 2010). By preventing water and solutes from the soil from flowing through the apoplastic layer, these structures guard the plant roots from nanomaterials and their entrance into the central cylinder through the xylem tissue. This natural barrier is bypassed by the formation of the new lateral roots, which lead to rupturing of the cortical area, permitting the passage of the apoplastic bypass and access of the NMs into the vascular bundle, particularly the xylem through the cortex into the central cylinder (Faiyue et al. 2010). Plant cells have irregular sized holes or pores in the cellulose-based cell wall with dimensions ranging from 3.5–3.8 nm, particularly for root hair cells, to 4.5–5.2 nm in the cells of palisade parenchyma (Carpita et al. 1979). The size dimensions of the NPs play a significant role in ascertaining the extent of transportation across various tissues in the plant. In an NP uptake study, TiO_2 NPs tagged with alizarin red S (size 3 nm) were observed in several epidermal cell compartments and *Arabidopsis* seedlings (Kurepa et al. 2010). Likewise, NPs (<5 nm size) were easily able to penetrate through the intact cell membranes (Hischemöller et al. 2009). Apart from the size, pH and the concentration of divalent ions influence the porosity of cell walls (Fleischer et al. 1999).

The NPs can be internalized through specific extrinsic proteins, such as clathrins or adaptins, which form vesicles, thus permitting the NPs to attach to certain receptors on the cell membrane and move inside the cell cytoplasm. However, this mechanism only works with ultra-fine NPs as the diameter of the clathrin-coated vesicles ranges from 70–120 nm. Following exposure to 40 nm polystyrene nanoparticles labeled with Texas red, *Acer pseudoplatanus* (Sycamore maple) protoplasts mostly absorbed the particles via fluid-phase endocytosis that is receptor-mediated and clathrin-dependent (Moscatelli et al. 2007). Using the fluorescence tagging technique, another 20 nm-diameter polyethylene glycol-coated quantum dots (QDs) were observed to exist in spherical organelles (Etxeberria et al. 2006). Another unique way of NPs internalization is through the non-endocytic route that involves a variety of gated channels, such as the cell membrane aquaporins which are water channels, to enable the nonselective absorption of nanoparticles (Fitzpatrick and Reid 2009; Rico et al. 2011). According to Tyerman et al. (2002), these gated channels have a diameter of 0.8–0.25 nm. The NPs through these channels enter cells and can be transported in the plant system both through apoplast and symplast pathways. The symplast transport of NPs among various plant cells is predominantly assisted by the plasmodesmata. The aggregation or accumulation and transportation of NPs are two interdependent mechanisms which control the process of accumulation of the NPs at specific tissues or regions in the plant or their translocation and uniform distribution across different tissue types in the plant body. Furthermore, the flow and accumulation of NPs inside the plant tissues are impacted by the shape, size, chemical makeup, plant stage, nanoparticle habitat, plant species, concentration, and other similar parameters. On internalization, the NPs may alter the circadian patterns and programmed cell death (PCD) process in the plant tissues or organs (Yanik and Vardar 2018a, b). Exposure of eggplant root tissues to varying concentrations of cobalt oxide NPs resulted in programmed cell death (PCD), as evidenced by DNA fragmentation and excessive accumulation of reactive oxygen species (ROS) (Faisal et al. 2016).

10.3.2 Induction of Oxidative Stress and Reactive Oxygen Species Production

The ROS are produced by various metabolic activities that take place in the organelles of plant cells, such as peroxisomes, chloroplasts, and mitochondria (Møller et al. 2007). Cell death results from excessive production and accumulation of reactive oxygen moieties or species within the cell, despite the fact that plants have a mechanism to combat harmful compounds. Since oxidative damage results from an excessive ROS burst, plant hormones, including salicylic acid and abscisic acid, as well as enzymes (oxidoreductase, catalases), are involved in neutralization of the produced ROS molecules. This oxidative damage may be caused due to increased production of ROS moieties by interaction with a variety of biotic or abiotic stress agents, leading to cell death (Carocho and Ferreira 2013). The ROS moieties generated in response to NPs exposure include singlet oxygen ($^1O_2$), hydroxyl radical(HO·), superoxide ($O_2^{\cdot-}$), and hydrogen peroxide (H_2O_2) (Mourato et al. 2012). The main producers of (H_2O_2) are the epidermal cells, bundle sheath cells, and parenchyma tissue of the plant leaves (Zhao et al. 2012). Additionally, this process triggers the Fe^{2+} and Cu^{2+} catalyzed Fenton reaction, producing extremely reactive ROS (hROS) and hydroxyl radicals (HO·) that are intractable by any known enzymatic system. The nuclear component and cell membrane sustain irreparable damage caused by this hROS (Freinbichler et al. 2011; Mourato et al. 2012). While hydrogen peroxide (H_2O_2) levels are tracked to indirectly observe this molecule, metal-based nanoparticles actively create HO· (Kaur et al. 2022b). A significant contributing factor to nanotoxicity is oxidative stress. The degree of ROS generation inside cells is dictated by the chemical and physical characteristics of NPs, including their solubility, adsorption potential, and crystalline phase. Due to their inherent capacity to produce harmful radicals, metallic nanoparticles take part in redox processes, catalyze Fenton reactions, and alter cellular absorption, and can have a significant negative impact as oxidative stress. Moreover, NPs may have direct interactions with lipids, proteins, and DNA to cause further biological consequences as protein breakdown, mutagenesis, and disruption of the membrane (Nel et al. 2006).

Cellular oxidative stress is defined as an imbalance in the quantities of pro-oxidant and antioxidant substances inside the cell (Jones 2008). Therefore, when pro-oxidants are preferred over antioxidants, oxidative stress develops. The amount of ROS produced and antioxidant defense define the level of oxidative stress. The production of ROS, including hydrogen peroxide (H_2O_2), hydroxyl radical (-OH), nitric oxide (NO), peroxynitrite ($ONOO^-$), and superoxide radical ($O_2^{\bullet -}$), is one of the frequent causes of oxidative stress. The majority of ROS are created as byproducts of metabolic processes such as neutrophil-mediated phagocytosis, cytochrome P450 enzymatic metabolism, and mitochondrial respiration (Bast et al. 1991). MNPs/MONPs are major catalysts for the production of ROS, leading to various stress responses such as cell cycle disruption (Siddiqui 2024), DNA damage (Manuja et al. 2021; Solorio-Rodriguez et al. 2024), autophagy, and lysosomal dysfunction(Feng et al. 2024; Florance et al. 2024) primarily as a consequence of cellular NPs uptake. NPs have been demonstrated to display a hormetic response and can kill particular cells, ranging from modest degrees of oxidative stress to magnifying the impact (Jalal et al. 2021; Singh et al. 2025; Soni et al. 2024).

Nanomaterials may exhibit two specific forms of oxidative stress with the foremost considered as the primary oxidative stress caused due to the direct effect possibly by various mechanisms, such as the photocatalytic activity (e.g., TiO_2 NPs) and metal ions release from the NPs surface (Fallah et al. 2024; Madkour 2020). The secondary indirect oxidative stress involves the generation of these reactive species by the malfunction of the mitochondria in response to exposure and interactions with the NPs. The electron transport pathway in the mitochondria continuously produces O_2^-, H_2O_2, and hydroxyl radicals. Normally, these ROS are eliminated by an antioxidant system made up of glutathione, catalase, and superoxide dismutase (Rao et al. 2025). However, some intracellular metal ions that are released from internalized nanoparticles result in mitochondrial dysfunction and the formation of ROS. This has a detrimental effect on the redox balance within the cell by oxidizing proteins, lipids, and DNA. At the molecular level, oxidative stress mostly affects cellular components or macromolecules, including DNA, proteins, and lipids. Thus, NPs-induced oxidative stress expresses itself in several ways, including growth inhibition, autophagy, cell senescence, cell death, and proliferation. Increased lipid peroxidation and ROS production cause observable lysosomal membrane destabilization and cell nuclear condensation.

10.3.3 Disruption of Cellular Membranes and Membrane Integrity

Nanoparticles can impair DNA synthesis and enzyme activity, damage cell membrane integrity and function, and interfere with metabolic processes that generate reactive nitrogen species (RNS) and ROS (Thiruvengadam et al. 2024). When nanomaterials interact at the cellular level, plants show signs of nanotoxicity, which include growth, suppression, changed cell division, and structural changes (Gao et al. 2023). Following their entry into the xylem and stele tissues in plants, the nanomaterials travel along apoplast or symplast routes to access the aerial regions, where they interact directly with cellular and subcellular organelles (Usman et al. 2020). Certain metal/metal oxide nanoparticles can produce toxic metal ions that can accumulate in plant cellular and subcellular organelles, such as cell membranes, nuclei, vacuoles, and chloroplasts and tissues. This can result in an excess of reactive oxygen species (Wang et al. 2018b). Cytotoxicity and genotoxicity, including membrane damage, chlorophyll degradation, PSII dysfunction, vacuole shrinkage, mitochondrial dysfunction, DNA damage, and chromosome abnormalities, are caused by the direct interaction of nanoparticles and ROS production in plant cell organelles (Yan and Chen 2019). When large quantities of ZnO nanoparticles were applied, the stomata and trichome morphology was altered. The NPs also changed the cellular organization by disrupting the thylakoid and grana organization and causing irregularities in the chloroplasts. These alteration in the cellular organization ultimately suppressed the growth of *H. sativum* indicating a toxic response (Rajput et al. 2021b).

10.3.4 Alteration of Photosynthesis and Chlorophyll Content

The plant physiological characteristics spanning over photosynthesis rate, water potential, and absorption of micro- and macronutrients, and variable hormone levels are phytotoxicity indicators induced by the nanoparticles (Peharec et al. 2021; Yang et al. 2018). Among the fundamental processes supporting plant survival, photosynthesis plays a central role by generating oxygen and ATP, which drive energy-dependent metabolic activities (Tripathi et al. 2017). The functional impact of nanoparticles on controlling plant photosynthesis also depends critically on their size. Compared to larger NPs, smaller ones frequently have a stronger tendency to concentrate in plant tissues and have a greater effect on plant physiology (Alkhatib et al. 2021). It is well known that the amount of chlorophyll in plant leaves indicates how well it can perform photosynthesis (Liu et al. 2020). By lowering the quantities of chlorophyll and photosynthetic pigments and decreasing photosynthetic activity, NPs can indirectly impact the overall physiological state of plants (Peharec et al. 2021).

Chlorophyll and carotenoid levels were significantly decreased as nanoparticle concentrations increased (Abdel Latef et al. 2020; Yanik and Vardar 2018a, b). *Camelina sativa* leaf extract derived AgNPs when applied at higher concentrations (2, 3, and 4 g L^{-1}) enhanced AgNPs in shoot and root tissues resulting in detrimental effects on the development of the seedlings and photosynthetic pigments (Mirmoeini et al. 2021). The ability of NPs to interfere with the chlorophyll synthesis in leaves leads to a negative effect on the plant photosynthetic system. Chlorophyll florescence analysis has been suggested as a powerful and widely used method to determine how the application of NPs affects photosynthetic efficiency (Rajput et al. 2021b; Wang et al. 2018b). The quantum efficiency (Fv/Fm), a characteristic photosystem II attribute is greatly impacted by NPs (Liu et al. 2020). In *Lactuca sativa*, foliar exposure to CuO NPs reduced photosynthetic pigments, net photosynthesis, chlorophyll fluorescence, plant leaf area, and biomass, all of which had an impact on the plant performance (Xiong et al. 2021). Therefore, high NP concentrations affect plant photosynthetic processes, which are impacted by the kind of plant, the properties of NPs, and the environment. Similarly, disruption of thylakoid organization, plastoglobules, morphology, and starch accumulation in the chloroplasts are affected by the NPs. In exposed plants, the nanoparticles alter the leaf chloroplast ultrastructure and morphology (Liang et al. 2018; Liu et al. 2020). The toxicology effects of ZnO nanoparticles (300–2000 mg/L) on *Hordeum sativum* growth were demonstrated by perturbing grana and thylakoid organization with chloroplast irregularity, deforming stomatal and trichome morphology, and affecting chlorophyll fluorescence emission (Rajput et al. 2021b). CeO_2 nanoparticles predominantly decreased photosynthetic pigments in *Lemna minor*, deformed stomata, and killed cells at a dosage greater than 0.1 mg/L (Liu et al. 2023).

10.3.5 Genotoxic Effects and DNA Damage

Research experiments showcased a DNA-damaging effect of NMs. However, systematic investigations of plant genotoxicity are few. DNA damage is frequently measured through the comet test, representing one of the tell-tale signs of NP-induced genotoxicity. Treatment with copper oxide NPs led to DNA damage in test plants, including several grassland and agricultural plants (*Lolium perenne, Raphanus sativus, and Lolium rigidum*) (Atha et al. 2012). In *Allium cepa* root cells, ZnO NPs had genotoxic and cytotoxic effects, as shown by a rise in micronuclei and chromosomal abnormality and a drop in the mitotic index (Kumari et al. 2011). Other stress-related genes were also up-regulated in tomato plants exposed to carbon nanotubes (CNTs), in addition to the aquaporin genes. Microarray-based gene expression studies revealed that genotoxicity varies greatly and depends on the nanoparticle type (Francis et al. 2024). Likewise, published reports on maize plants revealed selective and biphasic growth of the root tissues on exposure to graphite-derived CNTs (Wu et al. 2023). According to the transcriptional study, seminal root gene expression is specifically modulated by single wall carbon nanotubes (SWCNTs) and root cell contact, leading to comparatively slower root development (Yan et al. 2013). Genotoxicity is manifested in response to ROS-mediated oxidative damage and altered molecules which build up in the nucleus, resulting in the production of mutagenic DNA lesions, and ultimately result in reduced plant development and even plant death (Tuteja et al. 2009). Genetic damage, including genetic mutations, DNA strand breaks, and DNA cross-linking, is brought on by NP-induced ROS (Fallah et al. 2024). The genotoxic potential of ZnO NPs was assessed in human liver HepG2 cells (Sharma et al. 2012). When these cells are exposed to ZnO NPs, oxidative stress-mediated DNA damage is caused. This was demonstrated by the rise in sites sensitive to formamidopyrimidine glycosylase (Fpg), a marker of DNA damage caused by oxidation. AgNPs may lead to altered gene expression, causing several epigenetic variations besides DNA damage in plants (Gowtham et al. 2024). Plant roots can readily uptake, and translocate AgNPs to above-ground tissues where these may also get deposited (Tripathi et al. 2024b). Similar effects in plant cells have been documented by exposure to gold (Au) NPs (Singh et al. 2024). The comet test and fluorescence in situ hybridization have been predominantly used to assess the extent of genotoxicity effect. As compared to AuNPs, the AgNPs exhibited comparatively higher ROS-mediated and DNA damage in the plant cells at a relatively lower dose of application, as studied in *Nicotiana tabacum* L. plants (Lovecká et al. 2021). Furthermore, the AuNPs exhibited a transient damaging effect (May et al. 2018). These studies demonstrated that nanoparticle-mediated genotoxicity tests in plants are now necessary to assess the possible influence on the health of plants.

10.3.6 Interaction of Nanoparticles with Plant Defense Mechanisms

It is essential to comprehend how plants protect themselves from oxidative stress brought on by nanoparticles. To defend itself against different biotic and abiotic stresses, the plant employs a number of defense mechanisms. The antioxidant glutathione (GSH) is essential for the body's defense against oxidative stress (Rao et al. 2025). It oxidized to a disulfide form, i.e., glutathione disulfide (GSSG) after scavenging ROS, and glutathione reductase subsequently recycle it. To reduce the stress load, several elements of plant cell machinery called phytochelatins (PCs) chelate heavy metals and function downstream to GSH (Chaudhary et al. 2018). These defense molecules and enzymes help to mitigate the heavy metal stress and, therefore, the metal/metal oxide NPs and other higher order NMs can elicit a similar response in plant cells/tissues (Landa et al. 2012). The carbon nanomaterials suppressed the transcription of genes involved in electron transport and energy pathways when compared —ZnO, TiO_2, and fullerene soot—in a nanotoxicity study performed on *Arabidopsis* (Landa et al. 2012). Furthermore, in comparison to the bulk heavy metal, the nanoparticles of the same chemistry exhibit enhanced production of ROS and membrane lipid peroxidation (Munir et al. 2023). When exposed to bulk Ag, Ag^+ ions, and Ag nanoparticles, sand-cultured wheat plants showed markedly increased levels of metallothionein (MT), a cysteine-rich protein, and GSSG. Wheat roots and shoots grown in CuO and ZnO nanoparticles with greater concentrations of GSSG exhibited similar tendencies (Dimkpa et al. 2013). Consequently, GSH plays a function in either the elimination of ROS generated in response to plant exposure to the nanoparticles or the detoxification of metal ions.

10.3.7 Effects of Nanoparticle Aggregation and Particle Size, Shape, and Coatings on Toxicity

A consorted interplay and synergy of a variety of factors contribute to phytotoxicity on nanomaterial application. The janus properties of nanomaterials may be greatly affected by application dosage and modification or functionalization of the nanoparticle surfaces. Alumina coating shielded the algal plant (*Pseudokirchneriella subcapitata*) against toxicity when compared to naked SiO_2 NPs (Van Hoecke et al. 2011). At mere 1 ppm concentrations, silver (Ag) nanoparticles between 20 and 80 nm in size significantly inhibited the development *Arabidopsis thaliana* seedlings. Ag nanoparticles may adsorb on the root surface either independently or in association with the root tips secondary metabolites, leading to browning of the principal root tips (cap and columella) following AgNPs treatment (Lin and Xing 2007). The specific physical interaction between the clay (bentonite), TiO_2 nanoparticles and the maize plant disrupted the flow of water and induced leaf responses. This effect was associated with a reduction in the permissible range of width of the cell wall pores from 6.6 to 3.0 nm in maize roots resulting in the decreased hydraulic conductivities (Asli and Neumann 2009). The surface characteristics of nanomaterials influence their ability to traverse across various plant system barriers. After being injected into pumpkin shoots, carbon-coated magnetic nanoparticles traveled from the application site into xylem vessels and stem parenchyma tissue. When applied to leaves, nanoparticles with limited translocation efficiency reached inside the epidermal cells of the petiole tissues (Corredor et al. 2009). The size and concentration of nanoparticles determine how they translocate and accumulate. Since the size exclusion limit of plasmodesmata between adjacent cells is necessary for cell-to-cell transfer, the 27 kDa, or approximately 3 nm, green fluorescent protein (GFP) travels across the cells. Because agglomeration is favored by higher concentrations, applying zinc oxide (ZnO) nanoparticles to soybean seeds at varying doses (500–4000 mg L^{-1}) exhibited noticeably greater absorption at 500 mg L^{-1} than at higher concentrations of 1000–4000 mg L^{-1} (López-Moreno et al. 2010). Integration of applied NMs reduces their absorption due to the difficulty of the agglomerated structures to pass through the hole in cell walls. The way the nanoparticles are applied is another intriguing feature. In a maize research, it was shown that, in contrast to root application, foliar spraying the plant with nano-TiO_2 particles helped to reduce the effects of heavy metal stress (cadmium ion toxicity) (Lian et al. 2020).

10.4 Plant Growth-Promoting Rhizobacteria: A Strategy for Alleviating Nanoparticle-Induced Stress in Plants

10.4.1 Plant Growth-Promoting Rhizobacteria and Their Mechanisms of Action

The plant-beneficial rhizobacteria were discovered and termed as plant growth-promoting rhizobacteria (PGPR) in the late twentieth century by Kloepper and Schroth (Ryu et al. 2024). PGPR are characteristically the essential component of the rhizosphere biota that can support the development of the host plant. The PGPRs exhibit physiological and metabolic versatility, probably due to environmental adaptation. The high growth rate, and metabolic plasticity enables PGPRs to metabolize a wide range of natural and xenobiotic substances. These microbes can directly promote plant growth through nitrogen fixation, solubilization of the soil minerals or sparingly soluble agri-inputs to enhance the availability of phosphorus, zinc, manganese, and potassium, enhancement of the iron acquisition through siderophore production, and the synthesis of growth-regulating substances (Tripathi et al. 2024a). Additionally, the absorption of nutrients, stress tolerance, and root health are all dependent on PGPRs (Liu et al. 2025) (Fig. 10.1). Besides these benefits, PGPRs may also produce low molecular mass compounds or enzymes that regulate the plant development and growth (Zhang et al. 2024).

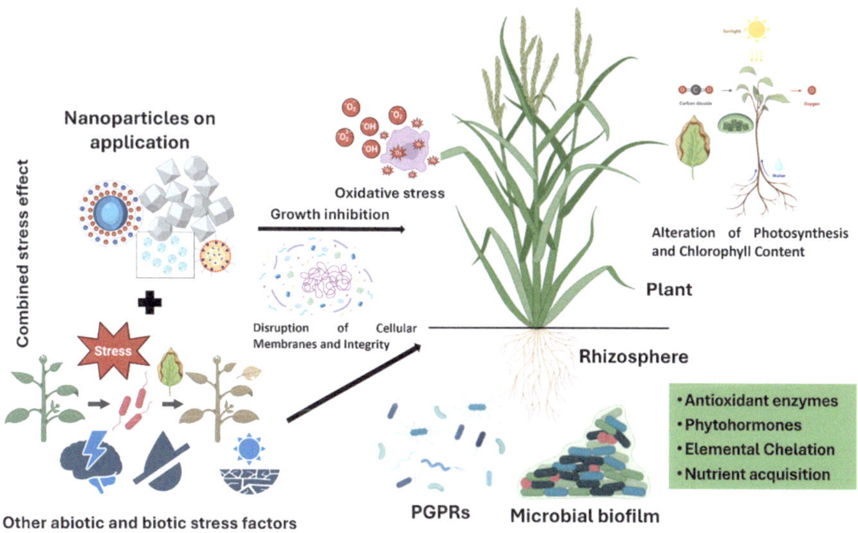

Fig. 10.1 Mechanistic interplay of nanomaterials with various types of stresses and plant beneficial rhizomicrobes for mitigation of nanomaterial-induced stress in plants

10.4.2 Mechanism of PGPR Activity

10.4.2.1 Nutrient Acquisition

Nitrogen (N) is the most crucial nutrient for plant growth and development. Even though the atmosphere contains roughly 78% N_2, developing plants cannot utilize this vast reservoir of nitrogen. Through the activity of nitrogenase enzyme complex, nitrogen fixing bacteria catalyze atmospheric nitrogen into ammonia, a bioavailable reduced form that plants can assimilate for growth (Kalia and Singh 2020). Biological nitrogen fixation (BNF) may be carried out by the N-fixers, often at moderate temperatures. Additionally, BNF is an affordable and environmentally responsible substitute for chemical fertilizers. The two main groups of nitrogen-fixing organisms are (a) symbiotic nitrogen-fixing bacteria, the rhizobiacea family members, which inhabit the root tissue/nodules of leguminous plants (e.g. R*hizobia*) and non-leguminous trees (e.g. *Frankia*); and (b) non-symbiotic nitrogen-fixing forms, which include cyanobacteria (*Anabaena*, *Nostoc*) and other loose associative or free-living N-fixers (*Azospirillum, Azotobacter, Gluconacetobacter diazotrophicus,* and *Azocarcus,* among others). All diazotrophs contain various genes needed to synthesize and for operations of the nitrogenase enzyme complex including the structural genes, genes involved in metal cofactor biosynthesis, electron donation, activators of the Fe protein, and regulatory genes necessary for the production and operation of the nitrogenase (*nif*) and *fix* enzyme complex genes (Li and Chen 2020). Phosphorus is the second most limiting nutrient for plants, following nitrogen in terms of availability and essentiality for plant growth. Phosphorus reserves in soil are not accessible in amounts that are appropriate for plants (Jahan et al. 2025). Only the soluble forms, such as mono- and dibasic phosphate, can be absorbed by plants (Pang et al. 2024). Of the total phosphorus in soil, between 30 and 50% is found in organic forms, with the remainder existing as inorganic bound-P. Microorganisms have the ability to mineralize this pool, releasing soluble phosphates that the plant can use. The plant unavailable forms of phosphate can be solubilized by a wide variety of bacterial genera, including *Aerobacter, Erwinia, Chryseobacterium, Achromobacter, Agrobacterium, Micrococcus, Psuedomonas, Bacillus, and Rhizobium* (Lei et al. 2025). Novel P-solubilizers have been isolated and identified from diverse niches which exhibit promising P-solubilization potentials (Visser et al. 2025). Microorganisms use two basic methods to solubilize phosphate: (i) the release of organic acids which interact with cations in the phosphate salts to mobilize phosphorus through ionic interactions, and (ii) the release of phosphate groups attached to organic materials through phosphatase enzyme activities (Ughamba et al. 2025). These mechanisms are typically more effective in alkaline soils such as calcareous soils (Barrow and Hartemink 2023; Hartemink and Barrow 2023).

10.4.2.2 Micronutrient/Trace Element Acquisition Through Microbial Metallophores

The siderophores function as a receptor moiety existing on the outer cell membrane that helps in the absorption of iron through the formation of Fe-siderophore complex (Wei et al. 2025). Siderophores are a category of microbial metallophores, including low molecular weight substances (0.4–1.5 kDa) that selectively chelate iron (Fe^{3+}) and transport it through the cell membrane inside the microbial cell. Both bacterial and fungal cells synthesize and produce a range of metallophores. Due to their affinity for conjugation with metal ions, metallophores may serve a dual purpose, i.e., acquisition of the trace element(s) from the rhizosphere even under low-availability conditions (Singh et al. 2022) and mitigation of the heavy metal stress by complexation with the metallophore moiety having promising potential in microbial remediation technology (Zelaya-Molina et al. 2025). After entering in the cell, the chelated metal is liberated and made accessible to the microorganism (Brick et al. 2025). Iron is an important micronutrient required by living cells, including bacteria, due to its critical role in metabolic activities. It cannot be directly assimilated by microbes in the soil because ferric iron (Fe^{3+}), though abundant in nature, is moderately soluble. Soil microorganisms produce and secrete low-molecular weight iron binding compounds to enhance their survival. By restricting or eliminating the fungal pathogen's access to available iron, the majority of Fe^{3+} in the rhizosphere is bound by siderophores. Iron shortage inhibits their growth, reduces nucleic acid synthesis, inhibits sporulation, and alters cell shape. These aspects lead to pathogen suppression. After being reduced and released into the periplasm and cytoplasm, iron is susceptible to oxidation (Ilbert and Bonnefoy 2013). In order to stop it from getting in the cell cytoplasm, this ferrous ion is readily trapped inside a carrier molecule and maintained in its new reduced state (Rajasekaran et al. 2022). Alternatively, the reduced ferrous form can be directly incorporated into metabolic processes before oxidization occurs (Chan et al. 2023). Although the general structure of microbial siderophores varies greatly, the majority have hydroxamate and catechol groups, which aid in chelating the ferric ion. *Pseudomonas* strains that produce siderophores quickly colonize the roots of various crops, resulting in improved yield. Furthermore, this bacterial genus exhibits both PGP and biocontrol potentials, thereby enhancing the growth and development of host plant, while also helping in decreasing the phytopathogenic population in rhizosphere and on the rhizoplane. Published research indicates that siderophore-mediated iron competition directly influences the biocontrol activity of these isolates (Deb and Tatung 2024). Another interesting upcoming research avenue is the use of siderophores and metallophores for bioremediation of heavy metals from different aquaculture and agricultural eco-niches (Gomes et al. 2024).

10.4.2.3 Production of Plant Growth Regulators

The exact synchronization of expansion, differentiation, and cell division processes is necessary for the growth and development of plants (Harashima and Schnittger 2010). Signaling molecules must be exchanged between roots and shoots for this to occur, and both biotic and abiotic variables can have an influence. PGPRs are capable of communication and interactions with other rhizosphere-dwelling microflora as well as plant cells using a cascade of signaling moieties or low molecular weight compounds (Mmotla et al. 2025). PGPR may promote plant development by producing and releasing phytohormones such as indole-3-acetic acid, cytokinin, gibberellin, and ethylene-degrading enzyme known as 1-aminocyclopropane (Stefen et al. 2022).

Among the various phytohormones produced by PGPRs, -auxins- particularly indole derivatives such as indole-3-butyric acid (IBA), indole-3-acetic acid (IAA), and their precursors- are the most commonly synthesized, with nearly 80% of the PGP bacteria producing it (Kejela 2024). These indole derivatives are used by the auxin-producing bacteria to coordinate several behavioral features through signaling mechanisms such as virulence factor synthesis, antibiotic resistance, biofilm formation, and metal stress adaption (Melander et al. 2014; Rolón-Cárdenas et al. 2022). Auxins are produced by a variety of biosynthetic routes in both plants and microorganisms. Auxins primarily influence the plant root system, leading to increased length, weight, branching, and surface area (Pappalettere et al. 2024). Better nutrition and growth potential result from these changes due to increase in the plant's capability to probe the soil for nutrient exchange (Yang et al. 2024). Inoculation of auxin-producing bacteria leads to emergence of adventitious roots from the stem (Tsavkelova et al. 2024). Plants and fungi synthesize gibberellins-the diterpenoid phytohormones through two independent but convergent evolution derived GA biosynthetic pathways (Goyal and Kalia 2020; Nett et al. 2017). In 1926, the first report of occurring in the culture filtrates of the fungus Fusarium moniliforme (also known as Gibberella fujikuroi in its sexual state) was made by the Japanese plant pathologist Dr. Eiichi Kurosawa (Rademacher 1994). Bacterial gibberellins are obtained through a third independent but convergent evolutionary pathway (Nett et al. 2017). The operon responsible for bacterial gibberellin biosynthesis includes several additional genes, suggesting their involvement in the availability of the isoprenoid precursors. Therefore, it remains uncertain whether these operons are strictly required for the gibberellin biosynthesis (Nett et al. 2017). These phytohormones are produced by a few bacterial genera, including the rhizobiaceae family genera, i.e., *Bradyrhizobium japonicum, Sinorhizobium fredii, Mesorhizobium loti,* and *Rhizobium etli,* which exhibit symbiotic interaction with different legume host plants. Also, two *Bacillus* genera, i.e., *Bacillus pumilus* and *Bacillus licheniformis* are also identified to contain the gibberellin biosynthesis operon (Salazar-Cerezo et al. 2018). Plant endogenous activities such as cell proliferation, germination, flowering, elongation of the stem, and the formation and initiation of root nodules in legume plants are all regulated by gibberellins (Keswani et al. 2022). Application of microbial-derived gibberellins in different test plants revealed that, in addition to the earlier-stated benefits, the plant defense system is activated, modulating the resistance and susceptibility responses of the plants to phytopathogens (Keswani et al. 2022). Another important phytohormone-Cytokinins are synthesized as well as metabolized by PGPRs. The primary significance of synthesis of a variety of cytokinin metabolites (highly conserved low molecular mass compounds) is to modulate the interactions with the host plant. As purine derivatives, cytokinins interact with the tRNA molecules to improve the efficiency and fidelity of mRNA translation of signal-producing pathways, both in plants and the PGPRs (Frébortová and Frébort 2021). For plant cells to divide and differentiate, cytokinins are necessary (Svolacchia and Sabatini 2023). These a stimulating effect on the cell division activity in young leaves and shoot meristems. It's the interplay of different types and concentration of ratio proportion of auxins and cytokinins which regulates various developmental processes, including root and shoot growth and apical dominance (Frebort et al. 2011).

Ethylene is a crucial plant hormone with a dual mode of action. It is a gaseous metabolite required for optimal plant development and growth, and it also act as a stress hormone as it invokes physiological changes in plants (Aman et al. 2025; Khan et al. 2024). Various abiotic and biotic stress agents lead to a rise in levels of endogenous ethylene, severely impacting the growth of the plant (Fortt et al. 2022). Rhizobacteria that produce 1-aminocyclopropane-1-carboxylate (ACC) deaminase promote the growth of the plant by reducing ethylene levels and enhancing salt tolerance and reducing drought and high temperature stress in plants (del Orozco-Mosqueda et al. 2020; McBride et al. 2025).

10.4.3 PGPR–Nanoparticle Interactions: Mechanical Insights

Applying different PGPRs to mitigate the negative impacts of the nanomaterial pollution has been proposed as a successful method for enhancing soil quality (Fig. 10.2). The inclusion of NPs may boost PGPR efficiency because they give PGPR a high surface area substrate and more sustenance for life.

10.4.3.1 Altered Plant–Microbe Interactions During Nanoparticle Stress

Plant stress can be effectively reduced by the synergy among the three components, i.e., rhizo-microbes, nanomaterials, and plants (Kalia and Sreelakshmi 2025; Kaur et al. 2022b). The interactions among these components may favor or promote direct delivery of nutrients and other molecules for improved nutrient absorption and use by plants and elicit an array of defense mechanisms to cope with the stress caused by the agents, leading to a reduction in the impact of the stress (Rajput et al. 2021a). The bioactive and signal molecules besides the phytohormones provided by the beneficial rhizo-microbes improve the hormonal balance of plants and increase the synthesis of several phytochemicals related to stress-reduction mechanisms

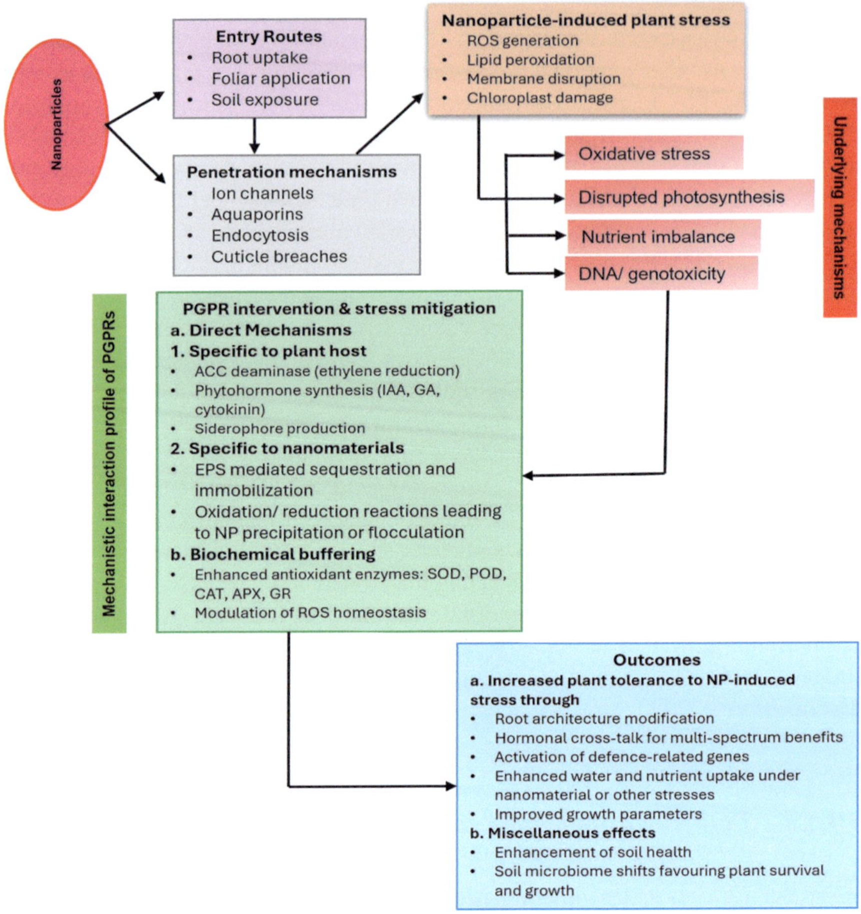

Fig. 10.2 Detailed direct and indirect processes involved in PGPR-mediated mitigation of the NP-induced stress in different crop plants

(Sodhi et al. 2025). By improving the hydraulic conductivity of the root which leads to osmotic adaption, and thus beneficial microbes help plants become more resilient to water deprivation. This can be achieved through ongoing advantages and a reduction in microbial efficacy, offering a workable framework for increasing plant production and resistance to environmental stressors (Sodhi et al. 2025). However, nanomaterials at a specific concentration or dose of application may cause nanotoxicity to the plant cell and tissues. The manifestations of nanotoxicity may be reduced or mitigated through the plant beneficial products synthesized and secreted by PGPRs (Solanki et al. 2024). A recent study in eggplant showcased the growth and biochemical responses facilitated by PGPRs under ZnO NPs stress through elicitation of the metabolic shifts in proline, malonaldehyde, and antioxidant synthesis, which led to alleviation of this abiotic stress (Solanki et al. 2024). The indirect negative impact on the host plant may arise from the nanotoxicity of the applied nanomaterials, which can alter the behavior of beneficial rhizomicroflora in the plant's rhizosphere (Sharma et al. 2023).

10.4.3.2 Biochemical Mechanisms of PGPR in Reducing Nanoparticle Toxicity

The PGPBs in the rhizosphere are, therefore, necessary for plants to reach maximum potential under stress conditions. Microbes use a variety of strategies, including antioxidant system enhancement, plant hormone modification, and nutrient uptake facilitation, to assist plants in withstanding stress. Nanomaterials exhibit janus properties; therefore, at higher concentrations, these are perceived by the plant cells as potential stress agents, causing a range of biochemical and physiological reactions (Kalia and Sreelakshmi 2025; Kaur et al. 2022b, 2023).

10.4.3.3 Antioxidant Potential for Modulation of Reactive Oxygen Species

Although plants naturally create reactive oxygen species (ROS) as byproducts of cell metabolism; nevertheless, abiotic stressors alter the subtle and fine balance of ROS produced and removed in plant cellular components (Mansoor et al. 2022). These radicals can potentially damage cellular macromolecules, including lipids, photosynthetic pigments, proteins, and eventually the cellular structure (Hasanuzzaman et al. 2020b).

Other antioxidant enzymes that plants utilize to reduce ROS include glutathione peroxidase (GPX), peroxidase (POD), catalase (CAT), ascorbate peroxidase (APX), and glutathione reductase (GR) (Hasanuzzaman et al. 2020a; Rao et al. 2025). Therefore, use of NMs that have antioxidant properties under stressful conditions increases the plant cell's capacity to scavenge ROS, increasing their chances of survival (Li et al. 2024; Rao et al. 2025). In order to help stop oxidative damage, the enzyme superoxide dismutase (SOD) transforms superoxide anions (O_2^-) into oxygen molecules (O_2) and hydrogen peroxide (H_2O_2). SOD is a vital enzyme system that eliminates free radicals (O_2^-) produced by stressed plants (Jomova et al. 2024; Mishra et al. 2023). Catalase (CAT) and peroxidase (POX) enzymes inhibit the ROS production from O_2^- and H_2O_2 via the Haber-Weiss process (Singh 2022). POX breaks down H_2O_2 and maintains redox balance. It also removes hydrogen peroxide by accelerating the oxidation of phenolic substrates, using H_2O_2 as an electron acceptor (Rajput et al. 2021a).

10.4.3.4 Role of Exopolysaccharides and Biofilm Formation

PGPR promotes the growth of plants directly by preventing diseases from harming plants, and indirectly by increasing the production of siderophores, phytohormones, exopolysaccharides, and biofilm formation (Yadav et al. 2021).

High-molecular-weight polymers, primarily polysaccharides, produced by PGPRs (Exopolysaccharides or EPS) enable bacteria to adhere to plant roots and form biofilms. These biofilms operate as a barrier to protect plant roots and microorganisms from environmental stressors, including those caused by nanoparticles (Abarquero et al. 2022). The bacterial EPS types include the slime and capsular EPS (Raj et al. 2023). Stress triggers bacteria to develop biofilm under unfavorable conditions. As a defense mechanism against the stress the nanoparticles generate, the bacteria respond by producing additional EPS (Kaur et al. 2024). Both within the bacterial cells and on the surface of the EPS matrix, the shape, surface charge, size, hydrophobicity, and functional groups of the nanoparticles influence their interactions with the components of the biofilm (Fetsiukh et al. 2021; Kaur et al. 2024). Nanoparticle-induced EPS generation has potential uses in biotechnology and medicine, among other domains. However, it is significant to take into account that the kind of nanoparticle, the bacterial species, and the ambient factors may all affect how the nanoparticles affect the formation of bacterial EPS. The breakdown of organic substances, desiccation resistance, and the transformation of heavy metals are all facilitated by the enzymatic activity of EPS (Balíková et al. 2022). PGPRs react to NP exposure by producing more EPS, which helps to bind and sequester NPs and lessen their phytotoxic effects (Agarwal et al. 2025). Because of this sequestration, NPs are unable to enter plant tissues and cause oxidative damage (Morcillo and Manzanera 2021). Bacteria normally create EPS as a defense mechanism when under stress. Additionally, heavy metals and other hazardous ions released by NPs can be chelated by EPS, mitigating the adverse effect they have on plants and microorganisms. The biofilm matrix limits the NP's mobility and bioavailability in the rhizosphere, which functions as a physical and chemical barrier creating a new hybrid niche, i.e., nanofouling (Deschênes and Ells 2020). A contrary impact of the nanomaterials on the bacterial biofilms has also been reported, which includes disruption of the cell membranes and quorum-sensing genes engaged in EPS secretion (Afrasiabi and Partoazar 2024; Ahmed et al. 2020). Nevertheless, generation of EPS and subsequent biofilm formation by the PGPRs are essential adaptive responses to stress brought on by nanoparticles. These processes demonstrate the promise of PGPRs in sustainable agricultural methods to combat the nanomaterial pollution of the land and water resources.

10.4.3.5 Role of PGPRs in Nanoparticle Transformation and Detoxification

The interaction between PGPRs and NPs significantly influences the transformation and detoxification of nanoparticles in the soil-plant system. By reducing NP-induced stress through a variety of methods, PGPRs improve the development of plants and overall soil health. The two main processes for detoxification of nanoparticle toxicity when it comes in contact with bacterial cells, are biosorption and bioaccumulation. Bioaccumulation, ion exchange, physical adsorption, and complexation are the primary mechanisms that control biosorption (Kaur et al. 2022a). The metal ions that are positively charged readily adsorb on the negatively charged bacterial cell wall, immobilizing the metal and lessening its harmful intracellular effects. Metal ions may interact with the cell's metabolic and physiological components once they're inside, causing oxidative stress, and affect the function, either directly or indirectly, of important bacterial metabolites such as proteins, DNA, and lipids. Cells have a number of strategies to combat these harmful effects, with the predominant mechanism being ion exclusion. In the presence of PGPR, nanoparticle immobilization in soil is necessary for boosting crop development and lowering metal bioaccumulation in crops (Bai et al. 2024). For example, *Enterobacter* sp. strain EG16 exhibits a Cd resistance (Cd^{2+} MIC, >250 mg L^{-1}), a 31% cell surface biosorption capacity, a strong ability to produce the plant hormone IAA and siderophores, and the ability to bind Cd intracellularly (Huang et al. 2018). By oxidizing Fe(II) and Mn(II), *Burkholderia* sp. D54 substantially generated Mn and Fe precipitates and limited metal absorption on *Sedum alfredii's* root surfaces (El-Maraghy et al. 2020; Guo et al. 2020). The negative effects of AgNPs on *Brassica juncea* seedlings were significantly mitigated by PGPR addition with AgNP, which also resulted in a lower level of nitrous oxide (Vishwakarma et al. 2017). Improved antioxidant enzyme activities may help detoxify reactive free radicals and lessen the detrimental effects of AgNPs stress on *B. juncea* seedlings, hence reducing biomass build-up in terms of weight and length.

10.5 Conclusion and Future Perspectives

This manuscript outlines the diverse interactions of nanomaterials with plants, their mechanisms of toxicity, the role of the PGPRs, and mechanisms of mitigation of the NPs-induced stress in plants. Integration of the application of beneficial microbes in conventional or existing agricultural practices offers a viable approach to alleviate the adverse effects of NPs-induced stress in plants. PGPRs augment plant resilience through mechanisms including phytohormone production, siderophores, and EPS generation, besides the modulation of the antioxidant enzyme activities. Together, these activities enhance plant growth and stress tolerance on exposure to nanomaterials.

To maximize their synergistic benefits, future studies must be focused on understanding the molecular interactions among the PGPRs and NPs. The stability and effectiveness of PGPR formulations might be further improved by employing advanced nano-encapsulation technologies, guaranteeing their continued action under a variety of environmental conditions. Application of PGPRs and NPs combined has the capacity to enhance nutrient availability, improve soil health, and support sustainable crop production. To assess the impact over time and ecological security of such integrated techniques, extensive field research is required. By utilizing the advantageous characteristics of PGPRs, novel approaches to deal with the problems caused by NP-induced stress, can be realized for advancing sustainable agriculture in the process.

References

Abarquero D, Renes E, Fresno JM, Tornadijo ME (2022) Study of exopolysaccharides from lactic acid bacteria and their industrial applications: a review. Int J Food Sci Technol 57:16–26. https://doi.org/10.1111/ijfs.15227

Abdel Latef AAH, Zaid A, Abu Alhmad MF, Abdelfattah KE (2020) The impact of priming with Al2O3 nanoparticles on growth, pigments, Osmolytes, and antioxidant enzymes of Egyptian Roselle (Hibiscus sabdariffa L.). Cultivar Agron 10:681. https://doi.org/10.3390/agronomy10050681

Afrasiabi S, Partoazar A (2024) Targeting bacterial biofilm-related genes with nanoparticle-based strategies. Front Microbiol 15. https://doi.org/10.3389/fmicb.2024.1387114

Agarwal H, Chaudhary D, Aggarwal H, Karala C, Purkait N, Sharma N, Mishra A, Mishra V, Kumar A, Singh P, Dufossé L, Joshi N (2025) Molecular underpinning of heavy metal sequestration through advanced remediation strategies in higher plants. Plant Stress 16:100881. https://doi.org/10.1016/j.stress.2025.100881

Ahmed B, Ameen F, Rizvi A, Ali K, Sonbol H, Zaidi A, Khan MS, Musarrat J (2020) Destruction of cell topography, morphology, membrane, inhibition of respiration, biofilm formation, and bioactive molecule production by nanoparticles of Ag, ZnO, CuO, TiO 2 , and Al 2 O 3 toward beneficial soil bacteria. ACS Omega 5:7861–7876. https://doi.org/10.1021/acsomega.9b04084

Alkhatib R, Alkhatib B, Abdo N (2021) Effect of Fe3O4 nanoparticles on seed germination in tobacco. Environ Sci Pollut Res 28:53568–53577. https://doi.org/10.1007/s11356-021-14541-x

Aman S, Swain S, Dutta E, Abbas S, Li N, Shakeel SN, Binder BM, Schaller GE (2025) Modulation of plant growth and development through altered ethylene binding affinity of the ethylene receptor ETR1. BMC Plant Biol 25:436. https://doi.org/10.1186/s12870-025-06469-y

Asli S, Neumann PM (2009) Colloidal suspensions of clay or titanium dioxide nanoparticles can inhibit leaf growth and transpiration via physical effects on root water transport. Plant Cell Environ 32:577–584. https://doi.org/10.1111/j.1365-3040.2009.01952.x

Atha DH, Wang H, Petersen EJ, Cleveland D, Holbrook RD, Jaruga P, Dizdaroglu M, Xing B, Nelson BC (2012) Copper oxide nanoparticle mediated DNA damage in terrestrial plant models. Environ Sci Technol 46:1819–1827. https://doi.org/10.1021/es202660k

Bai X, Bol R, Chen H, Cui Q, Qiu T, Zhao S, Fang L (2024) A meta-analysis on crop growth and heavy metals accumulation with PGPB inoculation in contaminated soils. J Hazard Mater 471:134370. https://doi.org/10.1016/j.jhazmat.2024.134370

Balíková K, Vojtková H, Duborská E, Kim H, Matúš P, Urík M (2022) Role of exopolysaccharides of pseudomonas in heavy metal removal and other remediation strategies. Polymers (Basel) 14:4253. https://doi.org/10.3390/polym14204253

Barrow NJ, Hartemink AE (2023) The effects of pH on nutrient availability depend on both soils and plants. Plant Soil 487:21–37. https://doi.org/10.1007/s11104-023-05960-5

Barthlott W, Neinhuis C (1997) Purity of the sacred lotus, or escape from contamination in biological surfaces. Planta 202:1–8. https://doi.org/10.1007/s004250050096

Bast A, Haenen GRMM, Doelman CJA (1991) Oxidants and antioxidants: state of the art. Am J Med 91:S2–S13. https://doi.org/10.1016/0002-9343(91)90278-6

Begum ZA, Rahman IMM, Tate Y, Sawai H, Maki T, Hasegawa H (2012) Remediation of toxic metal contaminated soil by washing with biodegradable aminopolycarboxylate chelants. Chemosphere 87:1161–1170. https://doi.org/10.1016/j.chemosphere.2012.02.032

Brick MB, Hussein MH, Mowafy AM, Hamouda RA, Ayyad AM, Refaay DA (2025) Significance of siderophore-producing cyanobacteria on enhancing iron uptake potentiality of maize plants grown under iron-deficiency. Microb Cell Factories 24:3. https://doi.org/10.1186/s12934-024-02618-4

Carocho M, Ferreira ICFR (2013) A review on antioxidants, prooxidants and related controversy: natural and synthetic compounds, screening and analysis methodologies and future perspectives. Food Chem Toxicol 51:15–25. https://doi.org/10.1016/j.fct.2012.09.021

Carpita N, Sabularse D, Montezinos D, Delmer DP (1979) Determination of the pore size of cell walls of living plant cells. Science (80-.) 205:1144–1147. https://doi.org/10.1126/science.205.4411.1144

Chan C, Ng D, Fraser ME, Schryvers AB (2023) Structural and functional insights into iron acquisition from lactoferrin and transferrin in Gram-negative bacterial pathogens. Biometals 36:683–702. https://doi.org/10.1007/s10534-022-00466-6

Chaudhary K, Agarwal S, Khan S (2018) Role of Phytochelatins (PCs), Metallothioneins (MTs), and heavy metal ATPase (HMA) genes in heavy metal tolerance. In: Mycoremediation and environmental sustainability, pp 39–60. https://doi.org/10.1007/978-3-319-77386-5_2

Corredor E, Testillano PS, Coronado M-J, González-Melendi P, Fernández-Pacheco R, Marquina C, Ibarra MR, de la Fuente JM, Rubiales D, Pérez-de-Luque A, Risueño M-C (2009) Nanoparticle penetration and transport in living pumpkin plants: in situsubcellular identification. BMC Plant Biol 9:45. https://doi.org/10.1186/1471-2229-9-45

Cui D, Zhang P, Ma Y, He X, Li Y, Zhao Y, Zhang Z (2014) Phytotoxicity of silver nanoparticles to cucumber (Cucumis sativus) and wheat (Triticum aestivum). J Zhejiang Univ Sci A 15:662–670. https://doi.org/10.1631/jzus.A1400114

Deb CR, Tatung M (2024) Siderophore producing bacteria as biocontrol agent against phytopathogens for a better environment: a review. S Afr J Bot 165:153–162. https://doi.org/10.1016/j.sajb.2023.12.031

del Orozco-Mosqueda MC, Glick BR, Santoyo G (2020) ACC deaminase in plant growth-promoting bacteria (PGPB): an efficient mechanism to counter salt stress in crops. Microbiol Res 235:126439. https://doi.org/10.1016/j.micres.2020.126439

Deschênes L, Ells T (2020) Bacteria-nanoparticle interactions in the context of nanofouling. Adv Colloid Interf Sci 277:102106. https://doi.org/10.1016/j.cis.2020.102106

Dimkpa CO, McLean JE, Martineau N, Britt DW, Haverkamp R, Anderson AJ (2013) Silver nanoparticles disrupt wheat (Triticum aestivum L.) growth in a sand matrix. Environ Sci Technol 47:1082–1090. https://doi.org/10.1021/es302973y

El-Maraghy SS, Tohamy TA, Hussein KA (2020) Expression of SidD gene and physiological characterization of the rhizosphere plant growth-promoting yeasts. Heliyon 6:e04384. https://doi.org/10.1016/j.heliyon.2020.e04384

Etxeberria E, Gonzalez P, Baroja-Fernández E, Romero JP (2006) Fluid phase endocytic uptake of artificial Nano-spheres and fluorescent quantum dots by sycamore cultured cells. Plant Signal Behav 1:196–200. https://doi.org/10.4161/psb.1.4.3142

Faisal M, Saquib Q, Alatar AA, Al-Khedhairy AA, Ahmed M, Ansari SM, Alwathnani HA, Dwivedi S, Musarrat J, Praveen S (2016) Cobalt oxide nanoparticles aggravate DNA damage and cell death in eggplant via mitochondrial swelling and NO signaling pathway. Biol Res 49:20. https://doi.org/10.1186/s40659-016-0080-9

Faiyue B, Al-Azzawi MJ, Flowers TJ (2010) The role of lateral roots in bypass flow in rice (Oryza sativa L.). Plant Cell Environ 33:702–716. https://doi.org/10.1111/j.1365-3040.2009.02078.x

Fallah S, Yusefi-Tanha E, Peralta-Videa JR (2024) Interaction of nanoparticles and reactive oxygen species and their impact on macromolecules and plant production. Plant Nano Biol 10:100105. https://doi.org/10.1016/j.plana.2024.100105

Feng Y, Fu H, Zhang X, Liu S, Wei X (2024) Lysosome toxicities induced by nanoparticle exposure and related mechanisms. Ecotoxicol Environ Saf 286:117215. https://doi.org/10.1016/j.ecoenv.2024.117215

Fetsiukh A, Conrad J, Bergquist J, Timmusk S (2021) Silica particles trigger the exopolysaccharide production of harsh environment isolates of growth-promoting Rhizobacteria and increase their ability to enhance wheat biomass in drought-stressed soils. Int J Mol Sci 22:6201. https://doi.org/10.3390/ijms22126201

Fitzpatrick K, Reid R (2009) The involvement of aquaglyceroporins in transport of boron in barley roots. Plant Cell Environ 32:1357–1365. https://doi.org/10.1111/j.1365-3040.2009.02003.x

Fleischer A, O'Neill MA, Ehwald R (1999) The pore size of non-Graminaceous plant cell walls is rapidly decreased by Borate Ester cross-linking of the Pectic polysaccharide Rhamnogalacturonan II. Plant Physiol 121:829–838. https://doi.org/10.1104/pp.121.3.829

Florance I, Cordani M, Pashootan P, Moosavi MA, Zarrabi A, Chandrasekaran N (2024) The impact of nanomaterials on autophagy across health and disease conditions. Cell Mol Life Sci 81:184. https://doi.org/10.1007/s00018-024-05199-y

Fortt J, González M, Morales P, Araya N, Remonsellez F, Coba de la Peña T, Ostria-Gallardo E, Stoll A (2022) Bacterial modulation of the plant ethylene signaling pathway improves tolerance to salt stress in Lettuce (Lactuca sativa L.). Front Sustain Food Syst 6. https://doi.org/10.3389/fsufs.2022.768250

Francis DV, Abdalla AK, Mahakham W, Sarmah AK, Ahmed ZFR (2024) Interaction of plants and metal nanoparticles: exploring its molecular mechanisms for sustainable agriculture and crop improvement. Environ Int 190:108859. https://doi.org/10.1016/j.envint.2024.108859

Frebort I, Kowalska M, Hluska T, Frebortova J, Galuszka P (2011) Evolution of cytokinin biosynthesis and degradation. J Exp Bot 62:2431–2452. https://doi.org/10.1093/jxb/err004

Frébortová J, Frébort I (2021) Biochemical and structural aspects of Cytokinin biosynthesis and degradation in bacteria. Microorganisms 9:1314. https://doi.org/10.3390/microorganisms9061314

Freinbichler W, Colivicchi MA, Stefanini C, Bianchi L, Ballini C, Misini B, Weinberger P, Linert W, Vareŝlija D, Tipton KF, Della Corte L (2011) Highly reactive oxygen species: detection, formation, and possible functions. Cell Mol Life Sci 68:2067–2079. https://doi.org/10.1007/s00018-011-0682-x

Gakis GP, Aviziotis IG, Charitidis CA (2023) Metal and metal oxide nanoparticle toxicity: moving towards a more holistic structure–activity approach. Environ Sci Nano 10:761–780. https://doi.org/10.1039/D2EN00897A

Gao M, Chang J, Wang Z, Zhang H, Wang T (2023) Advances in transport and toxicity of nanoparticles in plants. J Nanobiotechnology 21:75. https://doi.org/10.1186/s12951-023-01830-5

Gomes AFR, Almeida MC, Sousa E, Resende DISP (2024) Siderophores and metallophores: metal complexation weapons to fight environmental pollution. Sci Total Environ 932:173044. https://doi.org/10.1016/j.scitotenv.2024.173044

Gowtham HG, Shilpa N, Singh SB, Aiyaz M, Abhilash MR, Nataraj K, Amruthesh KN, Ansari MA, Alomary MN, Murali M (2024) Toxicological effects of nanoparticles in plants: mechanisms involved at morphological, physiological, biochemical and molecular levels. Plant Physiol Biochem 210:108604. https://doi.org/10.1016/j.plaphy.2024.108604

Goyal A, Kalia A (2020) Fungal Phytohormones: plant growth- regulating substances and their applications in crop productivity. In: Yadav AN, Mishra S, Kour D, Yadav N, Kumar A (eds) Agriculturally Important Fungi for Sustainable Agriculture. Springer Nature Switzerland AG, pp 143–169

Gulzar ABM, Mazumder PB (2022) Helping plants to deal with heavy metal stress: the role of nanotechnology and plant growth promoting rhizobacteria in the process of phytoremediation. Environ Sci Pollut Res 29:40319–40341. https://doi.org/10.1007/s11356-022-19756-0

Guo JK, Zhao J, Ren XH, Jia HL, Muhammad H, Lv X, Wei T, Hua L (2020) Effects of Burkholderia sp. D54 on growth and cadmium uptake of tomato, ryegrass and soybean plants. Int J Environ Sci Technol 17:1149–1158. https://doi.org/10.1007/s13762-019-02571-5

Hall JL (2003) Transition metal transporters in plants. J Exp Bot 54:2601–2613. https://doi.org/10.1093/jxb/erg303

Handa M, Kalia A (2024) Nanoparticle-plant-microbe interactions have a role in crop productivity and food security. Rhizosphere 30:100884. https://doi.org/10.1016/j.rhisph.2024.100884

Harashima H, Schnittger A (2010) The integration of cell division, growth and differentiation. Curr Opin Plant Biol 13:66–74. https://doi.org/10.1016/j.pbi.2009.11.001

Hartemink AE, Barrow NJ (2023) Soil pH - nutrient relationships: the diagram. Plant Soil 486:209–215. https://doi.org/10.1007/s11104-022-05861-z

Hasanuzzaman M, Bhuyan MHM, Zulfiqar F, Raza A, Mohsin S, Mahmud J, Fujita M, Fotopoulos V (2020a) Reactive oxygen species and antioxidant defense in plants under abiotic stress:

revisiting the crucial role of a universal defense regulator. Antioxidants 9:681. https://doi.org/10.3390/antiox9080681
Hasanuzzaman M, Bhuyan MHMB, Parvin K, Bhuiyan TF, Anee TI, Nahar K, Hossen MS, Zulfiqar F, Alam MM, Fujita M (2020b) Regulation of ROS metabolism in plants under environmental stress: a review of recent experimental evidence. Int J Mol Sci 21:8695. https://doi.org/10.3390/ijms21228695
Hischemöller A, Nordmann J, Ptacek P, Mummenhoff K, Haase M (2009) In-vivo imaging of the uptake of upconversion nanoparticles by plant roots. J Biomed Nanotechnol 5:278–284. https://doi.org/10.1166/jbn.2009.1032
Hu J, Wu X, Wu F, Chen W, Zhang X, White JC, Li J, Wan Y, Liu J, Wang X (2020) TiO2 nanoparticle exposure on lettuce (Lactuca sativa L.): dose-dependent deterioration of nutritional quality. Environ Sci Nano 7:501–513. https://doi.org/10.1039/C9EN01215J
Huang F, Wang Z-H, Cai Y-X, Chen S-H, Tian J-H, Cai K-Z (2018) Heavy metal bioaccumulation and cation release by growing Bacillus cereus RC-1 under culture conditions. Ecotoxicol Environ Saf 157:216–226. https://doi.org/10.1016/j.ecoenv.2018.03.077
Hussain HI, Yi Z, Rookes JE, Kong LX, Cahill DM (2013) Mesoporous silica nanoparticles as a biomolecule delivery vehicle in plants. J Nanopart Res 15:1676. https://doi.org/10.1007/s11051-013-1676-4
Ilbert M, Bonnefoy V (2013) Insight into the evolution of the iron oxidation pathways. Biochim Biophys Acta Bioenerg 1827:161–175. https://doi.org/10.1016/j.bbabio.2012.10.001
Jahan N, Mahmud U, Khan MZ (2025) Sustainable plant-soil phosphorus management in agricultural systems: challenges, environmental impacts and innovative solutions. Discov Soil 2:13. https://doi.org/10.1007/s44378-025-00039-2
Jalal A, de Oliveira Junior JC, Ribeiro JS, Fernandes GC, Mariano GG, Trindade VDR, dos Reis AR (2021) Hormesis in plants: physiological and biochemical responses. Ecotoxicol Environ Saf 207:111225. https://doi.org/10.1016/j.ecoenv.2020.111225
Jefferson DA (2000) The surface activity of ultrafine particles. Philos Trans R Soc London Ser A Math Phys Eng Sci 358:2683–2692. https://doi.org/10.1098/rsta.2000.0677
Jomova K, Alomar SY, Alwasel SH, Nepovimova E, Kuca K, Valko M (2024) Several lines of antioxidant defense against oxidative stress: antioxidant enzymes, nanomaterials with multiple enzyme-mimicking activities, and low-molecular-weight antioxidants. Arch Toxicol 98:1323–1367. https://doi.org/10.1007/s00204-024-03696-4
Jones DP (2008) Radical-free biology of oxidative stress. Am J Physiol Physiol 295:C849–C868. https://doi.org/10.1152/ajpcell.00283.2008
Kalia A, Singh J (2020) Plant-microbe interactions: applications for plant-growth promotion and in situ agri-waste management. In: Plant Microbiome Paradigm. Springer International Publishing, Cham, pp 49–69. https://doi.org/10.1007/978-3-030-50395-6_3
Kalia A, Sreelakshmi MV (2025) Nanoelicitors: a promising strategy for sustainable crop production under biotic stress conditions. In: Abd-Elsalam KA, Mohamed HI (eds) . Springer, Singapore, pp 125–142. https://doi.org/10.1007/978-981-96-6569-3_5
Kaur P, Sharma S, Albarakaty FM, Kalia A, Hassan MM, Abd-Elsalam KA (2022a) Biosorption and bioleaching of heavy metals from electronic waste varied with microbial genera. Sustainability 14:935. https://doi.org/10.3390/su14020935
Kaur H, Kaur J, Kalia A, Kuca K (2022b) The Janus face of nanomaterials: physiological responses as inducers of stress or promoters of plant growth? In: Plant and nanoparticles. Springer Nature Singapore, Singapore, pp 395–426. https://doi.org/10.1007/978-981-19-2503-0_16
Kaur J, Dhiman D, Kalia A (2023) Antioxidant activity of mycosynthesized nanoparticles. In: Myconanotechnology. CRC Press, pp 108–128
Kaur H, Kalia A, Manchanda P (2024) Elucidating the effect of TiO 2 nanoparticles on mung bean rhizobia via in vitro assay: influence on growth, morphology, and plant growth promoting traits. J Basic Microbiol. https://doi.org/10.1002/jobm.202300306
Kejela T (2024) Phytohormone-producing Rhizobacteria and their role in plant growth. In: New insights into Phytohormones. IntechOpen. https://doi.org/10.5772/intechopen.1002823

Keswani C, Singh SP, García-Estrada C, Mezaache-Aichour S, Glare TR, Borriss R, Rajput VD, Minkina TM, Ortiz A, Sansinenea E (2022) Biosynthesis and beneficial effects of microbial gibberellins on crops for sustainable agriculture. J Appl Microbiol 132:1597–1615. https://doi.org/10.1111/jam.15348

Khan S, Alvi AF, Khan NA (2024) Role of ethylene in the regulation of plant developmental processes. Stress 4:28–53. https://doi.org/10.3390/stresses4010003

Kumari M, Khan SS, Pakrashi S, Mukherjee A, Chandrasekaran N (2011) Cytogenetic and genotoxic effects of zinc oxide nanoparticles on root cells of Allium cepa. J Hazard Mater 190:613–621. https://doi.org/10.1016/j.jhazmat.2011.03.095

Kurepa J, Paunesku T, Vogt S, Arora H, Rabatic BM, Lu J, Wanzer MB, Woloschak GE, Smalle JA (2010) Uptake and distribution of Ultrasmall Anatase TiO 2 Alizarin Red S Nanoconjugates in Arabidopsis thaliana. Nano Lett 10:2296–2302. https://doi.org/10.1021/nl903518f

Landa P, Vankova R, Andrlova J, Hodek J, Marsik P, Storchova H, White JC, Vanek T (2012) Nanoparticle-specific changes in Arabidopsis thaliana gene expression after exposure to ZnO, TiO2, and fullerene soot. J Hazard Mater 241–242:55–62. https://doi.org/10.1016/j.jhazmat.2012.08.059

Larue C, Castillo-Michel H, Sobanska S, Trcera N, Sorieul S, Cécillon L, Ouerdane L, Legros S, Sarret G (2014) Fate of pristine TiO2 nanoparticles and aged paint-containing TiO2 nanoparticles in lettuce crop after foliar exposure. J Hazard Mater 273:17–26. https://doi.org/10.1016/j.jhazmat.2014.03.014

Lei Y, Kuai Y, Guo M, Zhang H, Yuan Y, Hong H (2025) Phosphate-solubilizing microorganisms for soil health and ecosystem sustainability: a forty-year scientometric analysis (1984–2024). Front Microbiol 16. https://doi.org/10.3389/fmicb.2025.1546852

Li Q, Chen S (2020) Transfer of nitrogen fixation (nif) genes to non-diazotrophic hosts. Chembiochem 21:1717–1722. https://doi.org/10.1002/cbic.201900784

Li K, Li H, Rahman S u, Rehman A, Khalid M (2024) Nanoparticle-mediated modulation of plant performance and microbiome dynamics: insights into interplay mechanisms. Biocatal Agric Biotechnol 61:103366. https://doi.org/10.1016/j.bcab.2024.103366

Lian J, Zhao L, Wu J, Xiong H, Bao Y, Zeb A, Tang J, Liu W (2020) Foliar spray of TiO2 nanoparticles prevails over root application in reducing Cd accumulation and mitigating Cd-induced phytotoxicity in maize (Zea mays L.). Chemosphere 239:124794. https://doi.org/10.1016/j.chemosphere.2019.124794

Liang L, Tang H, Deng Z, Liu Y, Chen X, Wang H (2018) Ag nanoparticles inhibit the growth of the bryophyte, Physcomitrella patens. Ecotoxicol Environ Saf 164:739–748. https://doi.org/10.1016/j.ecoenv.2018.08.021

Lin D, Xing B (2007) Phytotoxicity of nanoparticles: inhibition of seed germination and root growth. Environ Pollut 150:243–250. https://doi.org/10.1016/j.envpol.2007.01.016

Liu Y, Yue L, Wang C, Zhu X, Wang Z, Xing B (2020) Photosynthetic response mechanisms in typical C3 and C4 plants upon La 2 O 3 nanoparticle exposure. Environ Sci Nano 7:81–92. https://doi.org/10.1039/C9EN00992B

Liu Y, Zhao X, Ma Y, Dai W, Song Z, Wang Y, Shen J, He X, Yang F, Zhang Z (2023) Interaction of cerium oxide nanoparticles and ionic cerium with Duckweed (Lemna minor L.): uptake, distribution, and phytotoxicity. Nano 13:2523. https://doi.org/10.3390/nano13182523

Liu K, Deng F, Zeng F, Chen Z-H, Qin Y, Chen G (2025) Plant growth-promoting rhizobacteria improve drought tolerance of crops: a review. Plant Growth Regul 105:567–581. https://doi.org/10.1007/s10725-025-01300-y

López-Moreno ML, de la Rosa G, Hernández-Viezcas JA, Peralta-Videa JR, Gardea-Torresdey JL (2010) X-ray Absorption Spectroscopy (XAS) corroboration of the uptake and storage of CeO2 nanoparticles and assessment of their differential toxicity in four edible plant species. J Agric Food Chem 58:3689–3693. https://doi.org/10.1021/jf904472e

Lovecká P, Macůrková A, Záruba K, Hubáček T, Siegel J, Valentová O (2021) Genomic damage induced in Nicotiana tabacum L. plants by colloidal solution with silver and gold nanoparticles. Plants 10:1260. https://doi.org/10.3390/plants10061260

Madkour LH (2020) Pathways for nanoparticle (NP)-induced oxidative stress, pp 285–328. https://doi.org/10.1007/978-3-030-37297-2_7

Mansoor S, Ali Wani O, Lone JK, Manhas S, Kour N, Alam P, Ahmad A, Ahmad P (2022) Reactive oxygen species in plants: from source to sink. Antioxidants 11:225. https://doi.org/10.3390/antiox11020225

Manuja A, Kumar B, Kumar R, Chhabra D, Ghosh M, Manuja M, Brar B, Pal Y, Tripathi BN, Prasad M (2021) Metal/metal oxide nanoparticles: toxicity concerns associated with their physical state and remediation for biomedical applications. Toxicol Reports 8:1970–1978. https://doi.org/10.1016/j.toxrep.2021.11.020

May S, Hirsch C, Rippl A, Bohmer N, Kaiser J-P, Diener L, Wichser A, Bürkle A, Wick P (2018) Transient DNA damage following exposure to gold nanoparticles. Nanoscale 10:15723–15735. https://doi.org/10.1039/C8NR03612H

McBride S, Errickson W, Huang B (2025) Effects of chemical and biological inhibitors of ethylene on heat tolerance in annual bluegrass. HortScience 60:310–316. https://doi.org/10.21273/HORTSCI18179-24

Mekuye B, Abera B (2023) Nanomaterials: an overview of synthesis, classification, characterization, and applications. Nano Sel 4:486–501. https://doi.org/10.1002/nano.202300038

Melander RJ, Minvielle MJ, Melander C (2014) Controlling bacterial behavior with indole-containing natural products and derivatives. Tetrahedron 70:6363–6372. https://doi.org/10.1016/j.tet.2014.05.089

Mirmoeini T, Pishkar L, Kahrizi D, Barzin G, Karimi N (2021) Phytotoxicity of green synthesized silver nanoparticles on Camelina sativa L. Physiol Mol Biol Plants 27:417–427. https://doi.org/10.1007/s12298-021-00946-y

Mishra N, Jiang C, Chen L, Paul A, Chatterjee A, Shen G (2023) Achieving abiotic stress tolerance in plants through antioxidative defense mechanisms. Front Plant Sci 14. https://doi.org/10.3389/fpls.2023.1110622

Mmotla K, Sibanyoni NR, Allie F, Sitole L, Mafuna T, Mashabela MD, Mhlongo MI (2025) Exploring the intricacies of plant growth promoting rhizobacteria interactions: an omics review. Ann Microbiol 75:5. https://doi.org/10.1186/s13213-025-01793-y

Møller IM, Jensen PE, Hansson A (2007) Oxidative modifications to cellular components in plants. Annu Rev Plant Biol 58:459–481. https://doi.org/10.1146/annurev.arplant.58.032806.103946

Morcillo R, Manzanera M (2021) The effects of plant-associated bacterial exopolysaccharides on plant abiotic stress tolerance. Meta 11:337. https://doi.org/10.3390/metabo11060337

Moscatelli A, Ciampolini F, Rodighiero S, Onelli E, Cresti M, Santo N, Idilli A (2007) Distinct endocytic pathways identified in tobacco pollen tubes using charged nanogold. J Cell Sci 120:3804–3819. https://doi.org/10.1242/jcs.012138

Mourato M, Reis R, Louro L (2012) Characterization of plant antioxidative system in response to abiotic stresses: a focus on heavy metal toxicity. In: Advances in selected plant physiology aspects. InTech. https://doi.org/10.5772/34557

Munir N, Gulzar W, Abideen Z, Hasanuzzaman M, El-Keblawy A, Zhao F (2023) Plant–nanoparticle interactions: transcriptomic and proteomic insights. Agronomy 13:2112. https://doi.org/10.3390/agronomy13082112

Nadiminti PP, Dong YD, Sayer C, Hay P, Rookes JE, Boyd BJ, Cahill DM (2013) Nanostructured liquid crystalline particles As an alternative delivery vehicle for plant agrochemicals. ACS Appl Mater Interfaces 5:1818–1826. https://doi.org/10.1021/am303208t

Nel A, Xia T, Mädler L, Li N (2006) Toxic potential of materials at the Nanolevel. Science (80-) 311:622–627. https://doi.org/10.1126/science.1114397

Nett RS, Montanares M, Marcassa A, Lu X, Nagel R, Charles TC, Hedden P, Rojas MC, Peters RJ (2017) Elucidation of gibberellin biosynthesis in bacteria reveals convergent evolution. Nat Chem Biol 13:69–74. https://doi.org/10.1038/nchembio.2232

Nowack B, Bucheli TD (2007) Occurrence, behavior and effects of nanoparticles in the environment. Environ Pollut 150:5–22. https://doi.org/10.1016/j.envpol.2007.06.006

Pang F, Li Q, Solanki MK, Wang Z, Xing Y-X, Dong D-F (2024) Soil phosphorus transformation and plant uptake driven by phosphate-solubilizing microorganisms. Front Microbiol 15. https://doi.org/10.3389/fmicb.2024.1383813

Pappalettere L, Bartolini S, Toffanin A (2024) Auxin-producing bacteria used as microbial biostimulants improve the growth of tomato (Solanum lycopersicum L.) seedlings in hydroponic systems. Biotech 13:32. https://doi.org/10.3390/biotech13030032

Peharec Štefanić P, Košpić K, Lyons DM, Jurković L, Balen B, Tkalec M (2021) Phytotoxicity of silver nanoparticles on tobacco plants: evaluation of coating effects on photosynthetic performance and chloroplast ultrastructure. Nanomaterials 11:744. https://doi.org/10.3390/nano11030744

Pittol M, Tomacheski D, Simões DN, Ribeiro VF, Santana RMC (2017) Macroscopic effects of silver nanoparticles and titanium dioxide on edible plant growth. Environ Nanotechnol Monit Manag 8:127–133. https://doi.org/10.1016/j.enmm.2017.07.003

Rademacher W (1994) Gibberellin formation in microorganisms. Plant Growth Regul 15:303–314. https://doi.org/10.1007/BF00029903

Raj Y, Kumar A, Kumari S, Kumar R, Kumar R (2023) Comparative genomics and physiological investigations supported multifaceted plant growth-promoting activities in two Hypericum perforatum L.-associated plant growth-promoting Rhizobacteria for microbe-assisted cultivation. Microbiol Spectr 11. https://doi.org/10.1128/spectrum.00607-23

Rajasekaran MB, Hussain R, Siligardi G, Andrews SC, Watson KA (2022) Crystal structure and metal binding properties of the periplasmic iron component EfeM from Pseudomonas syringae EfeUOB/M iron-transport system. Biometals 35:573–589. https://doi.org/10.1007/s10534-022-00389-2

Rajput VD, Harish S, Verma KK, Sharma L, Quiroz-Figueroa FR, Meena M, Gour VS, Minkina T, Sushkova S, Mandzhieva S (2021a) Recent developments in enzymatic antioxidant defence mechanism in plants with special reference to abiotic stress. Biology (Basel) 10:267. https://doi.org/10.3390/biology10040267

Rajput VD, Minkina T, Fedorenko A, Chernikova N, Hassan T, Mandzhieva S, Sushkova S, Lysenko V, Soldatov MA, Burachevskaya M (2021b) Effects of zinc oxide nanoparticles on physiological and anatomical indices in spring barley tissues. Nano 11:1722. https://doi.org/10.3390/nano11071722

Rao MJ, Duan M, Zhou C, Jiao J, Cheng P, Yang L, Wei W, Shen Q, Ji P, Yang Y, Conteh O, Yan D, Yuan H, Rauf A, Ai J, Zheng B (2025) Antioxidant defense system in plants: reactive oxygen species production, signaling, and scavenging during abiotic stress-induced oxidative damage. Horticulturae 11:477. https://doi.org/10.3390/horticulturae11050477

Ravindran A, Prathna TC, Verma VK, Chandrasekaran N, Mukherjee A (2012) Bovine serum albumin mediated decrease in silver nanoparticle phytotoxicity: root elongation and seed germination assay. Toxicol Environ Chem 94:91–98. https://doi.org/10.1080/02772248.2011.617034

Rico CM, Majumdar S, Duarte-Gardea M, Peralta-Videa JR, Gardea-Torresdey JL (2011) Interaction of nanoparticles with edible plants and their possible implications in the food chain. J Agric Food Chem 59:3485–3498. https://doi.org/10.1021/jf104517j

Rolón-Cárdenas GA, Arvizu-Gómez JL, Soria-Guerra RE, Pacheco-Aguilar JR, Alatorre-Cobos F, Hernández-Morales A (2022) The role of auxins and auxin-producing bacteria in the tolerance and accumulation of cadmium by plants. Environ Geochem Health 44:3743–3764. https://doi.org/10.1007/s10653-021-01179-4

Ryu C-M, Nelson LM, De-Bashan L (2024) Editorial: highlights from the 12th plant growth-promoting rhizobacteria workshop. Front Plant Sci 15. https://doi.org/10.3389/fpls.2024.1470576

Salazar-Cerezo S, Martínez-Montiel N, García-Sánchez J, Pérez-y-Terrón R, Martínez-Contreras RD (2018) Gibberellin biosynthesis and metabolism: a convergent route for plants, fungi and bacteria. Microbiol Res 208:85–98. https://doi.org/10.1016/j.micres.2018.01.010

Schreiber L (2010) Transport barriers made of cutin, suberin and associated waxes. Trends Plant Sci 15:546–553. https://doi.org/10.1016/j.tplants.2010.06.004

Servin AD, Morales MI, Castillo-Michel H, Hernandez-Viezcas JA, Munoz B, Zhao L, Nunez JE, Peralta-Videa JR, Gardea-Torresdey JL (2013) Synchrotron verification of TiO 2 accumulation in cucumber fruit: a possible pathway of TiO 2 nanoparticle transfer from soil into the food chain. Environ Sci Technol 47:11592–11598. https://doi.org/10.1021/es403368j

Sharma V, Anderson D, Dhawan A (2012) Zinc oxide nanoparticles induce oxidative DNA damage and ROS-triggered mitochondria mediated apoptosis in human liver cells (HepG2). Apoptosis 17(8):852–870

Sharma MMM, Kapoor D, Rohilla R, Sharma P (2023) Nanomaterials and their toxicity to beneficial soil microbiota and fungi associated plants rhizosphere. In: Nanomaterials and nanocomposites exposures to plants. Springer Nature Singapore, Singapore, pp 353–380. https://doi.org/10.1007/978-981-99-2419-6_18

Siddiqui S (2024) DNA damage, cell death, and alteration of cell proliferation insights caused by copper oxide nanoparticles using a plant-based model. Biology (Basel) 13:805. https://doi.org/10.3390/biology13100805

Singh D (2022) Juggling with reactive oxygen species and antioxidant defense system—a coping mechanism under salt stress. Plant Stress 5:100093. https://doi.org/10.1016/j.stress.2022.100093

Singh V, Maharshi A, Singh DP, Upadhyay RS, Sarma BK, Singh HB (2018) Role of microbial seed priming and microbial Phytohormone in modulating growth promotion and defense responses in plants. In: Advances in seed priming. Springer Singapore, Singapore, pp 115–126. https://doi.org/10.1007/978-981-13-0032-5_8

Singh P, Chauhan PK, Upadhyay SK, Singh RK, Dwivedi P, Wang J, Jain D, Jiang M (2022) Mechanistic insights and potential use of Siderophores producing microbes in rhizosphere for mitigation of stress in plants grown in degraded land. Front Microbiol 13. https://doi.org/10.3389/fmicb.2022.898979

Singh MP, Kumar P, Singh AK, Prakash O, Parveen A, Gazali A, Sarangi PK, Sharma L, Prakash A (2024) Genotoxicity of gold nanoparticles in plants and underlying mechanisms. In: Husen A (ed) Plant response to gold nanoparticles. Springer Singapore, pp 79–95. https://doi.org/10.1007/978-981-97-7174-5_4

Singh A, Agrawal S, Rajput VD, Minkina T, Rensing C, Elshikh MS, Zhao Y, Alexiou A, Papadakis M, Ghazaryan K (2025) Biogenic nanoparticles for managing salinity stress-related crop and environmental risks: realistic applications and challenges. Discov Sustain 6:120. https://doi.org/10.1007/s43621-025-00855-0

Skiba E, Pietrzak M, Glińska S, Wolf WM (2021) The combined effect of ZnO and CeO2 nanoparticles on Pisum sativum L.: a photosynthesis and nutrients uptake study. Cells 10:3105. https://doi.org/10.3390/cells10113105

Soares EV, Soares HMVM (2021) Harmful effects of metal(loid) oxide nanoparticles. Appl Microbiol Biotechnol 105:1379–1394. https://doi.org/10.1007/s00253-021-11124-1

Sodhi GK, Wijesekara T, Kumawat KC, Adhikari P, Joshi K, Singh S, Farda B, Djebaili R, Sabbi E, Ramila F, Sillu D, Santoyo G, de los Santos-Villalobos S, Kumar A, Pellegrini M, Mitra D (2025) Nanomaterials–plants–microbes interaction: plant growth promotion and stress mitigation. Front Microbiol 15. https://doi.org/10.3389/fmicb.2024.1516794

Solanki B, Saleem S, Khan MS (2024) Amelioration of phytotoxic impact of biosynthesized zinc oxide nanoparticles: plant growth promoting rhizobacteria facilitates the growth and biochemical responses of Eggplant (Solanum melongena) under nanoparticles stress. Plant Physiol Biochem 211:108678. https://doi.org/10.1016/j.plaphy.2024.108678

Solorio-Rodriguez SA, Wu D, Boyadzhiev A, Christ C, Williams A, Halappanavar S (2024) A systematic genotoxicity assessment of a suite of metal oxide nanoparticles reveals their DNA damaging and Clastogenic potential. Nanomaterials 14:743. https://doi.org/10.3390/nano14090743

Soni S, Jha AB, Dubey RS, Sharma P (2024) Nanowonders in agriculture: unveiling the potential of nanoparticles to boost crop resilience to salinity stress. Sci Total Environ 925:171433. https://doi.org/10.1016/j.scitotenv.2024.171433

Stefen DLV, Nunes FR, Rodolfo GR, Segatto C, Anastácio TC, Lajus CR (2022) How phytohormones synthesized by PGPR affect plant growth? In: Secondary metabolites and volatiles of PGPR in plant-growth promotion. Springer International Publishing, Cham, pp 119–131. https://doi.org/10.1007/978-3-031-07559-9_7

Svolacchia N, Sabatini S (2023) Cytokinins. Curr Biol 33:R10–R13. https://doi.org/10.1016/j.cub.2022.11.022

Thiruvengadam M, Chi HY, Kim S-H (2024) Impact of nanopollution on plant growth, photosynthesis, toxicity, and metabolism in the agricultural sector: an updated review. Plant Physiol Biochem 207:108370. https://doi.org/10.1016/j.plaphy.2024.108370

Tripathi A, Liu S, Singh PK, Kumar N, Pandey AC, Tripathi DK, Chauhan DK, Sahi S (2017) Differential phytotoxic responses of silver nitrate (AgNO3) and silver nanoparticle (AgNps) in Cucumis sativus L. Plant Gene 11:255–264. https://doi.org/10.1016/j.plgene.2017.07.005

Tripathi A, Pandey VK, Jain D, Singh G, Brar NS, Taufeeq A, Pandey I, Dash KK, Samrot AV, Rustagi S (2024a) An updated review on significance of PGPR-induced plant signalling and stress management in advancing sustainable agriculture. J Agric Food Res 16:101169. https://doi.org/10.1016/j.jafr.2024.101169

Tripathi S, Mahra S, Sharma S, Mathew S, Sharma S (2024b) Interaction of silver nanoparticles with plants: a focus on the phytotoxicity, underlying mechanism, and alleviation strategies. Plant Nano Biol. 9:100082. https://doi.org/10.1016/j.plana.2024.100082

Tsavkelova EA, Volynchikova EA, Potekhina NV, Lavrov KV, Avtukh AN (2024) Auxin production and plant growth promotion by Microbacterium albopurpureum sp. nov. from the rhizoplane of leafless Chiloschista parishii Seidenf. orchid. Front Plant Sci 15. https://doi.org/10.3389/fpls.2024.1360828

Tuteja N, Ahmad P, Panda BB, Tuteja R (2009) Genotoxic stress in plants: shedding light on DNA damage, repair and DNA repair helicases. Mutat Res Mutat Res 681:134–149. https://doi.org/10.1016/j.mrrev.2008.06.004

Tyerman SD, Niemietz CM, Bramley H (2002) Plant aquaporins: multifunctional water and solute channels with expanding roles. Plant Cell Environ 25:173–194. https://doi.org/10.1046/j.0016-8025.2001.00791.x

Ughamba KT, Ndukwe JK, Lidbury IDEA, Nnaji ND, Eze CN, Aduba CC, Groenhof S, Chukwu KO, Anyanwu CU, Nwaiwu O, Anumudu CK (2025) Trends in the application of phosphate-solubilizing microbes as biofertilizers: implications for soil improvement. Soil Syst 9:6. https://doi.org/10.3390/soilsystems9010006

Usman M, Farooq M, Wakeel A, Nawaz A, Cheema SA, Rehman H, Ur Ashraf I, Sanaullah M (2020) Nanotechnology in agriculture: current status, challenges and future opportunities. Sci Total Environ 721:137778. https://doi.org/10.1016/j.scitotenv.2020.137778

Van Hoecke K, De Schamphelaere KAC, Ramirez-Garcia S, Van der Meeren P, Smagghe G, Janssen CR (2011) Influence of alumina coating on characteristics and effects of SiO2 nanoparticles in algal growth inhibition assays at various pH and organic matter contents. Environ Int 37:1118–1125. https://doi.org/10.1016/j.envint.2011.02.009

Vishwakarma K, Shweta, Upadhyay N., Singh J, Liu S, Singh VP, Prasad SM, Chauhan DK, Tripathi DK, Sharma S (2017) Differential phytotoxic impact of plant mediated silver nanoparticles (AgNPs) and silver nitrate (AgNO3) on Brassica sp. Front Plant Sci 8. https://doi.org/10.3389/fpls.2017.01501

Visser A-N, Zhang F, Guttman L, Masasa M, Wang S, Koedooder C, Shaked Y (2025) Unveiling the P-solubilizing potential of bacteria enriched from natural colonies of Red Sea Trichodesmium spp. Sci Total Environ 963:178446. https://doi.org/10.1016/j.scitotenv.2025.178446

Wang J, Koo Y, Alexander A, Yang Y, Westerhof S, Zhang Q, Schnoor JL, Colvin VL, Braam J, Alvarez PJJ (2013) Phytostimulation of Poplars and Arabidopsis exposed to silver nanoparticles and Ag+ at sublethal concentrations. Environ Sci Technol 47:5442–5449. https://doi.org/10.1021/es4004334

Wang C, Jiang K, Wu B, Zhou J, Lv Y (2018a) Silver nanoparticles with different particle sizes enhance the allelopathic effects of Canada goldenrod on the seed germination and seedling development of lettuce. Ecotoxicology 27:1116–1125. https://doi.org/10.1007/s10646-018-1966-9

Wang XP, Li QQ, Pei ZM, Wang SC (2018b) Effects of zinc oxide nanoparticles on the growth, photosynthetic traits, and antioxidative enzymes in tomato plants. Biol Plant 62:801–808. https://doi.org/10.1007/s10535-018-0813-4

Wei Z, Gu S, Vollenweider V, Zuo Y, Li Z, Kümmerli R (2025) Microbial siderophores for One Health. Trends Microbiol. https://doi.org/10.1016/j.tim.2025.05.002

Wu Q, Fan C, Wang H, Han Y, Tai F, Wu J, Li H, He R (2023) Biphasic impacts of graphite-derived engineering carbon-based nanomaterials on plant performance: effectiveness vs. nanotoxicity. Adv Agrochem 2:113–126. https://doi.org/10.1016/j.aac.2023.01.001

Xiong T, Zhang T, Xian Y, Kang Z, Zhang S, Dumat C, Shahid M, Li S (2021) Foliar uptake, biotransformation, and impact of CuO nanoparticles in Lactuca sativa L. var. ramosa. Hort Environ Geochem Health 43:423–439. https://doi.org/10.1007/s10653-020-00734-9

Yadav G, Srivastva R, Gupta P (2021) Endophytes and their applications as biofertilizers. In: Microbial Technology for Sustainable Environment. Springer Singapore, Singapore, pp 95–123. https://doi.org/10.1007/978-981-16-3840-4_7

Yan A, Chen Z (2019) Impacts of silver nanoparticles on plants: a focus on the phytotoxicity and underlying mechanism. Int J Mol Sci 20:1003. https://doi.org/10.3390/ijms20051003

Yan S, Zhao L, Li H, Zhang Q, Tan J, Huang M, He S, Li L (2013) Single-walled carbon nanotubes selectively influence maize root tissue development accompanied by the change in the related gene expression. J Hazard Mater 246–247:110–118. https://doi.org/10.1016/j.jhazmat.2012.12.013

Yang J, Jiang F, Ma C, Rui Y, Rui M, Adeel M, Cao W, Xing B (2018) Alteration of crop yield and quality of wheat upon exposure to silver nanoparticles in a life cycle study. J Agric Food Chem 66:2589–2597. https://doi.org/10.1021/acs.jafc.7b04904

Yang S, Yin R, Wang C, Yang Y, Wang J (2023) Phytotoxicity of zinc oxide nanoparticles and multi-walled carbon nanotubes, alone or in combination, on Arabidopsis thaliana and their mutual effects on oxidative homeostasis. PLoS One 18:e0281756. https://doi.org/10.1371/journal.pone.0281756

Yang P, Condrich A, Scranton S, Hebner C, Lu L, Ali MA (2024) Utilizing plant growth-promoting Rhizobacteria (PGPR) to advance sustainable agriculture. Bacteria 3:434–451. https://doi.org/10.3390/bacteria3040030

Yanik F, Vardar F (2018a) Mechanism and interaction of nanoparticle-induced programmed cell death in plants. In: Phytotoxicity of nanoparticles. Springer International Publishing, Cham, pp 175–196. https://doi.org/10.1007/978-3-319-76708-6_7

Yanik F, Vardar F (2018b) Oxidative stress response to aluminum oxide (Al_2O_3) nanoparticles in Triticum aestivum. Biologia (Bratisl) 73:129–135. https://doi.org/10.2478/s11756-018-0016-7

Zelaya-Molina LX, Chávez-Díaz IF, Urrieta-Velázquez JA, Aragón-Magadan MA, Puente-Valenzuela CO, Blanco-Camarillo M, de los Santos-Villalobos S, Ramos-Garza J (2025) Microbial Metallophores in the productivity of agroecosystems. Microbiol Res (Pavia) 16:67. https://doi.org/10.3390/microbiolres16030067

Zhang T, Jian Q, Yao X, Guan L, Li L, Liu F, Zhang C, Li D, Tang H, Lu L (2024) Plant growth-promoting rhizobacteria (PGPR) improve the growth and quality of several crops. Heliyon 10:e31553. https://doi.org/10.1016/j.heliyon.2024.e31553

Zhao L, Peng B, Hernandez-Viezcas JA, Rico C, Sun Y, Peralta-Videa JR, Tang X, Niu G, Jin L, Varela-Ramirez A, Zhang J, Gardea-Torresdey JL (2012) Stress response and tolerance of Zea mays to CeO 2 nanoparticles: cross talk among H_2O_2, heat shock protein, and lipid peroxidation. ACS Nano 6:9615–9622. https://doi.org/10.1021/nn302975u

Chapter 11
The Effects of Nanofungicides on Plant Microbiomes

Heba Yousef, Sozan E. El-Abeid, Reham I. AbdelHamid, and Kamel A. Abd-Elsalam

Abstract Nanotechnology has emerged as a transformative force in modern agriculture, offering innovative solutions to long-standing challenges in crop protection. This chapter explores the role of nanotechnology, particularly nanofungicides, in advancing sustainable plant disease management. Nanofungicides overcome the main drawbacks of traditional fungicides, such as environmental persistence, low bioavailability, and non-target toxicity, by utilizing the special physicochemical characteristics of materials at the nanoscale, such as increased surface area, enhanced reactivity, and controlled release. These nanoformulations enable precise targeting of fungal pathogens, improved adhesion to plant surfaces, and reduced chemical inputs, thereby enhancing effectiveness while minimizing ecological disruption. However, their integration into agroecosystems raises concerns about potential impacts on the plant microbiome, soil health, and non-target organisms. This chapter provides a comprehensive overview of nanofungicide classifications, mechanisms of action, and interactions with plant–microbe systems, alongside a critical examination of their benefits, risks, and future potential. Ultimately, nanotechnology holds significant promise for driving sustainable agriculture, but responsible development, regulatory oversight, and interdisciplinary research are essential to ensure its safe and equitable application.

Keywords Nanofungicides · Plant microbiome · Plant · pathogen interaction · Plant disease management

---

H. Yousef (✉) · S. E. El-Abeid · K. A. Abd-Elsalam
Plant Pathology Research Institute, Agricultural Research Center (ARC), Giza, Egypt
e-mail: heba.yousef@live.com

R. I. AbdelHamid
Sugar Crops Research Institute, Agriculture Research Center (ARC), Giza, Egypt

K. A. Abd-Elsalam (ed.), *Plant-Microbiome Nanotechnology*, Sustainability Sciences in Asia and Africa, https://doi.org/10.1007/978-981-95-5056-2_11

11.1 Introduction

The manipulation and application of matter at the nanoscale (1–100 nm) has significant implications for agriculture, particularly in the development of next-generation plant protection systems. Nanotechnology is a revolutionary finding in modern science. Materials at this scale display new and improved physicochemical characteristics that are not seen in their bulk equivalents, including greater surface area, elevated reactivity, and distinctive quantum effects. These qualities are currently being used in agriculture to solve long-standing problems with traditional agrochemicals, particularly fungicides. Despite their effectiveness, traditional fungicides frequently have drawbacks such as poor solubility, environmental persistence, quick breakdown, and non-target toxicity, which force repeated applications and leave chemical residues in food and soil systems.

As a result, formulations based on nanotechnology, particularly nanofungicides, are becoming more and more inventive substitutes that aim to enhance the specificity, effectiveness, and transport of active ingredients. By enabling targeted interactions with fungal pathogens, enhanced adherence to plant surfaces, and regulated and sustained release, these nanoformulations improve disease management while requiring less chemical input. Furthermore, the application of nanotechnology in agriculture raises significant concerns regarding its interactions with the plant microbiome and larger ecosystems while also opening up new possibilities for enhancing crop resilience, nutrient efficiency, and sustainability.

With an emphasis on nanofungicides, this chapter offers a fundamental review of nanotechnology in agriculture, including information on their classifications, mechanisms of action, physical and chemical characteristics, and interface with plant–microbe interactions. An in-depth understanding of how nanotechnology is changing sustainable crop protection and plant health management is provided by this examination of its advantages, effects on the environment, safety concerns, and future prospects.

11.2 Basics of Nanotechnology

Nanotechnology encompasses the approximation and usage of nanometer sized materials (1–100 nm), which exhibit novel physicochemical conduct properties as a result of enhanced surface and quantum outcomes (Kah et al. 2018). The emergence of nanotechnology in agriculture has created new horizons in the protection of crops with highly effective, targeted, and sustainable agrochemical formulations. It is one of the most promising applications of nanofungicides: nanotechnology-based products aiming to protect crops against fungal pathogens that can cause huge yield losses in the world (Fraceto et al. 2016).

These traditional fungicides are effective but suffer disadvantages such as being poorly soluble, rapidly degrading, having non-specific action, and so having to be applied repeatedly with the risk of chemical residues in food and soil (Kah et al. 2018). To handle challenges such as the chemical load and environmental impact, nanotechnology is developed to enhance the delivery, stability, and efficacy of active ingredients, which reduces the chemical load used (Chen and Yada 2011).

11.3 Types of Nanofungicides

The nanofungicides can broadly be formulated and act according to their mode of action:

11.3.1 Nanoparticles

Fungal properties of metallic and metal oxide nanoparticles, namely, silver (Ag), copper oxide (CuO), and zinc oxide (ZnO), are demonstrated in Table 11.1. Rai and Ingle (2012) demonstrate that these nanoparticles can cause disruption of fungal cell membranes, production of reactive oxygen species (ROS), and inhibition of cellular metabolism. Silver nanoparticles, for example, kill a broad spectrum of pathogens, such as *Fusarium* and *Botrytis* (Jo et al. 2009). Nanoparticles exert their effects through various mechanisms of action.

11.3.1.1 Disruption of Fungal Cell Membranes

11.3.1.1.1 Silver Nanoparticles (AgNPs)

AgNPs are able to bind with and enter fungal cell membranes, which leads to structural damage and high permeability of the cell membrane. This causes loss of cellular contents and eventual death of the cell (Matras et al. 2022).

Table 11.1 Key features of metallic and metal oxide nanoparticles as fungicides

Nanoparticle type	Main mechanisms of action	Target fungi examples	Notable features
Silver (AgNPs)	Membrane disruption, ROS generation, metabolic inhibition	Fusarium, Botrytis, Aspergillus, Penicillium	Broad-spectrum, surface charge affects efficacy (Matras et al. 2022)
Copper oxide (CuONPs)	Membrane disruption, ROS generation	Candida albicans	Synthesis conditions optimize activity (Imani et al. 2020)
Zinc oxide (ZnONPs)	ROS generation, membrane, and metabolic disruption	Botrytis cinerea, Penicillium expansum	It inhibits spore development and is effective at low concentrations (Zhang et al. 2019a, b)

11.3.1.1.2 Copper Oxide Nanoparticles (CuONPs)

The reaction between CuONPs introduces them into the fungal cell wall and membrane, due to which cells experience morphological alterations and membrane integrity to hinder cell survival (Imani et al. 2020).

11.3.1.1.3 Zinc Oxide Nanoparticles (ZnONPs)

ZnONPs deform fungal hyphae and inhibit the growth of reproductive structures, i.e., conidiophores and conidia, resulting in the death of fungi (He et al. 2011).

11.3.1.2 Generation of Reactive Oxygen Species (ROS)

One possible method for all three kinds of nanoparticle (Ag, CuO, and ZnO) is that they can generate oxidative stress in fungal cells through the generation of ROS. These reactive molecules cause damage to proteins, lipids, and DNA and affect vital cellular functions, resulting in cell death (Zhang et al. 2019a).

11.3.1.3 Inhibition of Cellular Metabolism

11.3.1.3.1 Silver Nanoparticles

AgNPs inhibit the expression of proteins associated with ATP production, disrupting energy metabolism in fungi.

11.3.1.3.2 Zinc Oxide Nanoparticles

Increased antioxidant activity (SOD, CAT) and elevated malondialdehyde (MDA) are indicators that ZnONPs as a whole disrupt physiological equilibrium within the cells. These disrupt metabolic pathways and are indicative of oxidative damage (Zhang et al. 2019a, b).

11.3.1.4 Spectrum of Antifungal Activity

11.3.1.4.1 Silver Nanoparticles

Managed to display a broad-spectrum against a series of plant pathogenic fungi, such as *Fusarium, Botrytis, Aspergillus,* and *Penicillium* species. Its effectiveness is reliant on the age of the particles, surface charge, and the concentration, whereby AgNPs that are positively charged have a higher fungicidal effect (Malik et al. 2024).

11.3.1.4.2 Copper Oxide Nanoparticles

They are efficient in resisting pathogens such as *Candida albicans*; optimum synthetic settings maximize antifungal effects (Imani et al. 2020).

11.3.1.4.3 Zinc Oxide Nanoparticles

It prevents the growth and development of post-harvest pathogens like *Botrytis cinerea* and *Penicillium expansum* so that it can be used in agriculture and food safety (He et al. 2011).

11.3.2 Nanocarriers

Organic nanocarriers (polymeric nanoparticles, nanocapsules, nanoliposomes) encapsulate conventional fungicides or natural antifungal compounds provided within nanocarriers (Table 11.2). These carriers serve to protect from degradation, allow for release, and improve uptake into the pathogen (Fraceto et al. 2016). Chitosan nanoparticles containing azoxystrobin have been tested for improved stability as well as prolonged activity in the field (Kah et al. 2018).

11.3.2.1 Types of Organic Nanocarriers

11.3.2.1.1 Polymeric Nanoparticles

Polymeric nanoparticles are biodegradable polymers like chitosan, polylactic acid (PLA), or poly (lactic-co-glycolic acid) (PLGA). They are protective agents of fungicides or natural compounds so that they are not swept away due to the ultraviolet (UV) light, hydrolysis, and enzyme degradation in the environment. Such nanoparticles facilitate the solubility and stability of fungicides with low water solubility

Table 11.2 Key features of organic nanocarriers in fungicide delivery

Nanocarrier type	Main functions	Example encapsulated fungicides	Notable benefits
Polymeric nanoparticles	Protection, controlled release, improved solubility	Azoxystrobin, natural extracts	Enhanced stability, bioavailability
Nanocapsules	Core-shell protection, targeted release	Carbendazim, essential oils	Reduced dosage, minimized toxicity
Nanoliposomes	Encapsulation, membrane fusion	Amphotericin B, plant extracts	Efficient delivery, reduced side effects

and strengthen bioavailability, as well as allowing a controlled or prolonged release at the site of infection (Fraceto et al. 2016).

11.3.2.1.2 Nanocapsules

Nanocapsules are based on core-shell systems, with the active ingredient encapsulated in a polymeric or lipid shell. The structure offers physical resistance to early degradation and controlled, sustained, and specific release of the fungicide. Nanocapsules also help decrease the dosage and frequency of their usage, thus minimizing pollution to the environment and possible toxicity (Fraceto et al. 2016).

11.3.2.1.3 Nanoliposomes

Nanoliposomes are spherical phospholipid vesicles that can accommodate hydrophilic and hydrophobic fungicides. They safeguard the unstable active components and ingestion of fungal tissues by fusing their membranes or bringing along membranes through endocytosis. This makes the delivery more efficient and decreases off-target effects to boost the overall efficacy of the fungicide (Fraceto et al. 2016).

Nanocarriers protect fungicides against environmental pressures like light, temperature, and enzymes, and this maintains the power of the fungicides during storage and/or use. The active ingredient is released gradually, which keeps effective concentrations over a long period and decreases the necessity of reapplication. Since the penetration and buildup of fungicides within the fungal cells or tissues has enhanced bioavailability and fungicidal efficacy, nanocarriers enhance these qualities. Nanocarriers help to address environmental problems in the management of plant diseases because they reduce leaching, volatilization, and non-target exposure, thereby making the management technique safer and sustaining the environment (Fraceto et al. 2016; Kah et al. 2018).

For example, chitosan nanoparticles containing azaxystrobin are frequently made from chitosan, a natural biopolymer, to encapsulate fungicides such as azaxystrobin. In addition to stabilizing Azoxystrobin and preventing its degradation, this chitosan nanoparticle provides a longer release of the product, resulting in continued field activity and improved disease control. Studies have shown that chitosan-based nanocarriers may reduce the quantity of fungicide required to achieve satisfactory outcomes, which is advantageous for sustainable agricultural practices (Kah et al. 2018).

11.3.3 Nanoemulsions

Nanoemulsions are fine oil-in-water or water-in-oil dispersions that are stabilized at the nanoscale. Their mechanism of action is to increase the solubility, penetration, and adherence of hydrophobic fungicides on the surfaces of plants and thereby

Table 11.3 Key advantages of nanoemulsions in fungicide delivery

Feature	Benefit
Increased solubility	Better dispersion and bioavailability of actives
Enhanced penetration	Improved uptake by plant and fungal cells
Rain fastness	Longer-lasting protection after rainfall
Controlled release	Sustained antifungal activity reduced leaching.
Lower phytotoxicity	Safer for plants, less off-target toxicity
Biofilm eradication	Effective against resistant fungal structures
Plant growth stimulation	Potential for improved crop vigor

enhance their efficacy and rain fastness (Campos et al. 2015). Nanoemulsion formulation of essential oils has been demonstrated to enhance their antifungal activity with lower phytotoxicity (Bocate et al. 2019) (Table 11.3).

11.3.3.1 Mechanisms of Action

11.3.3.1.1 Enhanced Solubility and Bioavailability

Nanoemulsions, which are hydrophobic fungicides or essential oils, are entrapped in the nanodroplets of the nanoemulsions, and this increases the apparent solubility of the hydrophobic sample in an aqueous system by many folds. This increased solubility means there is more efficient and uniform dispersion and availability of the active ingredient on plant surfaces (Prado et al. 2024).

11.3.3.1.2 Improved Penetration and Adherence

The nanoscale of the droplets provides closer contact with plants and better adhesion to plant surfaces, e.g., the cuticle and the stomata. It also potentiates the absorption of the active constituents into the plant and fungus cells, which makes them more effective even when the concentration is low (Mahajan et al. 2021).

11.3.3.1.3 Increased Rain Fastness

Nanoemulsions form a uniform, stretched crust on plant surfaces, which avoids elution of the active constituent by rain or irrigation. Rain fastness is a property that prevents elution, prolongs protection, and decreases frequent replenishment (Kutawa et al. 2021).

11.3.3.1.4 Targeted and Controlled Release

It is possible to tailor nanoemulsions to release fungicide gradually or, in one way or another, control it, which will guarantee a prolonged antifungal activity and reduce the chances of fungicide environmental losses through leaching or volatilization (Kutawa et al. 2021).

11.3.3.1.5 Enhanced Antifungal Activity of Essential Oils

Nanoemulsions of essential oils have demonstrated superior antifungal activity compared to non-nanoemulsions, primarily due to the enhanced efficacy of nanodroplets in contacting fungal cell membranes, leading to disruption of cell wall formation, inhibition of spore germination, and suppression of hyphal growth (Krishnamoorthy et al. 2021). Additionally, nanoemulsions are more effective than free essential oils in eradicating fungal biofilms, which are typically resistant to conventional treatments; for instance, nanoemulsions of *Lippia origanoides* essential oil eliminated up to 89.7% of *Candida albicans* biofilm mass, outperforming the free oil (Prado et al. 2024). These formulations also exhibit lower phytotoxicity than both essential oils and synthetic fungicides due to their ability to deliver lower effective concentrations with improved targeting and controlled release using biocompatible surfactants and carriers (Prado et al. 2024). Beyond disease control, nanoemulsions have been found to stimulate plant root and shoot growth by enhancing nutrient uptake and alleviating pathogen pressure. Moreover, by increasing fungicide efficacy and persistence, nanoemulsions contribute to reduced chemical input requirements, thereby minimizing environmental contamination and the likelihood of resistance development in pathogens (Prado et al. 2024).

11.4 Mechanisms of Action: How Nanofungicides Target Fungal Pathogens

Multiple mechanisms are used by Nanofungicides to control fungal pathogens.

1. Metallic nanoparticles can get attached to the fungal cell walls and also the fungal cell membranes, causing disruption of the structural integrity and also allowing increases in permeability, which finally results in the lysis of the fungal cell (Rai and Ingle 2012), as shown in Fig. 11.1.
2. Nanoparticles can produce ROS to damage cellular components, like proteins, lipids, and DNA, and to inhibit fungal growth (Jo et al. 2009).
3. Targeted delivery: Due to their functionalization, nanocarriers of active ingredients can be designed to recognize and bind to specific fungal biomarkers, which allows for precise delivery of the active ingredient at the site of infection (Fraceto et al. 2016).
4. Sustained release: The fungicides can be controlled through nanocarriers and nanoemulsions, maintaining effective concentrations for longer and seeking less frequent applications (Kah et al. 2018).

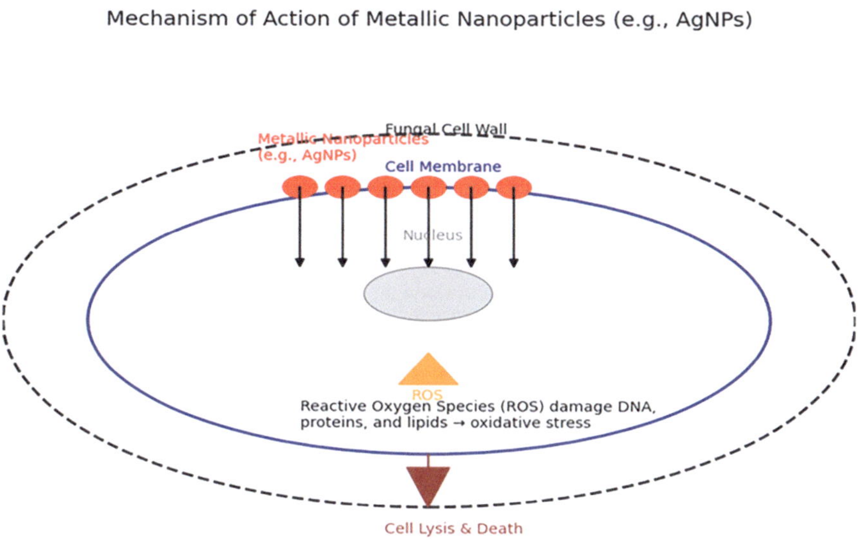

Fig. 11.1 Diagram of the mechanism of action of metallic nanoparticles (e.g., silver nanoparticles) on fungal pathogens. Considerable steps involve adherence of nanoparticles to the cell membrane, alteration of cellular membranes that lead to the formation of reactive oxygen species (ROS), and destruction of cellular organelles and terms with lysis

11.5 Properties of Nanofungicides

11.5.1 Size, Surface Area, and Reactivity

The small size of these materials gives them a high surface-to-volume ratio, thus making them more reactive and interacting with fungal cells (Kah et al. 2018). For instance, copper nanoparticles with diameters less than 50 nm have shown superior antifungal capacity compared to large particles due to the enlarged surface area, resulting in increased interaction with pathogens and the capability to disrupt cell processes better (Rai and Ingle 2012).

11.5.2 Target Specificity: Selective Action on Fungal Pathogens

A specific target for nanofungicides is one of the most inevitable benefits. By ligand or antibody modification of the surface of nanoparticles or nanocarriers that target unique fungal cell wall features, selective binding and delivery can be achieved (Fraceto et al. 2016). As a consequence, the impact on beneficial soil microbes and non-target organisms is reduced, which is a major limitation of conventional broad-spectrum fungicides (Chen and Yada 2011).

11.5.3 Potential for Reduced Chemical Load in Agriculture

Nanofungicides have proved to permit more effective use of active ingredients, often with lower doses needed to match, and often better than, control of fungal diseases of the traditional formulations (Kah et al. 2018). This allows controlled release of the fungicides from the nanocarriers, which maintain effective concentrations for extended periods with reduced frequency of applications and thus the total quantity of chemicals injected into the environment. For example, an attempt was made on the nanoencapsulated tricyclazole in which showed a 60% reduction in environmental leaching when compared to the conventional formulation (Fraceto et al. 2016). Not only that, but this also cuts their production costs and reduces the chance of the chemical residues finding their way into our food and environment.

Commonly, integration of nanotechnology in crop protection can be a paradigm change towards more sustainable, efficient, and environmentally friendly agricultural practices. Fungicides based on nanotechnology have superior chemical load, target specificity, and efficacy in comparison to conventional fungicides that address many of the shortcomings of the latter. The long-term environmental fate, toxicity, and regulatory aspects of these novel materials have not been further researched (Kah et al. 2018). Benefits from the next generation of smart crop protection products are coming from the advancement of biodegradable and stimuli-responsive nanocarriers (Table 11.4).

Table 11.4 Comparison of the traditional fungicides with the nanofungicides with respect to the main parameters of agriculture and the environment

Feature	Conventional fungicides	Nanofungicides
Solubility	Often poorly soluble; require surfactants or high concentrations	Enhanced solubility due to the nanoscale formulation
Target specificity	Broad-spectrum; may harm beneficial microorganisms	Can be functionalized for the selective targeting of pathogens
Application frequency	Require frequent applications due to rapid degradation and runoff	Sustained release formulations reduce application frequency
Environmental impact	High chemical load; risk of residue in soil and water	Reduced dosage and controlled release minimize environmental contamination
Stability and shelf life	Prone to degradation under UV light and moisture	Nanocarriers protect active ingredients and provide a longer shelf life
Efficacy	Moderate efficacy; may lead to resistance development over time	Enhanced penetration and interaction with fungal cells; higher antifungal activity
Phytotoxicity	May cause plant damage at high doses	Lower phytotoxicity due to targeted delivery and reduced dosage
Cost effectiveness	High input cost due to repeated use	Potentially more cost-effective due to reduced usage and improved performance

11.6 Components of the Plant Microbiome

In the case of plants, the plant microbiome is made up of very diverse microbial communities that interact with the plant, influencing its health and growth, and the plant's resilience. The interactions take place on separate niches, e.g., the rhizosphere (root zone) and phyllosphere (aerial plant surface), and have bacteria, fungi, viruses, and other microorganisms involved. This is a structured overview of the plant microbiome's components and functions.

The plant microbiome is the total collection of all microorganisms from plants. Bacteria, fungi, viruses, archaea, and protists are all involved in different ways to affect plant health and ecosystem functioning (Berg et al. 2020).

11.6.1 Bacteria

Within the plant microbiome, bacteria are most abundant and diverse. The total population of caulis, rosettes, and roots is represented by caulis, which inhabits all plant compartments (roots, stems, leaves, flowers, and seeds). Pseudomonas, Bacillus, Rhizobium, and the genus Azospirillum are all common bacterial genera. They can fix atmospheric nitrogen, solubilize phosphorus, and produce phytohormones as well as antibiotics and compete with the plants for resources for the support of plant growth (Compant et al. 2019).

11.6.2 Fungi

There are both beneficial and pathogenic fungi in the plant microbiome. Mycorrhizal fungi (e.g., Glomus, Rhizophagus) are beneficial fungi that take advantage of symbiotic associations with plant roots in return for carbohydrates for their support and treatment of increased nutrient uptake (e.g., phosphorus). Endophytic fungi that reside in the plant tissue do not cause disease, but can provide tolerance to abiotic stress to the plant (Yousef et al. 2025). Broadly, pathogenic fungi (e.g., Fusarium, Botrytis) can cause significant crop losses and are tempered by many of the microbes (Vandenkoornhuyse et al. 2015).

11.6.3 Viruses

Less studied but becoming an increasingly recognized microbiome member, plant-associated viruses are the focus of this dissertation. While a majority of viruses can be pathogenic, some can even modify the plant defenses or enhance certain advantageous responses, for example, increased drought tolerance, or to other pathogens (Roossinck 2015). The other microbiome members also interact with viruses, and the effects on community structure and function are influenced.

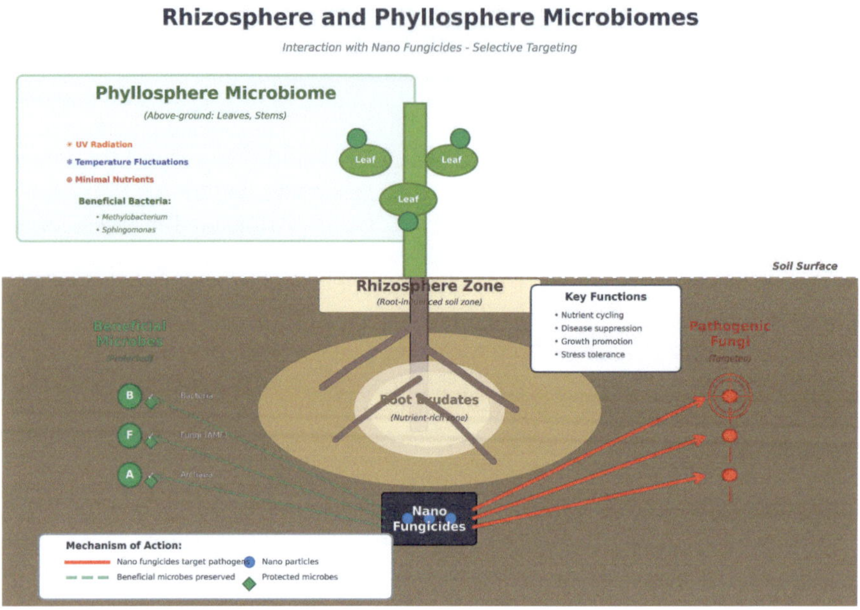

Fig. 11.2 Interaction between the rhizosphere microbiome and nanofungicides, showing selective targeting of pathogenic fungi while preserving beneficial microbial communities

11.6.4 Rhizosphere Microbiome

The rhizosphere is the zone directly influenced by root exudates and root-associated microorganisms in the narrow zone of soil. It is a hotspot for microbial activity with abundant, diverse, functionally and often more abundant populations of bacteria, fungi, or archaea compared to the bulk soil (Philippot et al. 2013). Specifically, rhizosphere microbes are responsible for important nutrient cycling, disease suppression, and plant growth promotion functions. Secreted specific exudates by plants actively shape the rhizosphere microbiome with recruits of beneficial microbes and repellents of pathogens (Sasse et al. 2018) (Fig. 11.2).

11.6.5 Phyllosphere Microbiome

The phyllosphere is the above-ground parts of plants, primarily the leaves, which are colonized by a multitude of bacteria, fungi, and yeasts. It is in a challenging phyllosphere through exposure to UV radiation, fluctuations in temperature, and very limited nutrients. Some bacteria (e.g., Methylobacterium, Sphingomonas) and some fungi are accordingly adapted to exist here. Some of these microbes can promote plant growth by producing growth-promoting compounds, improve stress tolerance, as well as provide a barrier against pathogens (Vorholt 2012).

11.7 Microbial Interactions with Plants

The cooperative relationships in which both microbes and plants benefit are known as mutualistic symbiosis. Typical examples include nitrogen-fixing bacteria like Rhizobium, which reside in legume root nodules and transform atmospheric nitrogen into a form that the plant can use, and mycorrhizal fungi, which improve plant nutrient uptake (particularly phosphorus) in return for carbohydrates. These interactions give bacteria nourishment and a conducive habitat for survival while also enhancing plant growth, stress tolerance, and ecosystem health overall.

11.7.1 Mycorrhizal Fungi

One of the most important mutualist relationships that plants have is with mycorrhizal (myco = fungus, rhizal = root) fungi. Arbuscular mycorrhizal fungi (AMF) invade the plant root cell in exchange for plant-derived carbohydrates to help the plant take up water and nutrients, especially phosphorus and micronutrients they cannot take up by themselves. It can benefit the plant with enhanced growth and good soil structure, and increased resistance to pathogens and other environmental stresses (Smith and Read 2010).

11.7.1.1 The Mechanisms and Interactions of Arbuscular Mycorrhizal Fungi (AMF)

Arbuscular mycorrhizal fungi (AMF) are a specific group of mycorrhizal fungi that penetrate the cortical cells of plant roots, forming intricate structures called arbuscular. These arbuscules are the primary sites of nutrient exchange between the fungus and the plant. In this symbiotic relationship:

11.7.1.1.1 Nutrient Uptake

AMF hyphae extend far into the soil, increasing the root's effective surface area. This allows the plant to access water and essential nutrients, especially phosphorus, zinc, and other micronutrients that are otherwise difficult for roots to absorb directly (Fig. 11.3).

11.7.1.1.2 Carbohydrate Exchange

In exchange, the plant furnishes the fungi with carbohydrates (principally sugars) synthesized by photosynthesis. Fungi do not have the means of photosynthetically producing their food; they rely on these carbohydrates.

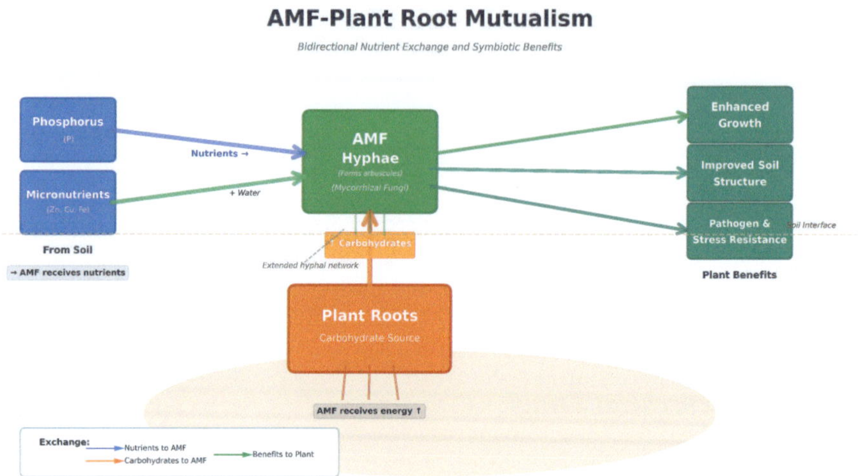

Fig. 11.3 The diagram illustrates the mutualistic relationship between AMF and plant roots, showing the exchange of nutrients and carbohydrates and the resulting benefits for the plant

11.7.1.1.3 Enhanced Growth

The improved nutrient and water uptake often results in increased plant growth and vigor, especially in nutrient-poor soils.

11.7.1.1.4 Additional Benefits to Plants

Beyond nutrient exchange, AMF provide several other advantages to their plant hosts:

11.7.1.1.5 Structure of Soil

Due to the ubiquitous nature of the fungal hyphae, soil particles are attached in clumps to increase their aggregation and stability. This increases soil aeration, water retention, and erosion.

11.7.1.1.6 Increased Resistance to Pathogens

AMF can enhance pathogen resistance in plants because a protective barrier is formed around the root of the plants by AMF, which increases their defense mechanism.

11.7.1.1.7 Resistance to Environmental Stresses

Plants that are related to the AMF have also been found to be more resistant to drought, salinity, and toxicity of heavy metals because the fungi play an indirect buffering role against these stresses.

11.7.1.1.8 Ecological and Agricultural Significance

Mycorrhizal associations are fundamental to ecosystem health, playing a crucial role in ecosystem productivity and biodiversity by influencing plant community structure and nutrient cycling. In agricultural settings, arbuscular mycorrhizal fungi (AMF) inoculation benefits many crops such as wheat, maize, and legumes by enhancing nutrient uptake, reducing reliance on chemical fertilizers, and sustainably improving crop yields.

11.7.2 Nitrogen-Fixing Bacteria

Rhizobium species in legumes and Frankia in actinorhizal plants are nitrogen-fixing bacteria that transform atmospheric nitrogen to ammonia (a form that can be taken up by the plant) within nodules formed on the plant roots. It helps to reduce the use of synthetic nitrogen fertilizers and assists in plant growth in nutrient-poor soils (Oldroyd and Dixon 2014).

11.8 Pathogen Suppression: Beneficial Microbes That Protect Plants from Disease

Plants coexist with many microorganisms, which play crucial roles in protecting plants from diseases. Beneficial bacteria and fungi are important components of the plant microbiome that operate as biocontrol agents, suppressing harmful pathogens through a variety of complimentary ways. These natural defenders help maintain plant health and reduce the reliance on chemical pesticides.

11.8.1 Mechanisms of Pathogen Suppression

Biocontrol agents are some of the members of the plant microbiome that suppress plant pathogens through multiple mechanisms.

11.8.1.1 Competition for Nutrients and Space

Beneficial microbes colonize the root surface and surrounding soil (the rhizosphere), effectively occupying ecological niches that might otherwise be available to pathogens. By rapidly consuming essential nutrients such as iron, carbon sources, and minerals, these microbes limit the resources available to pathogenic organisms. This competitive exclusion reduces the ability of pathogens to establish themselves and cause infection.

For example, Rhizobacteria such as Pseudomonas species efficiently colonize root surfaces, creating a physical and nutritional barrier against invading pathogens.

11.8.1.2 Antibiosis: Production of Antimicrobial Compounds

Many beneficial microbes produce many bioactive compounds that directly inhibit or kill pathogens. These include:

11.8.1.2.1 Antibiotics

Small molecules such as phenazines, pyoluteorin, and 2,4-diacetylphloroglucinol (DAPG) produced by bacteria like Pseudomonas fluorescens can suppress or kill pathogenic bacteria and fungi.

11.8.1.2.2 Siderophores

These are iron-chelating compounds secreted by microbes to scavenge iron from the environment. By sequestering iron, siderophores deprive pathogens of this essential nutrient, limiting their growth.

11.8.1.2.3 Lytic Enzymes

Some microbes secrete enzymes like chitinases and glucanases that degrade the cell walls of fungal pathogens.

11.8.1.3 Induced Systemic Resistance (ISR)

Certain beneficial bacteria and fungi can “prime” the plant’s immune system, a phenomenon known as Induced Systemic Resistance (ISR). Unlike direct antagonism, ISR enhances the plant’s ability to respond more quickly and robustly to pathogen attacks by activating defense signaling pathways.

Induced systemic resistance (ISR) typically involves signaling molecules such as jasmonic acid and ethylene, which activate the expression of defense-related genes throughout the plant. For example, *Pseudomonas fluorescens* not only produces antibiotics and siderophores but also stimulates ISR, thereby enhancing resistance against a wide range of pathogens. Similarly, *Trichoderma* species, widely recognized as fungal biocontrol agents, parasitize pathogenic fungi by coiling around their hyphae and secreting cell wall-degrading enzymes, while also inducing ISR in the host plant to bolster its defense mechanisms (Mahmoud et al. 2021).

Biocontrol agents exert synergistic impacts by employing multiple mechanisms simultaneously, which enhances their effectiveness in suppressing plant diseases. Their use supports sustainable agriculture by reducing reliance on chemical fungicides and bactericides, thereby promoting more environmentally friendly farming practices. In practical applications, strains of *Trichoderma* and *Pseudomonas* are widely used as commercial biofungicides to control soil-borne diseases in crops like tomatoes, cucumbers, and cereals.

11.8.2 Impact of Microbial Communities on Plant Growth, Stress Resistance, and Nutrient Cycling

Plant ecosystems and plant health depend on microbial communities. They range from facilitating plant growth and enhancing resistance to stress to stimulating the process of nutrient cycling that supports terrestrial and aquatic ecosystems.

11.8.2.1 Plant Growth Promotion

Rhizosphere bacteria and fungi produce phytohormones (e.g., auxins, gibberellins, cytokinins, and a variety of others) that promote root and shoot development.

11.8.2.1.1 Phytohormone Production

Rhizosphere bacteria and fungi synthesize phytohormones such as auxins, gibberellins, and cytokinins, which stimulate root and shoot development, leading to increased plant biomass and vigor (Xiao et al. 2022).

11.8.2.1.2 Increased Nutrient Availability

Microbial communities increase the availability of needed nutrients to the plants either through the secretion of chelators (siderophores) and other organic acids to solubilize soil nutrients or by pathways that enhance nutrient availability (Bernardin et al. 2024a).

11.8.2.1.3 Enhanced Plant Phenotype

The direct impact of microbial functions on plant growth has been expounded because functional meta-compositions of microbial communities could promote significant augmentation of plant biomass, contrasted to leaf size and nutrient content (Bernardin et al. 2024b).

11.8.2.2 Stress Resistance

Microbes help plants fight against abiotic stresses. Plants can resist abiotic stresses like drought, salinity, and heavy metals by the production of osmoprotectants, modifying the levels of hormones, or enhancing the levels of antioxidants (Xu et al. 2018).

11.8.2.2.1 Abiotic Stress Mitigation

Microbes help plants withstand abiotic stresses such as drought, salinity, and heavy metal toxicity by producing osmoprotectants, modulating plant hormone levels, and enhancing antioxidant defenses (Ge et al. 2023).

11.8.2.2.2 Recruitment of Beneficial Microbes

Under stress, plants can recruit specific soil microorganisms to the rhizosphere, forming stable communities that enhance plant tolerance through metabolic exchange and interdependent interactions (Ge et al. 2023).

11.8.2.2.3 Effects of Legacy

The presence of beneficial microbes following stress events can create a "Defense Biome," which not only aids the current plant but can also enhance the fitness of offspring by establishing a legacy of increased stress resistance.

11.8.2.3 Nutrient Cycling

By breaking down organic matter, mineralizing nutrients, and solubilizing infusible forms of phosphorus and other minerals, microbes enable plants to access these nutrients (Philippot et al. 2013), thus making the availability of these nutrients available for the plants.

Microbial communities are the primary drivers of nutrient cycling, orchestrating the transformation and movement of essential elements through various metabolic pathways:

11.8.2.3.1 Decomposition and Mineralization

Microbes play a crucial role in organic matter breakdown by decomposing plant litter and organic residues, transforming complex organic compounds into simpler forms and releasing essential nutrients like nitrogen (N), phosphorus (P), and sulfur (S) back into the soil for plant uptake. Through the process of mineralization, these microbes further convert organic nutrients into inorganic forms such as ammonium and phosphate, which are directly available to plants (Vietorisz et al. 2024).

11.8.2.3.2 Nitrogen Cycling

Certain bacteria, such as *Rhizobium* and *Azotobacter*, perform nitrogen fixation by converting atmospheric nitrogen into ammonia, supplying plants with a vital source of nitrogen (Vietorisz et al. 2024). Furthermore, through processes like nitrification, which converts ammonia into nitrate, and denitrification, which converts nitrate to gaseous nitrogen forms, microbial communities play a crucial role in the nitrogen cycle by controlling the availability and loss of nitrogen within ecosystems (Vietorisz et al. 2024).

11.8.2.3.3 Phosphorus Solubilization

Phosphate-solubilizing bacteria and fungi enhance phosphorus availability by secreting organic acids and enzymes that convert insoluble phosphorus compounds into soluble forms readily absorbable by plants (Vietorisz et al. 2024). Additionally, ectomycorrhizal fungi and other functional guilds contribute significantly to phosphorus cycling by producing oxidoreductase enzymes that facilitate the oxidation and reduction of phosphate in the soil (Vietorisz et al. 2024).

11.8.2.3.4 Sulfur and Other Elemental Cycles

Microbes play a vital role in sulfur cycling by mediating the reduction and oxidation of sulfur compounds, thereby regulating sulfur availability in soils and aquatic ecosystems (Vietorisz et al. 2024). This function, along with the broader functional diversity of microbial taxa, ensures the maintenance of all major nutrient cycles—including carbon, nitrogen, phosphorus, and sulfur ultimately supporting ecosystem productivity and resilience (Vietorisz et al. 2024).

11.8.2.3.5 Response to Environmental Change

Microbial communities can rapidly adjust their metabolic activities and composition in response to changes in temperature, moisture, pH, and nutrient availability, maintaining ecosystem stability and nutrient cycling under varying conditions (Vietorisz et al. 2024).

The plant microbiome is a diversified active community of bacteria, fungi, viruses, and other microorganisms residing in different plant niches. For example, through mutualistic relationships, pathogen suppression, and participation in nutrient cycling and stress resistance, the microbes are essential things in plant health and productivity. After highlighting the importance of their beneficial plant–microbe interactions, the authors offer promising strategies for harnessing these beneficial plant–microbe interactions to improve sustainable agriculture by reducing reliance on chemical inputs and improving crop resilience to environmental challenges.

11.9 Disruption of Microbial Balance

Bacteria's quick responses to their surroundings lead to rapid microbial transformation. Nature embodies the genetic composition of plants and their influence on their relationships with microbes, which can enhance nutrient absorption, disease resistance, and overall plant health. The environment significantly affects which bacteria are present and their interactions with plants (Astapati and Nath 2023). Agricultural methods that include biocontrol chemicals have created a crucial information gap regarding the consequences of introducing non-native compounds into the natural ecosystems of plants. Ranging from minor to significant impacts, various agents have been shown to affect soil and plant microbial communities, typically depending on factors such as soil depth, time since application, and the presence of organic amendments (Peterson et al. 2020). Widely used as a biocontrol fungus, Trichoderma species influence microbial populations differently depending on plant genotype and environmental conditions. Recent research on Trichoderma species has revealed that, while their use may have minimal impact on overall microbial diversity, they markedly alter the composition of microbial communities, particularly concerning bacteria and fungi (Moussa and Iasur Kruh 2025).

Nanofungicides provide novel approaches to crop protection, but in order to avoid detrimental impacts on microbial diversity and soil health, their potential to change the balance between beneficial and dangerous bacteria necessitates careful evaluation and additional research.

11.9.1 Effectiveness Against Fungal Infections

The substantial impact that fungal infections can have on agricultural productivity is highlighted by Chauhan et al. (2024). As an option for effectively managing these infections, bio-based nanofungicides have the potential to improve crop health and productivity. There are several modes of action of nanofungicides that improve the delivery of agrochemicals and micronutrients.

11.9.2 Specific Examples of Altered Microbial Populations

The use of zinc nanoparticles changed the composition of the microbial community in tea plants. With different concentrations of nanozinc, the analysis of the microbial community indicated that the use of zinc oxide nanoparticles did not significantly affect the diversity of the bacterial community in the leaves of tea plants. On the contrary, 100 mg/L of nanozinc reduced the diversity of fungi present in the plant (Chen et al. 2024). Thus, it was proven that the use of nanoparticles affects the presence of Marinococcus and Microbacterium, and this is consistent with the results of Dimkpa et al. (2013), where the results of Aconitum showed that zinc oxide nanoparticles affect the metabolism and the structure of the microbial community. A decrease in the abundance of pathogens and an increase in Bacillus were determined, thus improving growth parameters and enhancing gas exchange. In another study the microbial diversity after 20, and 40 days and found that AgNPs affect bacterial diversity, different concentrations change the functional bacterial diversity (Pallavi et al. 2016). Li et al. (2024) investigate the use of biogenic zinc oxide nanoparticles (ZnONPs) and *Pennisetum giganteum* in addressing soil cadmium contamination in plants. The findings reveal that ZnONPs considerably improve plant tolerance to Cd accumulation, boost antioxidative response, and increase soil enzyme activity. The study also looks at the bacterial community structure in plant rhizo-compartments, which is influenced by changes in soil enzymes caused by ZnONPs. Additionally, the study looks at how ZnO NP-induced changes in soil enzyme activity affect the bacterial community structure in plant rhizo-compartments. At low quantities, there were notable alterations in both the overall plant performance and the bacterial community structure associated with the rhizo-compartment (Fig. 11.4).

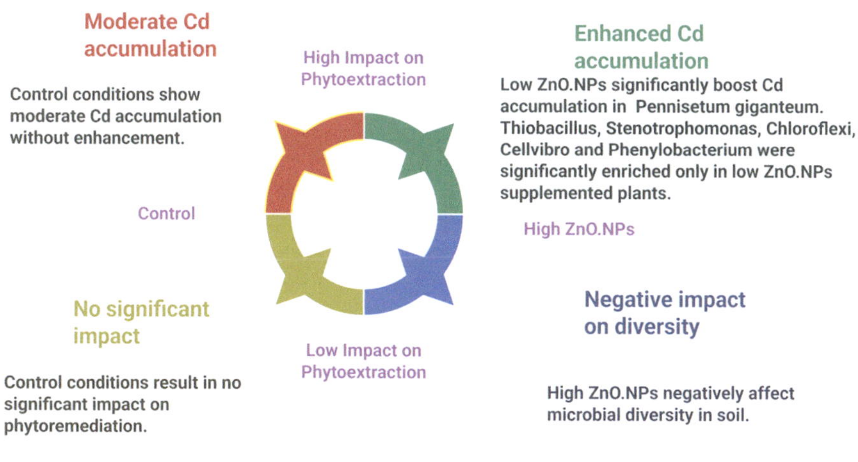

Fig. 11.4 Impact of ZnONPs on phytoremediation and microbial diversity

11.10 Selective Toxicity: A Double-Edged Sword

Design and properties of nanofungicides: Considerable work has been done to create nanofungicides with specific dimensions, surface properties, and shape. Improving their efficiency in delivering nutrients and managing pathogen infection depends on these attributes.

Toxicity considerations are an essential component of the application of nanofungicides and also address their toxicity. Knowing the toxicity levels helps in assessing the environmental and crop safety of different compounds (Wahab et al. 2024). Apart from helping in fungal disease management, nanofungicides are also being investigated for their use as nanosensors and in gene transfer. This increases their scope of use in agriculture and hence they are flexible tools for improving crop management (Kutawa et al. 2021). Their interactions with plant surfaces and fungal diseases. By enhancing these properties, adhesion and penetration can be improved and hence control pathogens (Łaźniewska et al. 2012).

The shape and size of nanofungicides are important for their delivery and absorption. While specific shapes may improve their ability to bind to fungal cells, thereby enhancing their efficacy, smaller nanoparticles can more easily pass through plant tissues and reach targeted infections (Ly et al. 2024).

Nanofungicides aim to help agrochemicals and micronutrients target more easily. Through this specific strategy, the active ingredients are released at the right time and place, thereby improving their efficacy in treating fungal diseases and supporting plant health (Karnawat et al. 2021). Controlled release mechanisms: Often, the design of nanofungicides includes controlled release systems to control the release of the active compound. This function reduces the environmental impact by helping to maintain effective fungicide concentrations over time, thereby reducing the demand for regular applications (Shang et al. 2019). Nanotechnology has improved pesticide application, facilitating safer management and enhancing efficacy. Researchers have created nanoparticulate systems using ionic gelation between chitosan and gum arabic for insulin loading. The insulin release profile in phosphate-buffered solutions (pH 6.5 and pH 7.2) differed from that in an acidic medium (pH 1.2). The enhanced solubility of chitosan in an acidic environment and the augmented swelling of arabic gum chains at pH 6.5 led to a decreased insulin release at that pH level. The nisin encapsulated tripolymeric nanoformulation comprising chitosan, sodium alginate, and pluronic F68 exhibited an initial burst release during the first week, followed by a steady release in the subsequent week. Guar gum was effectively acrylated to produce guar-gum-grafted polyacrylic acid, which increased stimulus response, reduced viscosity, and enhanced thermal stability (Kumar et al. 2022; Mosa et al. 2022). By enhancing biocompatibility, nanofungicides become more compatible with plants and their environments. This biocompatibility decreases the risk of harm to non-target species and encourages more environmentally friendly farming practices (Punniyakotti et al. 2024).

11.10.1 Targeting of Fungal Pathogens While Sparing Beneficial Microbes

Nano-biofungicides and bio-nanofungicides (Fig. 11.5) represent innovative approaches in agricultural pest management, offering distinct mechanisms and advantages. Nano-biofungicides are developed by encapsulating synthetic fungicides such as pyraclostrobin and epoxiconazole. This nanoencapsulation improves the stability of the active ingredients, enabling controlled release and reducing the frequency of application. Additionally, their enhanced surface-to-volume ratio facilitates better absorption by plant tissues while minimizing environmental impact. On the other hand, bio-nanofungicides are metallic nanoparticles synthesized through biogenic methods involving microorganisms, plant extracts, or enzymes (Peterson et al. 2020). The bio-corona, a dynamic, multilayered structure made of proteins, lipids, carbohydrates, or nucleic acids, surrounds the metallic core of these nanoparticles. In interactions with phytopathogens and the environment, the bio-corona is essentialt (Limo et al. 2018).

The mechanisms of action differ between the two types. For nano-biofungicides, the fungicidal activity is governed by the synthetic active ingredient encapsulated within the nanoparticles. In contrast, bio-nanofungicides derive their efficacy from the synergistic interactions between the metallic core and the bio-corona. Furthermore, the bio-corona in bio-nanofungicides enhances surface adhesion and compatibility with various leaf types, making them suitable for a wide range of crops. Both types of nanofungicides demonstrate promising capabilities, but understanding their structural properties and mechanisms of action is key to maximizing

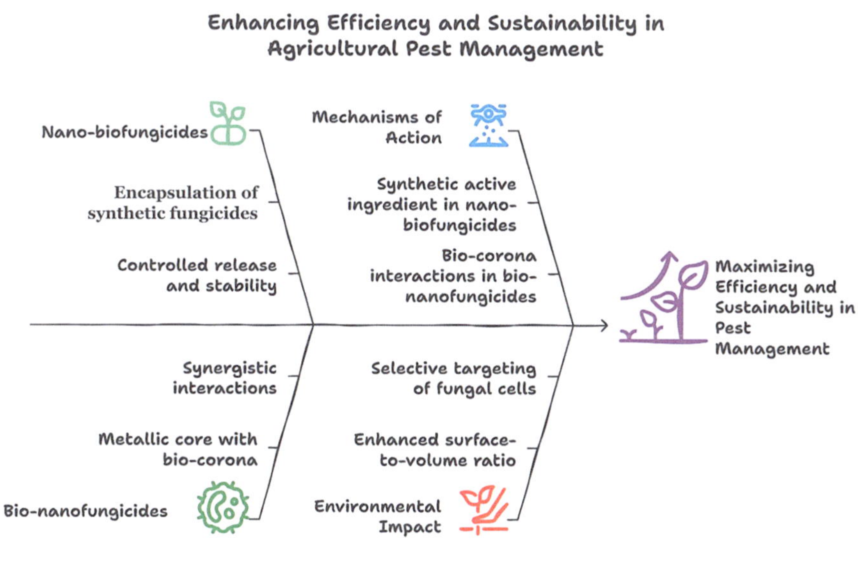

Fig. 11.5 Comparison between nano-biofungicides and bio-nanofungicides and their modes of action

their efficiency and sustainability in agricultural applications (Dávila Costa and Romero 2025).

Through their special modes of action, nanoparticles (NPs) present a viable method to target fungal infections while protecting useful bacteria. These systems let NPs target fungal cells only, therefore sparing non-target species from damage. NPs' uniqueness stems from their capacity to engage with fungal cell structures and processes unique from those of helpful bacteria. These are the main processes by which nanoparticles accomplish selective targeting.

11.10.2 Mechanisms of Action

Nanoparticles have the physical ability to break fungal cell walls and membranes, therefore causing cell lysis. This is accomplished by producing reactive oxygen species (ROS) and creating pores in the cell membrane, which are more successful against the strong cell walls of fungi than against other bacteria (Choudhary et al. 2024).

Nanoparticles can be designed to specifically reach the location of infection with antifungal drugs. This focused delivery lowers possible negative effects and minimizes non-target microorganism exposure. By means of a regulated release made possible by the encapsulation of antifungal medications inside NPs, the therapeutic efficacy against fungal infections is improved (Tarannum et al. 2024).

Some nanoparticles can disrupt quorum sensing, a vital communication mechanism for the production of fungal biofilm and virulence. By interfering with these signals, NPs can prevent the growth and spread of fungal infections without endangering beneficial bacteria that do not depend on similar communication systems (Cruz et al. 2024).

11.10.3 Safety and Selectiveness

Using microbial entities during the production of NPs guarantees their eco-friendliness and minimal toxicity to non-target species. This technique also improves the stability and biocompatibility of NPs, therefore making them safer for use in many surroundings, including agricultural situations (Choudhary et al. 2024; Cruz et al. 2024). Although nanoparticles show a good way to treat fungal infections, their long-term safety and environmental effects raise questions. More investigation is required to completely comprehend the interactions between NPs and non-target organisms, ensuring that their use does not unintentionally damage beneficial bacteria or cause resistance development in fungal populations (Slavin and Bach 2022).

11.10.4 Case Studies on Microbial Survival Post-Nanofungicide Application

Jayasree and Rajeshkumar (2024) describe how silver and gold nanoparticles fight fungal pathogens by damaging cell membranes, producing reactive oxygen species, and disrupting cellular functions. Their selectivity focuses on fungal cells, minimizing damage to helpful microbes while efficiently addressing infections. Pattanayak et al. (2023) reported that Nanotechnology has garnered considerable interest across multiple domains, including agriculture, the food industry, medicine, and cosmetics. Nanomaterials are consistently utilized throughout multiple sectors, including agribusiness, food production, pharmaceuticals, personal care products, and biotechnology. The interaction between nanoparticles and bacteria is essential in the treatment of numerous diseases, including the use of antimicrobial medicines. The creation and utilization of engineered agro-nanomaterials have been employed to mitigate the degradation of food crops caused by bacteria, fungi, viruses, and other pathogenic agents. The inhibitory capacity of nanoparticles on microbial proliferation entails the release and interaction of metal ions with cellular components, potentially resulting in the generation of reactive oxygen species, damage to cell walls and DNA, and disruption of the cell cycle. The essay examines many plant pathogens, their processes of phyto-pathogenesis, and their function in phytopathogen suppression.

Antimicrobial agents play a pivotal role in treating various diseases by preventing phytopathogen growth in host plants and early detection of pathogenic symptoms. Nanomaterials are being researched in agriculture for their potential to inhibit plant–microbe interactions, affecting the release and interaction of metal ions with cell components (Rajwade et al. 2020; Mgadi et al. 2024). This inhibition occurs through reactive oxygen species production, cell wall and DNA damage, and cell-cycle arrest. Mgadi et al. (2024) discussed mechanisms of phyto-pathogenesis, and the responses of nanoparticle interactions with different microbes, highlighting their role in phytopathogen suppression. The development and application of engineered agro-nanomaterials are crucial in overcoming these challenges and preserving food-crop produce.

11.11 Long-Term Ecological Consequences

11.11.1 Possible Shifts in Microbial Community Structure over Time

In comparison to the negative control, four general AgNP exposure reactions were noted: (1) intolerant, (2) impacted but recovering, (3) tolerant, and (4) stimulated phylotypes. Relationships between bacterial phylotypes and cell activity indicators revealed that rare bacterial phylotypes triggered by AgNPs did not seem to contribute much to cell activity, while tolerant and recovering bacteria were the most productive members of the community. Within five days of exposure, natural bacterial communities exhibited comparable activity levels to the negative control and were able to withstand single, low-level AgNP doses; however, the makeup of the bacterial communities differed from that of the control (Das et al. 2012). The environment will unavoidably be exposed to these materials as the availability of products based on nanoparticles increases. Therefore, wastewater, garbage, and other human-caused sources may introduce engineered nanoparticles (ENPs) into the soil environment; one type of ENP that may pose a risk to the environment is metallic oxide nanoparticles. It is crucial to investigate how ENP exposure affects soil bacteria since they play a significant role in providing services to both the ecosystem and humans (Hegde et al. 2016). Bacterial community activity, composition, and size were measured after exposure to nanosized (<50 nm) copper oxide (CuO) and magnetite (Fe3O4) particles in order to assess these impacts. Two types of dirt were investigated. Under two ENP concentrations (1%, 0.1%), the first type is a sandy loam (Bet-Dagan), and the second is a sandy clay loam (Yatir). The findings show that, in comparison to Yatir soil, the bacterial community in Bet-Dagan soil was more vulnerable to alteration as a result of exposure to certain ENPs. More precisely, CuO significantly impacted the size, composition, oxidative potential, and hydrolytic activity of the bacterial communities in Bet-Dagan soil. Although 1% CuO exposure significantly reduced oxidative potential and altered community composition, there were few other effects observed in the Yatir soil. Fe3O4 had no effect on the bacterial community in Yatir soil, but it altered the hydrolytic activity and composition of the bacterial community in Bet-Dagan soil. Additionally, once 0.1% CuO was added, the abundance of bacteria attributed to OTUs from the Bacilli class fell in Bet-Dagan soil, whereas the abundance increased with 1% CuO in Yatir soil. CuO addition to soil has a detrimental effect on other significant soil bacterial groupings, such as Rhizobiales and Sphingobacteriaceae. These findings suggest that soil habitats may be harmed by both ENPs. Additionally, it has been proposed that the organic matter and clay fraction in various soils interact with ENPs to lessen their toxicity (Frenk et al. 2013; Miao et al. 2019).

11.11.2 Impact on Microbial Diversity and Functional Potential

Microbial populations encounter engineered nanomaterials at varying concentrations in environmental matrices, including water, sewage, soil, and sediments. The toxicity of engineered nanomaterials on microbial communities remains inadequately comprehended; however, potential mechanisms encompass cell membrane damage, protein oxidation, genotoxicity, interaction with the respiratory chain, local proton depletion, and the production of reactive oxygen species (ROS) and/or apoptosis (Mgadi et al. 2024). The notion of engineered nanomaterials (ENM) toxicity to microorganisms mostly focuses on cellular processes, although there is a paucity of evidence at the molecular level. The interaction of engineered nanomaterials with Gram-positive and Gram-negative bacteria varies owing to the distinct compositions of their cell walls. This results in diverse interactions and varying degrees of toxicity. The primary problem is their toxicity to microorganisms that are advantageous to plants and those participating in nutrient mineralization, nitrogen cycle, and organic carbon decomposition in soil (Rajput et al. 2020). Engineered nanomaterials (ENMs), including TiO2, ZnO, CuO, Ag, carbon nanotubes (CNTs), and fullerenes, have shown a detrimental effect on soil microbial populations and plant growth-promoting microbial consortia, such as rhizobacteria and mycorrhiza. In some instances, the engineered nanomaterials (ENMs) dissolve in soil solution or saltwater, impacting the microbial population via the established mechanism of metal ion toxicity to cells. Efficient, dependable, and economical analytical techniques for identifying ENM-microbial interactions are essential for comprehending the mechanisms of interaction and ensuring the safe utilization and oversight of nanomaterial in the environment. Nonetheless, the majority of these techniques are labor-intensive, economically inefficient, and need specialized equipment and proficient personnel. Real-time monitoring of microbial interactions in soil and water may be infeasible owing to environmental variables affecting the bioavailability of engineered nanomaterial. Research on the toxicological effects of AgNPs on several bacteria has been conducted in pure culture medium. These bacteria include nitrifying bacteria (Nitrobacter and Silver nanoparticles are a popular antibacterial nanomaterial in consumer and medicinal goods. However, AgNPs may harm biogeochemical loop microorganisms in the environment. Numerous studies show that AgNPs damage microorganism cell membranes (Rajput et al. 2020; Hegde et al. 2016). Data in Table 11.5 show that the toxicological effects of silver nanoparticles (AgNPs) on a variety of microorganisms in pure culture media have been meticulously explored. The specific bacteria examined in these studies include Nitrobacter, Nitrosomonas, Escherichia coli, Pseudomonas aeruginosa, Staphylococcus aureus, and Bacillus subtilis, representing a diverse spectrum of microbial life. A notable investigation by Beddow et al. focused on the impact of both capped and uncapped nanosilver, along with silver sulfate (Ag_2SO_4), on different microbial species. This research encompassed *Nitrosomonas europaea, Nitrosospira multiformis, Nitrosococcus oceani, Escherichia coli,* and *Bacillus subtilis*.

The results were striking: all treatments that included AgNPs led to significant reductions in bacterial growth and nitrification capacity, highlighting the potent antimicrobial properties of silver nanoparticles. Among the various forms tested, capped nanosilver demonstrated superior effectiveness in inhibiting bacterial proliferation compared to its uncapped counterpart, with silver sulfate exhibiting a moderate level of efficacy (Hegde et al. 2016).

However, it is essential to approach these findings with caution. The controlled laboratory conditions in which these experiments were conducted may not accurately represent the complexities of real-world conditions. In ecological settings such as soil and sediment, engineered nanomaterials (ENMs) could act as sinks,

Table 11.5 The toxicological effects of different nanoparticles on a variety of microorganisms and their effect on carbon cycle

Type of nanoparticles	Test methods	Effect on microbial diversity/soil Examples	Impact on carbon cycle	Effects on microorganisms	References
AgNPs	On bacterial in media	Inhibit ammonia-oxidizing bacteria; Nitrobacter Inhibit nitrification in soil	Affects the ammoxidation process and disrupts the nitrogen cycle	High effects Inhibit the activity of ATP synthase and other proteins like hydroxylamine oxidoreductase and monooxygenase, generate ROS, destabilize outer membranes	Hegde et al. (2016)
AuNPs	On bacterial in media	No significant impact on ammoxidation	None	None	Hegde et al. (2016)
Carbonaceous nanomaterials	In vivo and in vitro	Escape wastewater treatments, influence microbial dynamics via intrinsic properties and environmental factors	Moderate potential disruption based on aggregation, interaction with organic matter and environmental pH affecting microbial processes	Alters microbial diversity and enzyme activities, affects nutrient cycling and plant–microbe interactions	Hegde et al. (2016)
Metal oxide nanoparticles (ZnO	Soil microorganisms, DNA based fingerprinting, bacterial growth assays	Reduces microbial diversity, toxic to nitrogen-fixing bacteria Bacillus subtillis, and Escherichia coli	Disrupts microbial processes critical for organic matter decomposition and carbon cycling	Dissolves into Zn ions, causing oxidative stress and membrane damage, inhibits enzyme activity	Hegde et al. (2016), Rajput et al. (2020), Ge et al. (2012)

CuO	Soil incubation studies, enzymatic activity assays, microbial community analysis	Alters microbial community composition, toxic to soil bacteria such as Rhizobiales, Bradyrhizobium	Impacts microbial respiration and nutrient cycling; indirectly affecting carbon turnover	Releases Cu ions, disrupting cellular processes and causing oxidative stress	Hegde et al. (2016), Rajput et al. (2020), Rajput et al. (2018)
TiO2	In vitro and in vivo studies and bacterial culture assays, soil respiration tests	Variable effects; minimal impact in vivo but inhibitory in vitro	Limited impact on carbon cycling, effects depend on environmental condition	Generates ROS, inhibits bacterial growth in vitro but shows no significant effects in vivo	Hegde et al. (2016), Ge et al. (2012), Rajput et al. (2018)

potentially modifying their bioavailability and interaction with microbial communities. Therefore, it is imperative to corroborate these toxicological assessments in more natural sediment and soil environments, as these systems play a crucial role in adsorbing and altering the behavior of contaminants like silver nanoparticles (Hegde et al. 2016; Rajput et al. 2020; Rajput et al. 2018).

11.11.3 Risks of Microbial Resistance

Especially in antimicrobial environments, the application of nanoparticles raises concerns such as resistance. Bacteria exposed to nanoparticles, inducing stress responses similar to those caused by medications, can coexpress antibiotic resistance genes and exhibit cross-adaptation to antibiotics (Zhang et al. 2020a). This may lead to the selection of resistant strains and horizontal gene transfer, spreading resistance genes across various bacterial populations. Moreover, the use of nanoparticles could unintentionally accelerate the development of antibiotic resistance by: Nanoparticles occasionally promote biofilm formation, which serves as a protective barrier for bacteria, thereby increasing their resistance to both antibiotics and nanoparticles (Markowicz 2023). Semi-lethal doses of nanoparticles may enable bacteria to adapt and develop resistance mechanisms such as efflux pumps or enzymatic degradation (Babita and Sehrawat 2024). Residual nanoparticles in the environment may create hotspots for bacterial interaction and gene transfer, further propagating resistance. Researchers are exploring methods to mitigate these risks (Wang et al. 2017). One of the many synergistic strategies being studied by scientists is the combination of nanoparticles with antibiotics to reduce the necessary doses and thereby prevent resistance development. Producing nanoparticles that safely dissolve using green technologies is beneficial for the environment. Designing nanoparticles that specifically target diseases without harming beneficial microbes or the environment (Zhang et al. 2020a).

11.12 Leaching and Soil Interaction

Copper nanoparticles significantly altered the bacterial populations present in biofilm systems and aquatic environments. Notably, copper nanoparticles reduced the presence of Acidobacteria, Aminicinants, Anaerulinaceae, and Anaerulina. Metagenomic functional analysis indicated that the bacterial group most affected by copper nanoparticles was the genera involved in carbohydrate and glycan synthesis, negatively impacting the activities of the extracellular enzymes beta-glucosidase (GLU) and E-leucine aminopeptidase (LAP). This clearly indicates that the introduction of copper oxide nanoparticles negatively affects biogeochemical processes and carbon and nitrogen cycling in biofilm systems. While these toxic effects of copper oxide nanoparticles can be mitigated when the aquatic environment is enriched with natural organic matter, such as humic acid, the use of copper oxide inhibits oxygen activity, thus reducing the activity of heterotrophic organisms (Miao et al. 2019). Soil bacteria may be poisoned by carbon nanomaterials (CNMs), and these same substances may change the bioavailability of nutrients or the toxicity of organic molecules. C60 fullerene, for instance, prevents the proliferation of typical water and soil microorganisms. C60 in its hydroxylated or polyvinyl pyrrolidone-coated forms is an effective oxidizing agent that produces a singlet oxygen species, which is harmful to cells. Research indicates that acetate oxidation by bacteria may be entirely blocked by C60 at a concentration of 140 μg per liter, whilst the process can be reduced by CdSe quantum dots (QD) at a concentration of 200 μg per liter. Quantum dots are poisonous to bacteria, according to research, because they produce toxic metals or ions. Unfortunately, there is a lack of data about the environmental stability and dissociation of quantum dots, as well as their harmful impacts on microbial populations. No matter their composition or functionalization, carbon nanotubes (CNTs) are known to have a negative impact on soil microbes and constitute a serious danger to microbial ecosystems. Scientific investigations have shown that microbes experience a decrease in cell viability and, in many cases, cell death, upon direct contact with highly pure CNT aggregates. Induction of oxidative stress responses and physical penetration of cellular components are the main underlying processes driving this cytotoxicity. The ability of CNTs to inhibit bacterial growth in pure cultures has been demonstrated in a number of investigations on their antibacterial characteristics. To illustrate the point, one study looked at how MWCNTs affected the number and activity of soil microbes (Lawrence et al. 2016). They discovered something surprising; most enzyme activities, including important ones like 1,4-β-glucosidase, cellobiohydrolase, xylosidase, 1,4-b-N-acetylglucosaminidase, and phosphatase, were significantly suppressed at a concentration of 500 μg MWCNTs per gram of soil. At this concentration, the toxicity of MWCNTs to microbes and their metabolic processes was demonstrated by a dramatic decrease in enzymatic activity and total microbial biomass (Chung et al. 2011; Zhang et al. 2020b). Several culture experiments have also shown that MWCNTs have antibacterial properties. Furthermore, studies examining the impact of SWCNTs on pure E. coli strains and soils with varying organic content have shown a concerning pattern: SWCNTs have the ability to disturb microbial communities and cause changes in soil metabolic activity, especially in low-organic-matter environments (Liu et al. 2020). Raw SWCNTs and their functionalized variants, such as polyethylene glycol and m-polyaminobenzene sulfonic acid, were the subject of a thorough toxicology study that found that six weeks of repeated exposure to raw SWCNTs severely reduced the metabolic activity of bacterial populations. Carbon nanotube (CNT) exposure has far-reaching consequences for several microbial species; for example, polymer nanocomposites containing PNC have shown toxicity to Pseudomonas aeruginosa. In light of these results, it is critical to evaluate the environmental effects of CNTs in both cultivated and uncultivated soils immediately (Verdugo 2015; Kharlamova et al. 2022).

11.13 Soil Health and Plant Growth

Nanotechnology in agriculture offers fascinating possibilities to improve plant growth. While the use of nanotechnology in agriculture is still in its early stages, a number of products have already been created and brought to market. Plant diseases can be controlled and growth development can be encouraged with the aid of fungicides that are based on copper (Cu) nanopesticides (Simonin et al. 2018; Adhikari and Ramana 2019). These positive effects demonstrate how nanotechnology may be used to enhance agricultural processes, which in turn leads to an increase in food production and stimulates farming methods that are also more ecologically friendly. Organic compounds, including polystyrene intercalated into kaolinite clay layers, also produce cementing nanocomposites able to regulate nutrient release from fertilizer capsules (Shang et al. 2019; Zhang et al. 2019a, b). Sulfate-modified polystyrene and other engineered nanomaterials can improve soil fertility and crop production even as they offer long-term plant protection as nanobactericides, nanofungicides, and nanoinsecticides (Kim et al. 2010). Direct contact with certain nanoparticles has been related to both beneficial and harmful effects on plant development. For example, CeO2 nanoparticles have been shown to lower nitrogen-fixing bacteria in soybean root nodules, hence lowering nitrogen fixation rates (Adhikari and Ramana 2019). In addition to maize biological output, zeolite fertilizers in Chernozem are good for mobile humus (Kulikova et al. 2020). Patented nanocomposites such as N, P, K, micronutrients, mannose, and amino acids also enhance nutrient absorption and usage in grain crops (Guha et al. 2020). Balanced NPK treatment coupled with Zn, S, and Mo nanoparticles improves grain production in pulses grown in red and lateritic soils under certain circumstances, as yields rise dramatically by 13% and 38% for green and black gram, respectively (Adhikari and Ramana 2019; Ghosh 2019; Baral et al. 2020; Dhlamini et al. 2020; Ahmed et al. 2017).

11.14 Benefits of Nanofungicides

As few agrochemicals as possible must be used in sustainable agriculture to protect the environment. There is no doubt that environmentally sustainable nanoformulations are necessary. Metallic complexes, natural or manufactured chemicals, elements, and primarily inorganic-based compounds are examples of nanoagents with fungicidal activity. In terms of controlling plant diseases, promoting plant growth, and inhibiting fungal growth, nano-biofungicides have demonstrated encouraging results. By facilitating the delayed and continuous release of active ingredients and enhancing penetration, nanoencapsulation can also enhance the usage of fungicides. In agriculture, nanofungicides can offer an economical, efficient, and ecologically safe method of controlling pests. Target species are greatly impacted by fungicide nanoformulation, which also fully shields the environment from premature degradation. Safe handling and less environmental pollution are two benefits of using nanofungicides. Hexaconazole NPs (100 nm) stabilized with polyethylene glycol (PEG) were more potent, safer, and stronger than bulk hexaconazole when compared to conventional fungicide formulations (Kumar et al. 2015). Polymeric nanocapsules and solid lipid nanoparticles are beneficial. These changes lessen toxicity to humans and the agroecosystem, decrease losses from leaching or degradation, and facilitate the release of bioactive compounds and their transportation to the site of action in the targeted plants.

11.14.1 Improved Efficiency

Compared to traditional fungicides, nanofungicides have shown noticeably higher efficacy, mostly because of their nanoscale size. Usually made up of nanoscale-engineered active components or carriers, these materials have a larger surface area, are more reactive, and have better penetration into plant tissues and pathogen cells (Kah et al. 2018; Chhipa 2017).

In order to increase the bioavailability and fungicidal capability of active compounds, especially those with limited water solubility, the nanoformulation enhances their solubility, dispersion, and stability (Rai and Ingle 2012). It has been demonstrated that metallic nanoparticles, specifically those of silver (AgNPs), copper (CuNPs), and zinc oxide (ZnONPs), have potent antifungal effects by disrupting fungal cell walls, producing reactive oxygen species (ROS), and interfering with metabolic processes (Fig. 11.6) (Liu et al. 2020).

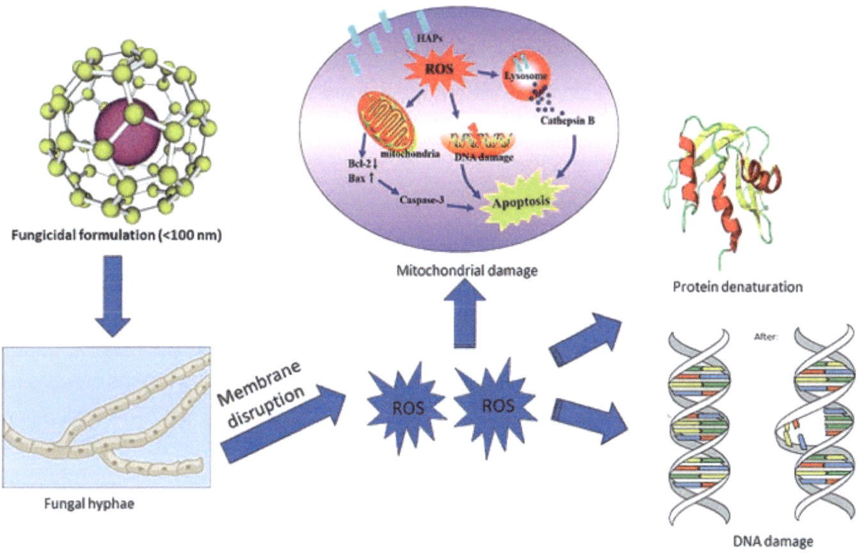

Fig. 11.6 Mechanism of antifungal activity of nanoparticles against fungi (Venugopal et al. 2024)

Additionally, the tiny particle size allows for more even and thorough penetration into the target regions, guaranteeing a more direct and efficient contact with fungal infections. By reducing losses from leaching or photodegradation, this targeted distribution improves the product's overall fungicidal efficacy (Fraceto et al. 2016).

According to recent research, nanofungicides can effectively control diseases at lower dosages, suggesting that they could eventually replace or minimize the need for conventional chemical fungicides while preserving or even enhancing crop protection results (Servin et al. 2015).

11.14.2 Lower Dosage and Frequency of Applications

The ability of nanofungicides to lower the dosage and frequency of treatment is one of their main advantages over traditional fungicides. The distinct physicochemical characteristics of nanoparticles, including their high surface-area-to-volume ratio, increased reactivity, and capacity to deliver regulated release of active substances, are primarily responsible for this enhancement (Kah et al. 2018).

In order to improve bioavailability and adherence to plant surfaces, nanoparticles can improve the stability and solubility of fungicides that are not very soluble in water. Compared to conventional formulations, these characteristics lead to increased efficacy at lower concentrations (Chhipa 2017). Additionally, the slow-release properties of nanocarriers prolong the active compounds' functional lifespan by sustaining effective concentrations for a longer amount of time, which lessens the need for frequent reapplication (Fraceto et al. 2016).

Reducing the frequency of applications helps farmers save money on labor and operating expenses while also lessening the environmental impact of excessive chemical use. Additionally, decreased chemical load in the environment lowers the chance of fungicide resistance in pathogenic fungi as well as non-target toxicity (Servin et al. 2015). Most of all, nanofungicides are a potential step toward crop protection techniques that are more accurate, sustainable, and efficient.

11.14.3 Sustainability of the Environment

As an environmentally benign substitute for traditional chemical fungicides, nanofungicides have demonstrated promise in improving environmental sustainability in agricultural practices. Several environmental benefits result from their increased effectiveness at lower application doses and decreased frequency of usage, which also helps to minimize total chemical inputs into the environment.

The fact that nanofungicides cause less environmental contamination is one of their most noteworthy benefits. Traditional chemical fungicides are frequently used in high quantities, which damages non-target organisms, degrades soil, and causes considerable runoff into water sources (Rai and Ingle 2012). In contrast, several nanofungicides have controlled release mechanisms that guarantee the progressive delivery of active ingredients over time, reducing leaching and minimizing the buildup of pesticides in the environment (Servin et al. 2015). The risk of environmental damage can be further decreased by this slow-release characteristic, which can further lessen the persistence of harmful leftovers.

Furthermore, nanofungicides can more precisely target particular infections. Because conventional fungicides frequently harm beneficial organisms, including pollinators, soil microorganisms, and aquatic life, our targeted administration reduces the unwanted impact on these organisms (Kah et al. 2018). Some nanoparticles also decompose more easily in the environment and leave behind fewer toxic residues, such as those derived from biodegradable materials (Chhipa 2017).

The broader objectives of sustainable agriculture, which emphasizes reducing environmental pollution while maintaining good crop yields, are in line with the decreased use of synthetic chemicals linked to nanofungicides. A more resource-efficient farming system is also facilitated by lower doses and fewer applications, which save farmers money and reduce agricultural energy consumption (Fraceto et al. 2016). As a result, nanofungicides present a viable approach to crop protection that is sustainable, with advantages ranging from less reliance on chemicals to a smaller ecological effect overall.

11.14.4 Reduced Developing Resistance

One of the greatest problems facing modern agriculture is the emergence of fungal disease resistance, especially when employing traditional chemical fungicides. Fungal species may develop defense mechanisms over time that make them less vulnerable to particular fungicides, resulting in decreased effectiveness and a greater demand for chemical applications (Gisi and Sierotzki 2008). Nonetheless, because of their distinct mode of action and increased efficacy at low dosages, nanofungicides offer a promising approach to reducing the development of resistance.

The fact that nanomaterials have numerous mechanisms of action is one of the main reasons why resistance develops more slowly. The evolutionary pressure that normally results in resistance is slowed by this multi-target impact, which makes it more difficult for fungi to build resistance mechanisms against all the effects at once (Rai and Ingle 2012; Kümmerer et al. 2019).

Furthermore, the improved penetration and bioavailability of nanofungicides enable them to function more efficiently at lower concentrations. This lessens the total amount of time that fungi are exposed to high concentrations of fungicides, which is a significant factor in the development of resistance in the usage of traditional pesticides (Kah et al. 2018). Because nanofungicides are frequently more effective at lower dosages, fewer applications mean fewer chances for fungal resistance to develop (Fraceto et al. 2016).

Furthermore, some nanomaterials' controlled release characteristics guarantee that the fungicide is applied gradually in a precise and consistent way, lowering the possibility that resistance may emerge as a result of fungicide concentration spikes. Better pathogen targeting and less frequent application lessen the selective pressure that conventional fungicide treatments frequently impose (Gandhi et al. 2018). A promising approach to addressing the rising issue of fungicide resistance in agricultural systems is provided by the special qualities of nanofungicides, including their capacity to target several facets of fungal physiology and their controlled release mechanisms.

11.14.5 Increased Stability and Shelf Life

For fungicides to be effective and be used in agricultural systems for an extended period of time, they must be stable in storage and able to retain their effectiveness over time (Kah et al. 2018). Through the encapsulation of active components in nanoparticles, which shield them from environmental elements including light, heat, and moisture, nanotechnology improves the fungicides' physical and chemical stability and prevents deterioration (Chhipa 2017; Fraceto et al. 2016).

A protective barrier surrounding the active fungicidal chemicals can be created by using nanomaterials as carriers or encapsulating agents, greatly enhancing their oxidation resistance and thermal stability (Mishra et al. 2017). This is especially helpful for fungicides that are susceptible to environmental changes because the nanocarriers protect the active components from deterioration, extending their shelf life. According to Rai and Ingle (2012), many nanofungicides have slow-release characteristics that allow them to retain constant activity over an extended period of time. This minimizes the need for frequent reapplication and guarantees that the active ingredients are present when needed for pathogen control.

The active fungicide components' solubility and dispersibility can also be improved by nanoparticles, increasing their effectiveness in various environmental settings. Enhancing the homogeneity and uniformity of the formulation improves the stability of nanofungicides by lowering the possibility of precipitation, agglomeration, or settling that can often occur in conventional formulations (Gandhi et al. 2018). This enhanced stability ensures that the fungicide stays effective from production to application by resulting in more dependable performance across the product's shelf life.

Furthermore, nanoencapsulation not only extends shelf life but also permits targeted distribution, guaranteeing regulated release of fungicides and lowering the possibility of premature degradation or loss of activity in the environment (Fraceto et al. 2016). Because they provide long-lasting and reliable disease control with less chemical use, nanofungicides help to make agricultural operations more cost-effective and ecologically friendly. Therefore, one crucial advantage of nanofungicides is their increased stability and longer shelf life, which guarantee their continued effectiveness over time and in a variety of applications and storage scenarios.

11.15 Nanofungicide Hazards

11.15.1 Environmental Hazards

Although nanofungicides provide effective and targeted disease control in agriculture, their distinct nanoscale characteristics give rise to serious ecotoxicological issues. These modified nanoparticles can change vital ecosystem processes, including nitrogen cycling, disturb microbial diversity, and negatively impact non-target soil microorganisms (Kah et al. 2018). Furthermore, they might seep into aquatic environments, where substances like copper and silver nanoparticles are especially harmful to aquatic life, which could result in bioaccumulation and biomagnification in food chains (Neha et al. 2021). Further increasing the risk is their tenacity and capacity to change into more reactive or toxic forms in the environment, particularly as these modifications may increase toxicity in unpredictable ways through synergistic interactions with other agrochemicals (Gogos et al. 2012). Consequently, even if nanofungicides offer a contemporary approach to crop protection, their long-term ecotoxicological effects demand careful consideration and regulation.

11.15.2 Threats to Human Health

The capacity of nanoparticles to cross biological barriers and build up in human tissues makes nanofungicides potentially harmful to human health. Cellular toxicity, inflammation, and oxidative stress can result from ingestion, inhalation, or skin contact with nanoparticle residues, whether through contaminated food or during application (Fraceto et al. 2016). In human lung and liver cells, for example, metal-based nanoparticles such as copper and silver, which are frequently employed for their antifungal qualities, have been linked to harmful effects (Ahmed et al. 2017). According to Singh et al. (2019), the tiny size and high reactivity of nanoparticles can enhance their absorption and distribution within the body, raising concerns about long-term health effects such as neurotoxicity, genotoxicity, and risks associated with chronic exposure. Risk assessment is further complicated by the lack of comprehensive toxicological data and clearly defined safety standards, highlighting the need for rigorous research before widespread use (Singh et al. 2019).

11.15.3 Safety and Regulatory Challenges

The licensing, monitoring, and use of nanofungicides are problematic due to the lack of recognized regulatory and safety frameworks, despite their increasing use in agriculture. According to Kah et al. (2013), conventional pesticide laws frequently fall short when it comes to nanomaterials because they fail to take into consideration the distinct physicochemical characteristics, behaviors, and interactions of nanoparticles. Since established techniques for assessing the toxicity, environmental fate, and exposure pathways of nanoparticles are still developing, this regulatory vacuum makes risk assessment more difficult. Additionally, there are concerns about consumer safety and transparency due to the absence of required labeling and traceability of nanomaterials in agrochemical products (Servin et al. 2015). Inconsistencies in the oversight and approval procedures for nanofungicides are also caused by the lack of collaboration across international regulatory agencies. These difficulties highlight how urgently specialized regulations, strong safety testing procedures, and thorough guidelines are required to guarantee the appropriate development and use of nanofungicides.

11.15.4 Cost and Accessibility

Although nanofungicides provide innovative crop protection options, their high production costs and restricted availability pose serious problems, especially for smallholder and resource-poor farmers. In comparison to traditional fungicides, the production of nanoparticles frequently necessitates specialized tools, high-purity materials, and energy-intensive procedures, which raises the cost (Kah and Hofmann 2014). These costs may prevent nanofungicides from being widely used in developing nations, which could lead to a greater divide between industrial and subsistence farming practices. Furthermore, access may be impeded and reliance on a small number of international suppliers may result from intellectual property limitations and a lack of local manufacturing infrastructure (Parisi et al. 2015). The promise for wider agricultural development may be undermined if efforts are not made to lower costs and provide equal access, as the advantages of nanotechnology in agriculture might continue to be limited to large-scale, high-input farming operations.

11.15.5 Unintended Consequences

The complicated interactions between nanoparticles and biological or environmental systems make it difficult to forecast the unexpected consequences of using nanofungicides. For instance, although useful in managing diseases, the antibacterial properties of nanoparticles may unintentionally damage pollinators, beneficial soil organisms, or plant–microbe symbioses, upsetting the ecological balance (Khot et al. 2012). Furthermore, nanoparticles have the ability to combine with other agrochemicals or environmental contaminants to form novel compounds that may be more harmful than previously thought (Nuruzzaman et al. 2016). Prolonged usage of nanofungicides may eventually lead to the emergence of fungal strains that are resistant, much like traditional pesticides, which would reduce their long-term efficacy. The necessity for cautious approaches in their deployment is further highlighted by the fact that many of these repercussions remain hypothetical due to the lack of long-term field investigations (Kah et al. 2018).

11.16 Conclusion

The development of highly effective and focused crop protection systems made possible by nanotechnology is changing the future of sustainable agriculture. Nanofungicides, which are designed materials that combine strong antifungal action with nanoscale precision, are among the most promising uses. Many of the drawbacks of conventional fungicides, such as overuse, quick degradation, and environmental contamination, can be addressed by nanofungicides through mechanisms like improved adherence to plant tissues, targeted fungal pathogens, and sustained release of active ingredients.

Furthermore, the plant microbiome, a complex and vital network of microbes that affect plant growth, stress tolerance, nutrient cycle, and disease resistance, interacts dynamically with these formulations. Although the advantages of nanofungicides, such as fewer applications, less chemical load, and increased effectiveness, are obvious, caution is necessary due to the possibility that they could upset the microbial balance or have an adverse effect on organisms that are not their intended targets. According to studies, microbial diversity may change when nanoagents are introduced into agricultural ecosystems, which could result in unforeseen ecological changes.

In addition to controlling pathogens, nanotechnology has demonstrated potential in improving agricultural yields, nutrient uptake, and soil fertility through the use of biocompatible formulations, nanocarriers, and nanofertilizers. However, issues with production costs, accessibility for smallholder farmers, regulatory control, environmental buildup, and health risks to humans still exist.

In conclusion, the development of nanofungicides is a critical step in the direction of resource-efficient and environmentally conscious agriculture. To completely comprehend and control the risks connected with their use, it will be crucial to develop thorough safety evaluations, legal frameworks, and long-term ecological studies as research progresses. Through the promotion of responsible innovation and interdisciplinary collaboration, nanotechnology can be a potent instrument in revolutionizing agricultural methods to address the dual issues of environmental sustainability and food security.

References

Adhikari T, Ramana S (2019) Nano fertilizer: its impact on crop growth and soil health. J Res PJTSAU 47:1

Ahmed B, Zaidi A, Khan MS, Rizvi A, Saif S, Shahid M (2017) Perspectives of plant growth promoting rhizobacteria in growth enhancement and sustainable production of tomato. In Microbial strategies for vegetable production (pp. 125–149). Cham: Springer International Publishing

Astapati AD, Nath S (2023) The complex interplay between plant-microbe and virus interactions in sustainable agriculture: harnessing phytomicrobiomes for enhanced soil health, designer plants, resource use efficiency, and food security. Crop Des 2(1):100028

Babita SS, Sehrawat N (2024) Phage therapy reimagined: viral. In: Emerging paradigms for antibiotic-resistant infections: Beyond the Pill, p 449. Springer, Singapore. https://doi.org/10.1007/978-981-97-5272-0_21

Baral K, Kumar D, Shivay YS (2020) Effect of zinc oxide nanoparticles embedded NPK fertilizer on growth, yield and zinc concentration and uptake of aromatic rice (Oryza sativa). Indian J Agric Sci 90(8):1626–1630

Berg G, Rybakova D, Fischer D, Cernava T, Champomier Vergès M-C, Charles T, Chen X et al (2020) Microbiome definition re-visited: old concepts and new challenges. Microbiome 8:1–22

Bernardin JR, Young EB, Gray SM, Bittleston LS (2024a) Microbial communities increase host plant leaf growth in a pitcher plant experimental system. bioRxiv, 2024-01

Bernardin JR, Young EB, Gray SM, Bittleston LS (2024b) Bacterial community function increases leaf growth in a pitcher plant experimental system. mSystems 9:e01298-24

Bocate KP, Reis GF, de Souza PC, Junior AGO, Durán N, Nakazato G et al (2019) Antifungal activity of silver nanoparticles and simvastatin against toxigenic species of Aspergillus. Int J Food Microbiol 291:79–86

Campos EVR, de Oliveira JL, Fraceto LF, Singh B (2015) Polysaccharides as safer release systems for agrochemicals. Agron Sustain Dev 35:47–66

Chauhan D, Talreja N, Ashfaq M (2024) Biogenic nanofungicides: next-generation anti-microbial materials. In: Novel applications in plant pathology. Nanobiotechnology for plant protection

Chen H, Yada R (2011) Nanotechnologies in agriculture: new tools for sustainable development. Trends Food Sci Technol 22(11):585–594

Chen H, Song Y, Wang Y, Wang H, Ding Z, Fan K (2024) Zno nanoparticles: improving photosynthesis, shoot development, and phyllosphere microbiome composition in tea plants. J Nanobiotechnol 22(1):389

Chhipa H (2017) Nanofertilizers and nanopesticides for agriculture. Environ Chem Lett 15:15–22

Choudhary A, Musheer N, Ashraf S, Saeed S (2024) Microbe-mediated nanoparticles: potential nanobiofungicides. In K. A. Abd-Elsalam (Ed.), Nanobiotechnology for plant protection: Nanofungicides (pp. 65–84). Elsevier. https://doi.org/10.1016/B978-0-323-95305-4.00004-0

Chung H, Son Y, Yoon TK, Kim S, Kim W (2011) The effect of multi-walled carbon nanotubes on soil microbial activity. Ecotoxicol Environ Saf 74(4):569–575

Compant S, Samad A, Faist H, Sessitsch A (2019) A review on the plant microbiome: ecology, functions, and emerging trends in microbial application. J Adv Res 19:29–37

Cruz JN, Muzammil S, Ashraf A, Ijaz MU, Siddique MH, Abbas R et al (2024) A review on mycogenic metallic nanoparticles and their potential role as antioxidant, antibiofilm and quorum quenching agents. Heliyon 10:e29500

Das P, Williams CJ, Fulthorpe RR, Hoque ME, Metcalfe CD, Xenopoulos MA (2012) Changes in bacterial community structure after exposure to silver nanoparticles in natural waters. Environ Sci Technol 46(16):9120–9128

Dávila Costa JS, Romero CM (2025) Nano-biofungicides and bio-nanofungicides: state of the art of innovative tools for controlling resistant Phytopathogens. Biophysica 5(2):15

Dhlamini B, Paumo HK, Katata-Seru L, Kutu FR (2020) Sulphate-supplemented NPK nanofertilizer and its effect on maize growth. Mater Res Express 7(9):095011

Dimkpa CO, Latta DE, McLean JE, Britt DW, Boyanov MI, Anderson AJ (2013) Fate of CuO and ZnO nano-and microparticles in the plant environment. Environ Sci Technol 47(9):4734–4742
Fraceto LF, Grillo R, de Medeiros GA, Scognamiglio V, Rea G, Bartolucci C (2016) Nanotechnology in agriculture: which innovation potential does it have? Front Environ Sci 4:20
Frenk S, Ben-Moshe T, Dror I, Berkowitz B, Minz D (2013) Effect of metal oxide nanoparticles on microbial community structure and function in two different soil types. PLoS One 8(12):e84441
Gandhi PR, Jayaseelan C, Kamaraj C, Rajasree SR, Mary RR (2018) In vitro antimalarial activity of synthesized TiO2 nanoparticles using Momordica charantia leaf extract against Plasmodium falciparum. J Appl Biomed 16(4):378–386
Ge Y, Schimel JP, Holden PA (2012) Identification of soil bacteria susceptible to TiO2 and ZnO nanoparticles. Appl Environ Microbiol 78(18):6749–6758
Ge J, Li D, Ding J, Xiao X, Liang Y (2023) Microbial coexistence in the rhizosphere and the promotion of plant stress resistance: a review. Environ Res 222:115298
Ghosh GK (2019) Red and lateritic soils and agri-productivity: issues and strategies. J Indian Soc Soil Sci 67(4):S104–S121
Gisi U, Sierotzki H (2008) Resistance of plant pathogens to fungicides: a global overview and trends. Phytopathology 98(1):1–13
Gogos A, Knauer K, Bucheli TD (2012) Nanomaterials in plant protection and fertilization: current state, foreseen applications, and research priorities. J Agric Food Chem 60(39):9781–9792
Guha T, Gopal G, Kundu R, Mukherjee A (2020) Nanocomposites for delivering agrochemicals: a comprehensive review. J Agric Food Chem 68(12):3691–3702
He L, Liu Y, Mustapha A, Lin M (2011) Antifungal activity of zinc oxide nanoparticles against Botrytis cinerea and Penicillium expansum. Microbiol Res 166(3):207–215
Hegde K, Brar SK, Verma M, Surampalli RY (2016) Current understandings of toxicity, risks and regulations of engineered nanoparticles with respect to environmental microorganisms. Nanotechnol Environ Eng 1:1–12
Imani MM, Safaei M, Moradpoor H, Rezaei R, Golshah A, Rezaei F, Jamshidy L (2020) Optimum synthesis of CuO nanoparticles with the highest antifungal activity against oral pathogen Candida albicans. J Appl Pharma Sci 10(2):021–025
Jayasree EA, Rajeshkumar AS (2024) Mechanism of silver and gold nanoparticles against infectious disease-causing pathogens. In: Smart nanomaterials for infectious diseases, (Vol. 62, pp. 151–177). Royal Society of Chemistry. https://doi.org/10.1039/BK9781837672813-00151
Jo YK, Kim BH, Jung G (2009) Antifungal activity of silver ions and nanoparticles on phytopathogenic fungi. Plant Dis 93(10):1037–1043
Kah M, Hofmann T (2014) Nanopesticide research: current trends and future priorities. Environ Int 63:224–235
Kah M, Kookana RS, Gogos A, Bucheli TD (2013) Exploring the use of engineered nanoparticles in agriculture: a risk-based perspective. Crit Rev Environ Sci Technol 43(24):2740–2781
Kah M, Kookana RS, Gogos A, Bucheli TD (2018) A critical evaluation of nanopesticides and nanofertilizers against their conventional analogues. Nat Nanotechnol 13(8):677–684
Karnawat M, Meena RK, Nagar D, Nagar R (2021) A review: applications of nanotechnology in plant growth and crop protection. Curr Res Agric Farm 2(1):40–45
Kharlamova MV, Paukov M, Burdanova MG (2022) Nanotube functionalization: investigation, methods and demonstrated applications. Materials 15(15):5386
Khot LR, Sankaran S, Maja JM, Ehsani R, Schuster EW (2012) Applications of nanomaterials in agricultural production and crop protection: a review. Crop Prot 35:64–70
Kim S, Pyo HB, Ko SH, Ah CS, Kim A, Kim WJ (2010) Fabrication of anionic sulfate-functionalized nanoparticles as an immunosensor by protein immobilization. Langmuir 26(10):7355–7364
Krishnamoorthy R, Gassem MA, Athinarayanan J, Periyasamy VS, Prasad S, Alshatwi AA (2021) Antifungal activity of nanoemulsion from Cleome viscosa essential oil against food-borne pathogenic Candida albicans. Saudi J Biol Sci 28(1):286–293

Kulikova A, Zakharov N, Yashin E, Khairtdinova N (2020) Efficiency of zeolite as maize fertilizer of Middle Volga chernozem. In: BIO web of conferences, vol 27. EDP Sciences, p 00058
Kumar S, Ahlawat W, Kumar R, Dilbaghi N (2015) Graphene, carbon nanotubes, zinc oxide and gold as elite nanomaterials for fabrication of biosensors for healthcare. Biosens Bioelectron 70:498–503
Kümmerer K, Dionysiou DD, Olsson O, Fatta-Kassinos D (2019) Reducing aquatic micropollutants–Increasing the focus on input prevention and integrated emission management. Sci Total Environ 652:836–850
Kumar R, Kumar N, Rajput VD, Mandzhieva S, Minkina T, Saharan BS et al (2022) Advances in biopolymeric nanopesticides: a new eco-friendly/eco-protective perspective in precision agriculture. Nanomaterials 12(22):3964
Kutawa AB, Ahmad K, Ali A, Hussein MZ, Abdul Wahab MA, Adamu A et al (2021) Trends in nanotechnology and its potentialities to control plant pathogenic fungi: a review. Biology 10(9):881
Lawrence JR, Waiser MJ, Swerhone GDW, Roy J, Tumber V, Paule A et al (2016) Effects of fullerene (C60), multi-wall carbon nanotubes (MWCNT), single wall carbon nanotubes (SWCNT) and hydroxyl and carboxyl modified single wall carbon nanotubes on riverine microbial communities. Environ Sci Pollut Res 23:10090–10102
Łaźniewska J, Macioszek VK, Kononowicz AK (2012) Plant-fungus interface: the role of surface structures in plant resistance and susceptibility to pathogenic fungi. Physiol Mol Plant Pathol 78:24–30
Li H, Rehman A, ur Rahman S, Li K, Yang T, Akuetteh P, Khalid M (2024) Biosynthesized zinc oxide nanoparticles modulate the phytoremediation potential of Pennisetum giganteum and its rhizocompartments associated microbial community structure. J Clean Prod 434:140346
Limo MJ, Sola-Rabada A, Boix E, Thota V, Westcott ZC, Puddu V, Perry CC (2018) Interactions between metal oxides and biomolecules: from fundamental understanding to applications. Chem Rev 118(22):11118–11193
Liu J, Wang C, Wang H, Du X (2020) Antifungal activity of metal-based nanoparticles: a review. J Nanomater 2020:Article ID 4294625. https://doi.org/10.1155/2020/4294625
Ly PD, Ly KN, Phan HL, Nguyen HH, Duong VA, Nguyen HV (2024) Recent advances in surface decoration of nanoparticles in drug delivery. Front Nanotechnol 6:1456939
Mahajan R, Tandon R, Kalia A, Mahajan BVC (2021) Nanoemulsion formulation of Ocimum gratissimum essential oil and its antifungal activity against Penicillium digitatum. J Nanosci Nanotechnol 21(6):3556–3565
Mahmoud EY, Hussien ZN, Ibrahim MM, Yousef H (2021) Using of certain biotic and abiotic inducers on controlling peanut Cercospora leaf spot. Curr Sci Int 10(1):18–28
Malik MA, Wani AH, Bhat MY, Siddiqui S, Alamri SA, Alrumman SA (2024) Fungal-mediated synthesis of silver nanoparticles: a novel strategy for plant disease management. Front Microbiol 15:1399331
Markowicz A (2023) The significance of metallic nanoparticles in the emerging, development and spread of antibiotic resistance. Sci Total Environ 871:162029
Matras E, Gorczyca A, Przemieniecki SW, Oćwieja M (2022) Surface properties-dependent antifungal activity of silver nanoparticles. Sci Rep 12(1):18046
Mgadi K, Ndaba B, Roopnarain A, Rama H, Adeleke R (2024) Nanoparticle applications in agriculture: overview and response of plant-associated microorganisms. Front Microbiol 15:1354440
Miao L, Wang P, Hou J, Yao Y, Liu Z, Liu S (2019) Low concentrations of copper oxide nanoparticles alter microbial community structure and function of sediment biofilms. Sci Total Environ 653:705–713
Mishra S, Singh BR, Naqvi AH, Singh HB (2017) Potential of biosynthesized silver nanoparticles using Stenotrophomonas sp. BHU-S7 (MTCC 5978) for management of soilborne and foliar phytopathogens. Sci Rep 7(1):45154

Mosa MA, El-Abeid SE, Khalifa MMA, Elsharouny TH, El-Baz SM, Ahmed AY (2022) Smart pH responsive system based on hybrid mesoporous silica nanoparticles for delivery of fungicide to control Fusarium crown and root rot in tomato. J Plant Pathol 104(3):979–992

Moussa S, Iasur Kruh L (2025) Balancing nature and nurture: the role of biocontrol agents in shaping plant microbiomes for sustainable agriculture. Microorganisms 13(2):323

Neha R, Adithya S, Jayaraman RS, Gopinath KP, Pandimadevi M, Praburaman L, Arun J (2021) Nano-adsorbents an effective candidate for removal of toxic pharmaceutical compounds from aqueous environment: a critical review on emerging trends. Chemosphere 272:129852

Nuruzzaman MD, Rahman MM, Liu Y, Naidu R (2016) Nanoencapsulation, nano-guard for pesticides: a new window for safe application. J Agric Food Chem 64(7):1447–1483

Oldroyd GE, Dixon R (2014) Biotechnological solutions to the nitrogen problem. Curr Opin Biotechnol 26:19–24

Pallavi MCM, Srivastava R, Arora S, Sharma AK (2016) Impact assessment of silver nanoparticles on plant growth and soil bacterial diversity. 3 Biotech 6:1–10

Parisi C, Vigani M, Rodríguez-Cerezo E (2015) Agricultural nanotechnologies: what are the current possibilities? Nano Today 10(2):124–127

Pattanayak SP, Bose P, Sunita P (2023) Interaction between nanoparticles and phytopathogens. In: Agricultural and environmental nanotechnology: novel technologies and their ecological impact. Springer Nature Singapore, Singapore, pp 169–220

Peterson PG, Merrett MF, Fowler SV, Barrett DP, Paynter Q (2020) Comparing biocontrol and herbicide for managing an invasive non-native plant species: efficacy, non-target effects and secondary invasion. J Appl Ecol 57(10):1876–1884

Philippot L, Raaijmakers JM, Lemanceau P, Van Der Putten WH (2013) Going back to the roots: the microbial ecology of the rhizosphere. Nat Rev Microbiol 11(11):789–799

Prado JCS, de Aguiar FLL, Prado GM, Nascimento JFD, de Sousa NV, Barbosa FCB et al (2024) Development and characterization of nanoemulsions containing Lippia origanoides Kunth essential oil and their antifungal potential against Candida albicans. J Appl Microbiol 135(11):lxae271

Punniyakotti P, Vinayagam S, Rajamohan R, Priya SD, Moovendhan M, Sundaram T (2024) Environmental fate and ecotoxicological behaviour of pesticides and insecticides in non-target environments: nanotechnology-based mitigation strategies. J Environ Chem Eng 12:113349

Rai M, Ingle A (2012) Role of nanotechnology in agriculture with special reference to management of insect pests. Appl Microbiol Biotechnol 94:287–293

Rajput VD, Minkina T, Sushkova S, Tsitsuashvili V, Mandzhieva S, Gorovtsov A et al (2018) Effect of nanoparticles on crops and soil microbial communities. J Soils Sediments 18:2179–2187

Rajput V, Minkina T, Sushkova S, Behal A, Maksimov A, Blicharska E et al (2020) ZnO and CuO nanoparticles: a threat to soil organisms, plants, and human health. Environ Geochem Health 42:147–158

Rajwade JM, Chikte RG, Paknikar KM (2020) Nanomaterials: new weapons in a crusade against phytopathogens. Appl Microbiol Biotechnol 104(4):1437–1461

Roossinck MJ (2015) Plants, viruses and the environment: ecology and mutualism. Virology 479:271–277

Sasse J, Martinoia E, Northen T (2018) Feed your friends: do plant exudates shape the root microbiome? Trends Plant Sci 23(1):25–41

Servin A, Elmer W, Mukherjee A, De la Torre-Roche R, Hamdi H, White JC et al (2015) A review of the use of engineered nanomaterials to suppress plant disease and enhance crop yield. J Nanopart Res 17:1–21

Shang Y, Hasan MK, Ahammed GJ, Li M, Yin H, Zhou J (2019) Applications of nanotechnology in plant growth and crop protection: a review. Molecules 24(14):2558

Simonin M, Colman BP, Tang W, Judy JD, Anderson SM, Bergemann CM et al (2018) Plant and microbial responses to repeated Cu (OH) 2 nanopesticide exposures under different fertilization levels in an agro-ecosystem. Front Microbiol 9:1769

Singh AV, Laux P, Luch A, Sudrik C, Wiehr S, Wild AM et al (2019) Review on the toxicity of engineered nanomaterials: an update. Arch Toxicol 93:3035–3091
Slavin YN, Bach H (2022) Mechanisms of antifungal properties of metal nanoparticles. Nano 12(24):4470
Smith SE, Read DJ (2010) Colonization of roots and anatomy of arbuscular mycorrhizas. In S. E. Smith & D. J. Read (Eds.), Mycorrhizal symbiosis (3rd ed., pp. 42–90). Academic Press. https://doi.org/10.1016/B978-012370526-6.50004-0
Tarannum N, Pooja K, Jakhar S, Mavi A (2024) Nanoparticles assisted intra and transdermic delivery of antifungal ointment: an updated review. Discover Nano 19(1):11
Venugopal E, Krishna SBN, Golla N, Patil SJ (2024) Nanofungicides: a promising solution for climateresilient plant disease management. In: plant quarantine challenges under climate change anxiety. Cham: Springer Nature Switzerland, pp 513–532
Vandenkoornhuyse P, Quaiser A, Duhamel M, Le Van A, Dufresne A (2015) The importance of the microbiome of the plant holobiont. New Phytol 206(4):1196–1206
Verdugo EM (2015) Reaction of carbon nanotubes with chemical disinfectants: byproduct formation and implications for nanotube environmental fate and transport. The University of Iowa. https://doi.org/10.17077/ETD.LRMHDB97
Vietorisz C, Policelli N, Li A, Adams LA, Bhatnagar JM (2024) Microbial functional guilds and genes are key to explaining soil nutrient cycling alongside soil and plant variables. bioRxiv, 2024–12
Vorholt JA (2012) Microbial life in the phyllosphere. Nat Rev Microbiol 10(12):828–840
Wahab A, Muhammad M, Ullah S, Abdi G, Shah GM, Zaman W, Ayaz A (2024) Agriculture and environmental management through nanotechnology: eco-friendly nanomaterial synthesis for soil-plant systems, food safety, and sustainability. Sci Total Environ 926:171862
Wang L, Hu C, Shao L (2017) The antimicrobial activity of nanoparticles: present situation and prospects for the future. Int J Nanomedicine 12:1227–1249
Xiao C, Xu C, Zhang J, Jiang W, Zhang X, Yang C et al (2022) Soil microbial communities affect the growth and secondary metabolite accumulation in Bletilla striata (Thunb.) Rchb. f. Front Microbiol 13:916418
Xu L, Naylor D, Dong Z, Simmons T, Pierroz G, Hixson KK et al (2018) Drought delays development of the sorghum root microbiome and enriches for monoderm bacteria. Proc Natl Acad Sci U S A 115(18):E4284–E4293
Yousef H, Abd-Elsalam KA, Abdel-Momen SM (2025) Endophytic fungi for reduction of mycotoxin contamination in crops. In: Fungal endophytes volume II: applications in agroecosystems and plant protection. Springer Nature Singapore, Singapore, pp 185–224
Zhang F, Liu X, Pentok M, Sauli E, He N, Zen X, Li X, Liu T (2019a) Molecular mechanism and changes of antioxidant enzyme in ZnO nanoparticles against fungus. J Biomed Nanotechnol 15(4):647–661. https://doi.org/10.1166/jbn.2019.2718
Zhang ZY, Huang L, Liu F, Wang MK, Ndzana GM, Liu ZJ (2019b) Transformation of clay minerals in nanoparticles of several zonal soils in China. J Soils Sediments 19:211–220
Zhang C, Sun R, Xia T (2020a) Adaption/resistance to antimicrobial nanoparticles: will it be a problem? Nano Today 34:100909
Zhang W, Zeng Z, Liu Z, Huang J, Xiao R, Shao B et al (2020b) Effects of carbon nanotubes on biodegradation of pollutants: positive or negative? Ecotoxicol Environ Saf 189:109914

Chapter 12
Impact of Nanoparticle-Supported Epiphytic Microorganisms on Plant Growth, Adaptation, and Disease Resistance

Elif Özlem Arslan Aydoğdu and Gülnihan Selim

Abstract Many biotic and abiotic environmental factors influence plant growth. One of the essential components supporting plants is the epiphytic flora. Epiphytic microorganisms, which can establish symbiotic, neutral, and mutualistic relationships with the plant, colonize the entire surface of the plant, from roots to flower organs and seeds. They protect their hosts against plant pathogens and predators, stress resistance, promote growth and development, and virtually all metabolic events. Although the importance of a healthy microbiome in soil and plants has been proven, the organisms involved and their responses to environmental dynamics are still the subject of research. Studies on the agricultural potential of nanoparticles have shown that, while supporting plant growth, they also increase plant resistance to biotic and abiotic stressors. The relationship of these molecules, which can be synthesized in an eco-friendly way by green synthesis and whose numerous benefits have been identified, with epiphytic flora is still a new and incomplete field. The effects of nanoparticles change when synthesized by members of the epiphytic flora, and the effects of nanoparticles synthesized in any way on epiphytic flora are still unclear. Studies show that after the application of various metal or metal oxide nanoparticles, changes occur in the composition of epiphytic microflora in plants and an increase in the adaptation of beneficial microorganisms, but there is also evidence that this flora may lose some of its members. In this chapter, the effect of nanoparticles synthesized by epiphytic flora and produced by green synthesis on plant growth, as well as recent studies on the effects of these particles on epiphytic flora, are included.

Keywords Epiphytic flora · Nanoparticles · Phyllosphere · Rhizoplane

E. Ö. Arslan Aydoğdu (✉)
Istanbul University, Faculty of Science, Department of Biology, Istanbul, Turkey
e-mail: eoarslan@istanbul.edu.tr

G. Selim
Istanbul University, Institute of Graduate Studies in Science, Istanbul, Turkey
e-mail: gulnihan.selim@istanbul.edu.tr

K. A. Abd-Elsalam (ed.), *Plant-Microbiome Nanotechnology*, Sustainability Sciences in Asia and Africa, https://doi.org/10.1007/978-981-95-5056-2_12

12.1 Introduction

While many abiotic and biotic factors affect the development of plants, in recent years, stress factors such as climate change, water shortage, nutrient imbalances in the soil due to improper agricultural and industrial practices, salinity, pH, and pressure of pathogens force the adaptation mechanisms of plants (Berendsen et al. 2012; Ding et al. 2022). Plants are generally supported against these stress factors by organisms in the rhizosphere and by their epiphytic and endophytic flora. Plant epiphytic flora mainly hosts prokaryotic and eukaryotic microorganisms that establish symbiotic or mutualistic relationships with the plant. They play critical roles such as increasing tolerance to both biotic and abiotic stress factors, and at the same time, plant growth (Butani et al. 2021; Das et al. 2023).

Microbial applications are becoming increasingly widespread in various fields, such as increasing agricultural yields, reducing pesticide and chemical fertilizers, and reclaiming infertile and/or contaminated soils. However, in recent years, the potential of epiphytic flora members to be used in all these activities has become more evident. Another technological breakthrough in agriculture in recent years is nanoparticles. Nanoparticles offer versatile benefits such as optimizing nutrient uptake, supporting plant growth due to their high surface area/volume ratios and physicochemical properties, helping to regulate plant resistance mechanisms against many abiotic stress factors, and providing protection against pathogens (Tsekhmistrenko et al. 2020; Hazarika et al. 2022). Green synthesis protocols have been rapidly developing in recent years as an alternative to chemical processes in nanoparticle synthesis. Among these methods, methods using fungi or bacteria attract much attention because of their eco-friendly properties, relatively low cost, and sustainability. Accordingly, we are encountering studies on nanoparticles derived from epiphytic bacteria and their benefits with more and more significant results (Bahrulolum et al. 2021; Jadoun et al. 2022; Sundar et al. 2025).

In addition to this, the effects of nanoparticle treatments on epiphytic microbiota are still not fully elucidated. In particular, changes in the composition and functions of epiphytic microflora in nanoparticle-treated plants raise the question of how these microorganisms alter the beneficial effects provided to the plant. Studies have shown that the composition of epiphytic microflora in plants treated with ZnO and Ag nanoparticles changed significantly, and the adaptation of beneficial microorganisms to the site increased (Chen et al. 2024; Przemieniecki et al. 2024). However, some studies have shown that nanoparticles, especially those with antimicrobial activity, change the flora of the soil, and that epiphytic flora can also be negatively affected. In other words, a nanoparticle can either support or harm the flora while providing various benefits to the plant (Jampílek and Králová 2025). Further studies on the effects of nanoparticles on biota that play a role in plant development, especially epiphytic and endophytic flora, should be prioritized to fill the knowledge gap.

12.2 Nanoparticles and Bacterially Originated/Related Nanoparticles

The term nano is derived from the Greek word "nanos," meaning dwarf, and is used to describe particles with dimensions of 1–100 nm. Nanoparticles in this size range are often much smaller than a protein or virus (Hulkoti and Taranath 2014). The concept of nanotechnology, which forms the basis of nanoparticles, was introduced by Richard Feynman in 1959, and particles with very small dimensions are called nanoparticles under this concept. Nanoparticles are frequently researched in the fields of medicine, biology, agriculture, biomedicine, and electronics (Hulla et al. 2015). In order for nanoparticles to be studied in these areas, they must have various properties as shown in Fig. 12.1 (Tsekhmistrenko et al. 2020).

Nanoparticles can be synthesized using chemical, physical, and biological methods. Examples of chemical synthesis include photochemical reduction and ultrasonic field techniques. Physical synthesis includes aerosol technologies, ultraviolet irradiation, laser ablation, and lithography. Finally, biological methods include synthesis using yeast, bacteria, and viruses. However, the production of nanoparticles by chemical and physical methods can both harm the environment and require the use of advanced equipment and methods due to hazardous substances in production. The biological method is more economical, time-saving, and eco-friendly compared to these methods. Therefore, it is also called green synthesis (Bahrulolum et al. 2021; Andaç et al. 2022). The fact that microorganisms can be effective in different parameters such as pH, pressure, temperature, or resistance to extreme conditions makes them much more attractive in the biological synthesis of nanoparticles (Jadoun et al. 2022) (Fig. 12.2).

Various fertilizers, pesticides, insecticides, or other industrial products are used to support plant growth in agriculture. However, these substances can lead to harmful results in terms of ecological balance. Therefore, it is preferred to use eco-friendly methods. For this purpose, nanotechnology has been used as one of the

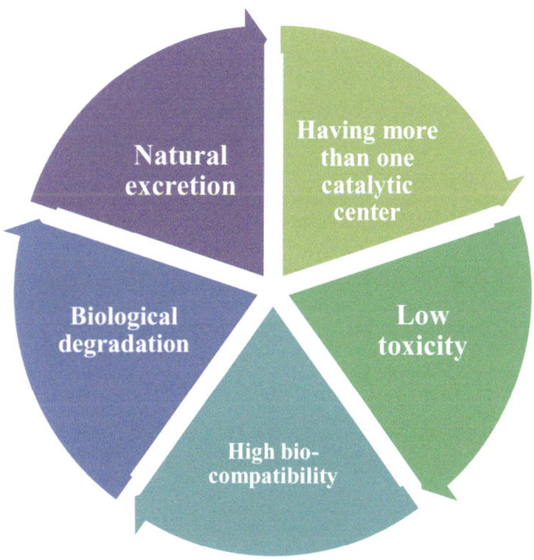

Fig. 12.1 Properties required for the use of nanoparticles in agriculture. The surface area/volume ratio of nanoparticles causes these particles to interact more easily with other parts. These features cause nanoparticles to absorb the substance more and make them the subject of research for different study topics (Tsekhmistrenko et al. 2020)

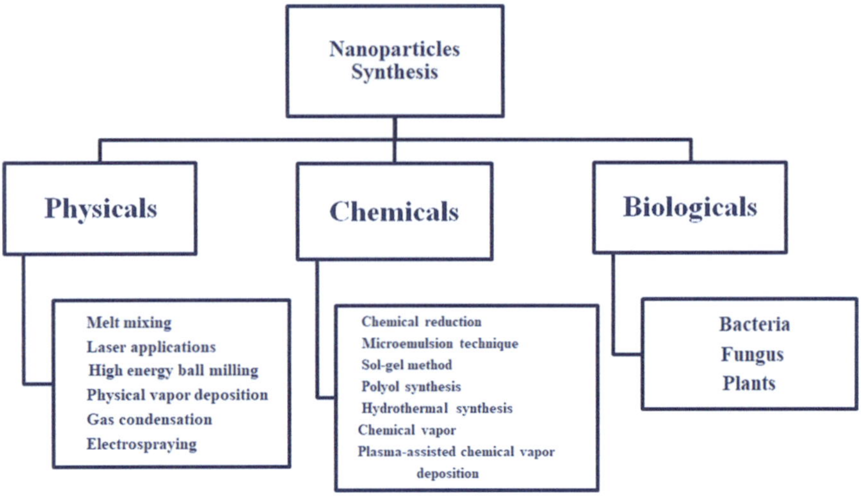

Fig. 12.2 Synthesis methods of nanoparticles (Andaç et al. 2022)

advantageous methods in recent years (Hazarika et al. 2022). Nanoparticles can be preferred in agriculture for reasons such as their compact size, easy transportation, handling, long-term storage, high efficiency, and non-toxicity (Singh et al. 2021). Nanoparticles are used in agriculture in areas such as plant growth promoters, nano-biosensors, nano-fertilizers, plant nano-pesticides, growth regulators, and against pathogens (Fig. 12.3) (Hazarika et al. 2022).

12.3 Important Nanoparticles in Agriculture

Studies have shown that nanoparticles can be synthesized from microorganisms such as fungi and microalgae, as well as a wide variety of bacterial species (Hosea et al. 1986; He et al. 2009; Binupriya et al. 2010; Shivaji et al. 2011; Srivastava et al. 2013; Hulkoti and Taranath 2014). Silver (Ag), gold (Au), copper (Cu), iron (Fe), titanium (Ti), and zinc (Zn) nanoparticles are among the most common and preferred nanoparticles in agricultural practices (Duhan et al. 2017; Singh et al. 2021).

12.3.1 Silver Nanoparticles (AgNP)

Silver is one of the nanoparticles that exhibit high stability and low chemical reactivity, which is why it is frequently synthesized (Mustapha et al. 2022). Mishra et al. (2014), it was reported that AgNPs obtained from *Serratia* sp. BHU-S4 bacteria have antifungal activity against the plant pathogen *Bipolaris sorokiniana*, which is called spot disease in wheat plants.

Fig. 12.3 Potential use areas of nanoparticles in agriculture

12.3.2 Gold Nanoparticles (AuNP)

Gold nanoparticles were first studied by synthesizing from *Bacillus subtilis* 168 bacteria (Mandal et al. 2006). In a study conducted with AuNPs synthesized from the *Laccaria fraterna* EM-1083 fungus, it was determined that they showed antibacterial activity against *Pseudomonas aeruginosa*. In addition, the nanoparticle inhibited biofilm development and pyocyanin production of *P. aeruginosa* (Samanta et al. 2017).

12.3.3 Zinc Oxide Nanoparticles (ZnONP)

In addition to prokaryotic cells such as bacteria, eukaryotic cells such as fungi and yeast can synthesize ZnO nanoparticles by using enzymes, proteins, and other biomolecular compounds via an intracellular or extracellular pathway. In a study, it was determined that ZnONPs obtained from rhizospheric fungi such as *Aspergillus tubulin, Aspergillus niger, Aspergillus fumigatus, Fusarium oxysporum,* and

Penicillium citrinum had antibacterial effects against *Salmonella enterica, P. aerogenosa, Escherichia coli, Staphylococcus aureus, Bacillus cereus,* and *Listeria monocytogenes* (Hefny et al. 2019).

12.3.4 Titanium Nanoparticles (TiNP)

Titanium nanoparticles are preferred in the agricultural, cosmetic, and biomedical fields. Their high chemicalstability, photocatalytic properties, and low toxicity are the main reasons for their preference (Ilyas et al. 2021). A study in 2018, it was determined that titanium nanoparticles obtained from *Pseudomonas putida* (PpFT1), *P. fluorescens* (PfBN1), and *B. subtilis* (BsBN3) bacteria, and *Trichoderma harzianum* (ThFT1), *T. atroviride* (TvGK2), and *T. hamatum* (TmSA2) fungi all reduced powdery mildew disease in wheat plants at different rates (Farhat et al. 2018).

12.3.5 Iron Nanoparticles (FeNP)

Iron-NPs are generally smaller than 20 nm in size. FeNPs are important due to their potential to replace traditional Fe-containing fertilizers (Singh et al. 2021). Tarafdar and colleagues synthesized iron nanoparticles from the fungus Aspergillus oryzae TFR9. Researchers reported that the nanoparticle obtained from this fungus, which uses $FeCl_3$ as a precursor metal salt, has valuable potential in the field of agriculture due to its low cost, eco-friendly, and advantageousness compared to other chemical methods (Tarafdar and Raliya 2013).

12.3.6 Copper Nanoparticles (CuNP)

Copper is an important nutrient for plant growth. In addition, nanoparticles obtained from copper have effective antibacterial activity, clean, cost-effective, and efficient nanoparticle properties for agriculture (Shobha et al. 2014). Copper and silica nanoparticles obtained from *P. fluorescens*, *T. atroviride,* and *Streptomyces griseus* microorganisms have been shown to be significantly effective against infections affecting tea plants caused by *Poria hypolateritia* and *Phomopsis theae* fungi. The obtained nanoparticles have the potential to be used in the management of plant root and stem diseases as well as for improving tea quality (Natesan et al. 2021).

12.4 Epiphytic Microbial Flora and Its Effect on Plant Development

Soil is an extremely rich environment in terms of both major and minor elements and microbial load and diversity. In this way, it supports plant development both mechanically and chemically. Plants, like all other living things, have a flora dominated by microorganisms that provide benefits for them. Although the plant-flora relationship in healthy plants supports both sides, this balance is disrupted negatively in the presence of microbial pathogens. Plants obtain their flora from the soil, which is also defined as the largest reservoir of microbial diversity (Butani et al. 2021). The soil layer that is closely associated with the plant is the region called the rhizosphere. Although this region refers to a limited number of soil masses that establish close connections with plant roots, it expresses a high microbial richness with a prokaryotic diversity of more than 30,000 (Berendsen et al. 2012). A limited part of this diversity appears as flora in plants.

The plant microbiome represents different biotas with mycorrhizal, rhizobial, epiphytic, and endophytic natures. These communities, which generally have similar tasks to the microbial flora of all other living things, are shaped according to the physicochemical and biological parameters of the environment (Ding et al. 2022; Koprivova and Kopriva 2022). Microorganisms colonizing the outer surfaces of plants, such as leaves, stems, roots, and stalks, form the epiphytic flora. Primarily composed of bacteria and fungi, epiphytic flora can also include other eukaryotic microorganisms, depending on the living environment. They protect the plant against pathogens with a mechanical barrier and also defend it with the antimicrobial agents they produce. It is also known that these organisms support plant reproduction and the plant's abiotic stress response. Epiphytic microorganisms are evaluated under two headings: rhizobiome and phillobiome. Although this distinction was deemed necessary because the diversity of epiphytes colonizing the above- and below-ground parts of the plant is shaped by abiotic and biotic factors, the relationships of these two different microbiomes with the plant are similar (Butani et al. 2021; Das et al. 2023) (Fig. 12.4).

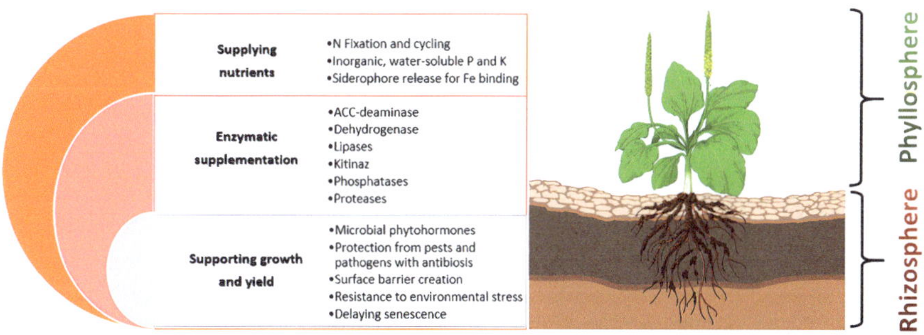

Fig. 12.4 Epiphytic flora communities and their basic functions

12.4.1 Formation of Epiphytic Flora

Root epiphytes located in the rhizoplane area are important components that play an active role in both plant development and protection. Some of the microorganisms responsible for the healthy execution of cycles in the soil settle in the rhizoplane and develop a close relationship with the plant. Although positive interactions are the focal point for plant development, it should not be forgotten that some parasitic epiphytes also tend to settle in this area. For this reason, the soil microbiome is of vital importance for plant health. This flora shows high diversity under the influence of various factors such as water content of the environment, amount of organic matter and pH (Berendsen et al. 2012; Yang et al. 2024). However, not all of the soil flora is transported to the rhizosphere. Many factors are involved in the inclusion of a microorganism on a plant surface, such as the organism's ability to adhere to any surface, the force of chemotaxis pulling it to the surface, and the microbial load of the adhesion surface. Primary metabolites such as sugars and amino acids secreted by plants, secondary metabolites such as flavonoids and terpenoids, and some volatile organic compounds serve to attract beneficial microorganisms (Siebers et al. 2018; Harbort et al. 2020; Liu et al. 2024; Pacheco-Moreno et al. 2024). Additionally, although less commonly used, specific polymers and other high-molecular-weight compounds may also serve as chemotaxis agents released by plants into the soil (Goswami and Deka 2022; Yang et al. 2024). These compounds attract some microorganisms toward the plant and especially aim to keep pathogens away.

Compounds released by the plant play a fundamental role in chemotaxis and colonization. In this way, the flora of plants in the same environment may change or differ depending on the type of plant or at different developmental stages (O'Neal and Alexandre 2020; Tsai et al. 2023). Since rhizobiome mechanisms are best clarified on bacteria, when we evaluate the adhesion phenomenon on the basis of these microorganisms, it can be said that epiphytic colonization is a kind of biofilm formation mechanism. Bacteria moving toward the root surface with the effect of chemotaxis lose their ability to move when they reach the region they are attracted to and stick to the surface. After weak bonding occurs by electrostatic or chemical means, the population in the environment increases logarithmically through the proliferation of bacteria adhering to the surface (Yang et al. 2024). In this process, many compounds play a role, such as bacterial surface polymers such as lipopolysaccharides (LPS) (Knights et al. 2021), bacterial signaling molecules such as di-AMP (Townsley et al. 2018), and plant surface molecules such as arabinogalactan (Imberty and Varrot 2008).

Weak adhesion does not mean that epiphytic settlement is complete. Plants have developed some mechanisms to protect themselves against pathogens. Overcoming these mechanisms and permanent settlement actually involves a very complex adaptation process. Many highly conserved microbial receptors, such as lipopolysaccharides (LPS), phospholipids, flg22, cold shock proteins (CSP), elongation factor

Tu (EF-Tu), chitin, and Nep1-like proteins, are recognized by plants and cause defense system activation (Pel et al. 2014; Zipfel and Oldroyd 2017).

The basic strategy is based on the microorganism changing the structure of these molecules or, if possible, destroying them. Although these escape mechanisms are more rapidly and effectively developed in pathogenic microorganisms, they are also found in beneficial bacteria (Yang et al. 2024). The best examples of this are seen in bacteria that break down their flagella with the proteases they secrete during adhesion and thus avoid defense stimulation via flg22. These bacteria can be pathogens such as *Pseudomonas syringae* and *Xanthomonas campestris*, as well as beneficial bacteria such as *Sinorhizobium meliloti* and *B. subtilis* (Felix et al. 1999; Trdá et al. 2014).

Another strategy is to prevent the defense from recognizing target molecules with some compounds such as organic acids and Nod factors released by bacteria (Zipfel and Oldroyd 2017; Yu et al. 2019). Another escape mechanism is the tolerance developed by the microorganism against various defense molecules released by the plant. Here, microorganisms either develop active neutralization strategies and inactivate the defense agents released by the plant or develop a passive adaptation strategy and express modifications that will not be affected by the released substances (Paulus et al. 2020; Xia et al. 2020).

When microorganisms do not stimulate the plant defense system, they begin to colonize and multiply on the roots without being exposed to the antimicrobial agents released by the plant. At this stage, new organisms that join the community can no longer be perceived by the plant, and the formation of a developed flora is paved. Epiphytic flora elements may change during and after these formation stages; new members may be added to the population; the dominant species distribution in the population may change for some organisms. This active process continues throughout the life of the plant and supports the plant.

12.4.2 Benefits of Epiphytic Flora to the Plant

Plants are covered with microorganisms from root to leaf and even to the fruit surface. While some of these organisms colonize the plant surface, some of them live inside the plant, such as in arbuscular mycorrhizas or endophytic bacteria, in the intercellular space or tissue. In addition, the existence and functions of microorganisms living in the rhizoplane and rhizosphere without establishing a tight bond with the plant should not be ignored. In all these points mentioned, microorganisms are actually flexible enough to adapt by sometimes passing from a very rigid base to its value. For this reason, the microbial flora of plants hosts an active population. Epiphytic flora is attributed to a very diverse community that colonizes not only the roots of the plant but also all surfaces, and can include photosynthetic microorganisms and plants that use the plant as hosts, depending on the presence of light. In the formation of this flora, organisms that can establish neutral and positive relationships with the plant are selected (Berendsen et al. 2012; Afzal et al. 2019; Butani et al. 2021; Das et al. 2023).

Epiphytic flora primarily involves the uptake of nutrients in the environment by the plant, as well as water uptake, stress resistance, and protection against diseases. Although epiphytic flora contains a high diversity, the best-known groups are bacteria. In particular, PGPB (plant growth-promoting bacteria) or systemic resistance-promoting bacteria have been the subject of many studies. The bacterial root microbiota is rich in members of *Proteobacteria* (mainly *Gammaproteobacteria* and *Alphaproteobacteria* classes) and *Actinobacteria*, as well as *Acidobacteria, Verrucomicrobia, Bacteroidetes,* and *Planctomycetes* (Reinhold-Hurek et al. 2015; Compant et al. 2019; Butani et al. 2021). Fungi represent the best-understood microbial group after bacteria. It is known that members of the *Ascomycota* phylum, such as *Trichoderma, Fusarium, Penicillium*, and *Scedosporium*, are dominant in root ecosystems (Salazar-Cerezo et al. 2018).

Perhaps the most significant supportive effect of epiphytic microorganisms on plants is their growth and metabolism supporting functions. Here, compounds released by bacteria and similar to plant hormones (phytohormones) and bacterial volatile organic compounds (VOCs) play a role. Phytohormones produced by plants are chemical compounds that promote plant growth (auxins, gibberellins (GA), cytokinins (CK) and ethylene (ET)) and are responsible for stress tolerance (brassinosteroids (BR), abscisic acid (ABA), salicylic acid (SA), jasmonic acid (JA) and strigolactone) (Bhatt et al. 2020). Auxins are the most important phytohormones produced by bacteria and fungi. Bacteria break down tryptophan to produce Indole-3-acetic acid (IAA), which plays the same role as an auxin in the plant. It is estimated that IAA, which can be synthesized in different ways, is synthesized by 80% of the rhizosphere flora (Butani et al. 2021). Information on bacteria that produce CKs, which are effective in cell division and root development, is much more limited. However, studies have shown that plants grown in soil to which these organisms were added have a noticeable difference in root and shoot development compared to the control group (Akhtar et al. 2020).

GAs, which are effective in new tissue development and reproduction, are hormones that are functional in almost all tissues and organs of the plant and regulate both plant development and physiology. It is known that this hormone, which is also produced by epiphytic and endophytic bacteria and fungi, plays a role in both development and adaptation to various stress factors as a plant homolog (Timofeeva et al. 2024). It is known that SA, another stress regulator, is synthesized in different environments by many well-known PGPBs such as *Pseudomonas, Achromobacter, Bacillus, Salmonella, Azospirillum, Yersinia, Vibrio,* and *Mycobacteria,* and that they support plant growth (Mishra and Baek 2021). Another regulator phytohormone that was not known to be produced by epiphytic or endophytic flora until recently is ABA. This hormone plays a role in the regulation of tolerance to stress via the catabolic pathway and in senescence. Studies have shown that various bacteria, such as *B. licheniformis, B. pumilus, Azospirillum lipoferum, Achromobacter xylosoxydans, Arthrobacter koreensis,* and *Brevibacterium halotolerans,* can produce this hormone (Timofeeva et al. 2024).

One of the most significant mechanisms in microbial regulation of plant growth is the blocking of ethylene production by ACC (1-aminocyclopropane-1-carboxylate) deaminase. Ethylene is present in very small amounts under normal conditions, but it is produced in excess under various stress conditions. Ethylene supports plant growth, serving a wide range of functions from fruit ripening to root development. Because it is a gas, it rapidly diffuses throughout plant tissues. When secreted in excess, it accelerates the plant's aging process. It is known that under some abiotic stress conditions, such as salinity, plant microflora produce ACC deaminase and thus suppress ethylene production. Examples of bacteria with this mechanism include *Bacillus filamentosus, P. migulae, P. fluorescens, Mesorhizobium, Burkholderia phytophirmans, Rhizobium leguminosarum Pseudomonas migulae, Janibacter indicus,* and *Brevibacterium casei* (Khalifa et al. 2020; Riyazuddin et al. 2020; Orozco-Mosqueda et al. 2023).

Another significant role of epiphytes is plant protection. Plants are under attack by many plant pathogens, from microorganisms to nematodes and insects. These can target roots and above-ground structures. The plant forms a strong line of defense against these pathogens with both its own metabolites and its microbiota. Epiphytic flora uses different mechanisms to protect the plant. These can be considered as contact inhibition, VOC, surfactants, and antimicrobial agents. Epiphytic bacteria such as *B. subtilis, Bacillus amyloliquefaciens, Enterobacter asburiae, Wautersia paucula, Serratia marcescens, Bacillus simplex, B. weihenstephanensis, B. subtilis,* and epiphytic fungi such as *F. oxysporum, Fusarium fujikuroi,* and *Rhizoctonia solani* are examples of these. These microorganisms exhibit antimicrobial, nematocidal, and insecticidal effects in different ways (Boer et al. 2019; Timofeeva et al. 2024; Xu et al. 2024).

Another function of the epiphytic flora in supporting plant growth is to regulate nutrient cycling and provide the plant with the major and minor elements it needs. Elements such as nitrogen, iron, and potassium are at the top of this list. In addition, the pH-regulating effect is another factor provided during elemental cycle regulation and is important for plant growth (Hirsch et al. 2024; Moreau et al. 2019; Tang et al. 2024). Although root epiphytes play an active role in these cycles, atmospheric N-fixing bacteria in the leaf flora have also been proven to support the plant (Zhu et al. 2023). The phyllosphere flora also provides nutritional support to the plant with functions such as anti-pesticide action, abiotic stress resistance, phytohormone secretion, and secretion of enzymes to degrade reactive oxygen compounds. Biological nitrogen fixation is a process described in bacteria and archaea and is generally inactivated by oxygen. Nitrogen fixation releases ammonia (NH_3). However, nitrogen sources available to plants are nitrate (NO_3^-) or ammonium (NH_4^+). The resulting product is assimilated into glutamate or can be converted into plant-available compounds via nitrogenase. *Azotobacter vinelandii, Rhodopseudomonas palustris, E. coli, Azospirillum, Azoarcus, Bacillus, Burkholderia, Klebsiella, Derxia, Actinomycetes, Herbaspirillum, Gluconacetobacter, Paenibacillus polymyxa, Pseudomonas*, and many cyanobacteria species are known to be involved in the nitrogen cycle, which is a highly complex cycle (Grzyb and Szulc 2024).

After nitrogen, the other crucial major element is phosphorus. Plants can take up phosphorus as orthophosphate, but even with external additions, phosphate in the soil is quickly converted to other forms that cannot be assimilated. Phosphate-solubilizing bacteria (PSB) synthesize acid phosphatases, phytases, phosphonatases, and C-P lyases to provide phosphorus in the form required by the plant. *Bacillus, Xanthomonas, Enterobacter, Acinetobacter, Pseudomonas, Alcaligenes, Actinobacteria, Lactococcus, Paenibacillus, Klebsiella, Serratia*, and *Burkholderia* are some of the rhizosphere PSBs (Li et al. 2021; Wu et al. 2022; Scott et al. 2024).

A part of the most commonly used chemical fertilizer combinations (NPKs) and involved in metabolic events related to growth and development, potassium plays a key role in many events, from homeostasis to protein synthesis and photosynthesis. K ions, which exist in four different forms depending on the degree of dissolution in the soil, are readily taken up by the plant, especially in the dissolved form. However, this constitutes only 1–2% of the K reserves in the soil. The rest is in silica forms such as mica, biotite, orthoclase, feldspar, and vermiculite. Potassium-solubilizing bacteria (PSB) are involved in the dissolution of potassium, silicon, and aluminum from these crystalline minerals. Some members of *Bacillota, Pseudomonadota, Actinomycetota,* and *Bacteroidota* (especially families of *Bacillaceae, Paenibacillaceae; Rhizobiaceae, Alcaligenaceae, Burkholderiaceae, Pseudomonadaceae, Enterobacteriaceae,* and *Lysobacteraceae; Micrococcaceae, Microbacteriaceae; Flavobacteriaceae, Cytophagaceae*, respectively) are KSB (Jain et al. 2022; Nassef et al. 2025).

A final feature of epiphytic microorganisms that support plant growth is their water-soluble siderophores. These proteins, which have an iron-binding function, are released into the environment, especially by bacteria, and catalyze iron transfer from the soil to the cells. Iron is vital for chloroform production and plant health. It also plays a regulatory role in non-green parts, from seed development to the formation of root shoots. Although relatively abundant in the earth's crust, the elemental form of iron taken up by organisms is very limited. Siderophore shows an exceptionally high affinity for Fe3+ ions. Siderophores of microbial origin are generally classified such as catecholate, phenolate, hydroxamate, (α-hydroxy) carboxylate, and mixed type according to their ligands. These compounds, classified by differences in molecular structure and iron-binding active sites, are also synthesized by members of the epiphytic flora. Bacteria have become so specialized that they have developed various siderophores to give them an advantage over other species they compete with. This structure, found in most bacteria, is also typical in the *Actinomycetota, Pseudomonadota, Cyanobacteriota*, and *Bacillota*, which harbor members of the epiphytic flora. Studies have shown that epiphytic microflora not only provides iron for plant growth; this Fe provided through siderophores increases the plant's resistance to abiotic stresses and weakens the disease-causing abilities of pathogens by causing iron deficiency (Grzyb and Szulc 2024; Riseh et al. 2025; Wang and Li 2025).

12.5 The Importance of Nanoparticle-Supported Epiphytic Flora in Plant Growth and Development

Microorganisms can develop different adaptations to survive in their environment and compete with other microorganisms. Microorganisms associated with plants can show similar characteristics. Therefore, the possibility of positive effects of nanoparticle studies related to these epiphytic microorganisms on the plant increases. Therefore, in this part of the chapter, researches revealing that nanoparticles associated with epiphytic microorganisms may be important in plant cultivation will be review.

12.5.1 Antimicrobial Effects of Nanoparticles

In agriculture, the infection of plants with various microorganisms causes crop losses. Therefore, preventing the infection of plants or controlling the infection is important for the sustainability of agriculture. Studies have shown that various microorganisms defined as endophytic can exhibit antimicrobial activity. In one study, AuNP produced by the endophytic bacterium *Pantoea ananatis* were found to exhibit effective antimicrobial activity, especially against pathogens such as *Bacillus cereus* and *Candida albicans* (Monowar et al. 2018). In another study, the antimicrobial activity of AgNPs obtained from the bacterium *Streptomyces* sp. MI-3 (isolated from the plant Mentha longifolia) was investigated. The results showed that the nanoparticles had antimicrobial activity against bacteria such as *Bacillus dimenuta*, *B. subtilis*, *S. aureus*, *P. aeruginosa*, and *E. coli*, and fungi such as *C. albicans* and *A. brasiliensis* (El-Gamal et al. 2018). In a study on CuONPs synthesized from Streptomyces zaomyceticus Oc-5 and Streptomyces pseudogriseolus Acv-11 bacteria (isolated from the plant *Oxalis corniculata*), it was determined that they have antimicrobial activity against the microorganisms *B. dimenuta, B. subtilis, S. aureus, P. aeruginosa, E. coli, C. albicans, A. brasiliensis, F. oxysporum, Pythium ultimum, A. niger* and *Alternaria alternata*. In addition, it was determined that the nanoparticles had larvicidal effects against *Musca domestica* and *Culex pipiens* (Hassan et al. 2019). In another study by the same team, it was revealed that AgNPs of these bacteria also have antibacterial, antifungal, and antilarvacidal effects (Fouda et al. 2020). In a study conducted with AgNPs synthesized by bacteria isolated from *Azadirachta indica*, it was determined that the B4 isolate, which was determined to be *B. subtilis*, had antibacterial activity on *K. pneumonia, P. aeruginosa, E. coli,* and methicillin-resistant *S. aureus* pathogens (Kattali et al. 2024). In another study, it was determined that AgNPs of the endophytic *Pseudomonas poae* CO strain isolated from *Allium sativum* had antifungal activity against *F. graminearum* (Ibrahim et al. 2020). A study, an endophytic strain, *B. siamensis*, was isolated from the *Coriandrum sativum* plant. It was revealed that AgNP synthesized from this bacterium protected rice plants (*Xanthomonas oryzae* and *Acidovorax oryzae*) from bacterial infections and supported plant growth (Ibrahim et al. 2019). Abdel-Maksoud et al. (2024) determined that TiO_2NPs synthesized from *Lactobacillus rhamnosus* bacteria exhibited antifungal activity against plant pathogens such as *A. flavus, A. versicolor, A. ustus, A. chinensis, P. citrinum,* and *P. chrysogenum.*

Similar properties are also observed in fungi. AgNPs were synthesized from an endophytic *Phomopsis helianthi* fungus isolated from the leaves of the *Nyctanthes arbor-tristis* plant. It was determined that the NPs obtained in this study showed antibacterial activity against the pathogens *E. coli, Staphylococcus* sp., *Morganella morgenii, Proteus vulgaris, P. aeruginosa, S. enteritidis,* and *Shigella boydii.* In addition, it has been discussed in the study that *Phomopsis* species may have anti-malarial effects in addition to their antimicrobial effects (Gond et al. 2020). Ponmurugan (2017), gold and silver nanoparticles synthesized from the *T. atroviride* fungus have been shown to have an antifungal effect against *Phomopsis theae*, a fungal infection agent that causes economic losses in tea plants. These NPs have the potential to be used as biological control agents. In a study, Ag and Au NPs were synthesized from the *Chaetomium globosum* endophytic fungus isolated from the *Tectona grandis* plant leaf. While no antibacterial effect could be detected in AuNP, AgNP showed significant activity against *E. coli, S. aureus, P. mirabilis,* and *K. pneumoniae* pathogens (Singh et al. 2018). Ag NPs obtained from *Cryptococcus laurentii* and *Rhodotorula glutinis* yeasts isolated from apple showed antifungal activity against phytopathogenic fungi (*Botrytis cinerea, P. expansum, A. niger, Alternaria* sp., and *Rhizopus* sp). Since the studied yeasts are GRAS microorganisms that are safe for use in food, the nanoparticles obtained from them have the potential to be used both in the prevention of plant diseases and in food packaging (Fernández et al. 2016).

12.5.2 Nanoparticles Support Plant Growth

In agriculture, plant development plays an important role in increasing product quality. For this reason, various substances such as chemical fertilizers are frequently used. However, the use of nanoparticles to support plant growth is seen as a more eco-friendly and economical method (Singh et al. 2023). Studies have shown that nanoparticles produced by bacteria can support plant growth in different ways. In a study conducted in 2022, it was determined that *Rhizobium pusense* (MS-1), *B. cereus* MS-2, *B. flexus* (MS-3), *Methylophilus flavus* (MS-4), and *P. aeruginosa* (MS-5) endophytic bacterial strains increased the growth of tomatoes. It was determined that plant growth occurred in the form of increased root and shoot length, biomass, and leaf chlorophyll content. In addition, it has been determined that nanoparticles have antibacterial effects against pathogenic bacteria such as *B. thuringiensis, P. putida, Azotobacter chroococcum, E. coli, B. licheniformis,* and *Rhizobium* sp. (Singh et al. 2022).

A 2022 study reported that ZnO nanoparticles induced siderophore production in the epiphytic *P. fluorescens* NK4 isolate. Furthermore, the study indicated that the presence of these nanoparticles increased siderophore production, supported plant growth, and suppressed the growth of the *P. viridiflava* NK2 strain, which causes fruit rot. In addition to the use of ZnO NPs as nano-fertilizers, it has been demonstrated that they also have the potential to be used as biocontrol agents (Al-Karablieh et al. 2022). Panichikkal et al. (2019) investigated the effects of nanoparticles synthesized using the *B. subtilis* SJ15 strain, isolated from soil, on plant growth and the endophytic Pseudomonas monteilii strain, known to produce IAA. According to the results obtained, it was determined that *P. monteilii* culture treated with AuNPs developed the plant root and shoot system and also increased its biomass (Panichikkal et al. 2019). In another study, SiNPs were synthesized from various endophytic bacteria isolated from various *Paspalum* spp. species. *S. marcescens* was isolated from *P. lenticulare, Enterobacter asburiae* from *P. rojasii*, and *Alcaligenes faecalis* bacteria were isolated from *P. compressifolium*. SiNPs belonging to these bacteria caused positive effects such as increased shoot length, increased root length, increased seed germination, high shoot dry weight, and seedling growth in *P. notatum* plant (de Moraes et al. 2024). A study conducted with chitosan nanoparticles (CNP) and AuNPs synthesized by *Pseudomonas* spp. bacteria isolated from *Cyclea peltata* found that nanoparticles increased the shoot length and fresh weight of *Vigna unguiculata* plant (Panichikkal and Krishnankutty 2022).

A study has shown that TiO_2NPs synthesized from *A. flavus* TFR 7 fungus isolated from the plant rhizosphere can be used as plant growth promoters on mung beans (*Vigna radiata* L.). In this study, it was determined that nanoparticles increased shoot length, root length, root area, root nodule, chlorophyll content, and total soluble leaf protein. In addition, an increase in acid phosphatase, alkaline phosphatase, phytase, and dehydrogenase enzymes was detected (Raliya et al. 2015). Positive effects of ZnO nanoparticles obtained from *A. fumigatus* TFR-8 fungus were observed on plant biomass in root stem, shoot length, root length, root area, chlorophyll content, total soluble leaf protein, rhizospheric microbial population, acid phosphatase, alkaline phosphatase, and phytase (Raliya and Tarafdar 2013; Mohd Yusof et al. 2019). FeNPs and MnNPs synthesized from endophytic *Paenibacillus polymyxa* DEBB B-358 bacteria, which were isolated from eucalyptus leaves and known to synthesize IAA, showed positive results in germination rate, root growth, and fresh weight in corn plants. The results obtained reveal that these nanoparticles have the potential to be used as nano-fertilizers (de França Bettencourt et al. 2020).

12.5.3 Nanoparticles Confer Resistance to Plants

Cadmium (Cd) is a toxic heavy metal that can be easily absorbed by various plants. Since the consumption of these plants by humans can cause the emergence of many diseases, it is considered a global problem. Different FeNPs (Fe_2O_3NP and Fe_3O_4NP) were synthesized from the *Methylobacterium currus* DSD isolate isolated from rice. The results obtained revealed that FeNPs play a role in reducing the toxicity and accumulation of Cd in plants. It was also determined that the mechanisms effective in the complexation process of Cd are affected by the chemical components of EPS formed by epiphytic microorganisms (Ding et al. 2024).

In a study in which CdONP was synthesized from *Cupriavidus metallidurans* CML2 bacteria isolated from *O. sativa*, the resistance of this nanoparticle to Cd was investigated. The nanoparticle showed high tolerance to heavy metals such as Cd^{2+}, Hg^{2+}, Cu^{2+}, Ni^{2+}, Zn^{2+}, $Se^{4+,}$ and Mn^{2+}. The results obtained in the study show that CdONP obtained from this bacterium can provide heavy metal resistance to plants and can also be used in the remediation of soils carrying high metal concentrations. In addition, it was determined that the nanoparticle obtained in this study can produce plant growth promoters such as IAA, soluble phosphorus compounds, and siderophores in rice plants and provide an increase in root lengths in rice seedlings (Zhang et al. 2024). Recent studies have shown that nanoparticles can be resistant to different conditions in the environment in which plants are located.

12.6 Effects of Nanoparticles on Epiphytic Microbiota

Nanoparticles can be used for different purposes in agriculture. It is also possible to make different improvements in terms of agriculture by administering the synthesized nanoparticles to plants. Nanoparticles can support plant development by affecting endophytic microbial communities and affecting features such as regulation, community structure, and function. For example, in a study, 2024, it was determined that ZnONPs changed the composition of epiphytic microorganisms in tea plants. In addition, it was determined that microorganisms belonging to the genera *Rhodococcus, Salana, Truepera, Mycobacterium, Chryseobacterium, Aureimonas, Staphylococcus, Stenotrophomonas, Aquipuribacter, Patulibacter, Arthrobacter, Nocardioides, Paracoccus, Arsenophonus, Microbacterium,* and *Brevundimonas* became dominant in tea plant samples treated with ZnONP. Some of these microorganisms are known to synthesize substances that improve living conditions for plants and repair their damage, while others are known to have antagonistic properties against pathogens (Chen et al. 2024). Similarly, the effects of AgNPs on the development of *Triticum aestivum* were examined. According to the results obtained, it was determined that microorganisms important for plant development, especially rhizobacteria that promote plant growth, increased after treatment with AgNPs. These microorganisms are known for their plant growth-supporting properties, such as N-fixation, bacterial phytohormones, siderophore production, urease, and protease production (Przemieniecki et al. 2024). Studies show that treating plants with nanoparticles also supports epiphytic microorganisms and contributes to plant growth and development.

12.7 Conclusion and Future Trends

Many environmental conditions negatively affect plant development. Today, many studies are being conducted on supporting plant development using nanoparticles. Some of these studies are directly related to endophytic microorganisms associated with plants. Studies show that metal oxide, metal, carbon-based, and silica nanoparticles can also support epiphytic microorganisms that significantly contribute to plant growth and development. It has been revealed that this interaction has an enhancing effect on plant development. In addition, studies have proven that nanoparticles synthesized from these epiphytic microorganisms also have positive effects on plant development.

Research on the support of nanoparticles for both epiphytic microorganisms and plant development primarily aims to standardize protocols. Thus, the use of nanoparticles can become more commercial. In addition, nanoparticles above a certain level can have a toxic effect on the environment. Therefore, appropriate doses should be determined and applications should be made carefully. The interaction between nanoparticles and epiphytic microorganisms holds great promise in terms of preventing food shortages that we may be exposed to in the future.

References

Abdel-Maksoud G, Abdel-Nasser M, Hassan SED, Eid AM, Abdel-Nasser A, Fouda A (2024) Biosynthesis of titanium dioxide nanoparticles using probiotic bacterial strain, Lactobacillus rhamnosus, and evaluate of their biocompatibility and antifungal activity. Biomass Convers Biorefinery 14(19):23961–23983. https://doi.org/10.1007/s13399-023-04587-x

Afzal I, Shinwari ZK, Sikandar S, Shahzad S (2019) Plant beneficial endophytic bacteria: mechanisms, diversity, host range, and genetic determinants. Microbiol Res 221:36–49. https://doi.org/10.1016/j.micres.2019.02.001

Akhtar SS et al (2020) Role of Cytokinins for interactions of plants with microbial pathogens and Pest insects. Front Plant Sci 10:1777. https://doi.org/10.3389/fpls.2019.01777

Al-Karablieh N, Al-Shomali I, Al-Elaumi L, Hasan K (2022) Pseudomonas fluorescens NK4 siderophore promotes plant growth and biocontrol in cucumber. J Appl Microbiol 133(3):1414–1421. https://doi.org/10.1111/jam.15645

Andaç M, Dikbaş Ç, Akyüz G (2022) Nanopartiküllerin genel özellikleri, sentez ve karakterizasyon teknikleri. Turkiye Klinikleri Veter Sci Pharmacol Toxicol Spec Topics 8(1):1–10

Bahrulolum H, Nooraei S, Javanshir N, Tarrahimofrad H, Mirbagheri VS, Easton AJ, Ahmadian G (2021) Green synthesis of metal nanoparticles using microorganisms and their application in the agrifood sector. J Nanobiotechnol 19:1–26. https://doi.org/10.1186/s12951-021-00834-3

Berendsen RL, Pieterse CM, Bakker PA (2012) The rhizosphere microbiome and plant health. Trends Plant Sci 17(8):478–486. https://doi.org/10.1016/j.tplants.2012.04.001

Bhatt D et al (2020) Role of growth regulators and Phytohormones in overcoming environmental stress. In: Protective chemical agents in the amelioration of plant abiotic stress: biochemical and molecular perspectives. Wiley, p 254. https://doi.org/10.1002/9781119552154.ch11

Binupriya AR, Sathishkumar M, Vijayaraghavan K, Yun SI (2010) Bioreduction of trivalent aurum to nano-crystalline gold particles by active and inactive cells and cell-free extract of

Aspergillus oryzae var. viridis. J Hazard Mater 177(1–3):539–545. https://doi.org/10.1016/j.jhazmat.2009.12.066

Boer W et al (2019) Pathogen suppression by microbial volatile organic compounds in soils. FEMS Microbiol Ecol 95(8). https://doi.org/10.1093/femsec/fiz105

Butani N, Desai P, Trivedi S (2021) Epiphytic microbes of roots: diversity and significance. In: Nath M, Bhatt D, Bhargava P, Choudhary DK (eds) Microbial Metatranscriptomics belowground. Springer, Singapore. https://doi.org/10.1007/978-981-15-9758-9_17

Chen H, Song Y, Wang Y, Wang H, Ding Z, Fan K (2024) ZnO NPs enhanced photosynthetic capacity, promoted new shoot development, and improved the community composition of Phyllosphere epiphytic and endophytic microorganisms in tea plants. https://doi.org/10.21203/rs.3.rs-4019055/v1

Compant S, Samad A, Faist H, Sessitsch A (2019) A review on the plant microbiome: ecology, functions, and emerging trends in microbial application. J Adv Res 19:29–37. https://doi.org/10.1016/j.jare.2019.03.004

Das T, Bhattacharyya A, Bhar A (2023) Linking Phyllosphere and rhizosphere microbiome to the plant–insect interplay: the new dimension of tripartite interaction. Physiologia 3(1):129–144. https://doi.org/10.3390/physiologia3010011

de França Bettencourt GM, Degenhardt J, Torres LAZ, de Andrade Tanobe VO, Soccol CR (2020) Green biosynthesis of single and bimetallic nanoparticles of iron and manganese using bacterial auxin complex to act as plant bio-fertilizer. Biocatal Agric Biotechnol 30:101822. https://doi.org/10.1016/j.bcab.2020.101822

de Moraes ACP, Kingsley KL, Ribeiro LDS, Vigna BBZ, Camargo ERD, White JF et al (2024) Beneficial bacteria associated with silica nanoparticles for growth promotion of Paspalum notatum. Appl Environ Soil Sci 2024(1):9971370. https://doi.org/10.1155/aess/9971370

Ding H, Liu T, Hu Q, Liu M, Cai M, Jiang Y, Cao C (2022) Effect of microbial community structures and metabolite profile on greenhouse gas emissions in rice varieties. Environ Pollut 306:119365. https://doi.org/10.1016/j.envpol.2022.119365

Ding S, Hu Z, Liu Y, Li L, Zeng Y, Jin D et al (2024) Enhanced cadmium absorption and tolerance of rice epiphytic microbes by iron oxide nanoparticles. Int Biodeterior Biodegradation 190:105770. https://doi.org/10.1016/j.ibiod.2024.105770

Duhan JS, Kumar R, Kumar N, Kaur P, Nehra K, Duhan S (2017) Nanotechnology: the new perspective in precision agriculture. Biotechnol Rep 15:11–23. https://doi.org/10.1016/j.btre.2017.03.002

El-Gamal MS, Salem SS, Abdo AM (2018) Biosynthesis, characterization, and antimicrobial activities of silver nanoparticles synthesized by endophytic Streptomyces sp. J Biotechnol 56:69–85

Farhat MG, Haggag WM, Thabet MS, Mosa AA (2018) Efficacy of silicon and titanium nanoparticles biosynthesis by some antagonistic fungi and bacteria for controlling powdery mildew disease of wheat plants. Int J Agric Technol 14:661

Felix G, Duran JD, Volko S, Boller T (1999) Plants have a sensitive perception system for the most conserved domain of bacterial flagellin. Plant J 18:265–276. https://doi.org/10.1046/j.1365-313x.1999.00265.x

Fernández JG, Fernández-Baldo MA, Berni E, Camí G, Durán N, Raba J, Sanz MI (2016) Production of silver nanoparticles using yeasts and evaluation of their antifungal activity against phytopathogenic fungi. Process Biochem 51(9):1306–1313. https://doi.org/10.1016/j.procbio.2016.05.021

Fouda A, Hassan SED, Abdo AM, El-Gamal MS (2020) Antimicrobial, antioxidant and larvicidal activities of spherical silver nanoparticles synthesized by endophytic Streptomyces spp. Biol Trace Elem Res 195:707–724. https://doi.org/10.1007/s12011-019-01883-4

Gond SK, Mishra A, Verma SK, Sharma VK, Kharwar RN (2020) Synthesis and characterization of antimicrobial silver nanoparticles by an endophytic fungus isolated from Nyctanthes arbor-tristis. Proc Natl Acad Sci India Sect B Biol Sci 90:641–645. https://doi.org/10.1007/s40011-019-01137-2

Goswami M, Deka S (2022) Rhizodeposits: an essential component for microbial interactions in rhizosphere. In: Singh UB, Rai JP, Sharma AK (eds) Re-visiting the rhizosphere eco-system for agricultural sustainability. Springer Nature, Singapore, pp 129–151. https://doi.org/10.1007/978-981-19-4101-6_7

Grzyb T, Szulc J (2024) Deciphering molecular mechanisms and diversity of plant Holobiont bacteria: microhabitats, community ecology, and nutrient acquisition. Int J Mol Sci 25(24):13601. https://doi.org/10.3390/ijms252413601

Harbort CJ, Hashimoto M, Inoue H, Niu Y, Guan R, Rombolà AD et al (2020) Root-secreted coumarins and the microbiota interact to improve iron nutrition in Arabidopsis. Cell Host Microbe 28:825–837.e6. https://doi.org/10.1016/j.chom.2020.09.006

Hassan SED, Fouda A, Radwan AA, Salem SS, Barghoth MG, Awad MA et al (2019) Endophytic actinomycetes Streptomyces spp mediated biosynthesis of copper oxide nanoparticles as a promising tool for biotechnological applications. JBIC J Biol Inorg Chem 24:377–393. https://doi.org/10.1007/s00775-019-01654-5

Hazarika A, Yadav M, Yadav DK, Yadav HS (2022) An overview of the role of nanoparticles in sustainable agriculture. Biocatal Agric Biotechnol 43:102399. https://doi.org/10.1016/j.bcab.2022.102399

He W, Zhou W, Wang Y, Zhang X, Zhao H, Li Z, Yan S (2009) Biomineralization of iron phosphate nanoparticles in yeast cells. Mater Sci Eng C 29(4):1348–1350. https://doi.org/10.1016/j.msec.2008.10.030

Hefny ME, El-Zamek FI, El-Fattah A, Mahgoub SAM (2019) Biosynthesis of zinc nanoparticles using culture filtrates of Aspergillus, Fusarium and Penicillium fungal species and their antibacterial properties against gram-positive and gram-negative bacteria. Zagazig J Agric Res 46(6):2009–2021. https://doi.org/10.21608/zjar.2019.51920

Hirsch M, Burges PL, Migueliz L et al (2024) Isolation of beneficial bacteria from strawberry (Fragaria x ananassa, Duch). Potentialities for fungal disease control and plant growth promotion. Plant Growth Regul 102:135–152. https://doi.org/10.1007/s10725-023-00989-z

Hosea M, Greene B, Mcpherson R, Henzl M, Alexander MD, Darnall DW (1986) Accumulation of elemental gold on the alga Chlorella vulgaris. Inorg Chim Acta 123(3):161–165. https://doi.org/10.1016/S0020-1693(00)86339-2

Hulkoti NI, Taranath TC (2014) Biosynthesis of nanoparticles using microbes—a review. Colloids Surf B: Biointerfaces 121:474–483. https://doi.org/10.1016/j.colsurfb.2014.05.027

Hulla JE, Sahu SC, Hayes AW (2015) Nanotechnology: history and future. Hum Exp Toxicol 34(12):1318–1321. https://doi.org/10.1177/0960327115603588

Ibrahim E, Fouad H, Zhang M, Zhang Y, Qiu W, Yan C et al (2019) Biosynthesis of silver nanoparticles using endophytic bacteria and their role in inhibition of rice pathogenic bacteria and plant growth promotion. RSC Adv 9(50):29293–29299. https://doi.org/10.1039/C9RA04246F

Ibrahim E, Zhang M, Zhang Y, Hossain A, Qiu W, Chen Y et al (2020) Green-synthesization of silver nanoparticles using endophytic bacteria isolated from garlic and its antifungal activity against wheat Fusarium head blight pathogen Fusarium graminearum. Nanomaterials 10(2):219. https://doi.org/10.3390/nano10020219

Imberty A and Varrot A (2008) Microbial recognition of human cell surface glycoconjugates. Curr Opin Struct Biol 18(5):567–576.

Ilyas M, Waris A, Khan AU, Zamel D, Yar L, Baset A et al (2021) Biological synthesis of titanium dioxide nanoparticles from plants and microorganisms and their potential biomedical applications. Inorg Chem Commun 133:108968. https://doi.org/10.1016/j.inoche.2021.108968

Jadoun S, Chauhan NPS, Zarrintaj P, Barani M, Varma RS, Chinnam S, Rahdar A (2022) Synthesis of nanoparticles using microorganisms and their applications: a review. Environ Chem Lett 20(5):3153–3197. https://doi.org/10.1007/s10311-022-01444-7

Jain D, Saheewala H, Sanadhaya S, Joshi A, Bhojiya AA, Verma AK, Mohanty SR (2022) Potassium solubilizing microorganisms as soil health engineers: an insight into molecular mechanism. In: Rhizosphere engineering. Academic Press, pp 199–214. https://doi.org/10.1016/B978-0-323-89973-4.00007-7

Jampílek J, Králová K (2025) Nanotoxicity effects of metal oxide nanoparticles on aquatic and soil ecosystems. In: Nanoscience and soil-water interaction in agroecosystem, p 1737
Kattali M, Mampett KP, Kodoor HF, Ponnenkunnathu SG, Soman S, Kar Mahapatra D et al (2024) Biofabrication of silver nanoparticles by Azadirachta indica rhizosphere bacteria with enhanced antibacterial properties. Eng 5(2):600–613. https://doi.org/10.3390/eng5020034
Khalifa AY, Metwally A, Ammar RB, Farghaly FA (2020) ACC deaminase-containing rhizobacteria from rhizosphere of Zygophyllum coccineum alleviate salt stress impact on wheat (Triticum aestivum L.). Scientific Journal For King Faisal University 21(1):89–101. https://doi.org/10.37575/b/agr/1988
Knights HE, Jorrin B, Haskett TL, Poole PS (2021) Deciphering bacterial mechanisms of root colonization. Environ Microbiol Rep 13(4):428–444
Koprivova A, Kopriva S (2022) Plant secondary metabolites altering root microbiome composition and function. Curr Opin Plant Biol 67:102227. https://doi.org/10.1016/j.pbi.2022.102227
Li JT, Lu JL, Wang HY, Fang Z, Wang XJ, Feng SW et al (2021) A comprehensive synthesis unveils the mysteries of phosphate-solubilizing microbes. Biol Rev 96(6):2771–2793. https://doi.org/10.1111/brv.12779
Liu Y, Xu Z, Chen L, Xun W, Shu X, Chen Y et al (2024) Root colonization by beneficial rhizobacteria. FEMS Microbiol Rev 48:fuad066. https://doi.org/10.1093/femsre/fuad066
Mandal D, Bolander ME, Mukhopadhyay D, Sarkar G, Mukherjee P (2006) The use of microorganisms for the formation of metal nanoparticles and their application. Appl Microbiol Biotechnol 69:485–492. https://doi.org/10.1007/s00253-005-0179-3
Mishra AK, Baek K-H (2021) Salicylic acid biosynthesis and metabolism: a divergent pathway for plants and bacteria. Biomolecules 11(5):705. https://doi.org/10.3390/biom11050705
Mishra S, Singh BR, Singh A, Keswani C, Naqvi AH, Singh HB (2014) Biofabricated silver nanoparticles act as a strong fungicide against Bipolaris sorokiniana causing spot blotch disease in wheat. PLoS One 9(5):e97881. https://doi.org/10.1371/journal.pone.0097881
Mohd Yusof H, Mohamad R, Zaidan UH, Abdul Rahman NA (2019) Microbial synthesis of zinc oxide nanoparticles and their potential application as an antimicrobial agent and a feed supplement in animal industry: a review. J Anim Sci Biotechnol 10:1–22. https://doi.org/10.1186/s40104-019-0368-z
Monowar T, Rahman MS, Bhore SJ, Raju G, Sathasivam KV (2018) Silver nanoparticles synthesized by using the endophytic bacterium Pantoea ananatis are promising antimicrobial agents against multidrug resistant bacteria. Molecules 23(12):3220. https://doi.org/10.3390/molecules23123220
Moreau D, Bardgett RD, Finlay RD, Jones DL, Philippot L (2019) A plant perspective on nitrogen cycling in the rhizosphere. Funct Ecol 33:540–552. https://doi.org/10.1111/1365-2435.13303
Mustapha T, Misni N, Ithnin NR, Daskum AM, Unyah NZ (2022) A review on plants and microorganisms mediated synthesis of silver nanoparticles, role of plants metabolites and applications. Int J Environ Res Public Health 19(2):674. https://doi.org/10.3390/ijerph19020674
Nassef K, Sahli A, Bouhdid S, Mezzoug N, Abrini J, Khay EO (2025) Phosphate solubilizing microorganisms and their use in sustainable agriculture: a review. Geomicrobiol J:1–20. https://doi.org/10.1080/01490451.2025.2457660
Natesan K, Ponmurugan P, Gnanamangai BM, Manigandan V, Joy SPJ, Jayakumar C, Amsaveni G (2021) Biosynthesis of silica and copper nanoparticles from Trichoderma, Streptomyces and Pseudomonas spp. evaluated against collar canker and red root-rot disease of tea plants. Arch Phytopathol Plant Protect 54(1–2):56–85. https://doi.org/10.1080/03235408.2020.1817258
O'Neal LVL, Alexandre G (2020) Specific root exudate compounds sensed by dedicated chemoreceptors shape Azospirillum brasilense chemotaxis in the rhizosphere. Appl Environ Microbiol 86:e01026–20. https://doi.org/10.1128/AEM.01026-20
Orozco-Mosqueda MDC, Santoyo G, Glick BR (2023) Recent advances in the bacterial Phytohormone modulation of plant growth. Plants 12(3):606. https://doi.org/10.3390/plants12030606

Pacheco-Moreno A, Bollmann-Giolai A, Chandra G, Brett P, Davies J, Thornton O et al (2024) The genotype of barley cultivars influences multiple aspects of their associated microbiota via differential root exudate secretion. PLoS Biol 22:e3002232. https://doi.org/10.1371/journal.pbio.3002232

Panichikkal J, Krishnankutty RE (2022) Chitosan and gold nanoparticles supplementation for augmentation of indole-3-acetic acid production by rhizospheric Pseudomonas aeruginosa and plant growth enhancement. Curr Microbiol 79(6):185. https://doi.org/10.1007/s00284-022-02850-4

Panichikkal J, Thomas R, John JC, Radhakrishnan EK (2019) Biogenic gold nanoparticle supplementation to plant beneficial Pseudomonas monteilii was found to enhance its plant probiotic effect. Curr Microbiol 76:503–509. https://doi.org/10.1007/s00284-019-01649-0

Paulus JK, Kourelis J, Ramasubramanian S, Homma F, Godson A, Hörger AC et al (2020) Extracellular proteolytic cascade in tomato activates immune protease Rcr3. Proc Natl Acad Sci USA 117:17409–17417. https://doi.org/10.1073/pnas.1921101117

Pel MJC, van Dijken AJH, Bardoel BW, Seidl MF, van der Ent S, van Strijp JAG et al (2014) Pseudomonas syringae evades host immunity by degrading flagellin monomers with alkaline protease AprA. Mol Plant-Microbe Interact 27:603–610. https://doi.org/10.1094/MPMI-02-14-0032-R

Ponmurugan P (2017) Biosynthesis of silver and gold nanoparticles using Trichoderma atroviride for the biological control of Phomopsis canker disease in tea plants. IET Nanobiotechnol 11(3):261–267. https://doi.org/10.1049/iet-nbt.2016.0029

Przemieniecki SW, Ruraż K, Kosewska O, Oćwieja M, Gorczyca A (2024) The impact of various forms of silver nanoparticles on the rhizosphere of wheat (Triticum aestivum L.)–shifts in microbiome structure and predicted microbial metabolic functions. Sci Total Environ 914:169824. https://doi.org/10.1016/j.scitotenv.2023.169824

Raliya R, Tarafdar JC (2013) ZnO nanoparticle biosynthesis and its effect on phosphorous-mobilizing enzyme secretion and gum contents in Clusterbean (Cyamopsis tetragonoloba L.). Agric Res 2:48–57. https://doi.org/10.1007/s40003-012-0049-z

Raliya R, Biswas P, Tarafdar JC (2015) TiO2 nanoparticle biosynthesis and its physiological effect on mung bean (Vigna radiata L.). Biotechnol Rep 5:22–26. https://doi.org/10.1016/j.btre.2014.10.009

Reinhold-Hurek B, Bünger W, Burbano CS, Sabale M, Hurek T (2015) Roots shaping their microbiome: global hotspots for microbial activity. Annu Rev Phytopathol 53(1):403–424. https://doi.org/10.1146/annurev-phyto-082712-102342

Riseh RS, Fathi F, Vazvani MG, Tarkka MT (2025) Plant colonization by biocontrol bacteria and improved plant health: a review. Front Biosci Landmark 30(1):23223. https://doi.org/10.31083/FBL23223

Riyazuddin R, Verma R, Singh K, Nisha N, Keisham M, Bhati KK, Kim ST, Gupta R (2020) Ethylene: a master regulator of salinity stress tolerance in plants. Biomolecules 10(6):959. https://doi.org/10.3390/biom10060959

Salazar-Cerezo S et al (2018) Fungal diversity and community composition of culturable fungi in Stanhopea tigrina cast gibberellin producers. Front Microbiol 9:612. https://doi.org/10.3389/fmicb.2018.00612

Samanta S, Singh BR, Adholeya A (2017) Intracellular synthesis of gold nanoparticles using an ectomycorrhizal strain EM-1083 of Laccaria fraterna and its nanoanti-quorum sensing potential against Pseudomonas aeruginosa. Indian J Microbiol 57:448–460. https://doi.org/10.1007/s12088-017-0662-4

Scott BM, Koh K, Rix GD (2024) Structural and functional profile of phytases across the domains of life. Curr Res Struct Biol:100139. https://doi.org/10.1016/j.crstbi.2024.100139

Shivaji S, Madhu S, Singh S (2011) Extracellular synthesis of antibacterial silver nanoparticles using psychrophilic bacteria. Process Biochem 46(9):1800–1807. https://doi.org/10.1016/j.procbio.2011.06.008

Shobha G, Moses V, Ananda S (2014) Biological synthesis of copper nanoparticles and its impact. Int J Pharm Sci Invent 3(8):6–28

Siebers M, Rohr T, Ventura M, Schütz V, Thies S, Kovacic F et al (2018) Disruption of microbial community composition and identification of plant growth promoting microorganisms after exposure of soil to rapeseed-derived glucosinolates. PLoS One 13:e0200160. https://doi.org/10.1371/journal.pone.0200160

Singh DK, Kumar J, Sharma VK, Verma SK, Singh A, Kumari P, Kharwar RN (2018) Mycosynthesis of bactericidal silver and polymorphic gold nanoparticles: physicochemical variation effects and mechanism. Nanomedicine 13(2):191–207. https://doi.org/10.2217/nnm-2017-0235

Singh RP, Handa R, Manchanda G (2021) Nanoparticles in sustainable agriculture: an emerging opportunity. J Control Release 329:1234–1248. https://doi.org/10.1016/j.jconrel.2020.10.051

Singh M, Qureshi KA, Jaremko M, Rajput M, Singh SK, Kaushalendra et al (2022) Bioprospects of endophytic bacteria in plant growth promotion and Ag-nanoparticle biosynthesis. Plants 11(14):1787. https://doi.org/10.3390/plants11141787

Singh NA, Narang J, Garg D, Jain V, Payasi D, Suleman S, Swami RK (2023) Nanoparticles synthesis via microorganisms and their prospective applications in agriculture. Plant Nano Biology:100047. https://doi.org/10.1016/j.plana.2023.100047

Srivastava SK, Yamada R, Ogino C, Kondo A (2013) Biogenic synthesis and characterization of gold nanoparticles by Escherichia coli K12 and its heterogeneous catalysis in degradation of 4-nitrophenol. Nanoscale Res Lett 8:1–9. https://doi.org/10.1186/1556-276X-8-70

Sundar V, Tamil Selvan S, Arumugam M, Chinnaraj S (2025) Investigation and characterization of eco-technological synthesis of spherical TiO2 nanoparticles from Thalassia hemprichi and analysis of biomedical properties. Appl Biochem Biotechnol, 17:1. https://doi.org/10.1007/s12010-024-05143-7

Tang X, Liao C, Huang X, Chen C, Xu D, Chen C (2024) Epiphytic microbiota source stimulates the fermentation profile and bacterial community of alfalfa-corn mixed silage. Front Microbiol 14:1247254. https://doi.org/10.3389/fmicb.2023.1247254

Tarafdar JC, Raliya R (2013) Rapid, low-cost, and ecofriendly approach for Iron nanoparticle synthesis using Aspergillus oryzae TFR9. J Nanoparticles. https://doi.org/10.1155/2013/141274

Timofeeva AM, Galyamova MR, Sedykh SE (2024) How do plant growth-promoting bacteria use plant hormones to regulate stress reactions? Plants 13(17):2371. https://doi.org/10.3390/plants13172371

Townsley L, Yannarell SM, Huynh TN, Woodward JJ, Shank EA (2018) Cyclic di-AMP acts as an extracellular signal that impacts Bacillus subtilis biofilm formation and plant attachment. MBio 9(2):10–1128

Trdá L, Fernandez O, Boutrot F, Héloir M-C, Kelloniemi J, Daire X et al (2014) The grapevine flagellin receptor VvFLS2 differentially recognizes flagellin-derived epitopes from the endophytic growth-promoting bacterium Burkholderia phytofirmans and plant pathogenic bacteria. New Phytol 201:1371–1384. https://doi.org/10.1111/nph.12592

Tsai HH, Wang J, Geldner N, Zhou F (2023) Spatiotemporal control of root immune responses during microbial colonization. Curr Opin Plant Biol 74:102369. https://doi.org/10.1016/j.pbi.2023.102369

Tsekhmistrenko SI, Bityutskyy VS, Tsekhmistrenko OS, Horalskyi LP, Tymoshok NO, Spivak MY (2020) Bacterial synthesis of nanoparticles: a green approach. Biosyst Divers 28(1):9–17. https://doi.org/10.15421/012002

Wang M, Li H (2025) Structure, function, and biosynthesis of Siderophores produced by Streptomyces species. J Agric Food Chem. https://doi.org/10.1021/acs.jafc.4c08231

Wu X, Rensing C, Han D, Xiao KQ, Dai Y, Tang Z et al (2022) Genome-resolved metagenomics reveals distinct phosphorus acquisition strategies between soil microbiomes. Msystems 7(1):e01107–21. https://doi.org/10.1128/msystems.01107-21

Xia Y, Ma Z, Qiu M, Guo B, Zhang Q, Jiang H et al (2020) N-glycosylation shields Phytophthora sojae apoplastic effector PsXEG1 from a specific host aspartic protease. Proc Natl Acad Sci USA 117:27685–27693. https://doi.org/10.1073/pnas.2012149117

Xu H, Wang L, Zhu M, Chen X (2024) Epiphytic fungi can drive a trade-off between pathogen and herbivore resistance in invasive Ipomoea cairica. Microorganisms 12(11):2130. https://doi.org/10.3390/microorganisms12112130

Yang L, Qian X, Zhao Z, Wang Y, Ding G, Xing X (2024) Mechanisms of rhizosphere plant-microbe interactions: molecular insights into microbial colonization. Front Plant Sci 15:1491495. https://doi.org/10.3389/fpls.2024.1491495

Yu K, Liu Y, Tichelaar R, Savant N, Lagendijk E, van Kuijk SJL et al (2019) Rhizosphere-associated Pseudomonas suppress local root immune responses by gluconic acid-mediated lowering of environmental pH. Curr Biol 29:3913–3920.e4. https://doi.org/10.1016/j.cub.2019.09.015

Zhang Y, Zhou Q, Gao C, Lu Y, Sheng Y, Xiao M et al (2024) Endophytic bacteria for Cd remediation in rice: unraveling the Cd tolerance mechanisms of Cupriavidus metallidurans CML2. J Hazard Mater 469:133846. https://doi.org/10.1016/j.jhazmat.2024.133846

Zhu YG, Peng J, Chen C, Xiong C, Li S, Ge A et al (2023) Harnessing biological nitrogen fixation in plant leaves. Trends Plant Sci 28(12):1391–1405. https://doi.org/10.1016/j.tplants.2023.05.009

Zipfel C, Oldroyd GED (2017) Plant signaling in symbiosis and immunity. Nature 543:328–336. https://doi.org/10.1038/nature22009

Chapter 13
Harnessing Plant–Microbiome Interactions Through Nanotechnology: Innovations in Disease Management and Plant Health

Mohammed Radi, Mohammed Taoussi, Abdelaaziz Farhaoui, Wafae Sfa, Hamza Rhazal, Hakima Achetoui, and Rachid Lahlali

Abstract The microbiome of plants is important for its ability to assist with the growth of plants, increase tolerance against biotic and abiotic stress, and maintain the health of crops. In recent years, nanotechnology has gained attention as an effective tool for the targeted modulation of plant–microbe interactions. Engineered NPs offer unique physico-chemical properties used to modulate microbial communities, mediate the delivery of beneficial microorganisms, and inhibit phytopathogens. In this chapter, we will provide an in-depth analysis of the potential of nanotechnology to improve plant–microbiome interactions, and we will focus on specific applications such as nanoparticles as biofertilizers, bionematicides, bionics, and delivery agents. In this study, we discuss the mechanistic knowledge for the interactions between nanoparticle–microbe–plant, together with the considerations of environ-

---

M. Radi · M. Taoussi
Phytopathology Unit, Department of Plant Protection, Ecole Nationale d'Agriculture de Meknès, Meknès, Morocco

Environment and Valorization of Microbial and Plant Resources Unit, Faculty of Sciences, Moulay Ismail University, Meknes, Morocco

A. Farhaoui
Phytopathology Unit, Department of Plant Protection, Ecole Nationale d'Agriculture de Meknès, Meknès, Morocco

Equipe de Recherche en Didactique des Sciences et Sciences de l'Education et TICE. (DSSE_TICE), Centre Régional des Métiers de l'Education et de la Formation de Fès Meknès, Fès, Morocco

W. Sfa · H. Rhazal
Phytopathology Unit, Department of Plant Protection, Ecole Nationale d'Agriculture de Meknès, Meknès, Morocco

Laboratory of Functional Ecology and Environmental Engineering, Sidi Mohamed Ben Abdellah University, Fez, Morocco

H. Achetoui · R. Lahlali (✉)
Phytopathology Unit, Department of Plant Protection, Ecole Nationale d'Agriculture de Meknès, Meknès, Morocco
e-mail: rlahlali@enameknes.ac.ma

K. A. Abd-Elsalam (ed.), *Plant-Microbiome Nanotechnology*, Sustainability Sciences in Asia and Africa, https://doi.org/10.1007/978-981-95-5056-2_13

mental and biosafety. Finally, the future of nanotechnology integration with sustainable agriculture and microbiome engineering is discussed, discussing opportunities and challenges for the future.

Keywords Plant microbiome · Nanotechnology · Engineered nanoparticles · Microbiome engineering · Plant disease management · Bioinoculants · Nanofertilizers · Nanopesticides · Sustainable agriculture · Plant–microbe interactions

13.1 Introduction

Plants are continuously interacting with a large diversity of microorganisms that are now identified as the plant microbiome. This dynamic and complex microbial community is found in specialized plant-associated niches, including the rhizosphere (zone of root), phyllosphere (leaf surface), and endosphere (internal tissues), and its primary functions are to drive plant nutritional, developmental, and health benefits (Turner et al. 2013; Vandenkoornhuyse et al. 2015). Recent studies show that these microbial partners are not uninvited guests but are contributors to nutrient cycling, promoting growth and, importantly, suppressing plant diseases. Plant growth-promoting rhizobacteria (PGPR) and arbuscular mycorrhizal fungi (AMF), beneficial microorganisms, play an important role in inducing systemic resistance responses, providing plants with increased resilience against a wide range of pathogens (Berendsen et al. 2012; Mendes et al. 2011). Yet these interactions are delicate, and environmental stresses and intensive agricultural practices can disrupt them, resulting in lower microbial diversity and poorer plant health.

Plant-disease management remains one of the major challenges of modern agriculture. Fungal, bacterial, and viral pathogens account for the losses of food globally, with the annual economic impact of these pathogens reaching billions (Savary et al. 2019). Historically, conventional synthetic pesticides and agrochemicals have played a dominant role in pathogen and pest control. Although beneficial at shorter timescale, this strategy tended to raise serious dilemmas on human health, environmental pollution, pathogen resistance, and soil microbial community degradation (Aktar et al. 2009; Popp et al. 2013). Therefore, this creates the compelling need for cost-effective, sustainable, and eco-friendly approaches that can reinforce natural plant defense mechanisms, yet maintain microbial continuity.

The field of nanotechnology is the science and engineering devoted to materials in the size range of 1–100 nm; it encompasses new formats for manipulating materials at the nanometer scale, creating great potential for revolutionizing agricultural practices. Nanomaterials have unique physicochemical features, like extremely high surface area, reactivity, and penetration of biological barricades, which may lead them to a broad range of agricultural uses (Nair et al. 2010). Application of NPs in controlled-release fertilizers, pesticide delivery systems, early disease detection, and soil remediation has already been investigated (Kah et al. 2013; Wang et al. 2016). More recently, studies have examined the possibility of probiotics in manipulating plant microbiome, including selective antimicrobial effects, targeted delivery of beneficial microbes, and enhancement of microbial colonization and activity (Khot et al. 2012). Microbiome engineering strategies applied in the context of nanobiotechnology are emerging as a precision toolset for reprogramming plant–microbe interactions to promote plant health. Yet the potential benefits brought by these technologies are accompanied by concerns about their environmental fate, biosafety, and ecological long-term consequences. To exploit the full potential of nanotechnology in agriculture, it is a prerequisite to fully understand nanoparticle–microbe–plant interactions, and to formulate smart, targeted nano-products that promote beneficial microbial functions without impacting non-target organisms or ecosystems.

This chapter aims to review recent scientific literature and emerging technologies bridging nanotechnology and plant microbiome research, emphasizing potential applications in plant disease management and crop health. It discusses the biological foundation of plant–microbiome interplay, the material characteristics and categorization of agro-nanomaterials, the mechanisms underlying the nanoscale modulation of microbial communities, and the ecotoxicological implications and future challenges of utilizing nanotechnologies in sustainable agriculture.

13.2 Agricultural Applications of Nanotechnology

13.2.1 The Use of Nanoparticles as Fertilizers and Plant Growth Stimulants

Recently, nanotechnology has evolved significantly, moving from the experimental phase to a concrete application. The use of nanotechnology to create slow-release, controlled-release fertilizers, pesticides, and herbicides has become essential for promoting sustainable, eco-friendly agriculture. Nanotechnology makes it possible to use nanostructured materials as carriers for agricultural chemicals, enabling intelligent delivery systems to be put in place (Chinnamuthu and Boopathi 2009).

Encapsulation and entrapment, along with the use of surface ionic bonds and weak bonds, allow for the storage, protection, distribution, and release of active substances that form fertilizers or crop stimulation products. Nanoparticles can improve the stability of agrochemicals by limiting environmental degradation, thus increasing their effectiveness while reducing their use. This reduction consequently contributes to the leaching of agricultural chemicals and allows the mitigation of their environmental impacts (Johnston 2010). In addition, the controlled release of the active ingredient allows for slow absorption, and helps avoid chemical overdoses, reducing risks to non-target organisms, and minimizing inputs and waste (Panpatte et al. 2016).

13.2.2 Nanoparticles as Plant Growth Stimulators

The use of fumed silica nanoparticles has a positive impact on plant growth. When these nanoparticles were applied to young tomato (*Lycopersicon esculentum*) plants, they showed accelerated plant growth and also increased germination at concentrations of 8 g/L. Furthermore, tests revealed that these nanoparticles do not affect the stability of the soil microbial community and the germination process of Lactuca sativa seeds (Pestovsky and Martínez-Antonio 2017).

Similarly, the use of selenium nanoparticles has shown promise in improving crop yields. The application of these nanoparticles (0.1 g/ha) to *Zea mays* seeds led to an increase in yield, as well as an increase in the starch and protein content of the seeds (Gudkov et al. 2020). However, the use of a fertilizer containing selenium did not have any positive effects on maize, suggesting that the benefits observed could be due to the specific properties of the nanoparticles rather than the mere presence of selenium (Tremblay et al. 2015).

Gold nanoparticles promoted shoot elongation without disturbing *Lactuca sativa* germination or soil microbial balance at a concentration of 0.013% (w/w) (Shah and Belozerova 2009). In *Arabidopsis thaliana* cultivation, treatment with these nanoparticles (80 μg/mL) stimulated germination, improved seed yield, and modulated the expression of certain miRNAs, enhancing antioxidant potential and enzymatic activity. In *Brassica juncea*, these nanoparticles increased chlorophyll and sugar content, but at high concentrations (50–100 ppm), they can cause oxidative stress (Kumar et al. 2013).

Silver nanoparticles play a role in stimulating seed germination, growth, and yield of several species, such as *Triticum aestivum*, *Brassica juncea*, *Raphanus sativus,* and *Arabidopsis*, and constitute a promising alternative to increase the productivity of these crops (Iram et al. 2014; Matras et al. 2023; Sharma et al. 2012).

Copper nanoparticles have been shown to positively influence several plant growth parameters, including the shoot-to-root ratio in *Lactuca sativa*, as well as seed yield and protein and oil content in Helianthus annuus. They can also enhance phytohormone levels (e.g., gibberellins and indole-3-acetic acid), thereby promoting growth in *Vicia sativa* and *Triticum aestivum*. Nonetheless, at high concentrations, their application through irrigation can become toxic (Pestovsky and Martínez-Antonio 2017).

The palladium nanoparticles were shown not to interfere with the balance of the soil microbiome or with the germination of Lactuca sativa seeds and significantly influenced the shoot-to-root length ratio (after 15 days) (Pestovsky and Martínez-Antonio 2017). The use of manganese nanoparticles on Vigna radiata is significant in relation to nitrogen metabolism and photosynthesis. This has led to improved plant growth and promotes increased chlorophyll and carotenoid production. In addition, these nanoparticles have shown low cell toxicity and compatibility with beneficial soil microorganisms (Pradhan et al. 2014). According to Feizi et al. (2012), titanium dioxide nanoparticles showed beneficial effects on the germination and growth of *Triticum aestivum* and *Spinacia oleracea* by improving photosynthesis and nitrogen metabolism processes. In addition, they preserved chloroplasts against aging induced by prolonged exposure to light.

13.2.3 Nanoparticles as a Nanofertilizer

Fertilizer is vitally important for increasing agricultural production. Around 30–40% of the total amount of fertilizer applied is absorbed by crops to promote their growth, while the rest is lost to the environment through evaporation, leaching, or degradation. This loss reduces the effectiveness of fertilizer and increases production costs. In addition, excessive use of fertilizer leads to pollution of natural resources, jeopardizing the balance of the ecosystem, particularly the soil, water, and air.

Recently, a decrease in yields of many crops has been observed due to inadequate fertilization and degradation of soil organic matter. In an effort to improve nutrient utilization and address the pollution problem, nanofertilizers present a promising solution. These fertilizers, designed to control nutrient release, could offer an alternative to chemical fertilizers. Nanofertilizers are produced primarily by encapsulating nutrients within nanostructured materials. Three main approaches are used for this purpose: Encapsulating nutrients within nanoporous materials; Coating nutrients with a thin polymer film; Formulating nutrients as nanometric particles or emulsions (Rai et al. 2012).

The gradual release of nutrients contained in fertilizer can be achieved by coating fertilizer particles in nanoparticulate membranes. According to Chinnamuthu and Boopathi (2009), the use of nanofertilizer containing macronutrients (nitrogen, phosphorus, potassium) and micronutrients has led to better nutrient uptake by cereal crops. Furthermore, there is some addition of natural minerals like nanoclays and zeolite in order to enhance the efficiency of the fertilizers, as alleged by Chinnamuthu and Boopathi (2009). The minerals have a crystalline coating which is rich in nutrients; hence, they can release slowly as per the demands of the plants. As an illustration, the use of zeolites in capturing, storing, and releasing nitrogen slowly is usually done (Leggo 2000).

Nanofertilizers have helped in limiting pollution of soil, water, and air through the gradual manner in which nutrients are released to the soil. Nutrient absorption into zeolite form is lower as compared to what is found with the commonly used ionic fertilizers. Millán et al. (2008) further note that there is a slow release of nitrogen fertilizers when using zeolite chips that are filled with urea. Surfactant-modified zeolites have been shown to be efficient nitrate absorbents. They find application in the controlled release of nitrate and other anions as fertilizer carriers. Moreover, ammonium-loaded zeolite particles have been shown to enhance phosphorus absorption, improve phosphate solubilization, and increase crop productivity (Allen et al. 1993), and a blend of zeolite and natural phosphate is capable of ensuring slow-releasing fertilization via the process of solubility and exchanging ions.

13.2.4 Nanoparticles as Nanopesticides

Many pesticides and herbicides have active ingredients (APIs) with low water solubility, which requires the use of a delivery system to apply them. These substances can be harmful to non-target organisms and promote pathogen resistance. Although there are several formulations on the market to overcome these limitations, these products are unstable and easily degrade. By using techniques such as encapsulation and controlled release methods, nanoformulations offer a solution to these challenges (Panpatte et al. 2016).

Nanopesticide formulations are recognized for their superior performance compared to commercial formulations (Table 13.1). This effectiveness is due to better bioavailability and absorption of active ingredients containing nanoparticles. The bioavailability of nanopesticide active ingredients depends on the characteristics of

Table 13.1 Nanoparticles and their antifungal and insecticidal effects

Type of nanoparticle	Effect	References
Silica nanoparticles	Effective against weevils, with insecticidal action linked to the release of silicate ions or activation of the natural defenses of grains.	Debnath et al. (2011), Epstein (2009)
Sulfur nanoparticles	Inhibit the growth of fungi such as *Aspergillus niger* and *Fusarium oxysporum* at relatively low concentrations. Their efficacy exceeds that of sulfur microparticles.	Pestovsky and Martínez-Antonio (2017), Roy Choudhury et al. (2011)
Zinc oxide (ZnO) nanoparticles	ZnO nanoparticles at a concentration of 50 μg/ml showed activity against *Aspergillus flavus*, *Aspergillus niger*, *Aspergillus fumigatus*, *Fusarium culmorum*, and *Fusarium oxysporum*. Their activity was size-dependent, with increased efficacy against *Fusarium* sp. at a concentration of 0.1 M.	Pestovsky and Martínez-Antonio (2017)
Silver nanoparticles	Effective against insects such as *Hippobosca maculata* and *Rhipicephalus microplus*, as well as against several plant pathogenic fungi like *Alternaria alternata* and *Botrytis cinerea*.	Pestovsky and Martínez-Antonio (2017), Rajakumar and Abdul Rahuman (2012)
Nickel nanoparticles	Effective against tick larvae and certain mosquitoes, with higher activity than Ni-hydrazine complexes.	Latvala et al. (2016), Rajakumar et al. (2013)
Copper and CuO nanoparticles	Effective against various fungal species such as *Alternaria alternata* and *Fusarium oxysporum*, with stronger action than traditional nanoparticles.	Zabrieski et al. (2015)
Fe_3O_4/ZnO/AgBr nanocomposite	This nanocomposite reduced the germination capacity of *Fusarium graminearum* and *Fusarium oxysporum* spores, with antifungal activity increasing with the composite concentration and the weight ratio of Fe_3O_4 to ZnO/AgBr.	Pestovsky and Martínez-Antonio (2017)
Reduced graphene oxide (rGO) nanosheets	rGO nanosheets showed dose-dependent and specific antifungal activity against *Aspergillus niger* and *Fusarium oxysporum* between 1 and 500 μg/ml.	Pestovsky and Martínez-Antonio (2017)
Gold nanoparticles	Active against several fungal and bacterial species, including *Aspergillus flavus* and *Puccinia graminis*, likely due to cellular damage.	Jayaseelan et al. (2013)

the carrier used (Chhabra et al. 2025). Nanopesticides are designed to target specific organisms and are relatively larger in size, making them impossible to absorb directly (Wen et al. 2010).

13.2.5 Nanoparticles as Pesticide Carriers

Nanoparticles offer significant potential for improving pesticide delivery. This assists in controlling the release of pesticides, maximizing the use of plant protection products, and reducing the adverse effects of the products on the ecosystem. Various nanoparticles are also researched with regard to potential encapsulation and protection of pesticides against degradation and extension of their durability (Pestovsky and Martínez-Antonio 2017). Silica nanoparticles can protect insecticides such as abamectin and avermectin from photodegradation and extend their release period for up to 27 days. The encapsulation of uniconazole was found to enhance rice growth and upgrade its efficiency (Wang et al. 2014; Wen et al. 2005). Hou et al. (2016) stated that the use of gold nanoparticles promotes the absorption of the fungicide ferbam by plants. SILVER nanoparticles, on the other hand, show synergistic action against many fungus-causing pathogens in combination with ketoconazole and fluconazole (Gajbhiye et al. 2009). Zinc oxide nanoparticles demonstrate antifungal effects on *Phytophthora capsici* since they induce ROS formation and enhance photodynamic degradation of the pesticide (Pestovsky and Martínez-Antonio 2017). Pestovsky and Martínez-Antonio (2017) also state that ferrihydrite nanoparticles prevent the phytotoxicity of fungicides and help their efficiency in higher proportions, especially when the particles are enriched in cobalt or aluminum. Kumar et al. (2014) explain that with the incorporation of imidacloprid into calcium alginate nanoparticles, the insecticidal properties of the compound can be optimized with respect to its toxicity to mammalian cells.

Chitosan nanoparticles are used to encapsulate essential oils from *Zataria multiflora* and *Schinus molle* as well as herbicides (paraquat, imazapic, imazapyr), to promote gradual release Maruyama et al. 2016; Mohammadi et al. 2015). In addition, according to Pradhan et al. (2013), polyethylene glycol nanoparticles serve as a protection for acephate by mitigating its toxicity and enhancing its efficacy against *Spodoptera litura* and tea mites.

13.2.6 Interaction of Nanoparticles with Plant-Associated Microorganisms

The application of nanofertilizer and nanopesticides can have an impact on various plant growth characteristics (such as seed germination, root and shoot growth, chlorophyll content, photosynthesis, flowering, fruiting, and crop yield), depending on the genetic make-up of the plant, soil and plant microbiology, soil nutrients (macronutrients and micronutrients), soil pH, moisture, and other environmental factors (Bratovcic et al. 2021; Juo and Franzluebbers 2003; Okey-Onyesolu et al. 2021).

Nanoparticles introduced into soil and plants could directly or indirectly affect the type of micro-organisms present and alter their functions (Kibbey and Strevett 2019; Mosquera et al. 2018). The effects of nanoparticles on microbial communities associated with plants are highly dependent on the type of plant, the type of nanoparticles (physical characteristics and chemical composition), the properties of the soil (clay and organic matter content), as well as its physicochemical characteristics (texture, organic matter content, pH) (Kibbey and Strevett 2019; Kumar et al. 2018; Peng et al. 2020). In the rhizosphere, plants secrete different types of exudates that stimulate the development of microorganisms. Similarly, these microorganisms play a crucial role in mobilizing nutrient cycles (Zhang et al. 2014). Nanoparticles can be absorbed directly by the plant through root hairs and stomata present on leaves (Carpita et al. 1979; Eichert et al. 2008; Eichert and Goldbach 2008). They are then transported by plasmodesmata from cell to cell within the plant (Zambryski 2004).

The nanoparticle microorganism interactions in the soil and the plant area play a critical role in the control of the disease as well as the maximization of plant growth. Nevertheless, this activity is tempered by the adverse or positive effects of the nanoparticles, respectively, in the form of antimicrobial agents or microbial growth inducers. The mechanism of nanoparticles in preventing the proliferation of various microorganisms is due to the release of the metal ions that attack different parts of the cell through various mechanisms. These include the formation of reactive oxygen species, creation of pores in the cell membrane, harm to cell walls, and DNA damage, as well as cell cycle arrest. All these acting processes end up inhibiting cell expansion and phytopathogen inhibition (Singh et al. 2019).

The use of nanoparticles as antimicrobials in a variety of studies has also demonstrated that nanoparticles can be effectively used to have a positive effect on microbial metabolism and functioning. Positive properties of nanoparticle-microbe interactions are high nanoparticle bioavailability when the surface area is increased. This enhances the absorption of nutrients by the microbes since the nanoparticles supply the microorganisms with their required nutrients that trigger their development and metabolism. Nanoparticles may also serve as functional nanotools for processes such as electron transfer, chemotaxis, and nutrient storage (Mansor and Xu 2020).

Knowledge of the mechanism of interaction between microorganisms and nanoparticles may help in creating eco-friendly nanomaterials. This may involve coming up with a green synthesis method for making nanoparticles. In a broader sense, the application of nanoparticles as fertilizers is an area that needs to be investigated further, since it may have altering effects on the microflora balance, which would affect the microorganisms present in the plant life (Mgadi et al. 2024).

13.3 Interaction of Nanoparticles with the Plant Microbiome

13.3.1 Effect of Nanoparticles on Microbial Communities: Balance between Toxicity and Positive Interactions

Plant microbial community is significant in plant health in various ways, including their protection against phytopathogens, enhancing their ability to resist phytopathogens, and stimulating their growth (Escribano-Vazquez et al. 2020). Nevertheless, alterations in this microbiome can occur due to phytopathogen infections or unfavorable environmental conditions and significantly decrease crop production. In this regard, the related contemporary developments in nanotechnologies have offered promising solutions to selectively stimulate and manipulate these populations of microbes, either through the generation of beneficial relationships between the microorganisms and the plant or by the removal of phytopathogens (Khan et al. 2021). Such novel approaches open up the opportunity to enhance the health of crops through a controlled, fine-tuned approach to their microbiome (Fig. 13.1).

13.3.1.1 Selective Toxicity of Nanoparticles against Pathogens

Plant disease management has bright prospects with nanoparticles specifically designed for antimicrobial applications. Copper (CuNPs), silver (AgNPs), zinc oxide (ZnONPs), and titanium dioxide (TiO_2 NPs) are among the metallic nanoparticles that have been found effective against microbial growth and can inhibit the

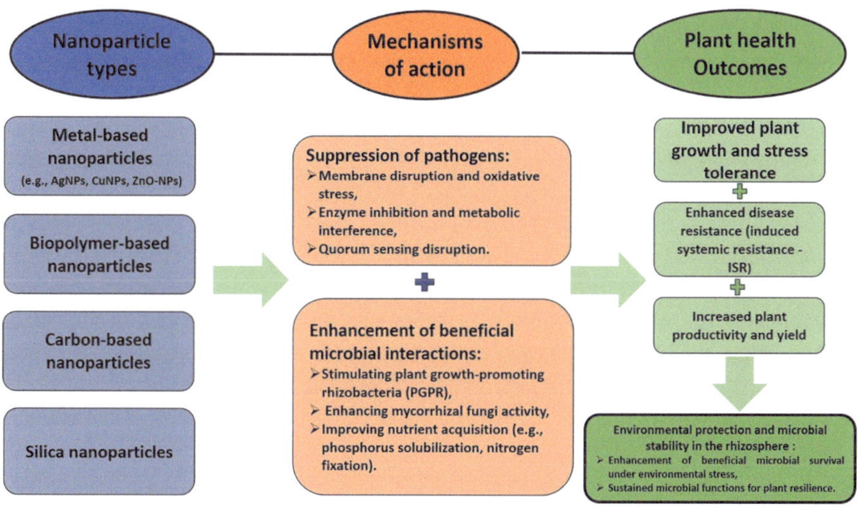

Fig. 13.1 Importance of nanomaterials in disease management and improving plant vigor

spread of plant pathogens. These structures produce the antimicrobial effect primarily by the formation of reactive oxygen species (ROS), the disturbance of the cell membrane structure, and the inhibition of several metabolic cascades. Application of metal-based nanomaterials has been shown to reduce the growth of different *Pythium* and *Fusarium* species, two leading plant pathogens, which cause a lot of harm in agriculture (Elshahawy et al. 2018).

Various mechanisms of action can be used to explain the antimicrobial effect of metal-based nanomaterials. The nanomaterials at this scale can also modify the structural arrangement of membranes that surround the microorganisms, resulting in leakage of cell components and generating high levels of oxidative stress (Tripathi et al. 2024). Most of the nanomaterials based on metals have the potential of hindering the growth of phytopathogens by inhibiting activities of key enzymes involved in various metabolic pathways, blocking respiration of the cells, and disturbing transport of nutrients required for growth (Hosseini et al. 2024). Moreover, certain nanomaterials are able to interfere with the cellular communication of bacteria, hampering the formation of biofilm and lowering the virulence of the pathogenic bacteria (Afrasiabi and Partoazar 2024). However, although nanoparticles have proven their efficacy against various infectious agents, concerns persist about their potentially harmful impacts on beneficial microbes. Oriented nanoparticle formulation technology has the potential to minimize these off-target impacts (Punniyakotti et al. 2024).

13.3.1.2 Role of Nanotechnology in Optimizing Beneficial Interactions in Favor of the Plant

Beyond their antimicrobial capacity, several nanomaterials also have the potential to encourage the emergence and proliferation of beneficial microbes, thereby strengthening interactions favorable to the plant (Fig. 13.1). Positive effects on the activity of mycorrhizal fungi as well as plant growth-promoting rhizobacteria (PGPR) have been recorded following the application of nanosilica and carbon-based nanoparticles (Guardiola-Márquez et al. 2023). The positive effect of nanoparticles on beneficial microorganisms is associated with positive impacts on plant growth and health. The stimulation of the population growth of beneficial microorganisms in the rhizosphere is linked to an improvement in atmospheric nitrogen fixation and phosphorus solubilization, which improves nutrient availability to plants, thus stimulating their growth and vigor (Francis et al. 2024). Nanoparticles promote biocontrol's efficacy by enhancing microbial antagonists' survival and their establishment in the rhizosphere. These nanoscale elements can also protect antagonists from environmental stress and optimize their activities against phytopathogens (Behl et al. 2024). In addition, NPs are known to induce beneficial microorganisms to release metabolites that activate immune responses, which induce systemic resistance in the plant (Imada et al. 2016). Therefore, it is crucial to optimize nanoparticle formulations to specifically minimize the growth of phytopathogens while stimulating the development of microbial antagonists that enter into beneficial interactions with the plant.

This is part of the sustainable management of the plant microbiome. Future studies should focus on creating functionalized nanoparticles that can target specific microbial populations while minimizing adverse effects on the agroecosystem (Ahmad et al. 2022).

13.3.2 Use of Nanotechnology in Delivering Beneficial Microorganisms

The use of nanomaterials to improve agricultural production offers optimistic prospects by exploiting them as structures to deliver beneficial microbes. These nanometric particles can safeguard these bioagents against various stresses, facilitate their installation in the rhizosphere, and ensure their precise delivery to targeted parts of the plant (Xu et al. 2022).

13.3.2.1 Optimization of Microbial Survival in the Face of Biotic and Abiotic Constraints

The viability of beneficial microorganisms can be seriously threatened by various environmental stresses, including drought, oxidative stress, and extreme temperatures. In the agroecosystem, such harsh conditions reduce the survival and activity of these microorganisms. Nanoscale engineering presents a tangible answer to this challenge by encapsulating these organisms in nanoscale structures. Therefore, the encapsulation of microorganisms by metal-based nanomaterial compounds can potentially improve the resilience of microorganisms by enhancing antioxidant defenses and ensuring the stability of membrane integrity. The use of nanotechnology in the encapsulation of microorganisms not only preserves the viability of microorganisms but also ensures their activity even in extreme climatic conditions. This strategy has already demonstrated encouraging results in agriculture, where it serves to preserve the persistence and proper functioning of beneficial microorganisms, even in situations of environmental stress (Moradipour et al. 2019).

13.3.2.2 Improved Targeted Colonization of Beneficial Microorganisms

To optimize their positive effect, it is crucial to introduce beneficial microorganisms into the plant environment in a targeted manner. The use of nanocarriers promotes the creation of conditions conducive to their attachment and installation on plant surfaces. Thus, some nanoparticles have demonstrated their ability to intensify the interaction between PGPRs and the plant root system, thus promoting their installation in the phyllosphere and rhizosphere. This is manifested by a significant improvement in the plant's vigor and a stimulation of its growth (Xu et al. 2022).

13.3.2.3 Targeted Releases of Beneficial Microorganisms for Optimal Efficiency

The design of functionalized nanoparticles allows for relevant targeting of beneficial microbes to the targeted parts of the plant where their intervention is expected. This precise method ensures a directed release of microbes at the targeted site, which increases their performance by hindering the development of phytopathogens and improving plant growth. Significant improvement in plant resistance to several pathogens by using nanocarriers to deliver beneficial microbes in a highly targeted manner in the rhizosphere has been demonstrated by recent research (Singh and Rai 2024).

13.3.3 Importance of Nanotechnology in Improving Plant Resistance and Activity of Beneficial Microbes

Nanocarriers are effective in enhancing the survival, stability, and delivery of beneficial bacteria and fungi, thereby enhancing plant resilience. In addition to their function as transporters, nanoparticles enhance plant defenses and control the release of active biological compounds that improve crop vigor (Behl et al. 2024).

13.3.3.1 Improving Microbial Survival and Functions Using Nanomaterials

One of the major inputs of nanocarriers for the agricultural sector is their ability to preserve beneficial microbes from abiotic stresses, including thermal variations, UV radiation, and drought (Handa and Kalia 2024). According to Behl et al. (2024), the encapsulation of beneficial microbes using nanomaterials enhances their organizational integrity while improving their biochemical activities, which creates favorable conditions for phosphate solubilization, nitrogen fixation, and nutrient availability within the rhizosphere. Research has proven that the integration of nanoformulated structures extends the lifespan of beneficial bacteria and strengthens their interrelationship with roots, which optimizes their agricultural performance (Behl et al. 2024).

13.3.3.2 Using Nanotechnology to Improve Plant Defense Systems

Nanocarriers have the potential to trigger plant immune system responses, in addition to preserving microbial activity (Fig. 13.1). Some nanomaterials promote the synthesis of microbial molecules such as volatile organic compounds, antibiotics, and siderophores, which prevent the development of plant pathogens (Choudhary et al. 2023). Systemic acquired resistance, as well as induced systemic resistance, can be induced using functionalized nanocarriers, which can optimize plant vigor and thus protect crops against phytopathogen invasions (Joshi 2025).

13.3.3.3 Importance of Nanocarriers in the Controlled Delivery of Bioactive Compounds

One of the major benefits of nanocarriers is their ability to control the gradual delivery of bioactive molecules and beneficial microbes, thereby ensuring a lasting impact on crop vigor. This carefully adjusted release mechanism promotes the sustainable implantation of antagonists, thereby reducing reliance on agrochemicals. Nanoencapsulated biofertilizers and biopesticides have been shown to promote the growth of several plants while reducing the negative impact on the agroecosystem associated with the use of traditional chemical inputs (Fu et al. 2020).

The use of nanocarriers in biotechnological strategies to deliver microorganisms and bioactive compounds is an encouraging option for sustainable control of phytopathogens while ensuring agricultural production that is respectful of the agroecosystem. Future research should focus on improving nanotechnological formulations to optimize their performance while ensuring their environmental compliance.

13.4 Nanotechnology-Based Approaches for Disease Management

13.4.1 Antimicrobial Nanomaterials for Plant Pathogen Control

Agriculture utilizes nanoscience for hormone distribution, seed germination, gene transfer, water resource management, nanobiosensors, nanobarcodes, and regulated agrochemical release (Dutta et al. 2023). Nanoparticles, created with specific characteristics like size, shape, and surface area, can transport active chemicals like fungicides to specific sites through conjugation, adsorption, or encapsulation (Khandelwal et al. 2016). Nano-based materials, including metallic oxides, nonmetals, metalloids, polymeric, and carbon nanostructures, can be used to treat plants' seeds, roots, soil, and leaves, preventing disease and promoting growth (Elmer et al. 2018). Nanoparticles, including silver, gold, copper, zinc oxide, and iron, are widely explored for their use as biosensors, protectants, therapeutants, and smart delivery vehicles for fungicides or actives (Dutta et al. 2023).

Nanoparticles have attractive potential in the fields of ultra-sensitive biomolecular detection, disease diagnosis, and the development of anti-microbial and therapeutic products. Integrating nanotechnology with biological science represents a promising new science of the present era (Dutta et al. 2022). In nanobiotechnology, biological agents like microbes, plant extracts, and enzymes have been combined into nanoparticles (1–100 nm in size) (Auría-Soro et al. 2019; Hyman and Denyes 2021). Overall, it represents a viable alternative for sustainable agriculture. In recent years, the use of chemical pesticides has provided very effective and convenient control over many plant diseases, but their growth has resulted in environmental disturbances with non-specific impacts posing a serious risk to beneficial microbes and human health (Deveci et al. 2021; Tang et al. 2021).

Nanoparticles exercising their suppressive activity against plant pathogens have included non-metals, metalloids, oxide metals, and nano-carbon materials (Elmer et al. 2018). Silver, titanium oxide, copper, gold, zinc oxide, iron, and silver have been the most extensively tested and widely used nanoparticles (Iravani et al. 2014; Kandi and Kandi 2015). Nanomaterials' mechanism of action vis-a-vis organisms comprises cell membrane perturbation, aggregation, suppression of H^+ATPase activity, inhibition of messenger RNA and protein synthesis, suppression of the production of toxins, blockage of the growth of microbes, and inhibition of nutritional flux (Fig. 13.2) (Dakal et al. 2016; Malerba and Cerana 2016).

NPs showed some efficacy on numerous pathogenic microorganisms (Table 13.2), notably *Rhizoctonia solani*, *Sclerotinia sclerotiorum*, *Fusarium oxysporum*, *Sclerotium rolfsii* (Abdelrhim et al. 2021; Kaman and Dutta 2019), *Colletotrichum* sp. (Goswami et al. 2020), *Sarocladium oryzae* (Das and Dutta 2021), *Xanthomonas oryzae pv. Oryzae* (Abdallah et al. 2019; Ogunyemi et al. 2020), and viruses such as bean yellow mosaic virus (BYMV) (Elbeshehy et al. 2015), potato virus Y (PVY) (Noha et al. 2018), and cucumber mosaic virus (CMV) (El-Sawy et al. 2017).

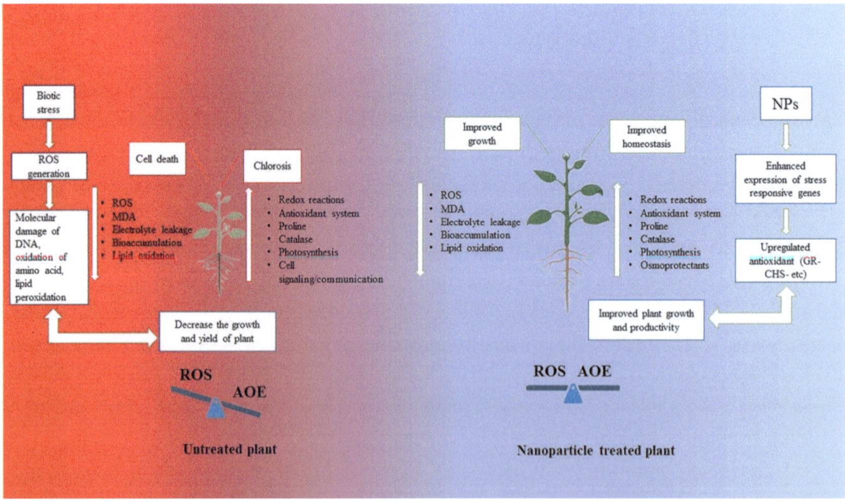

Fig. 13.2 Antimicrobial effect of nanomaterials for plant pathogen control

Table 13.2 Examples of the use of nanomaterials for plant pathogen control

Plant-pathogen type	Nanoparticle used	Target pathogen/disease	Mechanism of action	Efficacy/outcome	References
Nematodes	Green Silver Nanoparticles (GSN)	*Meloidogyne* spp. (root-knot nematodes)	Nematicidal effects via algal extract synthesis; reduces juvenile populations.	69.44% reduction in *M. javanica* at 17 mg/mL; no phytotoxicity.	Bhau et al. (2016), Elmer et al. (2018)
	Bacillus thuringiensis-nanoscale chitinases	*Caenorhabditis elegans*	Enhanced nematicidal delivery via immobilized enzymes.	Increased mortality by ~20% compared to non-nano formulations.	Qin et al. (2020)
	Cry protein nanocomposites	Nematodes (general)	Stabilized bioactive Cry proteins in nanoformulations (capsules, emulsions).	Improved stability and targeted delivery of nematicidal proteins.	Devi et al. (2019)
Fungi	Cyclic lipopeptide nanocrystals	*Aspergillus carbonarius, A. niger*	Controlled release of antifungal lipopeptides; inhibits spore germination.	Effective inhibition of hyphal growth and spore development.	Lin et al. (2020)
	Silver nanoparticles (AgNPs)	*Alternaria alternata, Penicillium* spp.	Disrupts cell membranes, inhibits spore germination.	Potent antifungal activity; size- and concentration-dependent efficacy.	Mansoor et al. (2021), Menichetti et al. (2023)
	Copper nanoparticles	*Phoma terrestris* (red root-rot)	Fungicidal properties via targeted delivery; enhances plant resistance.	Reduced fungal infection with lower environmental impact.	Ponmurugan et al. (2016)

(continued)

Table 13.2 (continued)

Plant-pathogen type	Nanoparticle used	Target pathogen/ disease	Mechanism of action	Efficacy/ outcome	References
Viruses	Layered double hydroxide (LDH) NPs	Cowpea mosaic virus	RNA-based gene silencing via nanoparticle carriers.	20-day protection post-application in cowpea plants.	Li et al. (2023)
	Chitosan/silica nanoparticles	Alfalfa mosaic virus, tobacco mosaic virus	Induces plant defense responses; inactivates viral replication.	Suppressed viral replication and systemic resistance activation.	Ahmed et al. (2023), Dutta et al. (2022)
	Silver nanoparticles (AgNPs)	Tobacco mosaic virus (TMV)	Physical deactivation of virus particles; high surface reactivity.	Reduced TMV infection and replication in tobacco plants.	Yadav and Yadav (2018)
Bacteria	Silver nanoparticles (AgNPs)	*Erwinia carotovora, Ralstonia solanacearum*	Disrupts bacterial membranes, inhibits DNA replication.	Efficacy comparable to chemical bactericides; concentration-dependent.	Menichetti et al. (2023)
	Zinc oxide nanoparticles (ZnO NPs)	*Xanthomonas citri* (citrus canker)	Targeted foliar delivery; suppresses bacterial growth.	Effective control of citrus canker in grapefruit trees.	Graham et al. (2016)
	Magnesium oxide nanoparticles (MgO)	*Ralstonia solanacearum* (bacterial wilt)	Activates salicylic acid/ jasmonic acid defense pathways.	Induced systemic resistance in tomato plants.	Imada et al. (2016)

13.4.2 *Nano-Delivery Systems and Mechanistic Aspects of Management of Plant Pathogens by NPs*

When herbicides are used in farming, conventional distribution systems are important. These systems need to focus on making products more efficient and controlling spray diversion (Eyhorn et al. 2015). The right use of controlled delivery methods is a potential alternative way to solve these issues. The managed delivery method involves releasing the right amount of pesticide ingredients for the crop, without giving it more than what it needs to get rid of the pest (Hou et al. 2021).

Several reports have discussed how nanotechnology can transform and improve traditional agrochemical applications, with the aim of enabling precision agriculture on a global scale (Ma 2019). Nanotechnology is the branch of science and technology that works with, creates, studies, controls, and/or uses materials that are on the order of nanometers. These are called nanomaterials and have a width or at least one of their measurements of 100 nm or less (Auffan et al. 2009). There is a lot to learn about nanostructured systems in agriculture (Ghormade et al. 2011). They can be made up of nanoparticles (nanocapsules, nanospheres, nanocrystals, nanocomposites, nanotubes, nanoneedles, etc.), micro/nanoemulsions, fullerenes, and biomimetic systems. Nanomaterials can be used in farming to protect crops (nanopesticides), feed plants (nanofertilizers), keep track of farming practices (nanobiosensors), and clean up pollution (pollutant cleanup nanostructured systems) (Ghormade et al. 2011). Due to their small size, high surface-to-volume ratio, ability to package actives in a core-shell diffusion system, and unique optical properties, nanostructured systems can be used to move chemical compounds for cultivation. These systems release the chemicals slowly and in a controlled way (Dutta et al. 2023).

A number of studies have shown that nanostructures that contain active substances work very well as controlled release devices for these actives. Researchers used micrometric (10–100 μm), sub-micrometric (1–10 μm), and nanometric (<1 μm) particles to help control the delivery of agrochemicals. They found that the nanometric particles were better than the others because they have more surface area per unit volume, are easier to fix, and speed up mass transfer (Ghormade et al. 2011). Quantum dots, metal oxides, biopolymers (natural or manufactured), clay minerals, emulsions, lipids, peptides, dendrimers, and other materials can be used as nanoparticles or to make nanostructured systems for use in the field (de Oliveira et al. 2020)

The primary actions that are made by NPs against microorganisms involve agglutination, cell membrane disruption, inhibition of RNA, protein, toxins, enzymes like H+-ATPase, and blockage of nutrient flow (Fig. 13.3) (González-Melendi et al. 2008). Nanoparticles act as carrier molecules for the targeted release of active substances into plant systems, reducing environmental impact. Benefits include higher solubility, site-specific administration, longer shelf life, less non-target species impact, and lingering environmental effects (Wang et al. 2023). Also, the stability and activity of nano-based formulations are greater as compared to conventional pesticides even under unfavorable environmental conditions (rainfall and UV exposure), thereby reducing the number of applications, toxicity, and overall costs (He et al. 2023). The antioxidant enzymes, such as superoxide dismutase (SOD), catalase, glutathione peroxidase, glutathione reductase, and ascorbate peroxidase, play a crucial role in plant growth and development. However, the relationship between plant adaptability and phytotoxicity remains debatable. The accumulation of metallic nanoparticles in ecosystems poses a threat to human and ecological health, as they can easily be absorbed into biological systems and enter the food chain, potentially affecting beneficial organisms and microbes (Rajput et al. 2021). Research on nanotoxicology has increased in recent years, but more

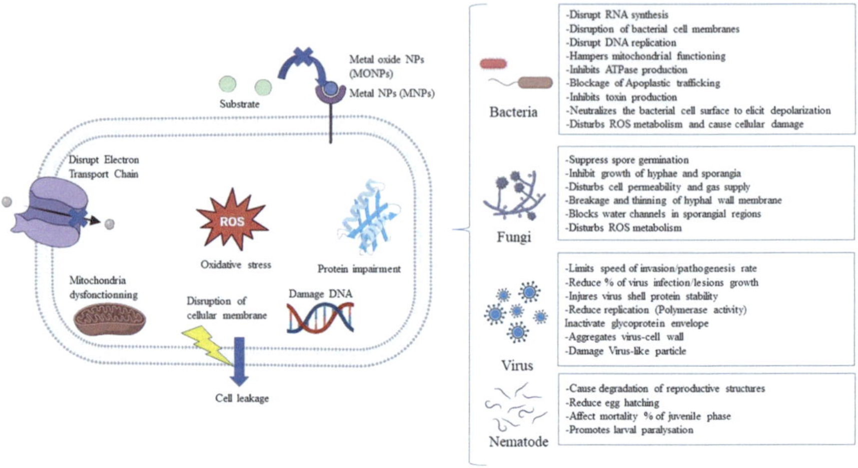

Fig. 13.3 Mechanistic aspects of management of plant pathogens by NPs

studies are needed to understand environmental fate, transformation, bioavailability, transport, and toxicity. However, polymeric nanoparticles like cellulose, lignin, and chitosan are antimicrobial compounds with antifungal, antibacterial, and plant growth-promoting properties, making them eco-friendly and abundant in nature (Dutta et al. 2023).

13.4.3 Case Studies and Recent Research Advances

Tomah et al. (2020) demonstrated that AgNPs, produced by *Trichoderma* spp., successfully inhibited Sclerotinia sclerotiorum in vitro, achieving 93.8%, 100%, and 100% inhibition at a dose of 200 μg/mL. The research also demonstrated direct interaction between AgNPs and fungal cells, encompassing AgNP contact, accumulation, micropore creation, and lamellar fragment generation. Also, Chen et al. (2020) discovered that Magnesium oxide nanoparticles (MgONPs) impeded fungal growth, spore germination, and sporangium development in vitro. The antifungal impact was ascribed to direct contact, adsorption by fungal hyphae, and alterations in cell morphology. In greenhouse circumstances, 500 μg/ml of MgONP decreased tobacco black root rot and black shank disease by 42.35% and 36.58%, respectively. Zaki et al. (2022) conducted another experiment revealing that the soil-borne pathogens *Macrophomina phaseolina, Rhizoctonia solani,* and *Fusarium fujikuroi* exhibited resistance to the antifungal properties of mycogenically synthesized zinc oxide nanoparticles (ZnONPs) generated by *Trichoderma* spp. A significant antifungal effect was seen in both cotton seedlings and in vitro. Similar outcomes were shown when copper oxide nanoparticles (CuONPs) were used to combat *Phytophthora nicotianae* (Chen et al. 2022). They found that 100 mg L-1 of CuONPs treatment under marijuana conditions resulted in a 33.69% increase in control effectiveness and the reduction of tobacco black shank disease without producing phytotoxicity (Chen et al. 2022). They also revealed that the utilized nanoparticles have improved SOD enzyme activity and intracellular ROS accumulation as antifungal strategies. Cu-content in the leaves and roots of treated plants, as well as a cascade of defense enzymes and resistance genes, were all considerably activated when tobacco plants were exposed to CuONPs (Dutta et al. 2023). In addition, Encinas Basurto et al. (2020) revealed that silver-chitosan nanoparticles inhibited mycelial growth of *Fusarium oxysporum* by up to 70% and reduced disease severity in tomato seedlings infected with Fusarium oxysporum 14 days post-inoculation. Furthermore, at a dose of 2000 ppm, nanoparticles had no adverse effects on the vegetative development of the seedlings. Further research is needed to determine the exact mode of action against soil-borne bacteria and viruses, and in vivo trials to validate in vitro results. Heavy metal nanoparticles raise toxicity concerns and could be a suitable alternative (Liu et al. 2023).

13.5 Environmental and Safety Considerations

Nanotechnology applied to the agricultural sector, particularly in the interaction with the plant–microbiome interface, offers new promising opportunities towards increasing productivity and disease resistance. In addition to these advancements, however, there is a growing need to urgently evaluate the environmental and health impacts of such developing technologies and applications of "nano-agriculture." Work should be done to mitigate any negative consequences, especially to the soil microbiome, which is foundational to sustainable plant health and nutrient cycling within the ecosystem (Kah et al. 2018).

Nitrogen-fixing bacteria, mycorrhizal fungi, and other plant growth-promoting rhizobacteria (PGPR) are critical in supporting essential processes such as nutrient solubilization, stress tolerance, and pathogen suppression (Gogos et al. 2012). Research studies indicate that metal-derived nanoparticles such as silver (AgNPs), zinc oxide (ZnO NPs), and copper oxide (CuO NPs) are non-target toxic to some microbial communities, even at minimal concentrations (Zhang et al. 2025; Simonin and Richaume 2015). These nanoparticles can impair the structure of microbial membrane integrity, generate oxidative molecules of oxygen species (ROS), and interfere with enzymatic activities, ultimately altering microbial diversity and function (Ge et al. 2011).

Also, the interaction between nanoparticles and soil matrices adds to another layer of complexity. The bioavailability, mobility, and reactivity of the nanoparticles can also be impacted by soil texture, pH, organic matter content, and cation exchange capacity For example, some nanoparticles that aggregate quickly may become inactive in heavily clayey soils, while others may become suspended and more mobile in sandy or less organic-rich soils, posing greater ecological risk (Cornelis et al. 2014). This highlights the importance of localized risk evaluations considering local soil microbiomes alongside tailored strategies for integrating nanotechnology into environmental solutions.

Many types of nanoparticles, in contrast to traditional pesticides and fertilizers, are intentionally engineered to be more resistant, which increases environmental risks (Keller et al. 2014). Organisms living in soil might provoke a chronic toxicity due to the accumulation of non-degradable and persistent nanoparticles from frequent applications (Vázquez Núñez and de la Rosa-Álvarez 2018). Even eco-friendly nanomaterials, such as those based on biopolymers (e.g., chitosan, PLA), must be examined for their breakdown residues, which may not always be safe (Parisi et al. 2015). The biodegradation speed depends on ecological factors, including humidity, temperature, and microbial activity. As a result, comprehensive life cycle assessments (LCA) are essential to determine their ecological impact, including energy consumption during synthesis, emissions, waste production, and their ultimate conversion in the environment (Rai et al. 2015).

Current regulations governing pesticides and fertilizers were developed for traditional chemicals and are often not specific enough to account for the distinctive physicochemical properties and behaviors of nanomaterials (Joseph et al. 2023). There is an urgent need for legal frameworks that are adapted to the nanoscale, including standards for nanoparticle characterization (size, shape, surface charge), uniform testing protocols for ecotoxicity, and requirements for environmental conditions (Sousa et al. 2018). Additionally, the danger assessment of nano-enabled agricultural products should adopt a precautionary and multi-tiered approach, integrating in vitro assays, soil microcosm studies, and long-term field trials (Handy et al. 2008). International bodies, including the OECD and the FAO, have initiated discussions on the governance of nano-agriculture, but harmonized and applicable regulations are still under development (Shah et al. 2024).

Farmers and consumers must be informed and educated about the potential benefits and risks associated with the use of nanoparticles (Parisi et al. 2015). This includes clear identification of the product, strategies of risk communication, and mechanisms for public participation. However, equitable opportunity to safe nano-agricultural innovations must be ensured, avoiding the creation of technological divides between developed and developing regions (He et al. 2019) (Fig. 13.4).

13.6 Future Perspectives and Challenges

Precision agriculture and microbiome engineering can greatly benefit from advancement in technology at the nanoscale. It's now possible to design and deliver inputs at the nanoscale level, unlocking precision in the modulation of microbiomes at the rhizosphere level, delivery of nutrients, and even controlled pesticide release (Parisi et al. 2015). Moreover, intelligent nanoscale sensors can monitor soil vitality, microbial activity, and physiological stress symptoms in plants in real time, making adaptive and data-driven crop management possible (Nair et al. 2010). Similarly, surface modification of nanoparticles can be created to precisely target either symbiotic microbe or pathogenic populations to bolster plant resilience and lessen the use of chemicals (Batacharyya et al. 1995). The combination of nanotechnology with

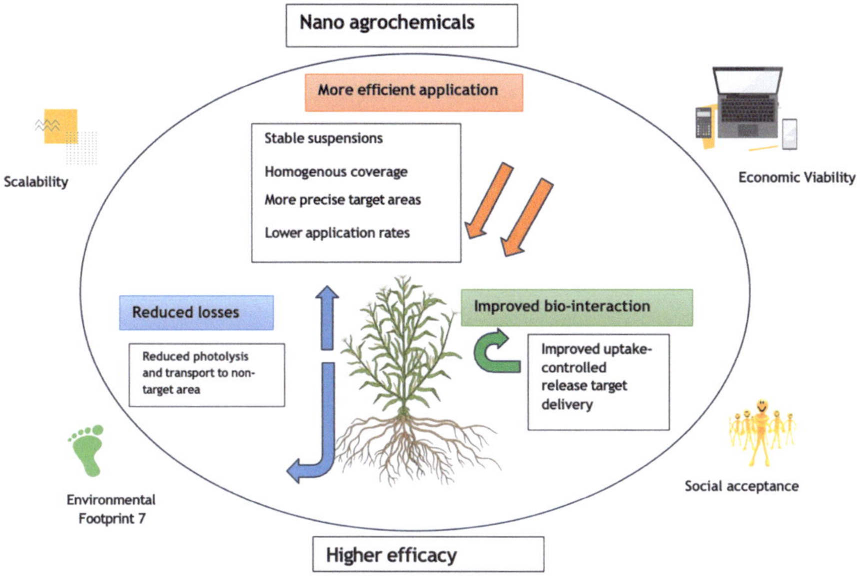

Fig. 13.4 Key drivers for applying nanotechnology to improve the efficacy of agrochemicals, along with associated socio-economic and environmental considerations

biology paves the way towards more advanced, sustainable, and efficient Agroecosystems.

Inasmuch as these advancements have exciting prospects, there remain a number of technological constraints and knowledge gaps. Most of the current understanding regarding the interactions between the complex soil-microbiome-plant system's nanomaterials lacks cohesiveness (Kah et al. 2018). Moreover, certain of these factors, such as nanoparticles clustering, their bioavailability, transport capability, and modification under diverse environmental conditions, are not yet fully understood (Servin et al. 2015). In addition, the scale-up potential of nanoformulations from lab synthesis to field application faces significant technical barriers, including cost-effectiveness, material stability, and compatibility with existing agricultural practices (Khot et al. 2012). High-scale field experiments and multi levels of biological systems approaches are urgently required to evaluate performance, safety, and specific sites outcomes of nano-interventions in real agricultural landscapes (Lowry et al. 2019).

Large-scale field experiments and multi-level biological systems approaches are urgently needed to evaluate the performance, safety, and site-specific outcomes of nano-interventions in real agricultural environments (He et al. 2019). Regulatory systems must develop to include nano-specific risk assessment protocols, standardized testing methods, and robust traceability systems (Peters et al. 2016). Furthermore, interdisciplinary dialogue involving scientists, policymakers, industry, and civil society is essential to ensure that innovation proceeds in a socially responsible and environmentally sound manner (Si et al. 2025).

13.7 Conclusion

Nanotechnology provides a powerful and effective strategy for enhancing the interaction between the plant–microbiome for better plant health in sustainable agriculture systems. This chapter covered the roles of beneficial microbial communities on disease resistance and plant resilience and how nanomaterials can modify these interactions through targeted delivery, enhanced colonization, and pathogen control.

Those strategies that make use of nanoscale features, including nanostructures and nanostrategies, help in managing precision plant diseases and are considered eco-friendly since they convert agricultural practices into green nano-enabled agriculture by reducing the use of chemicals. In ecology-conscious agriculture, the development of intelligent displayable or smart nanomaterials, along with real-time observation technologies, will be crucial. Altogether, the combination of nanotechnology with plant–microbiome science to research enables the development of robust, dynamic, and efficient systems for sustainable crop production.

Acknowledgements This study was funded by the Phytopathology Unit of the Ecole Nationale d'Agriculture de Meknès, Morocco.

References

Abdallah Y, Ogunyemi SO, Abdelazez A, Zhang M, Hong X, Ibrahim E, Hossain A, Fouad H, Li B, Chen J (2019) The green synthesis of MgO nano-flowers using Rosmarinus officinalis L.(Rosemary) and the antibacterial activities against Xanthomonas oryzae pv. oryzae. BioMed Res Int 2019

Abdelrhim AS, Mazrou YSA, Nehela Y, Atallah OO, El-Ashmony RM, Dawood MFA (2021) Silicon dioxide nanoparticles induce innate immune responses and activate antioxidant machinery in wheat against Rhizoctonia solani. Plants 10(12):2758

Afrasiabi S, Partoazar A (2024) Targeting bacterial biofilm-related genes with nanoparticle-based strategies. Front Microbiol 15:1387114

Ahmad F, Salem-Bekhit MM, Khan F, Alshehri S, Khan A, Ghoneim MM, Wu H-F, Taha EI, Elbagory I (2022) Unique properties of surface-functionalized nanoparticles for bio-application: functionalization mechanisms and importance in application. Nano 12(8):1333

Ahmed T, Luo J, Noman M, Ijaz M, Wang X, Masood HA, Manzoor N, Wang Y, Li B (2023) Microbe-mediated nanoparticle intervention for the management of plant diseases. Crop Health 1(1):1–11. https://doi.org/10.1007/s44297-023-00006-9

Aktar MW, Sengupta D, Chowdhury A (2009) Impact of pesticides use in agriculture: their benefits and hazards. Interdiscip Toxicol 2(1):1

Allen ER, Hossner LR, Ming DW, Henninger DL (1993) Solubility and cation exchange in phosphate rock and saturated clinoptilolite mixtures. Soil Sci Soc Am J 57(5):1368–1374. https://doi.org/10.2136/sssaj1993.03615995005700050034x

Auffan M, Rose J, Bottero J-Y, Lowry GV, Jolivet J-P, Wiesner MR (2009) Towards a definition of inorganic nanoparticles from an environmental, health and safety perspective. Nat Nanotechnol 4(10):634–641

Auría-Soro C, Nesma T, Juanes-Velasco P, Landeira-Viñuela A, Fidalgo-Gomez H, Acebes-Fernandez V, Gongora R, Almendral Parra MJ, Manzano-Roman R, Fuentes M (2019)

Interactions of nanoparticles and biosystems: microenvironment of nanoparticles and biomolecules in nanomedicine. Nanomaterials 9(10):1365
Batacharyya et al. (1995) Bhattacharyya et al.pdf. *9*(24):3489–3493
Behl K, Jaiswal P, Pabbi S (2024) Recent advances in microbial and nano-formulations for effective delivery and agriculture sustainability. *Biocatal Agric Biotechnol* 58:103180
Berendsen RL, Pieterse CMJ, Bakker PAHM (2012) The rhizosphere microbiome and plant health. Trends Plant Sci 17(8):478–486
Bhau BS, Phukon P, Ahmed R, Gogoi B, Borah B, Baruah J, Sharma DK, Wann SB (2016) A novel tool of nanotechnology: nanoparticle mediated control of nematode infection in plants. In: Microbial inoculants in sustainable agricultural Productivity: Vol. 2: Functional applications, pp 253–269
Bratovcic A, Hikal WM, Said-Al Ahl HAH, Tkachenko KG, Baeshen RS, Sabra AS, Sany H (2021) Nanopesticides and nanofertilizers and agricultural development: scopes, advances and applications. Open J Ecol 11(04):301–316. https://doi.org/10.4236/oje.2021.114022
Carpita N, Sabularse D, Montezinos D, Delmer DP (1979) Determination of the pore size of cell walls of living plant cells. Science 205(4411):1144–1147. https://doi.org/10.1126/science.205.4411.1144
Chen J, Wu L, Lu M, Lu S, Li Z, Ding W (2020) Comparative study on the fungicidal activity of metallic MgO nanoparticles and macroscale MgO against soilborne fungal phytopathogens. Front Microbiol 11:365
Chen J, Wu L, Kun S, Zhu Y, Wei D (2022) Nonphytotoxic copper oxide nanoparticles are powerful “nanoweapons” that trigger resistance in tobacco against the soil-borne fungal pathogen Phytophthora nicotianae. J Integr Agric 21(11):3245–3262
Chhabra P, Johari A, Tripathy DB, Gupta A (2025) Nano-pesticides, nano-herbicides and nano-fertilizers. In: Multifunctional materials. Wiley, pp 419–439. https://doi.org/10.1002/9781394234134.ch16
Chinnamuthu CR, Boopathi PM (2009) Nanotechnology and agroecosystem. 96(1/6):17–31. http://2828402980211270348-a-1802744773732722657-s-sites.googlegroups.com/site/10themadrasagriculturaljournal/home/vol-96-1-6/96-1-6-17-31.pdf?attachauth=ANoY7cq-9ptHvH-0vnEEbyFlY4jne-vy6Os90utZZxIq6glo1Gn_TeqwN2ld4RcvX79XmC9URaxnRZPvAmdVymMApIfIPzkrK3Sa3VDXTwXRlLsW2PXD7EwsknGi0wsoU_gw0wLLhShWIskaGyhz73TFcyWn9AzhPoxyVglq-sGbdQx18xZTgXwBJTvgMTScV3m9OOXtfInXvHsegFRtv52tbwaIb60-_VjedroJ6QGSDVcbv-kjCKDs9EXv_nmF-e8GTWQOh91q&attredirects=0
Choudhary N, Dhingra N, Gacem A, Yadav VK, Verma RK, Choudhary M, Bhardwaj U, Chundawat RS, Alqahtani MS, Gaur RK (2023) Towards further understanding the applications of endophytes: enriched source of bioactive compounds and bio factories for nanoparticles. Front Plant Sci 14:1193573
Cornelis G, Hund-Rinke K, Kuhlbusch T, Van Den Brink N, Nickel C (2014) Fate and Bioavailability of Engineered Nanoparticles in Soils: A Review,Interactions within natural soils have often been neglectedwhen assessing fate and bioavailability of engineered nanomaterials (ENM) in soils. Crit Rev Environ Sci Technol 44(24):2720–2764. https://doi.org/10.1080/10643389.2013.829767
Dakal TC, Kumar A, Majumdar RS, Yadav V (2016) Mechanistic basis of antimicrobial actions of silver nanoparticles. Front Microbiol 7:1831
Das A, Dutta P (2021) Antifungal activity of biogenically synthesized silver and gold nanoparticles against sheath blight of rice. J Nanosci Nanotechnol 21(6):3547–3555
de Oliveira C, Mulinari J, Reichert F, Júnior, A (2020) Nano-delivery systems of pesticides active agents for agriculture applications-an overview. In: Ciência, Tecnologia e Inovação: Do Campo à Mesa, Proceeding of International Agribusiness Congress, Recife, Brazil, pp 25–27
Debnath N, Das S, Seth D, Chandra R, Bhattacharya SC, Goswami A (2011) Entomotoxic effect of silica nanoparticles against Sitophilus oryzae (L.). J Pest Sci 84(1):99–105. https://doi.org/10.1007/s10340-010-0332-3

Deveci, H. A., Nur, G., Deveci, A., Kaya, I., Kaya, M. M., Kükürt, A., Gelen, V., Başer, Ö. F., & Karapehlivan, M. (2021). An overview of the biochemical and histopathological effects of insecticides. IntechOpen

Devi PSV, Duraimurugan P, Chandrika K (2019) Bacillus thuringiensis-based nanopesticides for crop protection. In: Nano-biopesticides today and future perspectives. Elsevier, pp 249–260

Dutta P, Kumari A, Mahanta M, Biswas KK, Dudkiewicz A, Thakuria D, Abdelrhim AS, Singh SB, Muthukrishnan G, Sabarinathan KG (2022) Advances in nanotechnology as a potential alternative for plant viral disease management. Front Microbiol 13:935193

Dutta P, Kumari A, Mahanta M, Upamanya GK, Heisnam P, Borua S, Kaman PK, Mishra AK, Mallik M, Muthukrishnan G (2023) Nanotechnological approaches for management of soil-borne plant pathogens. Front Plant Sci 14:1136233

Eichert T, Goldbach HE (2008) Equivalent pore radii of hydrophilic foliar uptake routes in stomatous and astomatous leaf surfaces—further evidence for a stomatal pathway. Physiol Plant 132(4):491–502. https://doi.org/10.1111/j.1399-3054.2007.01023.x

Eichert T, Kurtz A, Steiner U, Goldbach HE (2008) Size exclusion limits and lateral heterogeneity of the stomatal foliar uptake pathway for aqueous solutes and water-suspended nanoparticles. Physiol Plant 134(1):151–160. https://doi.org/10.1111/j.1399-3054.2008.01135.x

Elbeshehy EKF, Elazzazy AM, Aggelis G (2015) Silver nanoparticles synthesis mediated by new isolates of Bacillus spp ., nanoparticle characterization and their activity against Bean Yellow Mosaic Virus and human pathogens. Front Microbiol 6(May):1–13. https://doi.org/10.3389/fmicb.2015.00453

Elmer W, Ma C, White J (2018) Nanoparticles for plant disease management. Curr Opin Environ Sci Health 6:66–70

El-Sawy MM, Elsharkawy MM, Abass JM, Kasem MH (2017) Antiviral activity of 2–nitromethyl phenol, zinc nanoparticles and seaweed extract against cucumber mosaic virus (CMV) in eggplant. J Virol Antivir Res 6(2)

Elshahawy I, Abouelnasr HM, Lashin SM, Darwesh OM (2018) First report of Pythium aphanidermatum infecting tomato in Egypt and its control using biogenic silver nanoparticles. J Plant Prot Res 58(2)

Encinas Basurto DA, Alvarez Carvajal F, Armenta Calderon A, Gonzalez Soto TE, Esquer Miranda E, Juarez Onofre J, Mendez Ibarra R (2020) Silver nanoparticles coated with chitosan against Fusarium oxysporum causing the tomato wilt. Biotecnia 22:73

Epstein E (2009) Silicon: its manifold roles in plants. Ann Appl Biol 155(2):155–160. https://doi.org/10.1111/j.1744-7348.2009.00343.x

Escribano-Vazquez U, Beimfohr C, Bellet D, Thomas M, Zimmermann K, Langella P, Cherbuy C (2020) Symbioflor2® Escherichia coli genotypes enhance ileal and colonic gene expression associated with mucosal defense in gnotobiotic mice. Microorganisms 8(4):512

Eyhorn F, Roner T, Specking H (2015) Reducing pesticide use and risks-what action is needed. Briefing Paper: Helvetas, Swiss Intercooperation

Feizi H, Rezvani Moghaddam P, Shahtahmassebi N, Fotovat A (2012) Impact of bulk and Nanosized titanium dioxide (TiO_2) on wheat seed germination and seedling growth. Biol Trace Elem Res 146(1):101–106. https://doi.org/10.1007/s12011-011-9222-7

Francis DV, Abdalla AK, Mahakham W, Sarmah AK, Ahmed ZFR (2024) Interaction of plants and metal nanoparticles: exploring its molecular mechanisms for sustainable agriculture and crop improvement. Environ Int 190:108859

Fu L, Wang Z, Dhankher OP, Xing B (2020) Nanotechnology as a new sustainable approach for controlling crop diseases and increasing agricultural production. J Exp Bot 71(2):507–519

Gajbhiye M, Kesharwani J, Ingle A, Gade A, Rai M (2009) Fungus-mediated synthesis of silver nanoparticles and their activity against pathogenic fungi in combination with fluconazole. Nanomedicine 5(4):382–386. https://doi.org/10.1016/j.nano.2009.06.005

Ge Y, Schimel JP, Holden PA (2011) Evidence for negative effects of TiO_2 and ZnO nanoparticles on soil bacterial communities. Environ Sci Technol 45(4):1659–1664. https://doi.org/10.1021/es103040t

Ghormade V, Deshpande MV, Paknikar KM (2011) Perspectives for nano-biotechnology enabled protection and nutrition of plants. Biotechnol Adv 29(6):792–803

Gogos A, Knauer K, Bucheli TD (2012) Nanomaterials in plant protection and fertilization: current state, foreseen applications, and research priorities. J Agric Food Chem 60(39):9781–9792. https://doi.org/10.1021/jf302154y

González-Melendi P, Fernández-Pacheco R, Coronado MJ, Corredor E, Testillano PS, Risueño MC, Marquina C, Ibarra MR, Rubiales D, Pérez-de-Luque A (2008) Nanoparticles as smart treatment-delivery systems in plants: assessment of different techniques of microscopy for their visualization in plant tissues. Ann Bot 101(1):187–195

Goswami R, Bhattacharyya A, Dutta P (2020) Nanotechnological approach for management of anthracnose and crown rot diseases of banana

Graham JH, Johnson EG, Myers ME, Young M, Rajasekaran P, Das S, Santra S (2016) Potential of nano-formulated zinc oxide for control of citrus canker on grapefruit trees. Plant Dis 100(12):2442–2447

Guardiola-Márquez CE, López-Mena ER, Segura-Jiménez ME, Gutierrez-Marmolejo I, Flores-Matzumiya MA, Mora-Godínez S, Hernández-Brenes C, Jacobo-Velázquez DA (2023) Development and evaluation of zinc and iron nanoparticles functionalized with plant growth-promoting rhizobacteria (PGPR) and microalgae for their application as bio-nanofertilizers. Plants 12(20):3657

Gudkov SV, Shafeev GA, Glinushkin AP, Shkirin AV, Barmina EV, Rakov II, Simakin AV, Kislov AV, Astashev ME, Vodeneev VA, Kalinitchenko VP (2020) Production and use of selenium nanoparticles as fertilizers. ACS Omega 5(28):17767–17774. https://doi.org/10.1021/acsomega.0c02448

Handa M, Kalia A (2024) Exploring the dynamics of nanoparticle-plant-microbe interactions to realize the goal of improved crop productivity and food security. Rhizosphere:100884

Handy RD, Owen R, Valsami-Jones E (2008) The ecotoxicology of nanoparticles and nanomaterials: current status, knowledge gaps, challenges, and future needs. Ecotoxicology 17(5):315–325. https://doi.org/10.1007/s10646-008-0206-0

He X, Deng H, Hwang H m (2019) The current application of nanotechnology in food and agriculture. J Food Drug Anal 27(1):1–21. https://doi.org/10.1016/j.jfda.2018.12.002

He J, Li J, Gao Y, He X, Hao G (2023) Nano-based smart formulations: A potential solution to the hazardous effects of pesticide on the environment. J Hazard Mater 456:131599

Hosseini SN, Naghib SM, Kashani GK, Mozafari MR (2024) Chitosan-grafted graphene materials for drug delivery in wound healing. Curr Pharm Des

Hou R, Zhang Z, Pang S, Yang T, Clark JM, He L (2016) Alteration of the nonsystemic behavior of the pesticide ferbam on tea leaves by engineered gold nanoparticles. Environ Sci Technol 50(12):6216–6223. https://doi.org/10.1021/acs.est.6b01336

Hou Q, Zhang H, Bao L, Song Z, Liu C, Jiang Z, Zheng Y (2021) NCs-delivered pesticides: a promising candidate in smart agriculture. Int J Mol Sci 22(23):13043

Hyman P, Denyes J (2021) Bacteriophages in nanotechnology: history and future. In: Bacteriophages: Biology, technology, therapy, pp 657–687

Imada K, Sakai S, Kajihara H, Tanaka S, Ito S (2016) Magnesium oxide nanoparticles induce systemic resistance in tomato against bacterial wilt disease. Plant Pathol 65(4):551–560. https://doi.org/10.1111/ppa.12443

Iram F, Iqbal MS, Athar MM, Saeed MZ, Yasmeen A, Ahmad R (2014) Glucoxylan-mediated green synthesis of gold and silver nanoparticles and their phyto-toxicity study. Carbohydr Polym 104:29–33. https://doi.org/10.1016/j.carbpol.2014.01.002

Iravani S, Korbekandi H, Mirmohammadi SV, Zolfaghari B (2014) Synthesis of silver nanoparticles: chemical, physical and biological methods. Res Pharma Sci 9(6):385–406

Jayaseelan C, Ramkumar R, Rahuman AA, Perumal P (2013) Green synthesis of gold nanoparticles using seed aqueous extract of Abelmoschus esculentus and its antifungal activity. Ind Crop Prod 45:423–429. https://doi.org/10.1016/j.indcrop.2012.12.019

Johnston CT (2010) Probing the nanoscale architecture of clay minerals. Clay Miner 45(3):245–279. https://doi.org/10.1180/claymin.2010.045.3.245

Joseph TM, Kar Mahapatra D, Esmaeili A, Piszczyk Ł, Hasanin MS, Kattali M, Haponiuk J, Thomas S (2023) Nanoparticles: taking a unique position in medicine. Nano 13(3). https://doi.org/10.3390/nano13030574

Joshi H (2025) Exploring the efficacy of green nanoparticles in enhancing plant defense: a mechanistic investigation into immune response activation. J Nanopart Res 27(2):29

Juo ASR, Franzluebbers K (2003) Soil biology and microbiology. In: Tropical soils. Oxford University Press. https://doi.org/10.1093/oso/9780195115987.003.0008

Kah M, Beulke S, Tiede K, Hofmann T (2013) Nanopesticides: state of knowledge, environmental fate, and exposure modeling. Crit Rev Environ Sci Technol 43(16):1823–1867

Kah M, Kookana RS, Gogos A, Bucheli TD (2018) A critical evaluation of nanopesticides and nanofertilizers against their conventional analogues. Nat Nanotechnol 13(8):677–684. https://doi.org/10.1038/s41565-018-0131-1

Kaman PK, Dutta P (2019) Synthesis, characterization and antifungal activity of biosynthesized silver nanoparticle. Indian Phytopathol 72:79–88

Kandi V, Kandi S (2015) Antimicrobial properties of nanomolecules: potential candidates as antibiotics in the era of multi-drug resistance. Epidemiol Health 37

Keller AA, Vosti W, Wang H, Lazareva A (2014) Release of engineered nanomaterials from personal care products throughout their life cycle. J Nanopart Res 16(7). https://doi.org/10.1007/s11051-014-2489-9

Khan ST, Adil SF, Shaik MR, Alkhathlan HZ, Khan M, Khan M (2021) Engineered nanomaterials in soil: their impact on soil microbiome and plant health. Plants 11(1):109

Khandelwal N, Barbole RS, Banerjee SS, Chate GP, Biradar AV, Khandare JJ, Giri AP (2016) Budding trends in integrated pest management using advanced micro-and nano-materials: challenges and perspectives. J Environ Manag 184:157–169

Khot LR, Sankaran S, Maja JM, Ehsani R, Schuster EW (2012) Applications of nanomaterials in agricultural production and crop protection: a review. Crop Prot 35:64–70

Kibbey TCG, Strevett KA (2019) The effect of nanoparticles on soil and rhizosphere bacteria and plant growth in lettuce seedlings. Chemosphere 221:703–707. https://doi.org/10.1016/j.chemosphere.2019.01.091

Kumar V, Guleria P, Kumar V, Yadav SK (2013) Gold nanoparticle exposure induces growth and yield enhancement in Arabidopsis thaliana. Sci Total Environ 461–462:462–468. https://doi.org/10.1016/j.scitotenv.2013.05.018

Kumar S, Bhanjana G, Sharma A, Sidhu MC, Dilbaghi N (2014) Synthesis, characterization and on field evaluation of pesticide loaded sodium alginate nanoparticles. Carbohydr Polym 101:1061–1067. https://doi.org/10.1016/j.carbpol.2013.10.025

Kumar P, Burman U, Kaul RK (2018) Ecological risks of nanoparticles. In: Nanomaterials in plants, algae, and microorganisms. Elsevier, pp 429–452. https://doi.org/10.1016/B978-0-12-811487-2.00019-0

Latvala S, Hedberg J, Di Bucchianico S, Möller L, Odnevall Wallinder I, Elihn K, Karlsson HL (2016) Nickel release, ROS generation and toxicity of Ni and NiO micro- and nanoparticles. PLoS One 11(7):e0159684. https://doi.org/10.1371/journal.pone.0159684

Leggo PJ (2000) An investigation of plant growth in an organo-zeolitic substrate and its ecological significance. Plant Soil 219(1/2):135–146. https://doi.org/10.1023/A:1004744612234

Li M, Sun X, Yin M, Shen J, Yan S (2023) Recent advances in nanoparticle-mediated co-delivery system: a promising strategy in medical and agricultural field. Int J Mol Sci 24(6). https://doi.org/10.3390/ijms24065121

Lin N, Wang C, Ding J, Su L, Xu L, Zhang B, Zhang Y, Fan J (2020) Efficacy of nanoparticle encapsulation on suppressing oxidation and enhancing antifungal activity of cyclic lipopeptides produced by Bacillus subtilis. Colloids Surf B: Biointerfaces 193:111143

Liu L, Xie Y, Zhong X, Deng Q, Shao Q, Cai Z, Huang X (2023) Facilitating effects of the reductive soil disinfestation process combined with Paenibacillus sp. amendment on soil health and physiological properties of Momordica charantia. Front Plant Sci 13(January):1–14. https://doi.org/10.3389/fpls.2022.1095656

Lowry GV, Avellan A, Gilbertson LM (2019) Opportunities and challenges for nanotechnology in the agri-tech revolution. Nat Nanotechnol 14(6):517–522. https://doi.org/10.1038/s41565-019-0461-7

Ma Y (2019) Seed coating with beneficial microorganisms for precision agriculture. Biotechnol Adv 37(7):107423

Malerba M, Cerana R (2016) Chitosan effects on plant systems. Int J Mol Sci 17(7):996

Mansoor S, Zahoor I, Baba TR, Padder SA, Bhat ZA, Koul AM, Jiang L (2021) Fabrication of silver nanoparticles against fungal pathogens. Front Nanotechnol 3:679358

Mansor M, Xu J (2020) Benefits at the nanoscale: a review of <scp>nanoparticle-enabled</scp> processes favouring microbial growth and functionality. Environ Microbiol 22(9):3633–3649. https://doi.org/10.1111/1462-2920.15174

Maruyama CR, Guilger M, Pascoli M, Bileshy-José N, Abhilash PC, Fraceto LF, de Lima R (2016) Nanoparticles based on chitosan as carriers for the combined herbicides Imazapic and Imazapyr. Sci Rep 6(1):19768. https://doi.org/10.1038/srep19768

Matras E, Gorczyca A, Pociecha E, Przemieniecki SW, Żeliszewska P, Oćwieja M (2023) Silver nanoparticles affect wheat (Triticum aestivum L.) germination, seedling blight and yield. Funct Plant Biol 50(5):390–406. https://doi.org/10.1071/FP22086

Mendes R, Kruijt M, De Bruijn I, Dekkers E, Van Der Voort M, Schneider JHM, Piceno YM, DeSantis TZ, Andersen GL, Bakker PAHM (2011) Deciphering the rhizosphere microbiome for disease-suppressive bacteria. Science 332(6033):1097–1100

Menichetti A, Mavridi-Printezi A, Mordini D, Montalti M (2023) Effect of size, shape and surface functionalization on the antibacterial activity of silver nanoparticles. J Funct Biomater 14(5):244

Mgadi K, Ndaba B, Roopnarain A, Rama H, Adeleke R (2024) Nanoparticle applications in agriculture: overview and response of plant-associated microorganisms. Front Microbiol 15. https://doi.org/10.3389/fmicb.2024.1354440

Millán G, Agosto F, Vázquez M, Botto L, Lombardi L, Juan L (2008) Uso de clinoptilolita como un vehículo de fertilizantes nitrogenados en un suelo de la región Pampeana de Argentina. Ciencia e Investigación Agraria 35(3). https://doi.org/10.4067/S0718-16202008000300007

Mohammadi A, Hashemi M, Hosseini SM (2015) Nanoencapsulation of Zataria multiflora essential oil preparation and characterization with enhanced antifungal activity for controlling Botrytis cinerea, the causal agent of gray mould disease. Innovative Food Sci Emerg Technol 28:73–80. https://doi.org/10.1016/j.ifset.2014.12.011

Moradipour M, Saberi-Riseh R, Mohammadinejad R, Hosseini A (2019) Nano-encapsulation of plant growth-promoting rhizobacteria and their metabolites using alginate-silica nanoparticles and carbon nanotube improves UCB1 pistachio micropropagation. J Microbiol Biotechnol 29:1096

Mosquera J, García I, Liz-Marzán LM (2018) Cellular uptake of nanoparticles versus small molecules: a matter of size. Acc Chem Res 51(9):2305–2313. https://doi.org/10.1021/acs.accounts.8b00292

Nair R, Varghese SH, Nair BG, Maekawa T, Yoshida Y, Kumar DS (2010) Nanoparticulate material delivery to plants. Plant Sci 179(3):154–163

Noha K, Bondok AM, El-Dougdoug KA (2018) Evaluation of silver nanoparticles as antiviral agent against ToMV and PVY in tomato plants. Sciences 8(01):100–111

Ogunyemi SO, Zhang M, Abdallah Y, Ahmed T, Qiu W, Ali MA, Yan C, Li B (2020) The biosynthesis of three metal oxide nanoparticles (ZnO, MnO2, and MgO) and their antibacterial activity against the bacterial leaf blight pathogen. Front Microbiol 11:588326

Okey-Onyesolu CF, Hassanisaadi M, Bilal M, Barani M, Rahdar A, Iqbal J, Kyzas GZ (2021) Nanomaterials as nanofertilizers and nanopesticides: an overview. ChemistrySelect 6(33):8645–8663. https://doi.org/10.1002/slct.202102379

Panpatte DG, Jhala YK, Shelat HN, Vyas RV (2016) Nanoparticles: the next generation technology for sustainable agriculture. In: Microbial inoculants in sustainable agricultural productivity. Springer India, pp 289–300. https://doi.org/10.1007/978-81-322-2644-4_18

Parisi C, Vigani M, Rodríguez-Cerezo E (2015) Agricultural nanotechnologies: what are the current possibilities? Nano Today 10(2):124–127. https://doi.org/10.1016/j.nantod.2014.09.009

Peng C, Tong H, Shen C, Sun L, Yuan P, He M, Shi J (2020) Bioavailability and translocation of metal oxide nanoparticles in the soil-rice plant system. Sci Total Environ 713:136662. https://doi.org/10.1016/j.scitotenv.2020.136662

Pestovsky YS, Martínez-Antonio A (2017) The use of nanoparticles and nanoformulations in agriculture. J Nanosci Nanotechnol 17(12):8699–8730. https://doi.org/10.1166/jnn.2017.15041

Peters RJB, Bouwmeester H, Gottardo S, Amenta V, Arena M, Brandhoff P, Marvin HJP, Mech A, Moniz FB, Pesudo LQ, Rauscher H, Schoonjans R, Undas AK, Vettori MV, Weigel S, Aschberger K (2016) Nanomaterials for products and application in agriculture, feed and food. Trends Food Sci Technol 54:155–164. https://doi.org/10.1016/j.tifs.2016.06.008

Ponmurugan P, Manjukarunambika K, Elango V, Gnanamangai BM (2016) Antifungal activity of biosynthesised copper nanoparticles evaluated against red root-rot disease in tea plants. J Exp Nanosci 11(13):1019–1031

Popp J, Pető K, Nagy J (2013) Pesticide productivity and food security. A review. *Agron Sustain Dev* 33:243–255

Pradhan S, Roy I, Lodh G, Patra P, Choudhury SR, Samanta A, Goswami A (2013) Entomotoxicity and biosafety assessment of PEGylated acephate nanoparticles: A biologically safe alternative to neurotoxic pesticides. J Environ Sci Health B 48(7):559–569. https://doi.org/10.1080/03601234.2013.774891

Pradhan S, Patra P, Mitra S, Dey KK, Jain S, Sarkar S, Roy S, Palit P, Goswami A (2014) Manganese nanoparticles: impact on non-nodulated plant as a potent enhancer in nitrogen metabolism and toxicity study both in vivo and in vitro. J Agric Food Chem 62(35):8777–8785. https://doi.org/10.1021/jf502716c

Punniyakotti P, Vinayagam S, Rajamohan R, Priya SD, Moovendhan M, Sundaram T (2024) Environmental fate and ecotoxicological behaviour of pesticides and insecticides in non-target environments: nanotechnology-based mitigation strategies. J Environ Chem Eng:113349

Qin J, Tong Z, Zhan Y, Buisson C, Song F, He K, Nielsen-LeRoux C, Guo S (2020) A bacillus thuringiensis chitin-binding protein is involved in insect peritrophic matrix adhesion and takes part in the infection process. Toxins 12(4):252

Rai V, Acharya S, Dey N (2012) Implications of nanobiosensors in agriculture. J Biomater Nanobiotechnol 03(02):315–324. https://doi.org/10.4236/jbnb.2012.322039

Rai M, Ribeiro C, Mattoso L, Duran N (2015) Nanotechnologies in food and agriculture, pp 1–347. https://doi.org/10.1007/978-3-319-14024-7

Rajakumar G, Abdul Rahuman A (2012) Acaricidal activity of aqueous extract and synthesized silver nanoparticles from Manilkara zapota against Rhipicephalus (Boophilus) microplus. Res Vet Sci 93(1):303–309. https://doi.org/10.1016/j.rvsc.2011.08.001

Rajakumar G, Rahuman AA, Velayutham K, Ramyadevi J, Jeyasubramanian K, Marikani A, Elango G, Kamaraj C, Santhoshkumar T, Marimuthu S, Zahir AA, Bagavan A, Jayaseelan C, Kirthi AV, Iyappan M, Siva C (2013) Novel and simple approach using synthesized nickel nanoparticles to control blood-sucking parasites. Vet Parasitol 191(3–4):332–339. https://doi.org/10.1016/j.vetpar.2012.08.028

Rajput VD, Harish, Singh RK, Verma KK, Sharma L, Quiroz-Figueroa FR, Meena M, Gour VS, Minkina T, Sushkova S (2021) Recent developments in enzymatic antioxidant defence mechanism in plants with special reference to abiotic stress. Biology 10(4):267

Roy Choudhury S, Ghosh M, Mandal A, Chakravorty D, Pal M, Pradhan S, Goswami A (2011) Surface-modified sulfur nanoparticles: an effective antifungal agent against Aspergillus Niger and Fusarium oxysporum. Appl Microbiol Biotechnol 90(2):733–743. https://doi.org/10.1007/s00253-011-3142-5

Savary S, Willocquet L, Pethybridge SJ, Esker P, McRoberts N, Nelson A (2019) The global burden of pathogens and pests on major food crops. Nat Ecol Evol 3(3):430–439

Servin A, Elmer W, Mukherjee A, De la Torre-Roche R, Hamdi H, White JC, Bindraban P, Dimkpa C (2015) A review of the use of engineered nanomaterials to suppress plant disease and enhance crop yield. J Nanopart Res 17(2):1–21. https://doi.org/10.1007/s11051-015-2907-7

Shah V, Belozerova I (2009) Influence of metal nanoparticles on the soil microbial community and germination of lettuce seeds. Water Air Soil Pollut 197(1–4):143–148. https://doi.org/10.1007/s11270-008-9797-6

Shah MA, Shahnaz T, Zehab-ud-Din, Masoodi JH, Nazir S, Qurashi A, Ahmed GH (2024) Application of nanotechnology in the agricultural and food processing industries: A review. Sustain Mater Technol 39(December 2023):e00809. https://doi.org/10.1016/j.susmat.2023.e00809

Sharma P, Bhatt D, Zaidi MGH, Saradhi PP, Khanna PK, Arora S (2012) Silver nanoparticle-mediated enhancement in growth and antioxidant status of brassica juncea. Appl Biochem Biotechnol 167(8):2225–2233. https://doi.org/10.1007/s12010-012-9759-8

Si C, Liu X, Jiao J, Xu J, Ye L, Li J, Han Q (2025) Developing single ligand strategy for engineering a high-performance photocatalyst with the pore space partition and open-metal sites. Appl Catal B-Environ 362(October 2024):124687. https://doi.org/10.1016/j.apcatb.2024.124687

Simonin M, Richaume A (2015) Impact of engineered nanoparticles on the activity, abundance, and diversity of soil microbial communities: a review. Environ Sci Pollut Res 22(18):13710–13723. https://doi.org/10.1007/s11356-015-4171-x

Singh NK, Rai AK (2024) Enhancing crop productivity through nanotechnology: A comprehensive review of strategies and results. J Exp Agric Int 46(5):435–458

Singh J, Vishwakarma K, Ramawat N, Rai P, Singh VK, Mishra RK, Kumar V, Tripathi DK, Sharma S (2019) Nanomaterials and microbes' interactions: a contemporary overview. 3 Biotech 9(3):68. https://doi.org/10.1007/s13205-019-1576-0

Sousa GFM, Gomes DG, Campos EVR, Oliveira JL, Fraceto LF, Stolf-Moreira R, Oliveira HC (2018) Post-emergence herbicidal activity of nanoatrazine against susceptible weeds. Front Environ Sci 6(MAR):1–6. https://doi.org/10.3389/fenvs.2018.00012

Tang G, Tian Y, Niu J, Tang J, Yang J, Gao Y, Chen X, Li X, Wang H, Cao Y (2021) Development of carrier-free self-assembled nanoparticles based on fenhexamid and polyhexamethylene biguanide for sustainable plant disease management. Green Chem 23(6):2531–2540

Tomah AA, Alamer ISA, Li B, Zhang J-Z (2020) Mycosynthesis of silver nanoparticles using screened Trichoderma isolates and their antifungal activity against Sclerotinia sclerotiorum. Nano 10(10):1955

Tremblay GF, Bélanger G, Lajeunesse J, Chouinard PY, Charbonneau É (2015) Timothy response to increasing rates of selenium fertilizer in Eastern Canada. Agron J 107(1):211–220. https://doi.org/10.2134/agronj14.0397

Tripathi S, Tiwari K, Mahra S, Victoria J, Rana S, Tripathi DK, Sharma S (2024) Nanoparticles and root traits: mineral nutrition, stress tolerance and interaction with rhizosphere microbiota. Planta 260(2):34

Turner TR, James EK, Poole PS (2013) The plant microbiome. Genome Biol 14:1–10

Vandenkoornhuyse P, Quaiser A, Duhamel M, Le Van A, Dufresne A (2015) The importance of the microbiome of the plant holobiont. New Phytol 206(4):1196–1206

Vázquez Núñez E, de la Rosa-Álvarez G (2018) Environmental behavior of engineered nanomaterials in terrestrial ecosystems: uptake, transformation and trophic transfer. Curr Opin Environ Sci Health 6:42–46. https://doi.org/10.1016/j.coesh.2018.07.011

Wang Y, Cui H, Sun C, Zhao X, Cui B (2014) Construction and evaluation of controlled-release delivery system of Abamectin using porous silica nanoparticles as carriers. Nanoscale Res Lett 9(1):655. https://doi.org/10.1186/1556-276X-9-655

Wang P, Lombi E, Zhao F-J, Kopittke PM (2016) Nanotechnology: a new opportunity in plant sciences. Trends Plant Sci 21(8):699–712

Wang X, Xie H, Wang P, Yin H (2023) Nanoparticles in plants: uptake, transport and physiological activity in leaf and root. Materials 16(8):3097

Wen L, Li Z, Zou H, Liu A, Chen J (2005) Controlled release of avermectin from porous hollow silica nanoparticles. Pest Manag Sci 61(6):583–590. https://doi.org/10.1002/ps.1032

Wen Y, Yuan Y, Chen H, Xu D, Lin K, Liu W (2010) Effect of chitosan on the enantioselective bioavailability of the herbicide Dichlorprop to Chlorella pyrenoidosa. Environ Sci Technol 44(13):4981–4987. https://doi.org/10.1021/es100507p

Xu T, Wang Y, Aytac Z, Zuverza-Mena N, Zhao Z, Hu X, Ng KW, White JC, Demokritou P (2022) Enhancing agrichemical delivery and plant development with biopolymer-based stimuli responsive core–shell nanostructures. ACS Nano 16(4):6034–6048

Yadav A, Yadav K (2018) Nanoparticle-based plant disease management: tools for sustainable agriculture. In Nanotechnol Life Sci (Issue December). https://doi.org/10.1007/978-3-319-91161-8_2

Zabrieski Z, Morrell E, Hortin J, Dimkpa C, McLean J, Britt D, Anderson A (2015) Pesticidal activity of metal oxide nanoparticles on plant pathogenic isolates of Pythium. Ecotoxicology 24(6):1305–1314. https://doi.org/10.1007/s10646-015-1505-x

Zaki SA, Ouf SA, Abd-Elsalam KA, Asran AA, Hassan MM, Kalia A, Albarakaty FM (2022) Trichogenic silver-based nanoparticles for suppression of fungi involved in damping-off of cotton seedlings. Microorganisms 10(2):344

Zambryski P (2004) Cell-to-cell transport of proteins and fluorescent tracers via plasmodesmata during plant development. J Cell Biol 164(2):165–168. https://doi.org/10.1083/jcb.200310048

Zhang N, Wang D, Liu Y, Li S, Shen Q, Zhang R (2014) Effects of different plant root exudates and their organic acid components on chemotaxis, biofilm formation and colonization by beneficial rhizosphere-associated bacterial strains. Plant Soil 374(1–2):689–700. https://doi.org/10.1007/s11104-013-1915-6

Zhang H, Zhang M, Xu T, Wang X, Qi J, Wang Y, Liu W, Zhu L, Yuan Z, Si C (2025) Ions and electrons dual transport channels regulated by nanocellulose for mitigating dendrite growth of zinc-ion batteries. Chem Eng J 505(December 2024):159476. https://doi.org/10.1016/j.cej.2025.159476

Chapter 14
Innovative Approaches to Combat Plant Pathogens Through Rhizosphere Engineering with Metal Nanoparticles

Fahad Faheem, Noman Ijaz, Muhammad Ashfaq, Muhammad Shaaban, Muhammad Mahmood Ahmed, Samah Bashir Kayani, and Mirza Abid Mehmood

Abstract Nanotechnology research has dramatically increased manifold in the last decade. Nanoparticles (NPs) possess distinct characteristics with a large surface area-to-volume ratio and small size (less than 100 nm). Several factors affected the qualities and applications of nanoparticles, including the physical features and synthesis approaches, such as size and shape. The rhizosphere is made up of soil, microbes, and plants that can be altered in numerous ways to enhance the effectiveness of inoculants. The targeted manipulation of the root-soil interface is known as rhizosphere engineering. Rhizosphere engineering using MNPs has emerged as an effective method for controlling plant pathogens while improving plant and soil health. It may provide a more favorable environment for plant growth by altering the

F. Faheem · N. Ijaz · M. Ashfaq
Plant Pathology, Institute of Plant Protection, Muhammad Nawaz Shareef University of Agriculture, Multan, Punjab, Pakistan

M. Shaaban
College of Agriculture, Henan University of Science and Technology, Luoyang, China

M. M. Ahmed
Department of Bioinformatics, Institute of Biochemistry, Biotechnology and Bioinformatics, The Islamia University of Bahawalpur, Bahawalpur, Punjab, Pakistan

S. B. Kayani
Govt. Graduate College (W) Mumtazabad, Multan, Punjab, Pakistan

M. A. Mehmood (✉)
Plant Pathology, Institute of Plant Protection, Muhammad Nawaz Shareef University of Agriculture, Multan, Punjab, Pakistan

State Key Laboratory of Vegetation Structure, Function and Construction; School of Ecology and Environmental Science, Yunnan University, Kunming, People's Republic of China

Ministry of Education Key Laboratory for Transboundary Ecosecurity of Southwest China, Crop Disease & Pest Biocontrol Engineering Research Center of Yunnan Province, Kunming, Yunnan, China
e-mail: abid.mehmood@mnsuam.edu.pk

K. A. Abd-Elsalam (ed.), *Plant-Microbiome Nanotechnology*, Sustainability Sciences in Asia and Africa, https://doi.org/10.1007/978-981-95-5056-2_14

composition and activity of microbial communities, modifying root exudation patterns, or beneficial organisms, such as endophytes, bio stimulants, and nanomaterials. Rhizosphere engineering, including nanomaterials like MNPs, is proposed as a promising and sustainable approach to improve plant resilience. MNPs can increase plant systemic resistance and boost the absorption of nutrients and tolerate different stresses. Most MNPs such as zinc, silver, iron, and copper exhibit distinct antimicrobial activity, allowing specific inhibition of pathogenic organisms without interfering with beneficial microbial communities. MNPs show potent antimicrobial action through numerous mechanisms, including the development of reactive oxygen species, membrane disruption, inhibition of enzyme activity, regulated ion release, or cell death. Metal NPs can be employed via root dipping, transplant drench, seed coating, foliar spray, and organic amendments. Such techniques enhance root-nanoparticle interaction and microbial activity and manage the soil-borne pathogens. This chapter focuses on the rhizosphere engineering using metal nanoparticles, their antimicrobial mechanism, and application methods. Future innovations should emphasize producing biodegradable and environmentally sustainable nanomaterials for safe and long-term use across industries.

Keywords Nanoparticles · Green synthesis · Antimicrobial mechanism · Rhizosphere engineering

14.1 Introduction

The rhizosphere, a confined area of interaction between soil particles and roots, represents the initial plant-influenced environment endured by soil microorganisms (Dessaux et al. 2016). The rhizosphere contains three zones: endorhizosphere, which contains microorganisms and mineral ions in root cortex and epidermis; rhizoplane, which is situated next to the root epidermal cells and mucilage; and the ectorhizosphere, which extends from the rhizoplane to the bulk soil (Poria et al. 2021). The rhizosphere also acts as signaling molecules to induce the biological and physiological contact between the soil microbiome and plant roots via altering the chemical and physical characteristics of the soil by suppressing the growth of competing plant species, stimulating beneficial symbiosis, e.g., with nitrogen-fixing bacteria, mycorrhizal fungi and epiphytes, and hindering the pathogenic fungi, bacteria, and insect attack (Wen et al. 2022). However, the microbial community is greatly influenced by the surrounding environmental factors, as well as stress factors, such as biotic (pathogens), and abiotic (salinity, flood, drought, and pH) (Gargallo-Garriga et al. 2018). In a previous study, authors described that the rhizosphere has greater microbial diversity and abundance than bulk soil or which is not directly impacted by the plant roots (Xu et al. 2021). Over the last few years, the increasing effects of climate change, intensive farming methods, and appearance of new plant pathogens have been shown to necessitate innovative solutions towards plant protection (Prasad et al. 2017). MNPs are one of the intriguing tools due to their wide antimicrobial activities, ability to control release, and potential to change the plant–microbe association in the rhizosphere (Handa and Kalia 2024). Properly used, ENMs such as silver, copper, zinc, and gold or titanium dioxide NPs possess high antimicrobial activity and little phytotoxicity (Kumar et al. 2022). The association between the nanoparticles and root surface area and microbes can be added with some surface charges, coating, and functional molecules (Perea Velez et al. 2021). The size, shape, chemical composition, quantity, and numerous application strategies are some factors that influence the effectiveness of MNPs (El-Khawaga et al. 2023). A new strategy for improving plant immunity, regulating beneficial microbes, and inhibiting soil-borne diseases is rhizosphere engineering with metal nanoparticles (Ali et al. 2020). Through the applications of nanoparticles (NPs), especially at the soil–root interface, researchers have successfully activated the plant defense system and systemic acquired resistance, as well as altered microbial diversity to favor beneficial microbes such as mycorrhizal fungi and PGPR (Saeed et al. 2021).

14.1.1 Rhizosphere Engineering

Rhizosphere is a small region of soil near plant roots and the origin of various microorganisms. The plants influence the surrounding soil by emitting a variety of substances called rhizodeposits, such as carbohydrates, secondary metabolites, and organic acids (Hakim et al. 2021). Microbiomes play a significant role in plant health, metabolism, such as plant hormones production, defense responses, nutrient uptake, nitrogen fixation, phosphate solubilization, and disease suppression (Backer et al. 2018; Mitter et al. 2019; Ortiz and Sansinenea 2021). Rhizosphere engineering could promote plant health and growth, yield, and protection against biotic and abiotic stresses, among other processes; so, it is an eco-friendly approach to sustainable agriculture production. Engineering of the rhizosphere could lower its resilience to agrochemicals with positive microorganisms' transgenic plants, biostimulants, and eco-friendly inputs (Ryan et al. 2009). Rhizospheres consist of three primary factors, and they include soil, plants, and microorganisms. Over the last few years, the influence of climate change and global trade has resulted in the emergence of new pathogens and dissemination of diseases, which has now threatened agricultural production (Singh et al. 2023a). The conditions of the environment in the soil are changing over the time, and microbial activity is highly influenced by the external factors (Jansson and Hofmockel 2020). Conversely, microbial communities either thrive or fall due to responses of changing conditions in the environment (Wu et al. 2019). Plant diseases and pests cause major losses in the agricultural industry. According to Savary et al. (2019), showed that diseases and pests may cause average global losses in potato, wheat, soybean, and corn by 30%, 17.2%, 21.5%, 21.4%, and 22.5%, respectively. Traditional techniques such as antibiotics and pesticides, which are based on chemicals, can have adverse effects on human health and environment, which can result in resistance among pathogens of plants (Ahmed et al. 2023). One of the present management approaches is the use of crop rotation, resistant cultivars, and soil-applied agrochemicals. Nevertheless, these conventional methods do not fully protect crops, while new methods are cost-effective, less toxic, such as MNPs are to be identified and employed in combating plant disease (Gogos et al. 2012; Athawale et al. 2018).

14.1.2 Metal Nanoparticles

Nanomaterials (NMs) have become popular because of their increased use in various areas. MNPs have been shown to have potential use in agriculture, where they have boosted plant growth and food supply. NMs such as silver, zinc, gold, and copper have distinct chemical and physical properties that render them suitable for use with plants (Ashraf et al. 2021). For instance, MNPs have been shown to enhance the growth and development of plants through the induction of plant defense mechanism against pathogens, stress-responsive genes, and increased nutrient uptake (Abdelkader et al. 2023). Meanwhile, nanoscale agrochemicals and nano-enabled immunomodulation of engineered nanomaterials (ENMs) increase plant response to diseases to increase their resistance capabilities (Ahmed et al. 2024; Zhang et al. 2024). Moreover, nano-probe technology monitors plant immunity and stress responses (biotic and abiotic) in real time (Son et al. 2023). Recently, applications of ENMs have improved plant immunity and tolerance to environmental stress via stimulation of SAR and induced systemic resistance (ISR) and regulation of antioxidative defense. Nanoclay is a new and promising nanomaterial (NM) which has demonstrated performance against the biotic and abiotic stresses (Abd-Elsalam et al. 2024).

The green synthesis method uses various biological entities, such as capping, reducing, or stabilizing agents, instead of expensive, hazardous, and toxic substances (Dhillon et al. 2012). Therefore, researchers adopt biosynthetic or green synthesis methods, in which biological systems of nanoparticles display physicochemical properties. The green method is cheaper and eco-friendly than chemical methods that employ expensive and hazardous chemicals (Arumugam et al. 2015; Singh et al. 2015). However, NPs production by biological syntheses such as plants and microorganisms represents a green method of synthesis that is very easy, cost-effective, biocompatible, safe, and efficient compared to several chemical and physical processes (Ankamwar et al. 2005). The biological methods of green synthesis can be used with bacteria, yeast, fungi, algae, plants, or other products. Nanoparticles are synthesized using plant parts, including stems, leaves, roots, seeds, and fruits, as their extracts contain reducing or stabilizing (capping) agents (Narayanan and Sakthivel 2011). The major objective of this chapter is to review the role of the rhizosphere in regulating plant health and demonstrating its use in rhizosphere engineering and explore the modern issues and perspectives in the field of nano-enabled rhizosphere treatment.

14.2 Action Mechanism of Metal Nanoparticles

14.2.1 Antifungal Activity

Metallic nanoparticles (MNPs) are intriguing substances for controlling plant pathogenic fungi. The most common metal nanoparticles are zinc, silver, copper, and gold, and they are gaining popularity as antifungal agents due to their antimicrobial characteristics (Murphy et al. 2015; Zazo et al. 2016). Various MNPs, including nickel, iron, platinum, aluminum, titanium, magnesium, and palladium, have been examined against some fungi. MNPs have many effects on fungi, including destruction of the cell membranes, altering the color of filamentous fungi, reducing budding in yeast, and preventing sporulation and germination (Hameed et al. 2015; Ogar et al. 2015). Earlier studies reported that silver nanoparticles (AgNPs) may enter cell walls and fungal membranes, causing cell death through interaction with membrane protein functional groups. According to Lipovsky et al. (2011), zinc oxide nanoparticles produce reactive oxygen species (ROS) that extremely reduce the growth of *Candida albicans*. Xia et al. (2016) stated that AgNPs had multiple effects on the fungal pathogen *Trichosporon asahii*, such as hyphae distortion, organelle degeneration, and cytoplasmic leakage. However, a recent study revealed that the primary lethal mechanism of Ag ions and AgNPs against *S. cerevisiae* is not membrane permeabilization nor oxidative stress (Käosaar et al. 2016).

Exposure to nanoparticles causes alterations in the fungal cell wall, resulting in surface shrinkage, pit and pore development, cellular aggregation, and overall distortion (Kim et al. 2009; Selvaraj et al. 2014; Xia et al. 2016). Microscopy has shown that nanoparticles might have direct contact and attachment with fungal cell walls during absorption, which causes morphological alterations. Exposure to AgNPs might not always result in significant damage that causes cell wall rupture or pits. Energy-dispersive X-ray spectroscopy can identify silver (Ag) dots in the cytoplasm. Difficulty in distinguishing the plasma membrane and cytoplasm after absorption of nanoparticles suggests that NPs may damage these structures and blur their boundaries (Arciniegas-Grijalba et al. 2017; Athie-García et al. 2018). Silver nanoparticles change the ratio of phosphatidylcholine-to-phosphatidylethanolamine in treated cells, which results in cell activity and loss of membrane integrity. Moreover, treated cells have lower C18:2 phospholipids and less potential for desaturation (Różalska et al. 2018). Higher intercellular and intracellular levels of trehalose and glucose lead to cell linkage in *Candida albicans,* and these compounds can act as a carbon source that offers protection during stress (Alvarez-Peral et al. 2002). Furthermore, when all the fungal growth was prevented, levels of nucleic acids and carbohydrates were reduced (He et al. 2011). After absorption to AgNPs, both *C. albicans* and *Saccharomyces cerevisiae* exhibited higher glucose levels; the reduction in glucose was probably connected to G1 cell cycle delay or the generation of ROS. Thereby, the formation of ROS may lead to a decline in the uptake of glucose by *C. albicans* (Lee et al. 2019). Smaller NPs may have the capacity to

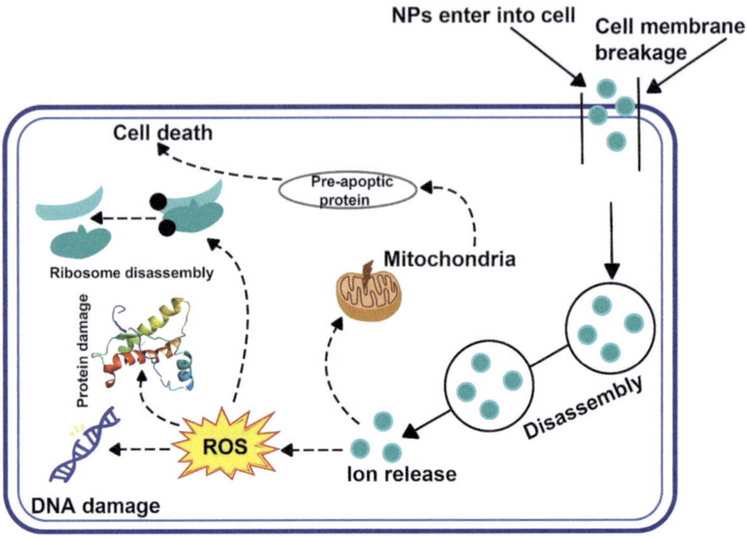

Fig. 14.1 Antifungal activity mechanism by NPs. NPs penetrate the cell by disrupting the cell membrane and lead to structural damage. They can release metal-ion within the cell interfering with the cellular activity, or destroy mitochondria, and trigger pre-apoptotic protein expression. These ions promote the production of ROS, which causes oxidative stress or protein's damage. It leads to protein malfunction, ribosome degradation, and impaired protein synthesis. All the effects lead to cellular dysfunction and cell death

induce fluid-phase endocytosis that passes through cell wall damage (Selvaraj et al. 2014). DNA fragmentation in the nucleus indicates that NPs can pass through the nuclear membrane or induce damage to the intracellular membrane, allowing them to connect with DNA. Ion release has improved antimicrobial activity with more release over time, as mentioned in Fig. 14.1 (Chen et al. 2016).

Few studies have suggested that ions are more harmful than nanoparticles (Pradhan et al. 2015). This could be related to size, which could increase cellular penetration or the capacity to form complexes with other biomolecules, including nucleic acids, negatively charged lipids, and proteins. NPs may have a significant influence on fungal hyphae and spores as the fungi treated with GO-Ag NPs, MgO NPs, Cu NPs, ZnO NPs, and Ag NPs depicted deformed, damaged and reduced hyphae (Kim et al. 2009; Ouda 2014; Chen et al. 2016; De La Rosa-García et al. 2018; Kumari et al. 2019). Most pathogenic fungi develop biofilms which increase their resistance to antifungal agents. Contrary to other fungal species, *S. cerevisiae* and few strains of *C. albicans* can produce more mature biofilms with extracellular matrix and stronger structural integrity. Both filamentous and yeast cells have the potential to form biofilms although it is not morphologically specific (Chandra et al. 2001). AgNPs binding to biofilms and cellular surfaces may result in cell damage and growth inhibition (Selvaraj et al. 2014). NPs possess fundamental antimicrobial properties, such as the production of ROS. These include metabolic by-products like hydroxyl radicals, superoxide anions, and hydrogen peroxide, which induce oxidative stress and damage cellular components i.e. DNA, proteins, and lipids (Das and Roychoudhury 2014). In a study, it was determined that superoxide radicals in *A. brassicicola* use a nitroblue tetrazolium reduction assay. Fungal mycelia treated with AgNPs elevated formazan, indicating that the superoxide radicals developed structural changes (Kumari et al. 2019). In some ways, it can be solved by simply altering its composition with other substances or integrating with other structures. An alternative method that could be used to improve antifungal activity of MNPs is nanofibers against fungi (Ifuku et al. 2015; Quirós et al. 2016).

14.2.2 Antibacterial Activity

Commonly used antibacterial agents are MNPs such as zinc oxide (ZnO), silver (Ag), gold (Au), copper oxide (CuO), magnesium oxide (MnO), and titanium oxide (TiO_2) (Beyth et al. 2015; Zhang 2015). Numerous methods may be utilized to describe MNPs and offer significant information about their electrical and physicochemical properties and morphology. Size, shape, roughness, and surface energy are the most important characteristics of NPs (Wang et al. 2017). The bactericidal impact of MNPs is described due to the small size of the MNPs with a large surface-to-volume ratio, which allows them to interact directly with the microbial membranes as opposed to metal ions in solution (Ruparelia et al. 2008). The antibacterial activity of MNPs is unclear in terms of the mechanism. Recent research provided several antibacterial mechanisms that may be explained by both chemical interactions and physical destruction. Such mechanisms include the generation of reaction oxygen species (ROS), and dissolution or leaching of ions that leads to the loss of cell wall and membrane integrity disrupting some of the metabolic pathways essential for maintaining bacterial viability (Kittler et al. 2010; Wang et al. 2014).

NPs can act upon the bacterial cells and affect their survival, and it is dependent on numerous factors, such as size, shape, concentration, stability, as well as surface charge. Antibacterial activity of NPs occurs through some of its mechanisms, such as ion release, oxidative stress, and non-oxidative stress, that cause a major mechanism (Azam et al. 2020). They also interfere with the bacterial cell wall, which serves as a barrier and helps maintain the cell wall. The action of NPs is based on the type of bacteria and composition of the cell wall (Linklater et al. 2020). Additionally, the interaction between bacteria and NPs is also influenced by the surface charge and size, where large hydrophilic NPs exclude the cell but remain on the surface of the bacteria cells, whereas the small hydrophobic NPs enter the bacteria cells (Jiang et al. 2022). It binds with the surface of the bacterial cell wall and causes damage by making minute holes in the cell membrane and making cracks (Gao et al. 2024). Small NPs interact with the proteins, making the cell membrane more permeable. They also cause leakage, injury, and death of the cell (Ahmed et al. 2020). Coulomb forces lead to the aggregation of large NPs on the cell surface and alteration of bacterial shape (structure) and augment the permeability of the membrane that leads to bacterial cell death (Timoszyk and Grochowalska 2022). NPs can alter the ion flow across the membrane, which results in a disturbance of Ca^{2+} ions in the cell. This leads to a possible process like apoptosis within the cell due to the concentration of calcium ions (Qiang et al. 2021). In Gram-negative bacteria, there is a negatively charged molecule, i.e., lipopolysaccharides (LPS), which is responsible for structural stability, membrane permeability, and protection. When NPs interact with LPS, the membrane becomes unstable and easy to break (Tavares et al. 2020). Many proteins in bacterial microbes help them in energy production and survival.

NPs attach to proteins by binding to nitrogen and sulfur atoms, altering protein shape or breaking them (Gahlawat and Choudhury 2019). Once NPs weaken the cell wall and enter the cell, they interact with DNA and block transcription (the process in which genetic instructions are copied). This inhibits the bacterial growth and further division, resulting in cell degradation and bacterial death (Rana et al. 2023). NPs also interfere with ribosomes that are required to make proteins, as the ribosomal subunit is not allowed to bind to tRNA, and no protein is produced. NPs reduce the ATP levels by inhibiting ATPase activity and make the survival of bacteria harder. Reduced levels of ATP stimulate the production of ROS, which damage proteins, DNA, the cell membrane, and other cellular structures (Ozdal and Gurkok 2022). ROS causes mutation in DNA, breaks the double helix, and disrupts the bacterial metabolism. Oxidative stress also increases when ROS generated by NPs inhibit the respiratory enzymes, which can lead to cell death. Conclusively, when NPs enter the cell, they reduce the production of ATP which slows down the metabolism and interfere with the binding of ribosomal subunits, hence stop protein synthesis, bacterial growth and division as well (Naz et al. 2020). Bacteria have some antioxidant defense enzymes such as superoxide dismutase (SOD) and glutathione peroxidase (GPx), to manage oxidative stress. However, increased levels of ROS generated by NPs can overpower these defense mechanisms and lead to dysfunction and DNA damage (Mourenza et al. 2020). In some cases, NPs directly destroy GPx without the production of increased levels of ROS, which triggers apoptosis like cell death (Fig. 14.2).

14.2.3 Antiviral Activity

Viruses are the smallest particles that consist of a capsid made up of proteins and either DNA or RNA (Krupovic et al. 2019). Viruses infect both eukaryotic and prokaryotic cells and can exist in various environments. The size of NPs can significantly impact their efficiency, as they interact directly with viral surface proteins and genomes, which are crucial for antiviral activity (Galdiero et al. 2011). However, smaller NPs can penetrate the host cell and pass through the viral genome, where they disrupt the viral components that aid in viral replication (Gurunathan et al. 2020). MNPs' antiviral activity is effective against both positive and negative-sense single-stranded RNA viruses in vitro and in vivo using different plants. According to Cai et al. (2019), SiO_2 and ZnO-NPs have antiviral activity against tobacco

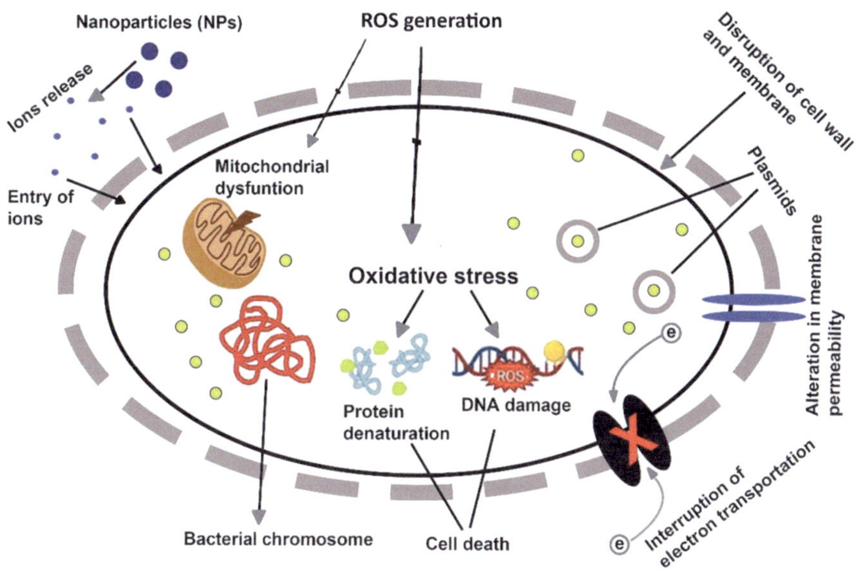

Fig. 14.2 Antibacterial activity mechanism by NPs. NPs bind and enter bacterial cells, disrupt cell walls and membranes, and penetrate ions into the cell. Within the cell, metal ions are released, causing mitochondrial dysfunction and the production of ROS. The result of ROS leads to oxidative stress that induces denaturation of proteins and DNA, which ultimately leads to cell death. Moreover, ROS also induces membrane permeability and damages plasmid and electron transport across the cell, which disrupts the essential functions of bacteria

mosaic virus in vitro. Similarly, Alkubaisi and Aref (2017) demonstrated that gold NPs induce harm to the virus-like particles of the BYDV-PAV, observing inflated and deteriorated virus-like particles (VLPs) displayed with AuNPs, alongside destroyed and missing particles, employing transmission electron microscopy (TEM).

NPs depict a broad-spectrum antiviral activity and target multiple stages of the virus life cycle. By interfering with viral components at different stages, these NPs can help prevent viral infection and reduce disease severity. One of the mechanisms by which NPs show their antiviral activity is inhibition of viral attachment and entry to the host cell (Gurunathan et al. 2020). Viruses rely on surface proteins for recognition and binding to specific receptors on host cells. These viral glycoproteins are neutralized by binding to the NPs and hence are unable to interfere with the host cell membrane. It prevents them from penetrating into the host cell and halts infection outside the cell (Najer 2024). They also alter the structure of viral proteins making them non-recognizable to host receptors and thus reducing the chance of the viral infection (Shah and Huseien 2020). They may also intercalate to viral nucleic acids such as DNA and RNA to prevent their replication and multiplication within the cell (Ghaffari et al. 2019). Since a virus is replicated by destroying the infected cells, it causes severe effects within the host cell and can be fatal. Hence, the killing of these cells boosts the chances of their proliferation in the body. NPs regulate cell

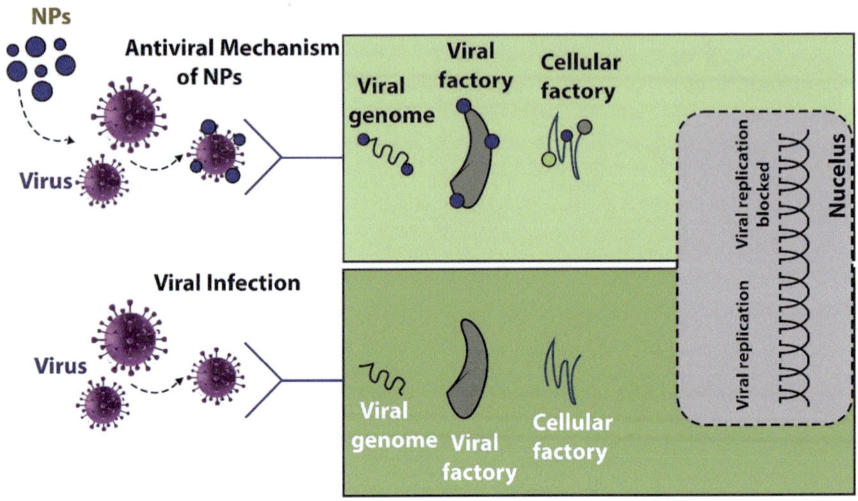

Fig. 14.3 Antiviral activity mechanism by NPs. Viral infection starts upon the entry of viral particles into the cell, in which viral and cellular factors aid viral genome replication. M-NPs prevent this process by directly interacting with viral particles, modifying their shape, and preventing their ability to enter host cells. After penetration, NPs combine with viral and cellular components, hindering their activity. In general, the antiviral activity of metal NPs impedes the spread of viruses in the host

apoptosis that involves depolarization of mitochondrial membrane potential (MMP). This pathway involves two apoptotic proteins: Bax and Bcl-2 (Liu and Tang 2020). Among these proteins, one is responsible for permeable mitochondrial membrane to promote cell death and the other prevents apoptosis. It helps to maintain cell integrity, survival, and reduce chances of virus-induced apoptosis (Suraweera et al. 2024). AgNPs release silver ions which decreases the pH of surrounding environment, creating acidic conditions that are unfavorable for the survival of virus and its replication as shown in Fig. 14.3 (Suraweera et al. 2024).

14.3 Application Methods of MNPs to Rhizosphere Engineering

An excellent approach to use MNPs strategically in the rhizosphere is to promote plant health, soil quality, and eliminate plant diseases (Kaningini et al. 2022). Rhizosphere engineering is one of the several processes that can deliver NPs and provide their physicochemical properties to optimize the interaction of the NPs with the plant roots and soil microorganisms (Hakim et al. 2021). The application procedures are crucial for boosting bioavailability, decreasing the environmental effects, and attaining desired objectives, such as disease management and better nutrient uptake.

14.3.1 Soil Amendment and Direct Soil Application

There are two most widely used methods (direct soil application and soil amendment) to introduce MNPs through the rhizosphere to defend plants against plant pathogens and increase plant health. Soil amendment is the process of increasing the quality of soil through the addition of substances that enhance the physical, chemical, and biological properties (Kv et al. 2019). MNPs such as silver, copper, iron, and zinc oxide can be applied to the soil, where they play multiple roles like increased nutrient availability and absorption, suppressing soil-borne pathogens by employing antimicrobial activity, enhancing the plant defense response, and increasing the soil microbial diversity, while selectively boosting beneficial microorganisms (Sun et al. 2021). In broadcasting and tillage methods, the MNPs are uniformly distributed across the field and mechanically mixed into the soil by plowing. NPs can be used alone with carriers such as compost, sand, and clay (Wahab et al. 2024). These approaches enhance the access of the NPs to the roots at an early stage of growth, or also suitable for extensive field applications. Another method is soil drenching, in which an MNP's suspension is prepared and added to the soil, usually around the plant base. It is also used in greenhouses, pots, or nursery beds (Ram et al. 2014).

14.3.2 Seed Treatment and Coating

Another new strategy in the field of rhizosphere engineering is the seed treatment and coating with MNPs, which stimulates seed germination, seedling vigor, and initial protection against soil-borne diseases. This approach implies the direct deposition of a thin layer of MNPs or the combination of MNPs with polymers or bioagents on the surface of the seed (Nile et al. 2022). There are many application methods, such as priming, dry coating, or film coating. In dry coating, NPs are mixed with carriers such as polymers or talc and applied as a dry coat. In film coating, MNPs are bound to the seed surface using binders such as polyvinyl alcohol (PVA, and starch) (Rhaman et al. 2022). AgNPs offer significant antimicrobial protection while ZnNPs promote germination and root elongation. MNPs are non-fertilizers, offering beneficial micronutrients such as Zn, Fe, and Cu, directly to the seedling to enhance nutrient absorption. Coated MNPs have the potential to stimulate beneficial microorganisms while lowering the negative ones (Fraceto et al. 2016).

14.3.3 Root Dipping and Transplant Drench

Root dipping and transplant drench are more specific approaches to introduce MNPs into the rhizosphere, encouraging early root development and protecting seedlings against soil-borne diseases. During transplantation, they stimulate a favorable plant–microbe relationship (Dutta et al. 2023). Normally, a nanoparticle solution is applied to the roots of plants before planting and left for 15–30 minutes (Sun et al. 2025). It provides defense against pathogens, including *Fusarium* and *Rhizoctonia,* at an early stage and induces early growth, root growth, and branching. When the plants are transplanted, a suitable quantity of metal nanoparticles is applied at the base of each plant. They are often employed in a nursery or greenhouse, especially for vegetables and fruits (Prasad et al. 2017). The main purpose of a transplant is to increase initial colonization by useful microbes and enhance the nutrient cycle in the soil.

14.3.4 Foliar Application and Systemic Translocation

Foliar application involves spraying MNPs on plant leaves, where the MNPs can be absorbed through stomata or cuticular channels (Hong et al. 2021). When the leaves absorbed the MNPs, they could penetrate the vascular system of the plant, specifically the phloem and xylem, and be circulated across the plant. Certain other NPs move downward via the leaves to the roots, into the rhizosphere (Wang et al. 2023b). Systemic modification in the plant also has the potential to alter the quantity and composition of root exudate amino acids and secondary metabolites, thereby influencing soil microbial communities by promoting beneficial microorganisms and reducing pathogens (Huang et al. 2014). Some MNPs, such as Zn, Mn, and Fe, may enhance micronutrient conditions in the roots via downward translocation, which may lead to root growth as well as the soil–plant relationship (Wang et al. 2023a).

14.3.5 Organic and Biological Amendments

Organic and biological amendments combined with MNPs is a new and promising method of increasing the rhizosphere health, plant growth, and pathogen resistance (Asadishad et al. 2018). This approach utilizes the benefit of combining with NPs with natural soil amendments, including compost, vermicompost, biochar, and beneficial microorganisms, in achieving the stability of MNPs, bioavailability, and efficacy, as well as controlling or reducing environmental risks presented by them (Rajput et al. 2023). These organic amendments include compost, biochar, and FYM that are rich in high-carbon and other important nutrients (Rahman et al. 2020). When combined with MNPs, they partially behave as carriers to ensure slow, controlled release of NPs in the rhizosphere (Wang et al. 2023a). This helps in reducing the leach down of nanoparticles and enhances their interaction with nutrients and the root systems. Meanwhile, MNPs can be used in combination with biological amendments such as PGPR, fungi (*Aspergillus, Trichoderma*) and mycorrhizal association to enhance the plant defense mechanisms (Ahmad and Ahmad 2024). For example, nano-copper was combined with *T. harzianum* to enhance antifungal effect, which results in systemic resistance in crops, such as tomato and cucumber (Kamel et al. 2022). This method is especially beneficial in stressful situations such as drought, salinity, or pathogen pressure, where traditional treatments are not very effective.

14.3.6 Precision Application Techniques

The combination of nanotechnology and precision agriculture provides a leading strategy to deliver agro-inputs, including MNPs in a smart, precise, and eco-friendly manner (Wahab et al. 2024). The application strategies include site-specific, controlled, and optimal delivery of MNPs to the rhizosphere according to real-time plant and soil requirements (Gehlout et al. 2022). However, to enhance their effectiveness and ensure the safety of their application, more advanced and complex approaches are required, particularly those that help to minimize losses and maximize contact with root zone.

14.3.6.1 Smart Sensors

One of the most precise agriculture techniques is sensor-guided application, where soil and plant sensors are employed in real time to provide location specificity with the delivery of MNPs into the rhizosphere (Nithinkumar et al. 2025). Different kinds of sensors are incorporated into this system, such as pH, soil-moisture sensors, electrical conductivity (EC), or nutrient sensors (nitrate, zinc, or iron) (Yin et al. 2021). Furthermore, these techniques increase sustainable agriculture because they can increase the usage efficiency of NPs and targeted NPs in the root zone more accurately and can be dynamically adapted to crops or environmental conditions.

14.3.6.2 Global Positioning System (GPS)

GPS plays the main role in precision agriculture and specifically in the targeted usage of MNPs in the rhizosphere engineering (Singh et al. 2024). GPS facilitates location-specific treatment, where farmers can meet variability in one field with the use of MNPs, such as AgNPs, ZnNPs, and CuNPs, only in areas that require a response (Yadav et al. 2023). However, a combination of GPS with GIS and VRT maps is computerized using soil health, lack of nutrients, pests or disease outbreaks, and agronomic variables (Singh et al. 2023a, b). This specific delivery increases the usage of NPs, lowering input costs, avoiding polluting the environment and overexposure, thus causing nanoparticle toxicity among plants.

14.3.6.3 IoT and AI

The combination of the Internet of Things (IoT) and Artificial Intelligence (AI) with precision agriculture is revolutionizing the use of the specific application of MNPs (Zn, Cu, Ag, and Fe), through its real-time, automated, and sustainable management of soil and crop health (Ashique et al. 2024). IoT includes the usage of interconnected sensors, devices, and networks that gather and relay information in the field to the cloud-based applications (Sharma et al. 2021). Many parameters can be measured through sensors installed in the soil and on plants, including soil pH, moisture content, EC, nutrients, temperature, and physiological response of plants (Postolache et al. 2022). AI systems can detect patterns and predict nutrient deficiencies, disease epidemics, and stress (Kumar et al. 2023). According to these forecasts, MNPs' site-specific applications can be mechanically activated by employing smart irrigation systems, drones, or precision sprayers. Additionally, the combination of IoT and AI enhances decision-making, lowers labor costs, and guarantees the sustainable use of NPs through continuous environmental condition adaptation (Miller et al. 2025).

14.3.6.4 Drone-Assisted Targeted Spraying

It is a modern precision agricultural technique that employs unmanned aerial vehicles to spray MNPs in a precise and efficient manner (Guebsi et al. 2024). Integrated with GPS navigation, multispectral imaging, and precision nozzles, agricultural drones can select regions in a field that should be treated, depending upon lack of nutrients, water stress, or pathogens (Olson and Anderson 2021). The primary advantage of using drone spraying is that it can cover areas that are not easily accessible or are delicate (steep ground, flooded regions, or overcrowded orchards) without causing damage to the crop (Das et al. 2024). Although drone-assisted spraying presents numerous advantages, it demands special attention to variables to promote the efficient and safe delivery, including nanoparticle dosage, drift rate, and the size of the droplet and wind (Guebsi et al. 2024).

14.4 Future Recommendations

Nanotechnology can change the universe because of its many applications in several fields. In the last few years, the field of nanotechnology has grown significantly, and in the future, it will have an enormous impact on the agriculture and food sector. Currently, many researchers are encouraging the green synthesis of NPs, as this approach will lead to the development of MNPs with specific size, morphology, and shape by utilizing reducing, capping, and stabilizing agents. The use of microorganisms in NPs biosynthesis is time-consuming while using plant extracts as compared to microbes is more reliable. There is a need to enhance the use of nano-fertilizers to increase crop productivity or stimulate sustainable farming while reducing the amount of fertilizer needed. Furthermore, nano-pesticides can be more targeted and decrease the usage of chemicals while reducing the need for huge quantities of pesticides. In precision agriculture, nano-biosensors also play a main role by allowing real-time detection of plant health, moisture levels, soil nutrients, and other components that can influence crop growth. As technology develops, these sensors in combination with AI help farmers improve their yields and promote efficient usage of resources. Plant stress tolerance is another important aspect; engineered nanoparticles can increase plant resilience by regulating stress-responsive genes and enhancing crop water retention. Nanotechnology also increases soil health by producing NPs that promote nutrient cycling. Moreover, nano-coatings on seeds promote germination, disease resistance, and early growth. Governments should adopt the main procedures for regulating the utilization of NPs in the agriculture sector. Safety evaluations and guidelines must confirm that nano-based products are not harmful to soil microorganisms, human health, and ecosystems. The combination of nanotechnology and smart farming techniques can make agriculture more productive, less hazardous, and better equipped to handle future limitations.

14.5 Conclusion

Nanotechnology enhances the performance and effectiveness of common objects. It also promotes a clean environment, air, water, and renewable energy for a sustainable future. According to the classification, there are parallel areas of applications with common goals, along with different applications that make the utilization of unique characteristics. Organic nanoparticles are non-toxic and biodegradable products commonly employed in food packaging, while inorganic NPs are reported in antibacterial studies. Organic NPs have unique properties and are utilized in industrial applications. Other vital techniques include green or biological methods such as using microbial biomolecules and plant extracts that facilitate in nanoparticle synthesis with low toxicity as compared to other conventional methods. NPs development and applications need to be controlled via characterization. Nanotechnology has a broad scope in the future, and it is governed by components, such as faster development, easier, and more effective materials characterization. Meanwhile, further research is needed to understand the mechanism of nanoparticle formation. Nano-fertilizers, nano-pesticides, nano-herbicides, nano-insecticides, and nanosensors are safe, have unique properties in agriculture, and help to promote sustainability.

The use of MNPs in agriculture provides novel responses to the industry's many problems. MNPs have the potential to modify or enhance these conventional practices, such as using NPs as biocontrol agents (BCAs) against chemical pesticides or enhancing nutrient delivery or plant growth with nano-fertilizers. These approaches increase nutrient management by improving plant nutrition and decreasing the demand for conventional fertilizers. Pests and plant diseases pose a significant threat to world food security, requiring new and sustainable methods to improve crop protection. NMs offer benefits over conventional approaches for plant disease management, including the potential to target certain pathogens. MNPs support the rhizosphere engineering in combating plant pathogens since they are less toxic and eco-friendly. The engineering of the rhizosphere provides a unique chance to enhance plant effectiveness through controlling root exudates, plant genotypes, microbiome engineering, and amendments of soil. The physicochemical properties of engineered NMs are unique and may impact the plant defense mechanism, making them a potential option to sustainably manage disease. Additionally, NMs have the potential to enhance plant activity and resistance to pathogens, contrary to crop diseases, which target molecular and biochemical processes. NMs help nutrient management, membrane stability, and protect the photosynthetic system from pathogens. Additionally, they enhance the plant defense mechanism by boosting stress protective compound accumulation and stress-responsive gene expression. Furthermore, many types of NMs, including nanotubes and nano-clays, provide specific characteristics, such as quick response time and increased sensitivity. Nano-sensors and nano-devices are new methods that show potential for real-time pathogen monitoring in plants with applications that include pre- and post-infection detection in lab and field conditions. Future studies should focus on understanding these complex aspects to fully use rhizospheric microbiomes. Understanding the hazards, advantages, and disadvantages of MNPs in agriculture can make them a safer and more sustainable method.

References

Abdelkader Y, Perez-Davalos L, LeDuc R, Zahedi RP, Labouta HI (2023) Omics approaches for the assessment of biological responses to nanoparticles. Adv Drug Deliv Rev 200:114992

Abd-Elsalam KA, Mehmood MA, Ashfaq M, Abdelkhalek TE, Hassan RK, Ravichandran M (2024) Liquid nanoclay: synthesis and applications to transform an arid desert into fertile land. Soil Systems 8:73

Ahmad F, Ahmad S (2024) NPs and soil microorganisms interactions in crop management—current status and future prospects. In: Industrial applications of soil microbes: volume 4, p 361

Ahmed B, Ameen F, Rizvi A, Ali K, Sonbol H, Zaidi A, Khan MS, Musarrat J (2020) Destruction of cell topography, morphology, membrane, inhibition of respiration, biofilm formation, and bioactive molecule production by nanoparticles of Ag, ZnO, CuO, TiO_2, and Al_2O_3 toward beneficial soil bacteria. ACS Omega 5:7861–7876

Ahmed T, Noman M, Gardea-Torresdey JL, White JC, Li B (2023) Dynamic interplay between nano-enabled agrochemicals and the plant-associated microbiome. Trends Plant Sci 28:1310–1325

Ahmed N, Alam M, Saeed M, Ullah H, Junaid M, Kanwal M, Ahmed S (2024) Role of plants in managing diseases. In: Ethnic knowledge and perspectives of medicinal plants. Apple Academic Press, pp 579–604

Ali MA, Ahmed T, Wu W, Hossain A, Hafeez R, Islam Masum MM, Wang Y, An Q, Sun G, Li B (2020) Advancements in plant and microbe-based synthesis of metallic nanoparticles and their antimicrobial activity against plant pathogens. Nano 10(6):1146

Alkubaisi NA, Aref NMA (2017) Dispersed gold nanoparticles potentially ruin gold barley yellow dwarf virus and eliminate virus infectivity hazards. Appl Nanosci 7:31–40

Alvarez-Peral FJ, Zaragoza O, Pedreno Y, Argüelles J-C (2002) Protective role of trehalose during severe oxidative stress caused by hydrogen peroxide and the adaptive oxidative stress response in Candida albicans. Microbiology 148:2599–2606

Ankamwar B, Chaudhary M, Sastry M (2005) Gold nanotriangles biologically synthesized using tamarind leaf extract and potential application in vapor sensing. Synth React Inorg Metal-Organic Nano-Metal Chem 35:19–26

Arciniegas-Grijalba PA, Patiño-Portela MC, Mosquera-Sánchez LP, Guerrero-Vargas JA, Rodríguez-Páez JE (2017) ZnO nanoparticles (ZnO-NPs) and their antifungal activity against coffee fungus Erythricium salmonicolor. Appl Nanosci 7:225–241

Arumugam A, Karthikeyan C, Hameed ASH, Gopinath K, Gowri S, Karthika V (2015) Synthesis of cerium oxide nanoparticles using Gloriosa superba L. leaf extract and their structural, optical and antibacterial properties. Mater Sci Eng C 49:408–415

Asadishad B, Chahal S, Akbari A, Cianciarelli V, Azodi M, Ghoshal S, Tufenkji N (2018) Amendment of agricultural soil with metal nanoparticles: effects on soil enzyme activity and microbial community composition. Environ Sci Technol 52:1908–1918

Ashique S, Raikar A, Jamil S, Lakshminarayana L, Gajbhiye SA, De S, Kumar S (2024) Artificial intelligence integration with nanotechnology: a new frontier for sustainable and precision agriculture. CNANO 21:242–273

Ashraf SA, Siddiqui AJ, Abd Elmoneim OE, Khan MI, Patel M, Alreshidi M, Moin A, Singh R, Snoussi M, Adnan M (2021) Innovations in nanoscience for the sustainable development of food and agriculture with implications on health and environment. Sci Total Environ 768:144990

Athawale V, Paralikar P, Ingle AP, Rai M (2018) Biogenically engineered nanoparticles inhibit *Fusarium oxysporum* causing soft-rot of ginger. IET Nanobiotechnol 12:1084–1089

Athie-García MS, Piñón-Castillo HA, Muñoz-Castellanos LN, Ulloa-Ogaz AL, Martínez-Varela PI, Quintero-Ramos A, Duran R, Murillo-Ramirez JG, Orrantia-Borunda E (2018) Cell wall damage and oxidative stress in Candida albicans ATCC10231 and Aspergillus niger caused by palladium nanoparticles. Toxicol In Vitro 48:111–120

Azam Z, Ayaz A, Younas M, Qureshi Z, Arshad B, Zaman W, Ullah F, Nasar MQ, Bahadur S, Irfan MM (2020) Microbial synthesized cadmium oxide nanoparticles induce oxidative stress and protein leakage in bacterial cells. Microb Pathog 144:104188

Backer R, Rokem JS, Ilangumaran G, Lamont J, Praslickova D, Ricci E, Subramanian S, Smith DL (2018) Plant growth-promoting rhizobacteria: context, mechanisms of action, and roadmap to commercialization of biostimulants for sustainable agriculture. Front Plant Sci 9:1473

Beyth N, Houri-Haddad Y, Domb A, Khan W, Hazan R (2015) Alternative antimicrobial approach: nano-antimicrobial materials. Evid Based Complement Alternat Med 2015:246012

Cai L, Liu C, Fan G, Liu C, Sun X (2019) Preventing viral disease by ZnONPs through directly deactivating TMV and activating plant immunity in Nicotiana benthamiana. Environ Sci Nano 6:3653–3669

Chandra J, Kuhn DM, Mukherjee PK, Hoyer LL, McCormick T, Ghannoum MA (2001) Biofilm formation by the fungal pathogen *Candida albicans*: development, architecture, and drug resistance. J Bacteriol 183:5385–5394

Chen J, Sun L, Cheng Y, Lu Z, Shao K, Li T, Hu C, Han H (2016) Graphene oxide-silver nanocomposite: novel agricultural antifungal agent against *Fusarium graminearum* for crop disease prevention. ACS Appl Mater Interfaces 8:24057–24070

Das K, Roychoudhury A (2014) Reactive oxygen species (ROS) and response of antioxidants as ROS-scavengers during environmental stress in plants. Front Environ Sci 2:53

Das A, Kadawla K, Nath H, Chakraborty S, Ali H, Singh S, Dubey VK (2024) Drone-based intelligent spraying of pesticides: current challenges and its future prospects. In: Chouhan SS, Singh UP, Jain S (eds) Applications of computer vision and drone technology in agriculture 4.0. Springer Nature Singapore, Singapore, pp 199–223

De La Rosa-García SC, Martínez-Torres P, Gómez-Cornelio S, Corral-Aguado MA, Quintana P, Gómez-Ortíz NM (2018) Antifungal activity of ZnO and MgO nanomaterials and their mixtures against *Colletotrichum gloeosporioides* strains from tropical fruit. J Nanomater 2018:1–9

Dessaux Y, Grandclément C, Faure D (2016) Engineering the rhizosphere. Trends Plant Sci 21(3):266–278

Dhillon GS, Brar SK, Kaur S, Verma M (2012) Green approach for nanoparticle biosynthesis by fungi: current trends and applications. Crit Rev Biotechnol 32:49–73

Dutta P, Kumari A, Mahanta M, Upamanya GK, Heisnam P, Borua S, Kaman PK, Mishra AK, Mallik M, Muthukrishnan G (2023) Nanotechnological approaches for management of soil-borne plant pathogens. Front Plant Sci 14:1136233

El-Khawaga AM, Zidan A, Abd El-Mageed AI (2023) Preparation methods of different nanomaterials for various potential applications: a review. J Mol Struct 1281:135148

Fraceto LF, Grillo R, de Medeiros GA, Scognamiglio V, Rea G, Bartolucci C (2016) Nanotechnology in agriculture: which innovation potential does it have? Front Environ Sci 4:20

Gahlawat G, Choudhury AR (2019) A review on the biosynthesis of metal and metal salt nanoparticles by microbes. RSC Adv 9:12944–12967

Galdiero S, Falanga A, Vitiello M, Cantisani M, Marra V, Galdiero M (2011) Silver nanoparticles as potential antiviral agents. Molecules 16:8894–8918

Gao Y, Dou H, Ma Y, Tian G, Weragoda DM, Li S, Yang X, Zhang Z, Fan G, Chen B (2024) Antibacterial performance effects of Ag NPs in situ loaded in MOFs nano-supports prepared by post-synthesis exchange method. J Environ Chem Eng 12:112133

Gargallo-Garriga A, Preece C, Sardans J, Oravec M, Urban O, Peñuelas J (2018) Root exudate metabolomes change under drought and show limited capacity for recovery. Sci Rep 8(1):12696

Gehlout S, Priyam A, Afonso L, Schultze GA, Singh PP (2022) Application of metallic nanoparticles as agri inputs: modulation in nanoparticle design and application dosage needed. In: Nanotechnology in agriculture and environmental science, pp 16–54

Ghaffari E, Rezatofighi SE, Ardakani MR, Rastegarzadeh S (2019) Delivery of antisense peptide nucleic acid by gold nanoparticles for the inhibition of virus replication. Nanomedicine (Lond) 14:1827–1840

Gogos A, Knauer K, Bucheli TD (2012) Nanomaterials in plant protection and fertilization: current state, foreseen applications, and research priorities. J Agric Food Chem 60:9781–9792

Guebsi R, Mami S, Chokmani K (2024) Drones in precision agriculture: a comprehensive review of applications, technologies, and challenges. Drones 8:686

Gurunathan S, Qasim M, Choi Y, Do JT, Park C, Hong K, Kim J-H, Song H (2020) Antiviral potential of nanoparticles—can nanoparticles fight against coronaviruses? Nano 10:1645

Hakim S, Naqqash T, Nawaz MS, Laraib I, Siddique MJ, Zia R, Mirza MS, Imran A (2021) Rhizosphere engineering with plant growth-promoting microorganisms for agriculture and ecological sustainability. Front Sustain Food Syst 5:617157

Hameed ASH, Karthikeyan C, Kumar VS, Kumaresan S, Sasikumar S (2015) Effect of Mg2+, Ca2+, Sr2+ and Ba2+ metal ions on the antifungal activity of ZnO nanoparticles tested against Candida albicans. Mater Sci Eng C 52:171–177

Handa M, Kalia A (2024) Exploring the dynamics of nanoparticle-plant-microbe interactions to realize the goal of improved crop productivity and food security. Rhizosphere:100884

He L, Liu Y, Mustapha A, Lin M (2011) Antifungal activity of zinc oxide nanoparticles against Botrytis cinerea and Penicillium expansum. Microbiol Res 166:207–215

Hong J, Wang C, Wagner DC, Gardea-Torresdey JL, He F, Rico CM (2021) Foliar application of nanoparticles: mechanisms of absorption, transfer, and multiple impacts. Environ Sci Nano 8:1196–1210

Huang X-F, Chaparro JM, Reardon KF, Zhang R, Shen Q, Vivanco JM (2014) Rhizosphere interactions: root exudates, microbes, and microbial communities. Botany 92:267–275

Ifuku S, Tsukiyama Y, Yukawa T, Egusa M, Kaminaka H, Izawa H, Morimoto M, Saimoto H (2015) Facile preparation of silver nanoparticles immobilized on chitin nanofiber surfaces to endow antifungal activities. Carbohydr Polym 117:813–817

Jansson JK, Hofmockel KS (2020) Soil microbiomes and climate change. Nat Rev Microbiol 18:35–46
Jiang Y, Zheng W, Tran K, Kamilar E, Bariwal J, Ma H, Liang H (2022) Hydrophilic nanoparticles that kill bacteria while sparing mammalian cells reveal the antibiotic role of nanostructures. Nat Commun 13:197
Kamel SM, Elgobashy SF, Omara RI, Derbalah AS, Abdelfatah M, El-Shaer A, Al-Askar AA, Abdelkhalek A, Abd-Elsalam KA, Essa T (2022) Antifungal activity of copper oxide nanoparticles against root rot disease in cucumber. J Fungi 8:911
Kaningini AG, Nelwamondo AM, Azizi S, Maaza M, Mohale KC (2022) Metal nanoparticles in agriculture: a review of possible use. Coatings 12:1586
Käosaar S, Kahru A, Mantecca P, Kasemets K (2016) Profiling of the toxicity mechanisms of coated and uncoated silver nanoparticles to yeast Saccharomyces cerevisiae BY4741 using a set of its 9 single-gene deletion mutants defective in oxidative stress response, cell wall or membrane integrity and endocytosis. Toxicol In Vitro 35:149–162
Kim K-J, Sung WS, Suh BK, Moon S-K, Choi J-S, Kim JG, Lee DG (2009) Antifungal activity and mode of action of silver nano-particles on Candida albicans. Biometals 22:235–242
Kittler S, Greulich C, Diendorf J, Köller M, Epple M (2010) Toxicity of silver nanoparticles increases during storage because of slow dissolution under release of silver ions. Chem Mater 22:4548–4554
Krupovic M, Dolja VV, Koonin EV (2019) Origin of viruses: primordial replicators recruiting capsids from hosts. Nat Rev Microbiol 17:449–458
Kumar A, Choudhary A, Kaur H, Guha S, Mehta S, Husen A (2022) Potential applications of engineered nanoparticles in plant disease management: a critical update. Chemosphere 295:133798
Kumar Y, Koul A, Singla R, Ijaz MF (2023) Artificial intelligence in disease diagnosis: a systematic literature review, synthesizing framework and future research agenda. J Ambient Intell Human Comput 14:8459–8486
Kumari M, Giri VP, Pandey S, Kumar M, Katiyar R, Nautiyal CS, Mishra A (2019) An insight into the mechanism of antifungal activity of biogenic nanoparticles than their chemical counterparts. Pestic Biochem Physiol 157:45–52
Usharani KV, Roopashree KM, Naik D (2019) Role of soil physical, chemical and biological properties for soil health improvement and sustainable agriculture. J Pharmacogn Phytochem 8(5):1256–1267
Lee B, Lee MJ, Yun SJ, Kim K, Choi I-H, Park S (2019) Silver nanoparticles induce reactive oxygen species-mediated cell cycle delay and synergistic cytotoxicity with 3-bromopyruvate in Candida albicans, but not in saccharomyces cerevisiae. Int J Nanomed 14:4801–4816
Linklater DP, Baulin VA, Le Guével X, Fleury J, Hanssen E, Nguyen THP, Juodkazis S, Bryant G, Crawford RJ, Stoodley P, Ivanova EP (2020) Antibacterial action of nanoparticles by lethal stretching of bacterial cell membranes. Adv Mater 32:2005679
Lipovsky A, Nitzan Y, Gedanken A, Lubart R (2011) Antifungal activity of ZnO nanoparticles—the role of ROS mediated cell injury. Nanotechnology 22:105101
Liu N, Tang M (2020) Toxic effects and involved molecular pathways of nanoparticles on cells and subcellular organelles. J Appl Toxicol 40:16–36
Miller T, Durlik I, Kostecka E, Kozlovska P, Łobodzińska A, Sokołowska S, Nowy A (2025) Integrating artificial intelligence agents with the internet of things for enhanced environmental monitoring: applications in water quality and climate data. Electronics 14:696
Mitter B, Brader G, Pfaffenbichler N, Sessitsch A (2019) Next generation microbiome applications for crop production—limitations and the need of knowledge-based solutions. Curr Opin Microbiol 49:59–65
Mourenza Á, Gil JA, Mateos LM, Letek M (2020) Oxidative stress-generating antimicrobials, a novel strategy to overcome antibacterial resistance. Antioxidants 9:361
Murphy M, Ting K, Zhang X, Soo C, Zheng Z (2015) Current development of silver nanoparticle preparation, investigation, and application in the field of medicine. J Nanomater 2015:696918

Najer A (2024) Pathogen-binding nanoparticles to inhibit host cell infection by heparan sulfate and sialic acid dependent viruses and protozoan parasites. Smart Med 3:e20230046

Narayanan KB, Sakthivel N (2011) Green synthesis of biogenic metal nanoparticles by terrestrial and aquatic phototrophic and heterotrophic eukaryotes and biocompatible agents. Adv Colloid Interf Sci 169:59–79

Naz S, Gul A, Zia M (2020) Toxicity of copper oxide nanoparticles: a review study. IET Nanobiotechnol 14:1–13

Nile SH, Thiruvengadam M, Wang Y, Samynathan R, Shariati MA, Rebezov M, Nile A, Sun M, Venkidasamy B, Xiao J, Kai G (2022) Nano-priming as emerging seed priming technology for sustainable agriculture—recent developments and future perspectives. J Nanobiotechnol 20:254

Nithinkumar K, Panchadi A, Bagrecha S, Jena T, Kumari M (2025) Precision farming for food security and agricultural land restoration. In: Babu S, Das A, Rathore SS, Singh R, Singh R (eds) Ecological solutions to agricultural land degradation. Springer Nature Singapore, Singapore, pp 341–371

Ogar A, Tylko G, Turnau K (2015) Antifungal properties of silver nanoparticles against indoor mould growth. Sci Total Environ 521:305–314

Olson D, Anderson J (2021) Review on unmanned aerial vehicles, remote sensors, imagery processing, and their applications in agriculture. Agron J 113:971–992

Ortiz A, Sansinenea E (2021) Recent advancements for microorganisms and their natural compounds useful in agriculture. Appl Microbiol Biotechnol 105:891–897

Ouda SM (2014) Antifungal activity of silver and copper nanoparticles on two plant pathogens, Alternaria alternata and Botrytis cinerea. Res J Microbiol 9:34

Ozdal M, Gurkok S (2022) Recent advances in nanoparticles as antibacterial agent. ADMET DMPK 10:115–129

Perea Velez YS, Carrillo-González R, Gonzalez-Chavez M d CA (2021) Interaction of metal nanoparticles–plants–microorganisms in agriculture and soil remediation. J Nanopart Res 23(9):206

Poria V, Singh S, Nain L, Singh B, Saini JK (2021) Rhizospheric microbial communities: occurrence, distribution, and functions. In: Nath M, Bhatt D, Bhargava P, Choudhary DK (eds) Microbial metatranscriptomics belowground. Springer Singapore, pp 239–271. https://doi.org/10.1007/978-981-15-9758-9_12

Postolache S, Sebastião P, Viegas V, Postolache O, Cercas F (2022) IoT-based systems for soil nutrients assessment in horticulture. Sensors 23:403

Pradhan A, Seena S, Schlosser D, Gerth K, Helm S, Dobritzsch M, Krauss G-J, Dobritzsch D, Pascoal C, Cássio F (2015) Fungi from metal-polluted streams may have high ability to cope with the oxidative stress induced by copper oxide nanoparticles. Environ Toxicol Chem 34:923–930

Prasad R, Bhattacharyya A, Nguyen QD (2017) Nanotechnology in sustainable agriculture: recent developments, challenges, and perspectives. Front Microbiol 8:1014

Qiang L, Jin H, Feng Y, Wu R, Song Y, Liu L (2021) Apoptosis-like bacterial death modulated by photoactive hyperthermia nanomaterials and enhanced wound disinfection application. Nanoscale 13:14785–14794

Quirós J, Gonzalo S, Jalvo B, Boltes K, Perdigón-Melón JA, Rosal R (2016) Electrospun cellulose acetate composites containing supported metal nanoparticles for antifungal membranes. Sci Total Environ 563:912–920

Rahman GKMM, Rahman MM, Alam MS, Kamal MZ, Mashuk HA, Datta R, Meena RS (2020) Biochar and organic amendments for sustainable soil carbon and soil health. In: Datta R, Meena RS, Pathan SI, Ceccherini MT (eds) Carbon and nitrogen cycling in soil. Springer Singapore, Singapore, pp 45–85

Rajput VD, Kumari A, Minkina T, Barakhov A, Singh S, Mandzhieva SS, Sushkova S, Ranjan A, Rajput P, Garg MC (2023) A practical evaluation on integrated role of biochar and nanomaterials in soil remediation processes. Environ Geochem Health 45:9435–9449

Ram P, Vivek K, Kumar SP (2014) Nanotechnology in sustainable agriculture: present concerns and future aspects. Afr J Biotechnol 13:705–713

Rana A, Pathak S, Lim D-K, Kim S-K, Srivastava R, Sharma SN, Verma R (2023) Recent advancements in plant- and microbe-mediated synthesis of metal and metal oxide nanomaterials and their emerging antimicrobial applications. ACS Appl Nano Mater 6:8106–8134

Rhaman MS, Tania SS, Imran S, Rauf F, Kibria MG, Ye W, Hasanuzzaman M, Murata Y (2022) Seed priming with nanoparticles: an emerging technique for improving plant growth, development, and abiotic stress tolerance. J Soil Sci Plant Nutr 22:4047–4062

Różalska B, Sadowska B, Budzyńska A, Bernat P, Różalska S (2018) Biogenic nanosilver synthesized in Metarhizium robertsii waste mycelium extract – As a modulator of Candida albicans morphogenesis, membrane lipidome and biofilm. PLoS ONE 13(3): e0194254. https://doi.org/10.1371/journal.pone.01942

Ruparelia JP, Chatterjee AK, Duttagupta SP, Mukherji S (2008) Strain specificity in antimicrobial activity of silver and copper nanoparticles. Acta Biomater 4:707–716

Ryan PR, Dessaux Y, Thomashow LS, Weller DM (2009) Rhizosphere engineering and management for sustainable agriculture. Plant Soil 321:363–383

Saeed Q, Xiukang W, Haider FU, Kučerik J, Mumtaz MZ, Holatko J, Naseem M, Kintl A, Ejaz M, Naveed M (2021) Rhizosphere bacteria in plant growth promotion, biocontrol, and bioremediation of contaminated sites: a comprehensive review of effects and mechanisms. Int J Mol Sci 22(19):10529

Savary S, Willocquet L, Pethybridge SJ, Esker P, McRoberts N, Nelson A (2019) The global burden of pathogens and pests on major food crops. Nat Ecol Evol 3:430–439

Selvaraj M, Pandurangan P, Ramasami N, Rajendran SB, Sangilimuthu SN, Perumal P (2014) Highly potential antifungal activity of quantum-sized silver nanoparticles against Candida albicans. Appl Biochem Biotechnol 173:55–66

Shah KW, Huseien GF (2020) Inorganic nanomaterials for fighting surface and airborne pathogens and viruses. Nano Express 1:032003

Sharma K, Anand D, Sabharwal M, Tiwari PK, Cheikhrouhou O, Frikha T (2021) A disaster management framework using internet of things-based interconnected devices. Math Probl Eng 2021:1–21

Singh A, Singh NB, Hussain I, Singh H, Singh SC (2015) Plant-nanoparticle interaction: an approach to improve agricultural practices and plant productivity. Int J Pharm Sci Invent 4:25–40

Singh BK, Delgado-Baquerizo M, Egidi E, Guirado E, Leach JE, Liu H, Trivedi P (2023a) Climate change impacts on plant pathogens, food security and paths forward. Nat Rev Microbiol 21:640–656

Singh M, Vermaa A, Kumar V (2023b) Geospatial technologies for the management of pest and disease in crops. In: Precision agriculture. Elsevier, pp 37–54

Singh A, Tomar B, Margaryan G, Rajput P, Minkina T, Mandzhieva S, ... Ghazaryan, K (2024) Emerging technologies for sustainable soil management and precision farming. Nanotechnology applications and innovations for improved soil health, 210–235

Son WK, Choi YS, Han YW, Shin DW, Min K, Shin J, Lee MJ, Son H, Jeong DH, Kwak S-Y (2023) In vivo surface-enhanced Raman scattering nanosensor for the real-time monitoring of multiple stress signalling molecules in plants. Nat Nanotechnol 18:205–216

Sun W, Dou F, Li C, Ma X, Ma LQ (2021) Impacts of metallic nanoparticles and transformed products on soil health. Crit Rev Environ Sci Technol 51:973–1002

Sun X-D, Ma J-Y, Feng L-J, Duan J-L, Yuan X-Z (2025) Precise tracking of nanoparticles in plant roots. Nat Protoc 20:248–271

Suraweera CD, Espinoza B, Hinds MG, Kvansakul M (2024) Mastering death: the roles of viral Bcl-2 in dsDNA viruses. Viruses 16:879

Tavares TD, Antunes JC, Padrão J, Ribeiro AI, Zille A, Amorim MTP, Ferreira F, Felgueiras HP (2020) Activity of specialized biomolecules against gram-positive and gram-negative bacteria. Antibiotics 9:314

Timoszyk A, Grochowalska R (2022) Mechanism and antibacterial activity of gold nanoparticles (AuNPs) functionalized with natural compounds from plants. Pharmaceutics 14:2599

Wahab A, Muhammad M, Ullah S, Abdi G, Shah GM, Zaman W, Ayaz A (2024) Agriculture and environmental management through nanotechnology: eco-friendly nanomaterial synthesis for soil-plant systems, food safety, and sustainability. Sci Total Environ 926:171862

Wang L, He H, Yu Y, Sun L, Liu S, Zhang C, He L (2014) Morphology-dependent bactericidal activities of Ag/CeO2 catalysts against Escherichia coli. J Inorg Biochem 135:45–53

Wang L, Hu C, Shao L (2017) The antimicrobial activity of nanoparticles: present situation and prospects for the future. Int J Nanomed 12:1227–1249

Wang Q, Zhang P, Zhao W, Li Y, Jiang Y, Rui Y, Guo Z, Lynch I (2023a) Interplay of metal-based nanoparticles with plant rhizosphere microenvironment: implications for nanosafety and nano-enabled sustainable agriculture. Environ Sci Nano 10:372–392

Wang X, Xie H, Wang P, Yin H (2023b) Nanoparticles in plants: uptake, transport and physiological activity in leaf and root. Materials 16:3097

Wen T, Yu G-H, Hong W-D, Yuan J, Niu G-Q, Xie P-H, Sun F-S, Guo L-D, Kuzyakov Y, Shen Q-R (2022) Root exudate chemistry affects soil carbon mobilization via microbial community reassembly. Fundam Res 2(5):697–707

Wu X, Li J, Ji M, Wu Q, Wu X, Ma Y, Sui W, Zhao L, Zhang X (2019) Non-synchronous structural and functional dynamics during the coalescence of two distinct soil bacterial communities. Front Microbiol 10:1125

Xia Z-K, Ma Q-H, Li S-Y, Zhang D-Q, Cong L, Tian Y-L, Yang R-Y (2016) The antifungal effect of silver nanoparticles on Trichosporon asahii. J Microbiol Immunol Infect 49:182–188

Xu H, Du H, Zeng F, Song T, Peng W (2021) Diminished rhizosphere and bulk soil microbial abundance and diversity across succession stages in Karst area, Southwest China. Appl Soil Ecol 158:103799

Yadav A, Yadav K, Ahmad R, Abd-Elsalam KA (2023) Emerging frontiers in nanotechnology for precision agriculture: advancements, hurdles and prospects. Agrochemicals 2:220–256

Yin H, Cao Y, Marelli B, Zeng X, Mason AJ, Cao C (2021) Soil sensors and plant wearables for smart and precision agriculture. Adv Mater 33:2007764

Zazo H, Colino CI, Lanao JM (2016) Current applications of nanoparticles in infectious diseases. J Control Release 224:86–102

Zhang X (2015) Gold nanoparticles: recent advances in the biomedical applications. Cell Biochem Biophys 72:771–775

Zhang P, Jiang Y, Schwab F, Monikh FA, Grillo R, White JC, Guo Z, Lynch I (2024) Strategies for enhancing plant immunity and resilience using nanomaterials for sustainable agriculture. Environ Sci Technol 58:9051–9060

Chapter 15
Microbial Nanoparticles for Boosting Plant Tolerance to Salinity and Drought Stress

Waqas Haider, Tahira Yasmeen, Malaika Ashiq, Fatima, Muhammad Saleem Arif, Mohsin Tariq, Faisal Islam, and Muhammad Riaz

Abstract Salinity and drought stress are among the most significant challenges confronting the world's semi-arid and arid regions, resulting in undesirable effects on agricultural output. Water stress has been recognized as the biggest threat to the global agriculture sector, which is closely linked to salinity, thus making it difficult to identify its distinct impacts. Research has shown that salinity levels often reflect the quality of water, and prolonged or severe drought can hasten soil salinization, which will ultimately lower crop yields and endanger agricultural sustainability. Though common practices of managing the salinity and drought stress in plants are substantial yet they consume a large volume of resources. The potential of nanotechnology as a biotechnological tool can be exploited for remodeling various aspects of agriculture under abiotic stresses. Nanoparticles (NPs) have the capability to greatly enhance a variety of scientific areas, including biological, medicinal, industrial, and agricultural sciences. Biosynthesized NPs can be applied to plants and soil to preserve ionic balance and enhance soil fertility, both of which can significantly increase crop yield by mitigating the hostile impacts of drought and salinity in different plants.

Keywords Metallic nanoparticles · Stress mitigation · Biosynthesized NPs · Plant tolerance · Soil health

W. Haider · T. Yasmeen (✉) · M. Ashiq · Fatima · M. S. Arif · M. Riaz
Department of Environmental Sciences, Government College University Faisalabad, Faisalabad, Pakistan
e-mail: tahirayasmeen@gcuf.edu.pk

M. Tariq
Department of Bioinformatics and Biotechnology, Government College University Faisalabad, Faisalabad, Pakistan

F. Islam
International Genome Center, Faculty of Life Sciences, Jiangsu University, Zhenjiang, China

K. A. Abd-Elsalam (ed.), *Plant-Microbiome Nanotechnology*, Sustainability Sciences in Asia and Africa, https://doi.org/10.1007/978-981-95-5056-2_15

15.1 Introduction

Soil salinity is one of the most critical threats to the arid and semi-arid regions for inducing undesirable impacts on agriculture. Greater concentrations of greenhouse gases (GHGs) in the atmosphere, in addition to the corresponding increase in air temperature and reduction in relative humidity, as well as severe precipitation events, are likely signs of climate change that drastically affect the level at which soil salinity emerges (Mukhopadhyay et al. 2021). Salinity-induced oxidative, ionic, and osmotic processes can drastically hinder crop growth and yield by hindering potassium (K^+) absorption and promoting its loss from plant cells. Salt stress further disrupts plant physiology by triggering lipid membrane damage, protein and nucleic acid degradation, excessive production of reactive oxygen species (ROS), and disturbance of cellular homeostasis (Taha et al. 2020). Besides salinity, drought is another abiotic stress that inhibits plant growth and cultivation by disrupting several agricultural, plant physiological, and metabolic processes. It is considered one of the most complex and difficult to monitor natural hazards due to its slow onset and long-term effects.

There are many common practices of managing the salinity and drought stress in plants, but they consume a large volume of resources. Nanoparticle applications are supposed to enhance plant performance in challenging circumstances. They are used to sense and monitor trace elements in soil, treat plant diseases and pathogens, and deliver nutrients to plants (Heikal et al. 2023). Different metallic NPs (Cu, ZnO, Se) have demonstrated excellent outcome when applied as nano-fertilizers in agricultural ecosystems (Dey et al. 2024). These NPs cause the transcriptional change in biosynthesis-related genes and signal transduction of phytohormones, thus influencing growth, development, and defense response in plants under stressful environments (Thilakarathne et al. 2025). Compared to chemical synthesis of NPs, biological synthesis of nanoparticles is the easiest, most sustainable, and eco-friendly way of producing NPs. Plants, bacteria, yeast, and fungi can synthesize NPs more conveniently. Moreover, unicellular and multicellular organisms can be used to synthesize intracellular and extracellular nanoparticles (Roopa et al. 2025).

15.2 Salinity and Drought Stress Impacts and Management Strategies

The continuous increase in salt-affected areas is considered a recurrent challenge to sustainable agriculture development (Corwin 2021). More than 800 million hectares of land worldwide are categorized as saline soil, either by salinity (397 Mha) or sodicity (434 Mha) (Hafez et al. 2020). Around one-third of the extensive irrigated areas in developing nations are either currently affected by salinity or may be at high risk of developing the same (Mohanavelu et al. 2021). According to estimates from the Food and Agriculture Organization (FAO) of the United Nations, innumerable salinity levels presently have damaged 11% of the world's irrigated area (34 Mha). Irrigation-induced salinity, especially where farmers rely on saline groundwater, restrict root growth and soil infiltration, ultimately raising crop water demand (Minhas et al. 2019). Additionally, these soil problems lead to a number of environmental problems, such as increased soil erosion from vegetation loss, groundwater contamination, ecosystem disruption, decreased natural nutrient cycling, soil fertility and organic matter loss, moderated crop productivity, and deprivation of general habitat quality (Mohanavelu et al. 2021). In addition, several observable symptoms, such as soil surface crusting, poor plant growth, and leaf discoloration, can serve as qualitative indicators of salinity and sodicity in soils. These parameters can be measured in a lab setting by measuring electric conductivity (EC) and sodium adsorption ratio (SAR) or exchangeable sodium percentage (ESP), respectively. Table 15.1 shows different prominent features based on which soils are categorized as normal, saline, or sodic soils.

Table 15.1 Criteria for salinity and sodicity classification (Alonge et al. 2018; Scherer et al. 1996)

Soil type	Electrical conductivity (EC)	Sodium adsorption ratio (SAR)	Exchangeable sodium percentage (ESP)	pH	Characteristics	References
Normal soil	<4 dS/m	<13	<15	6.5–8.5	No significant salinity or sodicity issues and high permeability, supporting optimal plant growth	Richards (1954)
Saline soil	>4 dS/m	<13	<15	6.5–8.5	High salinity but low sodicity which can restrict plant growth due to osmotic stress. However, soil structure remains intact	Rengasamy (2010)
Sodic soil	<4 dS/m	>13	>15	>8.5	High sodicity with poor structure, leading to low permeability and poor plant growth. Crust formation and waterlogging	Qadir et al. (2006)
Saline-sodic soil	>4 dS/m	>13	>15	6.5–8.5	Both high salinity and sodicity, affecting both water uptake and soil structure. Limiting plant growth and soil aeration	Malcolm and Sumner (1998)

In addition to soil salinity, drought stress represents another critical abiotic factor that hinders the overall growth of crops by constraining seedling germination, root physiology, leaf structure, nutrient intake, and photosynthetic activity (Zhang et al. 2019). Scientists have developed a range of mitigation and adaptation strategies to address the imposed challenges of salinity and drought stress.

Strategic management techniques, such as the choice of drought-tolerant genotypes, the best time to sow, the right plant density, and efficient soil and nutrient management, can reduce yield losses in crops that are stressed by drought (Adeyemi et al. 2020). Plants can be made resistant to drought and salinity stress in a variety of ways. As drought and salinity resistance in crop species arise either directly or indirectly through genetic variation of traits, breeding selection can increase this resistance. Another essential short-term strategy to lessen the detrimental effects of drought on plants is seed priming (Khan et al. 2019). Using natural and synthetic plant growth regulators can also increase a plant's resistance to drought stress. By improving the production of proteins and the retention of internal water balance, gibberellic acid (GA) can be used to restore the hypocotyl length and weight in water in drought and water-stressed seedlings (Seleiman et al. 2021). In a range of plant challenges that reduce plant growth and productivity, osmoprotectants are a class of solutes that regulate osmotic pressure and signaling. These substances function without interfering with the cell metabolism (Brito et al. 2019). Silicon, which is one of the most prevalent elements in the Earth's crust, can be added to minerals to improve the general strength of both stressed and non-stressed plants and increase their resistance to different stress levels (Yan et al. 2018; Hurtado et al. 2020). Plants also illustrate improved growth with lessened negative effects when potassium (K) is added during drought stress conditions. Under drought conditions, plants absorb a lot of potassium for their internal regulation processes (Hasanuzzaman et al. 2018). Similarly, recent studies highlighting the positive effects of numerous organic amendments on plant growth in saline/sodic soils suggest that microbial activities reduce oxidative and osmotic stress, enhance stomatal conductance and improve the rate of seed germination under stressed conditions. In dry conditions, plant growth-promoting rhizobacteria (PGPR) contribute to reducing the negative effects by forming colonies in the root zones that encourage plant growth to improve the resistance against drought conditions (Balci et al. 2025). On the other side, it has been proven that organic amendments, such as biochar, microbial compost, green manure, and farmyard manure, enhance nitrogen uptake in the soil as a vital component for protein synthesis and promote a number of plant activities, including metabolism, protein buildup, and antioxidant effectiveness, especially under abiotic stress conditions (Irin and Hasanuzzaman 2024). Nanoparticle-based fertilizers and insecticides are also used to boost crop productivity and yield in both normal and challenging conditions. Nanoparticles (NPs) exhibit an assortment of physicochemical characteristics, including large surface area, high reactivity, adaptable pore size, and an array of particle morphologies (Saxena et al. 2016). They can be an innovative and beneficial approach to mitigate drought stress and to combat soil salinity (Hakeem and Pirzadah 2020). Different stress management strategies, along with underlying mechanisms and positive outcomes, are summarized in Table 15.2.

Table 15.2 Salinity and drought stress mitigation strategies

Mitigation strategy	Mechanism	Stress factor	Advantages	References
Organic amendment	Improve biological and physical properties, boost water holding capacity, and accessibility of nutrients	Salinity stress	Easily available, fast action, eco-friendly	Ullah et al. (2021), Irin and Hasanuzzaman 2024)
Plant growth-promoting rhizobacteria (PGPRs)	Intriguing techniques for improving homogeneous seedling growth, quick seed germination, and seed health	Salinity stress	Low cost involved, long-term benefits	Ha-Tran et al. (2021), Al-Turki et al. (2023)
Biochar	Boost soil moisture, increase soil surface space, and lower bulk density	Drought stress	Long-term beneficial method, environmentally sound, sustain the crop production	Akhtar et al. (2015), Wu et al. (2023)
Genetically modified crops	Change genetic pathways to improve resilience (e.g., drought, stress, heat), transcription factors	Abiotic stresses	Long-term effects under stress-responsive conditions	KhokharVoytas et al. (2023)
Chemical fertilizers	Provide micro and macro-nutrients necessary for plant growth, boost the crop yield	Salinity and drought stress	Short-term strategy to enhance the yield	Yaghoubi Khanghahi et al. (2021)
Microbial nanoparticles	Biomass increase, induction of strong cellular oxidative defense, better nutrient use efficiency and regulation	Drought stress	More reliable action, environmentally safe	Ahmad et al. (2021)

15.3 Nanoparticle Types

The particles with a single dimension between 1 and 100 nm are categorized as nanoparticles (NPs). NPs display a range of characteristics according to their size and surface functions (Najahi-Missaoui et al. 2020). Green synthesis techniques utilizing biological processes provide a less detrimental, more economical, and sustainable alternative to chemical or physical methods for nanoparticle formation. Furthermore, biological synthesis allows precise control over the size and shape of nanoparticles designed for specific uses. Various organisms, including bacteria, actinomycetes, fungi, yeasts, viruses, and algae, are being explored as natural reducing or stabilizing agents to produce metal nanoparticles (Ghosh et al. 2021).

NPs are categorized into a number of classes based on their size, shape, and chemical makeup. Nanoparticles (NPs) are categorized into four classes based on their size and dimensional characteristics: (i) zero-dimension NPs having all dimensions in nanoscale, (ii) one-dimensional NPs having one dimension outside the nanoscale, (iii) two-dimensional NPs having two dimensions outside the nanoscale, and (iv) three-dimensional NPs having all dimensions non-confined to nanoscale.

Based on their chemical and physical properties, some of the most notable classes of NPs are carbon-based NPs, Inorganic NPs, and organic NPs. Carbon-based NPs (CNP) are made solely from carbon atoms (Ealia and Saravanakumar 2017). Fullerenes and nanotubes made of carbon are two primary types of carbon-based NPs. These carbon nanofibers (CNFs) can improve nutrient use efficiency by releasing nutrients in accordance with the plant's requirement (Ashfaq et al. 2025).

Inorganic materials such as metals and ceramics are used to make inorganic nanoparticles. Metal NPs are nanoscale particles created by mechanical or chemical processes from pure metals or their compounds. Their composition can be either monometallic, bimetallic, or polymetallic (Toshima and Yonezawa 1998; Nascimento et al. 2018). Chemical reduction, hydro-thermal breakdown of metal salts, seeding, microemulsion, coprecipitation, and sonoelectro-deposition techniques are used to produce metal-based nanoparticles (Nam and Luong 2019). To synthesize metallic NPs Al, Cd, Co, Cu, Ag, Au Fe, Pb, and Zn metals are used (Ijaz et al. 2020; Kumar et al. 2021). Metal oxide nanoparticles (MONPs) are another kind of nanoparticle that is widely produced and the subject of extensive research for having unique physical and chemical properties, such as thermal conductivity and heat transfer (Fedele et al. 2011; Keblinski et al. 2008; Yu et al. 2008). MONPs are a promising tool for a variety of applications due to their high stability, ease of incorporation into hydrophobic and hydrophilic systems, and ease of engineering to shape them into the required size (Sanchez-Moreno et al. 2016). A variety of metal oxide nanoparticles, including ZnO, NiO, MnO_2, TiO_2, Fe_2O_3, and CO_3O_4, have been used to study the electrochemical detection of biomolecules (Immanuel et al. 2019). Many studies showed beneficial effects on plants by foliar spray of TiO_2 NPs due to their photocatalytic properties. TiO_2 NPs also improved plant height, ear and seed attributes, total yield, and harvest index in drought-stressed wheat plants (Jaberzadeh et al. 2013). Nonmetallic inorganic solids of phosphates, carbonates, and carbides, which are used to make ceramic NPs, come in a variety of sizes and shapes and are found as amorphous, porous, and polycrystalline forms. Researchers are paying attention to these NPs for their usage in catalytic reactions, photocatalysis, and imaging processes (Zeng et al. 2021).

Organic NPs are made of carbohydrates, proteins, lipids, or any other organic material, which are mostly used for targeted drug delivery and cancer therapy (Kurul et al. 2025).

15.4 Biological Synthesis of NPs

When microorganisms react with the metal ion of their environment and reduce them to their elemental form, the process is referred as the biosynthesis of NPs (Li et al. 2011). The biosynthesis of metallic NPs through microbes provides a greener route with no toxicity levels (Sharma et al. 2019). Microbial cells contain biomolecules such as proteins, carbohydrates, and DNA, which act as reactants, reducing agents, templates, and capping agents. These bioconjugate nanoparticles consequently possess biocompatibility and bioactive qualities (Sharma et al. 2019). Many different types of natural organisms, including bacteria, fungi, viruses, algae, and plants, rely on their enzymes, proteins, DNA, lipids, carbohydrates, and other constituents to generate nanoparticles. For bioremediation and materials production, bacteria that metabolize metals are shown to be eco-friendly catalysts. Indeed, bacteria may use respiration activities to aid in the production of various metal oxides (Kim et al. 2018). As NPs are synthesized by different living systems, some of them are described below.

Figure 15.1 illustrates the generalized steps involved in the biological synthesis of nanoparticles using diverse organisms such as bacteria, fungi, algae, and plants. The process can follow either an intracellular or extracellular route, depending on whether the metal precursor is added to the biomass or its supernatant. Subsequent incubation allows for the reduction of metal ions into nanoparticles. These biogenic nanoparticles are then harvested, typically by centrifugation, and can be stabilized by natural biomolecules that also act as capping agents. This eco-friendly approach offers a sustainable alternative to conventional chemical synthesis methods and highlights the potential of biological systems in nanotechnology.

15.4.1 Bacteria-Mediated NP Synthesis

Nanoparticle synthesis using bacteria has increased in recent years due to its dynamic use in different sectors. Green chemistry principles are applied in the microbial production of nanoparticles (NPs) to reduce the magnitude of hazardous compounds released as byproducts and promote safe handling of NPs (Noah 2019). As bacteria have a tendency to replicate exponentially, microbial-based synthesis provides a direct, inexpensive, and eco-friendly pathway to produce NPs on a large scale. In the process of metal ion bioreduction, bacterial cells perform exceptionally well as they are easy to culture, have short doubling times, exhibit genetic flexibility, and can adapt quickly to changes in their environment (Annamalai et al. 2021). Various bacteria, including *Arthrobacter gangotriensis, Arthrobacter kerguelensis, Pseudomonas Antarctica, Pseudomonas proteolytica*, and *Pseudomonas meridiana*, serve as reducing agents and factories of microbes to synthesize metal NPs (Singh et al. 2015). The triangle CuO nanoparticles made with *Halomonas elongate* demonstrated antimicrobial action counter to *S. aureus* and *E. coli* (Rad et al. 2018).

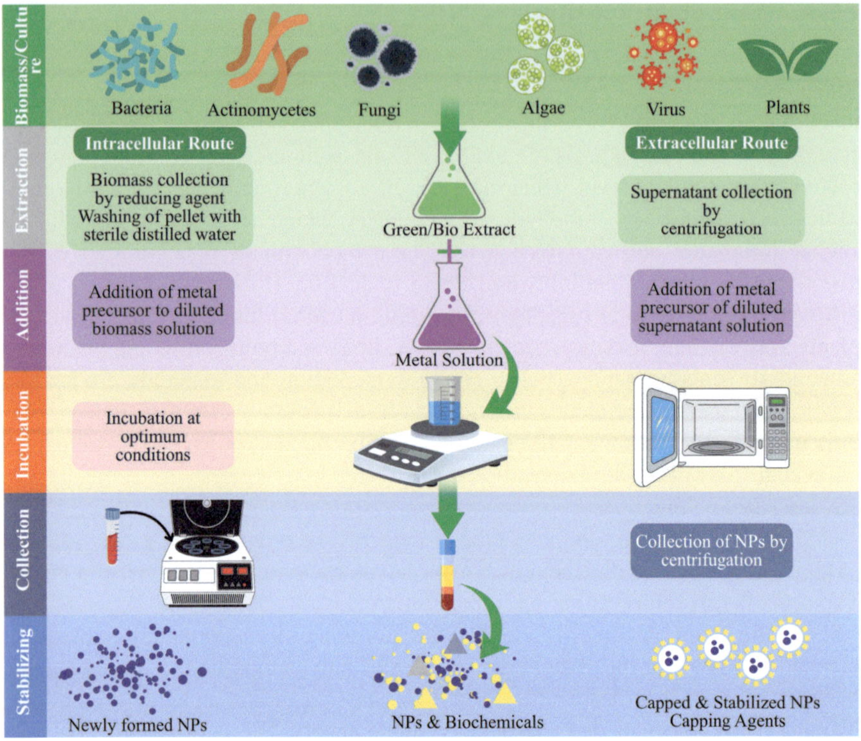

Fig. 15.1 Schematic representation of nanoparticle biosynthesis using intracellular and extracellular routes from various biological sources (Altammar 2023)

Recently, *Bacillus brevis*-synthesized silver nanoparticles (AgNPs) presented remarkable Antibacterial effectiveness against multidrug-resistant strains of *Salmonella typhi* and *Staphylococcus aureus* (Saravanan et al. 2018). Figure 15.1 illustrates bacteria-mediated nanoparticle synthesis through intracellular and extracellular mechanisms involving metal ion reduction and stabilization. In order to mitigate two concurrent abiotic stresses, such as drought and Cd, iron oxide (IONPs) were bio-fabricated using a locally identified *Bacillus* strain RNT1. When applied to rice plants, IONPs improved nutrient use efficiency, increased biomass, induced a strong cellular oxidative defense, and regulated the molecular mechanisms causing acropetal Cd translocation (Ahmad et al. 2021).

15.4.2 Fungi-Mediated NP Synthesis

Researchers believed that fungal species perform more efficiently in producing NPs than microbial cells. This could be explained by the rapid growth of fungal and mycelium biomass, large interactive surface area, and bio-reduction capacity of metal ions (Siddiqi and Husen 2016). Fungi are typically more effective and resourceful than bacteria because of their high aggregation, variety of bioactive metabolites, and faster production of nanoparticles (Alghuthaymi et al. 2015). The mycosynthesis of NPs involves the reduction of distinct metal ions to metal NPs while evaluating them for well-known applications. Fungal species such as *Aspergillus parasiticus, Penicillium parasiticus, Fusarium oxysporum, Mucor indicus,* and *Rhizopus stolonifer*, contribute to this process (Alshehri 2018; AbdelRahim et al. 2017). Ganachari et al. (2012) and Baskar et al. (2013) have described the production of silver and ZnO NPs by *Penicillium diversum* and *Aspergillus terreus*, respectively. Figure 15.1 presents how fungi can facilitate nanoparticle synthesis, either by secreting enzymes that reduce metal ions outside the cell or by taking up the ions and converting them into nanoparticles within their cells. Using fungi instead of bacteria to be used as natural nano-factories has an additional benefit of rendering, handling, processing, and biomass downstreaming.

15.4.3 Yeasts-Mediated NP Synthesis

The synthesis of nanoparticles by yeast strains pertaining to multiple genera is known to use unique methods, which lead to notable variations in mono-dispersity, particle size, position, and other characteristics (Ghosh et al. 2021). It has been demonstrated that yeasts, a class of ascomycetes prevalent in fungi, have more important capacities for acquiring NPs. Intracellular Gold nanoparticles (Au NPs) have been synthesized using the fungus *Verticillium luteoalbum,* while AgNPs are obtained from *Cladosporium cladosporioides* and *Penicillium fellutanum* using extracellular techniques (Karunakaran et al. 2023). Proteins, organic acids, and polysaccharides released by fungi also exhibit various shapes and are utilized to regulate the formation of spherical crystals. The process of nanoparticle synthesis from yeast is well displayed in Fig. 15.1.

15.4.4 Actinomycetes-Mediated NP Synthesis

Actinomycetes have been demonstrated to be important for the synthesis of metal nanoparticles (Manimaran and Kannabiran 2017). Actinomycetes contain numerous features with prokaryotes such as bacteria and fungi. It was found that proteins stabilize the Ag and Au NPs. Figure 15.1 shows how actinomycetes, known for their rich metabolic activity, can produce nanoparticles either inside their cells or in the surrounding medium by releasing natural compounds that help convert metal ions into stable nanoparticles. Proteins linked with Au NPs preserve free amino acids or cysteine (Ranjitha and Rai 2017). Moreover, only a small number of actinomycetes, particularly *Thermomonospora, Nocardia, Streptomyces,* and *Rhodococcus*, have demonstrated the ability to produce and characterize gold nanoparticles (El-Batal et al. 2015). Similarly, *Streptacidiphilus durhamensis, Streptomyces xinghaiensis,* and *Rhodococcus* sp. showed potential to synthesize spherical silver NPs (Buszewski et al. 2018; Otari et al. 2015; Wypij et al. 2018).

15.4.5 Algae-Mediated NP Synthesis

The trend of using algae for biosynthesis of NPs is increasing day by day. Algae hold high contents of secondary metabolites. They are fast grower, scalable, and easy to cultivate, eco-friendly and sustainable, which is suitable for NP synthesis. Moreover, synthesis of NPs using algae does not require an external reducing agent, which further reduces the NP synthesis cost. Both extracellular and intracellular mechanisms of NP synthesis are evident in algal cells, as shown in Fig. 15.1. In intracellular biosynthesis, NPs are formed inside the algal cells, while in extracellular biosynthesis, metal ions become attached to the surface of algal cells and cell metabolites reduce them at the cell surface (Chaudhary et al. 2020). Algal cells of *Cystoseira trinodis* have been evident to have potential of synthesizing Cu NPs (Gu et al. 2018). *Sargassum tenerrimum, Laminaria japonica,* and *Turbinaria conoides* can also be used for the development of Au NPs, as stated by Ramakrishna et al. (2016), Ghodake and Lee (2011), and Swaminathan et al. (2011). Algal nanoparticles (NPs) have shown great promise in environmental preservation and bioremediation. Silver nanoparticles (Ag NPs) produced by *Haematococcus candid* and *Chlorella* sp. have been demonstrated to suppress the growth of *Penicillium expansum*, a primary cause of fruit spoiling (El-Moslamy et al. 2016), whilst *Chlorella vulgaris*-mediated Ag NPs have potent antifungal action in plants. As powerful bioremediators, gold nanoparticles (Au NPs) derived from *Turbinaria conoides* and *Sargassum tenerrimum* have demonstrated efficacy in the catalytic reduction and degradation of organic dyes. *Scenedesmus obliquus* generates lipid cadmium sulfide (CdS) NPs, which, because of their high retention capability, effectively retain and remediate Cd^{2+} ions (Abideen et al. 2022).

15.5 Mechanism of Microbial Nanoparticles in Improving Plant Tolerance

The application of nano-biotechnology in sustainable agriculture has tremendous potential. However, some nanoparticles (NPs) with particular physicochemical characteristics naturally increase plant development and stress tolerance rather than serving as nano-carriers. The physicochemical characteristics, administration strategy (foliar delivery, hydroponics, soil), and applied concentration influence the biological role of nanoparticles to be used for abiotic stress management (Zhao et al. 2020).

Using microorganisms to synthesize nanoparticles (NPs) is becoming a viable method for creating nanomaterials that can improve plants' resistance to biotic and abiotic influences. In plants, several metal and metal oxide nanoparticles (NPs) have been shown to improve defenses against pests and diseases when applied in small amounts (Shaikh et al. 2024). During drought and salinity stress, NPs regulate signaling pathways and stress-responsive genes, thus permitting the plant to evolve and endure under stress conditions (Rehman et al. 2024). NPs significantly boost the processes of nutrient uptake and help plants to maintain the nutritional balance and their distribution across various plant tissues (Kopittke et al. 2019). A recent study demonstrated that the inclusion of magnetic nanoparticles (NPs) into a hydroponic growing system significantly enhanced the concentrations of photosynthetic pigments in *Hordeum vulgare* leaves during the seedling growth phase. An improved photosystem-related gene activity was also evident with the application of magnetic NPs (Tombuloglu et al. 2020). The application of NPs can noticeably boost plant tolerance against drought stress by decreasing the production of phenolic compounds in plants. Likewise, increased photosynthetic pigments, enhanced photosynthetic efficiency and improved yield attributes along with reduced activities of CAT, SOD, POD enzymes and lower MDA contents were reported in NPs subjected wheat plants (Hafez et al. 2021).

15.5.1 Influence on Plant Physiology

Nitric oxide (NO) has been linked to a group of physiological developments, most appropriately germination, fruit ripening, and response to biotic and abiotic adverse conditions. NPs may help in the transportation of NO to specific plants (Zuccarelli et al. 2020). Metal nanoparticles can move in a plant system through various channels and processes; however, different factors, including the nature and amount of nanoparticles, the type and growth phase of the plant, underlying mechanisms, and the ambient environment, can alter these pathways (Khan et al. 2022). Plants treated with NPs demonstrated high rates of photosynthesis, stomatal conductance, CO_2 integration, transpiration rate, water use capability, chlorophyll, and proline content, besides improving plant tolerance against abiotic stresses (Khan et al. 2016; Ahmad et al. 2021). It was also observed that ZnO NPs application enhanced chlorophyll synthesis, photosynthetic efficiency, fluorescence, and the activity of chlorophyll producing enzymes (chlorophyllase), enabling plants to better withstand drought or salinity stress (Abd El-Mageed et al. 2021).

15.5.2 Stress Mitigation

Drought is one of the common abiotic stressors that substantially reduces agricultural productivity in arid regions. The feasibility of plant and nutrient transport rely on water (Hasan et al. 2020a, 2020b). Stress-related genes entailed in ion transport, antioxidant defense, and stress signaling pathways with altered expression levels have been evident after seed priming with CeO_2 NPs. Numerous metabolites that are known to improve plant stress tolerance were also accumulated by CeO_2 NP penetration during seed priming. Additionally, under salinity stress conditions, CeO_2 NPs lowered oxidative stress in plant cells by acting as scavengers of reactive oxygen species (ROS) (Francis et al. 2024). In another study, the topical application of ZnO NPs on wheat leaves demonstrated a stress tolerance mechanism through reduced oxidative stress markers, which appeared as increased antioxidant enzyme activity (SOD and POD) under a stressful environment (Alabdallah et al. 2021). A type of photocatalyst called TiO_2 NPs has the capacity to hydrolyze light into protons, electrons, and oxygen. During the light reaction stage, the generated electrons and protons are subsequently incorporated into the plant's electron transport chain, accelerating up photosynthesis under stress and helping reduce the oxidative damage caused by abiotic stresses such as salinity and drought (Aghdam et al. 2015).

Salt-stressed plants accumulate more Na^+, while their K^+ levels drop. Excessive Na^+ and Cl^- ions in wheat during salinity are hostile to the uptake of essential K^+ ions. Janicka-Russak and Kabała (2014) confirmed that the application of Ag NPs downregulated the quantity of Na^+ and Cl^- while upregulated the concentration of K^+ ions in both the roots and the leaves. This may occur due to the high K^+/Na^+ ratio in plant tissues. A decrease in Na^+-induced K^+ efflux leads to a high intracellular K^+ retention. Ashraf et al. (2019) reported decreased germination rate and other growth characteristics of soybean seedlings under NaCl stress, while seeds soaking in ZnO-NPs improved these observed parameters.

15.5.3 Soil Health Improvement

The specific physical and chemical features of nanoparticles (NPs) at the nanoscale have displayed substantial potential in influencing the attributes and behaviors of the soil. NPs modify soil structure, nutrient availability, and water dynamics by integrating into soil, organic matter, and water-based constituents (Zaib et al. 2023). Nanoparticles can amend the structure of soil by improving soil compaction, porosity, and aggregation (Liu et al. 2015). Engineered nanoparticles can help in the transportation of essential nutrients by providing targeted delivery to plant tissues. Wang et al. (2018) reported improved nutrient uptake in plants under the influence of Ag NPs. Organic materials in the soil that have functional groups such as –OH and –COOH can work together with nanoparticles. Stable aggregate development is assisted by this interrelation, which supports the organic materials binding to mineral surfaces (Zaib et al. 2023). Nanomaterials have also been presented to support a few advantageous soil processes, notably the nitrogen cycle, increased enzyme activity, and the instigation of helpful microbes. The rhizosphere microbiota plays a key role by supplying bioavailable nutrients and producing phytohormones in regulating plant development, nutrient uptake, physiological processes, and endurance under abiotic stressors (Liu et al. 2020).

15.6 Microbial Nanoparticles for Boosting Salinity Stress Tolerance

Salinity is among various abiotic stresses that stand out as one of the most detrimental abiotic stresses that may extremely damage the crops and reduce crop efficiency (Hussain et al. 2016). Via lower osmotic potential, accumulation of toxic ions, and deficiency of the photosynthetic machinery, salinity may have an influence on seedling growth and germination (Farooq et al. 2015). Under both stress and non-stress conditions, one of the key adaptive strategies that plants use to cope with induced damage is the variation in endogenous phytohormone levels, which regulate a number of responses throughout the plant life cycle. A suppressive effect on abscisic acid (ABA) levels and a stimulatory effect on IAA, GA3, and total cytokinins were observed with foliar application of AgNPs in common bean plants (El-Batal et al. 2016). Under salt stress, AgNPs treated wheat plants showed improved performance compared to control plants. Modification in growth characteristics along with elevated levels of metabolic chemicals was also reported in silver nanoparticles (Ag NPs) primed plants (Asanova et al. 2019). Foliar application of Ag NPs lessened damage index parameters together with H_2O_2 and ROS during salinity stress by encouraging proline metabolism, nitrogen absorption, and antioxidative mechanism (Wahid et al. 2020). They also demonstrated how the exogenous application of nanoparticles affected the endogenous ABA concentration, which in turn positively controlled the plants' chlorophyll content. Abdoli et al. (2020) concluded that the application of Fe_2O_3 NPs enabled *Trachyspermum ammi* plants by enhancing the K^+/Na^+ ratio and increasing iron (Fe) content to minimize salinity stress. According to frequent studies on tomato plants, the Cu-NPs direct spray lowered the salt stress by fostering formation as well as keeping optimal health (Pérez-De-Luque 2017). Through enhancing the expression of the protein OsHMA3 and stimulating salt transport into the vacuole, silicon nanoparticles have been found to improve plant development and antioxidant enzymes' activity under salinity stress (Siddiqui et al. 2020). Mohamed et al. (2017) claimed that application of *Aspergillus terreus*-mediated silver NPs on salt-stressed wheat plants, can increase soluble sugar and proline content along with boosting shoot fresh and dry biomass. Similarly, Manzoor et al. (2021) observed enriched plant growth and increased N, P, and K uptake under Cd and salinity stress with *Pantoea ananatis*-mediated FeO-NPs application. The application of microbial nanoparticles has been developed as a promising strategy

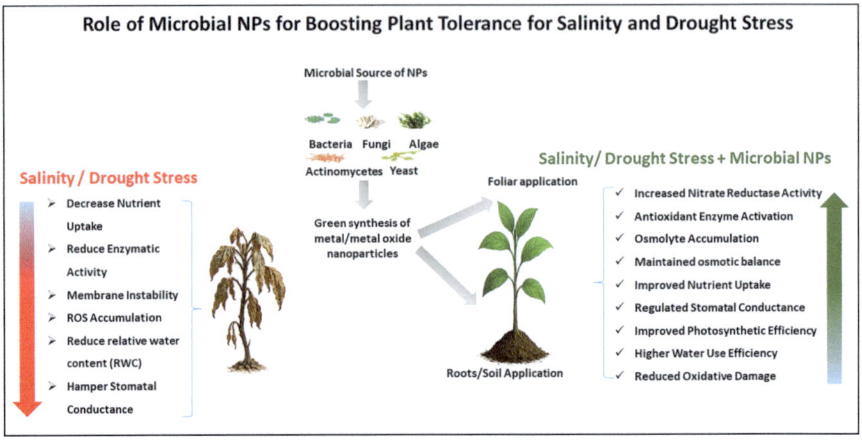

Fig. 15.2 Microbial synthesis of metal/metal oxide nanoparticles and their role in enhancing plant tolerance to salinity and drought stress through physiological and biochemical variations

to enhance plant resilience in salinity stress. NPs synthesized by various microbes modulate key physiological and biochemical pathways to diminish stress effects, as presented in Fig. 15.2.

15.7 Microbial Nanoparticles for Boosting Drought Stress Tolerance

Osmoregulation and single-cell growth across cellular membranes can be influenced by adequate soil water and short- to long-distance transport in order to produce healthy crops (Iwuala et al. 2019). Ahmad et al. (2021) established that the *Bacillus* strain RNT1 can be utilized to produce iron oxide nanoparticles, which trigger genetic pathways and boost nutrient uptake from the surroundings. Under Cd toxicity, these nanoparticles were applied to rice plants, which improved drought stress tolerance. Another study led by Hojjat and Ganjali (2016) illustrated that Ag NPs upgraded growth characteristics of lentils, enabling plants to regulate water balance and better grow under drought stress. Similarly, El-Zohri et al. (2021) reported that the use of NPs (ZnO) depressed the loss of vital osmolytes by preserving the strength of the membrane and considerably decreasing the development of MDA and H_2O_2 in tomato plants. Semida et al. (2021) claimed enhanced PS-II and metabolic efficiency by the exogenous application of NPs (ZnO) through maintaining cell water content and membrane stability in drought stress. In addition, SiO_2 NPs were found to considerably minimize the detrimental effects of drought stress by promoting photosynthesis, transpiration, relative water content (RWC), and water intake in *Pisum sativum L.* plants. SiO_2 NPs considerably upraised non-enzymatic processes in drought-stressed plants by upgrading total phenolic content (TPC),

2,2-diphenyl-1-picrylhydrazyl (DPPH) radical eliminating capacity, and ferric reducing antioxidant power (FRAP) (Sutulienė et al. 2021). The application of NPs (TiO_2) encouraged glycine betaine (GB) and proline synthesis, boosted nitrate reductase (NR) activity, improved nitric oxide (NO) production, and upregulated osmolyte formation, making plants tolerant against oxidative stress (Khan et al. 2020). Yasmeen et al. (2016) also reported considerably higher spike length, grain number per spike, and grain weight in water-stressed wheat plants exposed to Fe and Cu NPs. Microbial nanoparticles play an important role in improving plant drought tolerance by enhancing antioxidant defense, water retention, and nutrient intake systems. Their mode of action under water stress is illustrated in Fig. 15.2. Phytogenic or biogenic nanoparticles have been used to diminish the detrimental consequences of drought on plant health (Chakraborty et al. 2022). It has been reported that citrate-coated iron nanoparticles produced via *Chaetomorpha antennina* effectively increased drought tolerance in *Setaria italica* by up-regulating protective osmolyte levels and key antioxidant enzyme activity (Sreelakshmy et al. 2016).

15.8 Application Techniques of NPs in Agriculture

15.8.1 Foliar Application

NPs are usually applied topically in agriculture. Foliar-applied NPs are frequently utilized as biosensors, post-harvest preservers, fertilizers, herbicides, and nutritional additives for plants (Tighe-Neira et al. 2018). Plants treated with low-solubility fertilizers often exhibit poor growth due to insufficient uptake of essential trace elements (Li et al. 2016). Furthermore, several plant nutrients have the potential to precipitate or bind with organic matter and soil colloids, which would further limit their availability to plants. Therefore, slow-release nanofertilizers have been created using biodegradable natural macromolecules, such as carboxyl methyl cellulose polymers, kaolinite, mesoporous silica, and chitosan (Achari and Kowshik 2018). Foliar-applied nanofertilizers offer the advantages of being rapidly absorbed by plants, being more economical, and having a negligible influence on soil health when compared to conventional soil-applied fertilizers (Meier et al. 2020). Slow-release nanofertilizers can improve the uptake of N, P, and K in plants and accomplish biofortification quickly (Salem et al. 2016; Budke et al. 2020). However, a number of requirements must be met for leaf absorption, i.e., NPs must be in a form that the plant can use, have a long enough residence period, and be able to pass through the epidermis to reach the target organ. Diffusion into the leaves is one method through which NPs may enter the epidermis, depending on the ion gradient (Li et al. 2016). Improved yield, photosynthates, and Zn levels were recorded in the ZnO NPs sprayed rice (Zhang et al. 2025). Another study demonstrated that the topical application of fullerene nanoparticles can improve a plant's tolerance to water stress by enhancing its water retention capacity and reducing drought-induced oxidative stress (Borišev et al. 2016). The effectiveness of foliar-applied microbial nanoparticles in enhancing plant stress tolerance is supported by their ability to deliver nutrients directly, trigger antioxidant responses, and regulate stress-related hormones. A comprehensive overview of these mechanisms, along with application strategies and associated considerations, is illustrated in Fig. 15.3.

15.8.2 Soil Applications

Soil application is done by mixing NPs in the topsoil layer (Shah et al. 2022). Nanoparticles are easily integrated into the pore space between soil particles because of their small size, especially into fine soil particles that are not under a lot of pressure. Soil-applied microbial nanoparticles contribute to stress mitigation by improving nutrient availability, enhancing microbial interactions, and supporting soil structure. The broader impacts of microbial NPs on plant resilience and environmental sustainability are summarized in Fig. 15.3. The soil is impacted differently by two-dimensional nanoparticles (carbon nanotubes) and three-dimensional ones (colloidal silica). The detailed surface area of the nanoparticle increases as the particle size decreases. This suggests a bigger atomic surface, which raises contact with other particles and increases the ion exchange capacity. Microbe-mediated Ag, ZnO, FeO, CuO, and TiO_2 NPs are mostly used in agricultural systems, which have wide impacts on soil metabolic pathways, enzyme activities, phytohormones and antibiotics production, and soil–microbe interactions besides soil remediation (Chaudhary et al. 2020).

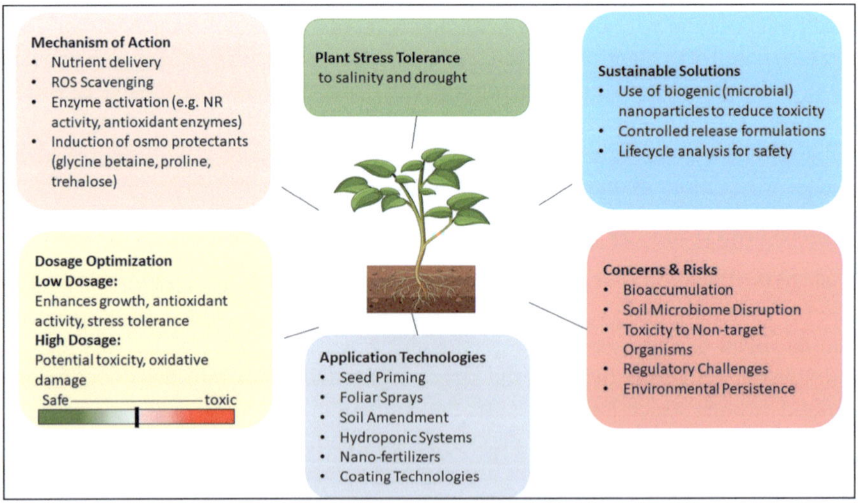

Fig. 15.3 An overview of microbial nanoparticles improving plant tolerance to salinity and drought via physiological mechanisms, optimized application, and sustainability considerations

15.9 Dosage and Safety

NPs with a smaller size range are rapidly absorbed by plants and may exhibit greater phytotoxicity even at low concentrations because it has been thoroughly reviewed that NPs smaller than 5 nm can easily translocate through cell wall pores and about 20 nm through plasmodesmata. However, the potential movement from roots to shoots is hindered, resulting in reduced bioaccumulation. For example, ZnO and CeO_2 NPs exhibited their harmful effects on cucumber and decreased the plant yield by 31.6% at higher concentrations (Zhao et al. 2013). Conversely, the application of 1000 ppm of ZnO NPs enhanced the amount of chlorophyll and encouraged seed germination and seedling vigor in peanut plants (Vasant and Raghav 2024). More precisely, metal NPs' primary mechanism of toxicity is due to their reliance, leading to the excessive production of reactive oxygen species (ROS), particularly due to ion release in the biological environment, which results in hazardous consequences. Additionally, it disrupts the mitochondrial and chloroplast electron transport chains. Higher concentrations of NPs also reduce auxins, cytokinins, and salicylic acid production, which may indicate a hormone imbalance in plants and impact metabolism. Photosynthesis may be negatively impacted by extremely high NPs concentrations, which could lead to plant death or growth inhibition (Mittal et al. 2020). Therefore, a more precise estimation of the applicable dose of NPs is crucial to get maximum positive outputs. The influence of nanoparticle concentration on plant responses ranging from beneficial to phytotoxic is a critical consideration. These dosage-related effects, alongside their mechanisms and risks, are outlined in Fig. 15.3.

15.10 Challenges and Limitations

15.10.1 Technical Challenges

Despite the fact that nanoparticles have many benefits, industrial production is challenging since the optimization factors needed to produce reproducible nanoparticles are not well defined. Aggregation, contamination, and nanoparticle degradation are among the challenging factors encountered during its manufacturing. Low yield also raises scalability concerns. NPs have a variety of possible applications in industries. However, there are still multiple barriers. The processing of metal nanoparticles with exact control over size and form is a significant problem. High temperatures and harsh chemical conditions are used in several metal NP synthesis techniques, which might make it difficult to scale up for large-scale manufacturing. Furthermore, metal nanoparticles' properties and possible uses can be greatly influenced by their size and form; thus, it's critical to synthesize NPs with exact control over size and shape (Fernandes et al. 2023).

15.10.2 Environmental and Ecological Concerns

Extensive research has been done to explore the specific effects of NPs, yet only limited progress has been achieved their associated health and safety risks. The clear and precise structure for NP use in the agri-food businesses and risk management throughout the various sectors is largely vague. In addition to demonstrating advancements in nano-safety research, risk assessment, risk management, and policymaking, as well as facilitation for risk communication among various stakeholders, including researchers, policymakers, insurance companies, and the general public, must also be established. Environmental health risk is a significant area of concern. Therefore, the deliberate or direct exposure of food crops to NPs (such as nanopesticides, nanofertilizers, and nanoherbicides), along with indirect or incidental exposure, has been the main focus of plant-based nano-safety research (Stone et al. 2018).

Physicochemical characteristics of NPs are the main issue that must be resolved before using NPs in any field because they may impose environmental risks. The most significant attributes of the NPs related to toxicity are their size and concentration. It can be linked to a larger surface area with a smaller size, which ultimately leads to increased absorption and interaction with the plant system, thereby raising the risk of adverse effects (Mittal et al. 2020). Nevertheless, introducing nanoparticles into the environment may have unforeseen repercussions that could result in dangers and toxicity. Nanoparticles are easily transferred by wind or water, contaminating soil, water, and unexpected habitats. Ecosystem equilibrium could be disturbed by the interaction of NPs with naive species or bioaccumulate in the food chain because of their high surface reactivity and large surface area-to-volume ratio. Soil health has been negatively affected by NPs because they have changed the composition of the microbial population. Soil microbiota considerably disturbs the decomposition of organic matter, nutrient cycling, and plant health (Zhu et al. 2022). Few microbes may find NPs toxic that could obstruct these activities. Since agricultural products become a basic part of the human diet, there is a potential risk that nanoparticles used in agriculture could pose health problems on ingestion through the food chain could build up in plant or animal tissues that are edible (Karumannil et al. 2023). Therefore, the possible benefits of disbursing nanotechnology in agriculture should be sensibly weighed against any unsuspected hitches (Thabet and Alqudah 2024). These possible concerns are assimilated into the wider context of nanoparticle benefits and risks in Fig. 15.3.

References

Abd El-Mageed TA, Shaaban A, Abd El-Mageed SA, Semida WM, Rady MO (2021) Silicon defensive role in maize (*Zea mays* L.) against drought stress and metals-contaminated irrigation water. SILICON 13:2165–2176

AbdelRahim K, Mahmoud SY, Almarry KS, Mustafa AEZ, Husseiny SM (2017) Extracellular biosynthesis of silver nanoparticles using *Rhizopus stolonifer*. Saudi J Biol Sci 24:208–216
Abdoli S, Ghassemi-Golezani K, Alizadeh-Salteh S (2020) Responses of ajowan (*Trachyspermum ammi* L.) to exogenous salicylic acid and iron oxide nanoparticles under salt stress. Environ Sci Pollut Res 27(29):36939–36953
Abideen Z, Waqif H, Munir N, El-Keblawy A, Hasnain M, Radicetti E et al (2022) Algal-mediated nanoparticles, phycochar, and biofertilizers for mitigating abiotic stresses in plants: a review. Agronomy 12(8):1788
Achari GA, Kowshik M (2018) Recent developments on nanotechnology in agriculture: plant mineral nutrition, health, and interactions with soil microflora. J Agric Food Chem 66(33):8647–8661
Adeyemi O, Keshavarz-Afshar R, Jahanzad E, Battaglia ML, Luo Y, Sadeghpour A (2020) Effect of wheat cover crop and split nitrogen application on corn yield and nitrogen use efficiency. Agronomy 10(8):1081
Aghdam MTB, Mohammadi H, Ghorbanpour M (2015) Effects of nanoparticulate anatase titanium dioxide on physiological and biochemical performance of *Linum usitatissimum* (Linaceae) under well-watered and drought stress conditions. Rev Bras Bot 39(1):139–146
Ahmad A, Hashmi SS, Palma JM, Corpas FJ (2021) Influence of metallic, metallic oxide, and organic nanoparticles on plant physiology. Chemosphere 290:133329
Akhtar SS, Andersen MN, Liu F (2015) Residual effects of biochar on improving growth, physiology and yield of wheat under salt stress. Agric Water Manag 158:61–68
Alabdallah NM, Hasan MM, Hammami I, Alghamdi AI, Alshehri D, Alatawi HA (2021) Green synthesized metal oxide nanoparticles mediate growth regulation and physiology of crop plants under drought stress. Plants 10(8):1730
Alghuthaymi MA, Almoammar H, Rai M, Said-Galiev E, Abd-Elsalam KA (2015) Myconanoparticles: synthesis and their role in phytopathogens management. Biotechnol Biotechnol Equip 29:221–236
Alonge TA, Ojo OI, Adejumobi MA (2018) Electrical conductivity based classification and mapping of salt affected soils in kampe-omi irrigation scheme. GSJ 6(7):1075
Alshehri ANZ (2018) Using a novel *mucorindicus* CBS 226.29 ET for biosynthesis of gold nanoparticles and applying them in nanoremediation of azo dyes. Res J Appl Sci Eng Technol 15:197–205
Altammar KA (2023) A review on nanoparticles: characteristics, synthesis, applications, and challenges. Front Microbiol 14:1155622
Al-Turki A, Murali M, Omar AF, Rehan M, Sayyed RZ (2023) Recent advances in PGPR-mediated resilience toward interactive effects of drought and salt stress in plants. Front Microbiol 14:1214845
Annamalai J, Ummalyma SB, Pandey A, Bhaskar T (2021) Recent trends in microbial nanoparticle synthesis and potential application in environmental technology: a comprehensive review. Environ Sci Pollut Res 28(36):49362–49382
Asanova AA, Yashin SE, Trofimova TV, Polonskiy VI (2019) Application of silver nanoparticles to improve wheat seedlings growth. IOP Conf Ser Earth Environ Sci 315:052041
Ashfaq M, Gupta G, Verma N (2025) Carbon-based nanocarriers for plant growth promotion: Fuelling when needed. Nanoscale 17(2):616–634
Ashraf MY, Tariq S, Saleem M, Khan MA, Hassan SWU, Sadef Y (2019) Calcium and zinc mediated growth and physio-biochemical changes in mungbean grown under saline conditions. J Plant Nutr 43(4):512–525
Balci M, Arikan-Abdulveli B, Yildiztugay E, Ozfidan-Konakci C, Uysal A (2025) An effective sustainable strategy: plant growth promoting rhizobacteria (PGPR), *Bacillus atrophaeus*, in maize plants against challenging environments. J Environ Chem Eng 13(3):16778
Baskar G, Chandhuru J, Fahad KS, Praveen AS (2013) Mycological synthesis, characterization and antifungal activity of zinc oxide nanoparticles. Asian J Pharm Technol 3(4):142–146

Borišev M, Borišev I, Župunski M, Arsenov D, Pajević S, Ćurčić Ž, Vasin J, Djordjevic A (2016) Drought impact is alleviated in sugar beets (Beta vulgaris L.) by foliar application of fullerenol nanoparticles. PLoS One 11(11):e0166248

Brito N, Dinis N, Moutinho-Pereira N, Correia N (2019) Drought stress effects and olive tree acclimation under a changing climate. Plants 8(7):232

Budke C, thor Straten S, Mühling KH, Broll G, Daum D (2020) Iodine biofortification of field-grown strawberries–approaches and their limitations. Sci Hortic 269:109317

Buszewski B, Rațiu IA, Milanowski M, Pomastowski P, Ligor T (2018) The effect of biosilver nanoparticles on different bacterial strains' metabolism reflected in their VOCs profiles. J Breath Res 12:027105

Chakraborty S, Singh A, Roychoudhury A (2022) Biogenic nanoparticles and generation of abiotic stress-resilient plants: a new approach for sustainable agriculture. Plant Stress 6:100117

Chaudhary R, Nawaz K, Khan AK, Hano C, Abbasi BH, Anjum S (2020) An overview of the algae-mediated biosynthesis of nanoparticles and their biomedical applications. Biomolecules 10(11):1498

Corwin DL (2021) Climate change impacts on soil salinity in agricultural areas. Eur J Soil Sci 72(2):842–862

Dey S, Ghosh N, Nath S, Gopal G, Paul S, Mukherjee A, Paul S, Kundu R (2024) Application of multi-metallic nanoparticles in agriculture: the more, the better? Biocatal Agric Biotechnol 58:103238

Ealia SAM, Saravanakumar MP (2017) A review on the classification, characterisation, synthesis of nanoparticles and their application. IOP Conf Ser: Mater Sci Eng 263(3):032019

El-Batal A, Mona S, Al-Tamie M (2015) Biosynthesis of gold nanoparticles using marine *Streptomyces cyaneus* and their antimicrobial, antioxidant and antitumor (*in vitro*) activities. J Chem Pharm Res 7:1020–1036

El-Batal AI, Gharib FAE, Ghazi SM, Hegazi AZ, Hafz AGMAE (2016) Physiological responses of two varieties of common bean (Phaseolus VulgarisL.) to foliar application of silver nanoparticles. Nanomater Nanotechnol 6:13

El-Moslamy S, Kabeil S, Hafez EJO (2016) Bioprocess development for Chlorella vulgaris cultivation and biosynthesis of anti-phytopathogens silver nanoparticles. J Nanomater Mol Nanotechnol 5: 1. J Nanomater Mol Nanotechnol 9:2

El-Zohri M, Al-Wadaani NA, Bafeel SO (2021) Foliar sprayed green zinc oxide nanoparticles mitigate drought-induced oxidative stress in tomato. Plants 10(11):2400

Farooq M, Hussain M, Wakeel A, Siddique KHM (2015) Salt stress in maize: effects, resistance mechanisms, and management. Agron Sustain Dev 35:461–481

Fedele L, Colla L, Bobbo S, Barison S, Agresti F (2011) Experimental stability analysis of different water-based nanofluids. Nanoscale Res Lett 6:1–8

Fernandes C, Jathar M, Sawant BKS, Warde T (2023) Scale-up of nanoparticle manufacturing process. In: Pharmaceutical process engineering and scale-up principles. Springer, pp 173–203

Francis DV, Abdalla AK, Mahakham W, Sarmah AK, Ahmed ZF (2024) Interaction of plants and metal nanoparticles: exploring its molecular mechanisms for sustainable agriculture and crop improvement. Environ Int 190:108859

Ganachari SV, Bhat R, Deshpande R, Venkataraman A (2012) Extracellular biosynthesis of silver nanoparticles using fungi Penicillium diversum and their antimicrobial activity studies. BioNanoScience 2(4):316–321

Ghodake G, Lee DS (2011) Biological synthesis of gold nanoparticles using the aqueous extract of the brown algae Laminaria japonica. J Nanoelectron Optoelectron 6(3):268–271

Ghosh S, Ahmad R, Zeyaullah M, Khare SK (2021) Microbial nano-factories: synthesis and biomedical applications. Front Chem 9:626834

Gu H, Chen X, Chen F, Zhou X, Parsaee Z (2018) Ultrasound-assisted biosynthesis of CuO-NPs using brown alga *Cystoseira trinodis*: characterization, photocatalytic AOP, DPPH scavenging and antibacterial investigations. Ultrasonics Sonochem 41:109–119

Hafez EM, Omara AED, Alhumaydhi FA, El-Esawi MA (2020) Minimizing hazard impacts of soil salinity and water stress on wheat plants by soil application of vermicompost and biochar. Physiol Plant 172(2):587–602

Hafez EM, Osman HS, Gowayed SM, Okasha SA, Omara AED, Sami R et al (2021) Minimizing the adversely impacts of water deficit and soil salinity on maize growth and productivity in response to the application of plant growth-promoting rhizobacteria and silica nanoparticles. Agronomy 11(4):676

Hakeem KR, Pirzadah TB (2020) Nanobiotechnology in agriculture. Springer International Publishing

Hasan MM, Alharby HF, Uddin MN, Ali MA, Anwar Y, Fang X, Hakeem KR, Alzahrani Y, Hajar AS (2020a) Magnetized water confers drought stress tolerance in moringa biotype via modulation of growth, gas exchange, lipid peroxidation and antioxidant activity. Polish J Environ Stud 29(2):1625–1636

Hasan MM, Ali MA, Soliman MH, Alqarawi AA, Abd_Allah EF, Fang X (2020b) Insights into 28-homobrassinolide (HBR)-mediated redox homeostasis, AsA–GSH cycle, and methylglyoxal detoxification in soybean under drought-induced oxidative stress. J Plant Interact 15(1):371–385

Hasanuzzaman M, Bhuyan M, Nahar K, Hossain M, Mahmud J, Hossen M, Masud A, Moumita N, Fujita M (2018) Potassium: a vital regulator of plant responses and tolerance to abiotic stresses. Agronomy 8(3):31

Ha-Tran DM, Nguyen TTM, Hung SH, Huang E, Huang CC (2021) Roles of plant growth-promoting rhizobacteria (PGPR) in stimulating salinity stress defense in plants: a review. Int J Mol Sci 2(6):3154

Heikal YM, El-Esawi MA, El-Ballat EM, Abdel-Aziz HM (2023) Applications of nanoparticles for mitigating salinity and drought stress in plants: an overview on the physiological, biochemical and molecular genetic aspects. New Zealand J Crop Hortic Sci 51(3):297–327

Hojjat SS, Ganjali A (2016) The effect of silver nanoparticle on lentil seed germination under drought stress. Int J Farm Allied Sci 5(3):208–212

Hurtado AC, Chiconato DA, De Mello Prado R, Da Silveira Sousa G, Junior GSS, Gratão PL, Felisberto G, Viciedo DO, Santos DMMD (2020) Different methods of silicon application attenuate salt stress in sorghum and sunflower by modifying the antioxidative defense mechanism. Ecotoxicol Environ Safe 203:110964

Hussain MI, Lyra DA, Farooq M, Nikoloudakis N, Khalid N (2016) Salt and drought stresses in safflower: a review. Agron Sustain Dev 36(1):4–13

Ijaz I, Gilani E, Nazir A, Bukhari A (2020) Detail review on chemical, physical and green synthesis, classification, characterizations and applications of nanoparticles. Green Chem Lett Rev 13(3):223–245

Immanuel S, Aparna TK, Sivasubramanian R (2019) Graphene–metal oxide nanocomposite modified electrochemical sensors. In: Pandikumar A, Rameshkumar P (eds) Graphene-based electrochemical sensors for biomolecules. Elsevier, pp 113–138

Irin IJ, Hasanuzzaman M (2024) Organic amendments: enhancing plant tolerance to salinity and metal stress for improved agricultural productivity. Stress 4(1):185–209

Iwuala E, Odjegba V, Sharma V, Alam A (2019) Drought stress modulates expression of aquaporin gene and photosynthetic efficiency in *Pennisetum glaucum* (L.) R. Br genotypes. Curr Plant Biol 21:100131

Jaberzadeh A, Moaveni P, Moghadam HRT, Zahedi H (2013) Influence of bulk and nanoparticles titanium foliar application on some agronomic traits, seed gluten and starch contents of wheat subjected to water deficit stress. Notulae Botanicae Hortic Agrobotanici Cluj-Napoca 41(1):201

Janicka-Russak M, Kabała K (2014) The role of plasma membrane H+-ATPase in salinity stress of plants. Prog Bot:77–92

Karumannil S, Khan TA, Kappachery S, Gururani MA (2023) Impact of exogenous melatonin application on photosynthetic machinery under abiotic stress conditions. Plants 12(16):2948

Karunakaran G, Sudha KG, Ali S, Cho E (2023) Biosynthesis of nanoparticles from various biological sources and its biomedical applications. Molecules 28(11):4527

Keblinski P, Prasher R, Eapen J (2008) Thermal conductance of nanofluids: is the controversy over? J Nanopart Res 10:1089–1097

Khan MN, Mobin M, Abbas ZK, AlMutairi KA, Siddiqui ZH (2016) Role of nanomaterials in plants under challenging environments. Plant Physiol Biochem 110:194–209

Khan MN, Zhang J, Luo T, Liu J, Rizwan M, Fahad S, Xu Z, Hu L (2019) Seed priming with melatonin coping drought stress in rapeseed by regulating reactive oxygen species detoxification: antioxidant defense system, osmotic adjustment, stomatal traits and chloroplast ultrastructure perseveration. Ind Crops Prod 140:111597

Khan MN, AlSolami MA, Basahi RA, Siddiqui MH, Al-Huqail AA, Abbas ZK, Siddiqui ZH, Ali HM, Khan F (2020) Nitric oxide is involved in nano-titanium dioxide-induced activation of antioxidant defense system and accumulation of osmolytes under water-deficit stress in Vicia faba L. Ecotoxicol Environ Saf 190:110152

Khan Y, Sadia H, Shah SZA, Khan MN, Shah AA, Ullah N, Ullah MF, Bibi H, Bafakeeh OT, Khedher NB, Eldin SM, Fadhl BM, Khan MI (2022) Classification, synthetic, and characterization approaches to nanoparticles, and their applications in various fields of nanotechnology: a review. Catalysts 12(11):1386

KhokharVoytas A, Shahbaz M, Maqsood MF, Zulfiqar U, Naz N, Iqbal UZ et al (2023) Genetic modification strategies for enhancing plant resilience to abiotic stresses in the context of climate change. Funct Integr Genom 23(3):283

Kim T, Kim MG, Lee J, Hur H (2018) Biosynthesis of nanomaterials by shewanella species for application in lithium ion batteries. Front Microbiol 9:2817

Kopittke PM, Lombi E, Wang P, Schjoerring JK, Husted S (2019) Nanomaterials as fertilizers for improving plant mineral nutrition and environmental outcomes. Environ Sci Nano 6(12):3513–3524

Kumar JA, Krithiga T, Manigandan S, Sathish S, Renita AA, Prakash P, Prasad BN, Kumar TP, Rajasimman M, Hosseini-Bandegharaei A, Prabu D, Crispin S (2021) A focus to green synthesis of metal/metal based oxide nanoparticles: various mechanisms and applications towards ecological approach. J Clean Prod 324:129198

Kurul F, Turkmen H, Cetin AE, Topkaya SN (2025) Nanomedicine: how nanomaterials are transforming drug delivery, bio-imaging, and diagnosis. Next Nanotechnology 7:100129

Li X, Xu H, Chen ZS, Chen G (2011) Biosynthesis of nanoparticles by microorganisms and their applications. J Nanomater:270974

Li P, Du Y, Huang L, Mitter N, Xu ZP (2016) Nanotechnology promotes the R&D of new-generation micronutrient foliar fertilizers. RSC Adv 6(73):69465–69478

Liu X, Li Q, Wang O, Zhao Q, Zhang Z, Zhou Y (2015) Influence of TiO2 nanoparticles on the bioavailability of nutrients in Chinese cabbage. J Environ Sci 30:153–158

Liu Y, Li Y, Pan B, Zhang X, Zhang H, Steinberg CE, Qiu H, Vijver MG, Peijnenburg WJ (2020) Application of low dosage of copper oxide and zinc oxide nanoparticles boosts bacterial and fungal communities in soil. Sci Total Environ 757:143807

Malcolm E, Sumner RN (1998) Sodic soils-distribution, properties, management, and environmental consequences. Oxford University Press

Manimaran M, Kannabiran K (2017) Actinomycetes-mediated biogenic synthesis of metal and metal oxide nanoparticles: progress and challenges. Lett Appl Microbiol 64(6):401–408

Manzoor N, Ahmed T, Noman M, Shahid M, Nazir MM, Ali L et al (2021) Iron oxide nanoparticles ameliorated the cadmium and salinity stresses in wheat plants, facilitating photosynthetic pigments and restricting cadmium uptake. Sci Total Environ 769:145221

Meier S, Moore F, Morales A, González M-E, Seguel A, Meriño-Gergichevich C, Rubilar O, Cumming J, Aponte H, Alarcón D (2020) Synthesis of calcium borate nanoparticles and its use as a potential foliar fertilizer in lettuce (*Lactuca sativa*) and zucchini (*Cucurbita pepo*). Plant Physiol Biochem 151:673–680

Minhas P, Qadir M, Yadav R (2019) Groundwater irrigation induced soil sodification and response options. Agric Water Manag 215:74–85
Mittal D, Kaur G, Singh P, Yadav K, Ali SA (2020) Nanoparticle-based sustainable agriculture and food science: recent advances and future outlook. Front Nanotechnol 2:579954
Mohamed AKS, Qayyum MF, Abdel-Hadi AM, Rehman RA, Ali S, Rizwan M (2017) Interactive effect of salinity and silver nanoparticles on photosynthetic and biochemical parameters of wheat. Arch Agron Soil Sci 63(12):1736–1747
Mohanavelu A, Naganna SR, Al-Ansari N (2021) Irrigation induced salinity and sodicity hazards on soil and groundwater: an overview of its causes, impacts and mitigation strategies. Agriculture 11(10):983
Mukhopadhyay R, Sarkar B, Jat HS, Sharma PC, Bolan NS (2021) Soil salinity under climate change: challenges for sustainable agriculture and food security. J Environ Manag 280:111736
Najahi-Missaoui W, Arnold RD, Cummings BS (2020) Safe nanoparticles: are we there yet? Int J Mol Sci 22(1):385
Nam NH, Luong NH (2019) Nanoparticles: synthesis and applications. In: Mat Biomed Engineering. Elsevier, pp 211–240
Nascimento MA, Cruz JC, Rodrigues GD, de Oliveira AF, Lopes RP (2018) Synthesis of polymetallic nanoparticles from spent lithium-ion batteries and application in the removal of reactive blue 4 dye. J Clean Prod 202:264–272
Noah N (2019) Green synthesis: characterization and application of silver and gold nanoparticles. Elsevier eBooks, pp 111–135
Otari SV, Patil RM, Ghosh SJ, Thorat ND, Pawar SH (2015) Intracellular synthesis of silver nanoparticle by actinobacteria and its antimicrobial activity, Spectrochimica Acta Part A: Mol Biomol Spectroscopy 136:1175–1180
Pérez-De-Luque A (2017) Interaction of nanomaterials with plants: what do we need for real applications in agriculture? Front Environ Sci 5:12
Qadir M, Noble AD, Schubert S, Thomas RJ, Arslan A (2006) Sodicity-induced land degradation and its sustainable management: problems and prospects. Land Degrad Devel 17(6):661–676
Rad M, Taran M, Alavi M (2018) Effect of incubation time, $CuSO_4$ and glucose concentrations on biosynthesis of copper oxide (CuO) nanoparticles with rectangular shape and antibacterial activity: taguchi method approach. Nano Biomed Eng 10:25–33
Ramakrishna M, Rajesh Babu D, Gengan RM, Chandra S, Nageswara Rao G (2016) Green synthesis of gold nanoparticles using marine algae and evaluation of their catalytic activity. J Nanostructure Chem 6:1–13
Ranjitha VR, Rai VR (2017) Actinomycetes mediated synthesis of gold nanoparticles from the culture supernatant of *Streptomyces griseoruber* with special reference to catalytic activity. 3 Biotech 7(5):299
Rehman A, Khan S, Sun F, Peng Z, Feng K, Wang N, Jia Y, Pan Z, He S, Wang L, Qayyum A, Du X, Li H (2024) Exploring the nano-wonders: unveiling the role of nanoparticles in enhancing salinity and drought tolerance in plants. Front Plant Sci 14
Rengasamy P (2010) Soil processes affecting crop production in salt-affected soils. Funct Plant Biol 37(7):613–620
Richards LA (ed) (1954) Diagnosis and improvement of saline and alkali soils (No. 60). US Government Printing Office
Roopa MC, Thirumala S, Kallimani S, Manohara BM, Basavarajaiah SM (2025) Green synthesis of nanoparticles for enhanced wastewater treatment and other applications: a review and future perspectives. Next Mater 8:100664
Salem NM, Albanna LS, Awwad AM (2016) Green synthesis of sulfur nanoparticles using Punica granatum peels and the effects on the growth of tomato by foliar spray applications. Environ Nanotechnol Monit Manag 6:83–87
Sanchez-Moreno P, Ortega-Vinuesa JL, Peula-Garcia JM, Marchal JA, Boulaiz H (2016) Smart drug-delivery systems for cancer nanotherapy. Curr Drug Targets 19(4):339–359

Saravanan M, Barik SK, MubarakAli D, Prakash P, Pugazhendhi A (2018) Synthesis of silver nanoparticles from *Bacillus brevis* (NCIM 2533) and their antibacterial activity against pathogenic bacteria. Microb Pathog 116:221–226

Saxena R, Tomar RS, Kumar M (2016) Exploring nanobiotechnology to mitigate abiotic stress in crop plants. J Pharma Sci Res 8(9):974

Scherer TF, Seelig B, Franzen D (1996) Soil, water and plant characteristics important to irrigation. NDSu Extension Circular

Seleiman MF, Al-Suhaibani N, Ali N, Akmal M, Alotaibi M, Refay Y, Dindaroglu T, Abdul-Wajid HH, Battaglia ML (2021) Drought stress impacts on plants and different approaches to alleviate its adverse effects. Plants 10(2):259

Semida WM, Abdelkhalik A, Mohamed GF, El-Mageed T a A, El-Mageed S a A, Rady MM, Ali EF (2021) Foliar application of zinc oxide nanoparticles promotes drought stress tolerance in eggplant (*Solanum melongena* L.). Plants 10(2):421

Shah GM, Amin M, Shahid M, Ahmad I, Khalid S, Abbas G et al (2022) Toxicity of ZnO and Fe2O3 nano-agro-chemicals to soil microbial activities, nitrogen utilization, and associated human health risks. Environ Sci Europe 34(1):106

Shaikh A, Jamla M, Joshi S, Patil S, Oak U, Kumar V (2024) Microbial nanotechnology for producing stress smart crops. Plant Nano Biol 7:100063

Sharma D, Kanchi S, Bisetty K (2019) Biogenic synthesis of nanoparticles: a review. Arab J Chem 12:3576–3600

Siddiqi KS, Husen A (2016) Fabrication of metal nanoparticles from fungi and metal salts: scope and application. Nanoscale Res Lett 11:1–15

Siddiqui H, Ahmed KBM, Sami F, Hayat S (2020) Silicon nanoparticles and plants: current knowledge and future perspectives. Sustain Agric Rev 41:129–142

Singh R, Shedbalkar UU, Wadhwani SA, Chopade BA (2015) Bacteriagenic silver nanoparticles: synthesis, mechanism, and applications. Appl Microbiol Biotechnol 99:4579–4593

Sreelakshmy V, Deepa MK, Mridula P (2016) Green synthesis of silver nanoparticles from Glycyrrhiza glabra root extract for the treatment of gastric ulcer. J Dev Drugs 5(2):152

Stone V, Führ M, Feindt PH, Bouwmeester H, Linkov I, Sabella S, Murphy F, Bizer K, Tran L, Ågerstrand M (2018) The essential elements of a risk governance framework for current and future nanotechnologies. Risk Anal 38(7):1321–1331

Sutulienė R, Ragelienė L, Samuolienė G, Brazaitytė A, Urbutis M, Miliauskienė J (2021) The response of antioxidant system of drought-stressed green pea (Pisum sativum L.) affected by watering and foliar spray with silica nanoparticles. Horticulturae 8(1):35

Swaminathan S, Murugesan S, Damodarkumar S, Dhamotharan R, Bhuvaneshwari S (2011) Synthesis and characterization of gold nanoparticles from alga *acanthophora specifera* (VAHL) boergesen. Int J Nanosci Nanotechnol 2:85–94

Taha RS, Seleiman MF, Alotaibi M, Alhammad BA, Rady MM, Mahdi AHA (2020) Exogenous potassium treatments elevate salt tolerance and performances of Glycine max L. by boosting antioxidant defense system under actual saline field conditions. Agron 10(11):1741

Thabet SG, Alqudah AM (2024) Unraveling the role of nanoparticles in improving plant resilience under environmental stress condition. Plant Soil:1–18

Thilakarathne AS, Liu F, Zou Z (2025) Plant signaling hormones and transcription factors: key regulators of plant responses to growth, development, and stress. Plants 14(7):1070

Tighe-Neira R, Carmora E, Recio G, Nunes-Nesi A, Reyes-Diaz M, Alberdi M, Rengel Z, Inostroza-Blancheteau C (2018) Metallic nanoparticles influence the structure and function of the photosynthetic apparatus in plants. Plant Physiol Biochem 130:408–417

Tombuloglu H, Slimani Y, Tombuloglu G, Alshammari T, Almessiere M, Korkmaz AD, Baykal A, Samia ACS (2020) Engineered magnetic nanoparticles enhance chlorophyll content and growth of barley through the induction of photosystem genes. Environ Sci Pollut Res 27(27):34311–34321

Toshima N, Yonezawa T (1998) Bimetallic nanoparticles-novel materials for chemical and physical applications. New J Chem 22(11):1179–1201

Ullah N, Ditta A, Imtiaz M, Li X, Jan AU, Mehmood S et al (2021) Appraisal for organic amendments and plant growth-promoting rhizobacteria to enhance crop productivity under drought stress: a review. J Agron Crop Sc 207(5):783–802

Vasant GH, Raghav R (2024) Co-application of metal oxide nanoparticle (s) and plant growth promoting rhizobacteria on the growth of groundnut plant *(Arachis hypogaea L.)*. Doctoral dissertation, Department of Microbiology Faculty of Science Atmiya University

Wahid I, Kumari S, Ahmad R, Hussain SJ, Alamri S, Siddiqui MH, Khan MIR (2020) Silver nanoparticles regulate salt tolerance in wheat through changes in aba concentration, ion homeostasis, and defense systems. Biomolecules 10(11):1506

Wang D, Shi Y, Samake A, Fu S, Chen S, Yang M (2018) Silver nanoparticles enhance plant growth in a saline environment. J Nanopart Res 20(5):141

Wu Y, Wang X, Zhang L, Zheng Y, Liu X, Zhang Y (2023) The critical role of biochar to mitigate the adverse impacts of drought and salinity stress in plants. Front Plant Sci 14:1163451

Wypij M, Czarnecka J, Świecimska M, Dahm H, Rai M, Golinska P (2018) Synthesis, characterization and evaluation of antimicrobial and cytotoxic activities of biogenic silver nanoparticles synthesized from Streptomyces xinghaiensis OF1 strain. World J Microbiol Biotechnol 34(2):23

Yaghoubi Khanghahi M, Leoni B, Crecchio C (2021) Photosynthetic responses of durum wheat to chemical/microbiological fertilization management under salt and drought stresses. Acta Physiol Planta 43(8):123

Yan G, Nikolic M, Ye M, Xiao Z, Liang Y (2018) Silicon acquisition and accumulation in plant and its significance for agriculture. J Integr Agric 17(10):2138–2150

Yasmeen F, Raja NI, Razzaq A, Komatsu S (2016) Proteomic and physiological analyses of wheat seeds exposed to copper and iron nanoparticles. Biochim Biophys Acta Proteins Proteom 1865(1):28–42

Yu W, France DM, Routbort JL, Choi SU (2008) Review and comparison of nanofluid thermal conductivity and heat transfer enhancements. Heat Transfer Eng 29(5):432–460

Zaib M, Raza I, Zubair M, Arif Z, Mumtaz MM, Abbas MQ et al (2023) Nano-enabled soil amendments for improved soil structure and water holding capacity: an in-depth review. Int Res J Educ Technol 5(08):344–357

Zeng X, Li E, Xia G, Xie N, Shen Z, Moskovits M, Yu R (2021) Silica-based ceramics toward electromagnetic microwave absorption. J Europ Ceramic Soc 41(15):7381–7403

Zhang Z, Zhang J, Xu G, Zhou L, Li Y (2019) Arbuscular mycorrhizal fungi improve the growth and drought tolerance of Zenia insignis seedlings under drought stress. New For 50(4):593–604

Zhang H, Chen J, Liu X, Wang R, Zhang H, Yang Y (2025) Enhancing rice (*Oryza sativa* L.) yield and quality by improving photosynthesis with foliar application of zinc oxide nanoparticles. Environ Sci Nano 12:2331–2342

Zhao L, Sun Y, Hernandez-Viezcas JA, Servin AD, Hong J, Niu G et al (2013) Influence of CeO2 and ZnO nanoparticles on cucumber physiological markers and bioaccumulation of Ce and Zn: a life cycle study. J Agric Food Chem 61(49):11945–11951

Zhao L, Lu L, Wang A, Zhang H, Huang M, Wu H, Xing B, Wang Z, Ji R (2020) Nano-biotechnology in agriculture: use of nanomaterials to promote plant growth and stress tolerance. J Agric Food Chem 68(7):1935–1947

Zhu F, Yan Y, Doyle E, Zhu C, Jin X, Chen Z, Wang C, He H, Zhou D, Gu C (2022) Microplastics altered soil microbiome and nitrogen cycling: the role of phthalate plasticizer. J Hazard Mater 427:127944

Zuccarelli R, Rodríguez-Ruiz M, Lopes-Oliveira PJ, Pascoal GB, Andrade SCS, Furlan CM, Purgatto E, Palma JM, Corpas FJ, Rossi M, Freschi L (2020) Multifaceted roles of nitric oxide in tomato fruit ripening: NO-induced metabolic rewiring and consequences for fruit quality traits. J Exp Bot 72(3):941–958

Chapter 16
Plant Microbiome: Composition and Applications

Rayane Saifi, Hadjer Saifi, Hysen Kokiçi, and Bekri Xhemali

Abstract Plant health is now influenced not just by soil, nutrients, and climate but also by a microscopic ecosystem coexisting with them. The plant microbiome is an essential and unseen partner of plants, primarily located in the rhizospheres, phyllospheres, and endospheres, with a smaller presence in the spermospheres and carpospheres. The plant microbiome serves as a vital living barrier, functioning as an external immune system crucial for phytoprotection processes (interaction, anticipation, and response), occupying the plant niche prior to invaders, releasing antimicrobial bioactive compounds, and initiating internal defenses upon the onset of threats. The plant microbiome functions as a bio-intelligent interface between the host plant and its surrounding ecosystem, which serves as a critical agronomic mediator. This mediator mobilizes essential nutrients, such as nitrogen, iron, and phosphorus from soil and atmospheric sources, while also promoting plant growth through the biosynthesis of phytohormones and enhances plant resilience to pedoclimatic stresses. The plant microbiome has become a key component of scientific research due to its ecological and physiological functions, which have the potential to enhance the sustainability, productivity, and resilience of agricultural ecosystems. Maintaining it is essential for preserving plant health and a strategic investment in food security and agricultural sustainability. This chapter explains the

R. Saifi (✉)
Research Laboratory of Sciences and Environment: Bioresources, Geochemical-Physics, Legislation and Socio-Economic Development, University of Tamanrasset, Tamanrasset, Algeria
e-mail: rayanesaifi@univ-tam.dz

H. Saifi
Biogeochemistry of Desert Environments Laboratory, Scientific and Technical Research Center in Physicochemical Analysis, Department of Biological Sciences, Kasdi Merbah University, Ouargla, Algeria

H. Kokiçi
Institute of Plant Genetic Resources, Agricultural University of Tirana, Tirana, Albania

B. Xhemali
Laboratory of Plant Protection, Kosovo Institute of Agriculture, Pejë, Kosovo

K. A. Abd-Elsalam (ed.), *Plant-Microbiome Nanotechnology*, Sustainability Sciences in Asia and Africa, https://doi.org/10.1007/978-981-95-5056-2_16

concept and diversity of the plant microbiome and its fundamental importance for plant health. It focuses on innovative approaches to altering and improving plant microbiome functions, identifying factors that influence its balance, such as agricultural practices and climatic conditions.

Keywords Plant Health · Soil Microbiome Engineering · Nutrients · Climate Change

16.1 Introduction

Healthy plant generally indicates an autotrophic organism exhibiting optimum physiological performance, defined by effective nutrient absorption, active photosynthesis, sustained growth, successful reproduction, and resistance against disease and pests (Fadiji et al. 2024; Srivastav et al. 2024), but due to the innovation of high-throughput biomolecular sequencing technologies and the rise of bioinformatics and artificial intelligence, the concept of plant health has profoundly evolved, particularly in recent years, abandoning the perspective focused solely on the morphology and physiology of the plant in favor of a more holistic approach, integrating its biotic microenvironment and fundamentally its microbiome by elucidating that obtaining a healthy, fruitful and productive plant does not depend exclusively on the optimization of pedoclimatic factors (soil texture and its organic and inorganic matter profile, temperature, pH, air humidity, wind, rainfall, texture, etc.) and the fight against harmful biotic factors (pathogens, pests, weeds, etc.), but also on the permanent, intimate and complex interaction between the plant and the microbial communities cohabiting its ecosystems (in the vicinity, on or inside plant tissues), refuting the classic reductive version which considers them as unwanted companions, epiphytes, opportunists or strictly pathogens for the plant (Fadiji et al. 2024; Singh and Das 2024).

Scientific evidence has demonstrated that the phytomicrobiome is a critical factor influencing plant health, development, functionality, secondary metabolite synthesis, fruit quality, yield, resilience, and environmental adaptation. This has led to the emergence of the concept of "plant holobiont," which posits that the plant–microbiome partnership constitutes a complex, symbiotic, and inseparable biounion, essentially forming an integrated, functional, resistant, resilient, and interdependent superorganism, particularly at the trophic and immune levels. The vitality of this phytomicrobial biological network fundamentally relies on the equilibrium and stability of its many symbiotic connections (Compant et al. 2025). Joint progress in phytobiology, phytopathology, eco-microbiology and eco-agronomy has highlighted that the plant microbiome is involved in several key and decisive biofunctions associated with plant survival, health and adaptation, such as the improvement of nutrient availability and root absorption (solubilization of phosphorus, atmospheric nitrogen fixation, desorption of clay potassium, sequestration of edaphic iron, etc.); the biosynthesis and functional modulation of certain phytohormones (auxin, cytokinin, gibberlin, abscisic acids, etc.) which behave as external regulators stimulating aerial growth and proliferation and root elongation, thus leading to increased agricultural yield; resilience to biotic disturbances and strengthening of plant immune defense via the secretion of antimicrobial metabolites (antibiotics, siderophores, volatile toxic substances, hydrolytic enzymes, biosurfactant lipopeptides, bacteriocins, organic acids, etc.), competitive expansion in the root, foliar, floral and cauline niche to hinder the establishment of pathogenic strains, in addition the phytomicrobiome is the pivotal actor in the development of induced and acquired plant systemic resistance by behaving like an ecological vaccine; It also contributes to increasing the physiological tolerance of the plant to withstand abiotic stresses (salinity, acidity, extreme temperatures, heavy metal pollution, excess or deficit water, excessive UV radiation, etc.) through the modulation of the genetic expression of specific stress genes involved (Ge and Wang 2025; Compant et al. 2025; Zeng et al. 2025; Vincze et al. 2024).

Despite the remarkable and beneficial metabolic and agroecological power of the plant microbiome, a physiological imbalance of the holobiont can stimulate the appearance of deleterious behavior in certain microorganisms by becoming a significant stress factor for plants. This dysbiosis is often associated with a high risk of phytopathology (crown gall, angular leaf spot of cotton, bacterial wilt of tomato, etc.), nutritional deficiencies, and low agricultural productivity. In this context, plants often seek to control the dimensions, taxonomic diversity, stability, and functionality of their microbiomes via molecular signals and specific volatile compounds (Guo et al. 2025; Wang et al. 2024). Managing the dynamic interactions between the plant and its microbiome, which vary from mutualism to parasitism depending on ecological parameters, is a strategic pillar to improve phytohomeostasis related to food security, climate change, and the agroecological transition (Du et al. 2024; Saifi et al. 2024). Advancements in genomic, metabolomic, proteomic, transcriptomic, and genetic technologies have facilitated the development of new and sustainable approaches to enhance plant health and production. Introducing genetically modified microbial strains into soil is a viable method to promote plant development and achieve microbiome equilibrium. This method mitigates the threat of pathogen intrusion while diminishing dependence on traditional agrochemicals, recognized for their environmental longevity and toxicity (Noman et al. 2024; Fadiji et al. 2024).

This chapter seeks to elucidate current scientific advancements in the comprehension of plant–microbiome interactions by examining the origins of the plant microbiome, its ecological niches, and the characteristics of its relationships with the host plant over its life cycle. It underscores the crucial function of the microbiome in sustaining plant health and bolstering its immunity. This chapter analyzes the determinants of homeostasis, evaluates contemporary methodological techniques in microbiome research, and explores novel functional modification methods and advancements in plant microbiome engineering.

16.2 The Plant Microbiome

The interaction between the plant kingdom and microorganisms remained largely overlooked until 1888, when Dutch microbiologist Martinus Willem Beijerinck successfully isolated the first microbial germ (Rhizobium) from the plant microbiome (Mukherjee et al. 2024). This microorganism facilitates the fixation of atmospheric nitrogen, benefiting plants. This finding sparked extensive research on the subject, which continues to evolve to this day. The plant microbiome comprises a huge, intricate, and dynamic assemblage of bacterial, protistic, fungal, and archaeal communities, together with a virome that regulates their biological functions inside the host plant. The latter has this microbiome alone due to its inherent connection to sustaining metabolic homeostasis, immunity, and overall health (Mukherjee et al. 2024; Afridi et al. 2022).

The advancement of omics techniques has shown that the preferred plant habitat of microbial communities of the phytomicrobiome is located mainly at the level of rhizospheres (soil surrounding the root system), phyllospheres (leaf surface), endospheres (internal plant tissues), and to a lesser extent at the level of spermospheres (seed tissues) and carpospheres (fruit tissues); and that the plant host does not recruit its microbial partners randomly, on the contrary the structure, dimensions, diversity, trophic network and interaction mechanisms (ranging from mutualism to parasitism) of the plant microbiome are designated with unprecedented precision through a universal chemomolecular dialogue which involves the genotype and phenotype of the plant and microbial germs, the chemical profile of root exudates and volatile compounds ejected by plant organs (contents of amino acids, organic acids, sugars, vitamins, polyphenols, trace elements, etc.), intermicrobial interactions, cultural practices (types of fertilizers, pesticides, plowing, crop rotation, etc.), and climatic and edaphic conditions (Compant et al. 2025; Afridi et al. 2022; Santoyo 2022).

Due to its advanced metabolic capabilities and significant ecological functions, the plant microbiome serves as a vital physiological ally to the host plant, akin to its intrinsic systems. It acts as a functional extension of the plant genome (a second plant genome), contributing to bioregulation in development, environmental adaptation, and overall well-being, thereby designating this superorganism as a "holobiont" (Matthews 2024).

Plant architecture provides various passive entry points for microorganisms that do not require ATP, including stomata, lenticels, hydathodes, and root hairs. Tropism, physiological traits, adaptation mechanisms, and the invasion strategies of microbial pathogens influence the selection of these openings. Plant tissues subjected to mechanical trauma (from agricultural implements, climatic conditions, insect infestations, etc.) and germination, senescence, or growth regions demonstrate a notable susceptibility to pathogenic and opportunistic microorganisms (Berg et al. 2021). Particular microbial species (*Fusarium oxysporum*, *Magnaporthe oryzae*, *Erwinia amylovora*, *Agrobacterium tumefaciens*, etc.) effectively penetrate plant host tissues by secreting toxins and a range of hydrolytic enzymes, such as cellulase, hemicellulase, protease, and pectinase, to depolymerize plant cell walls, thereby promoting either beneficial or detrimental microbial advancement. Conversely, several species compromise and dismantle plant cell walls by inducing turgor pressure via the development of an appressorium, a specialized structure for adherence and penetration, to traverse the cuticle and epidermis of plants (Mukherjee et al. 2024; Berg et al. 2021).

The plant microbiome is a critical factor influencing plant health and productivity. It operates via interconnected biochemical networks to facilitate the coevolution of the plant–microbiome partnership by modulating several growth and resilience characteristics (Mukherjee et al. 2024).

The terms microbiota and microbiome must not be conflated. The former encompasses all microorganisms residing in a certain ecological niche. At the same time, the latter comprises not only microbial entities (the microbiota) but also their genetic composition, interactions among themselves and with their host, and their influence on the environment (Choi et al. 2021).

16.3 Origin of the Plant Microbiome

Plant niches acquire their microbial flora through two primary, different, and complementary processes. The first mechanism, horizontal transmission, enables the plant to acquire nearly its entire microbiome through dynamic interactions with its biotic and abiotic surroundings. In contrast, the second mechanism, active vertical transmission, guarantees partial retention of the microbiome across generations of plants (Tanger et al. 2024).

16.3.1 Microbiome of Horizontal Origin (Exogenous, Environmental)

The horizontal microbial associations of the plant microbiome are created on the plant's surface due to its interaction with the biocenosis or biotope of its surrounding environment. This process, which stays active throughout the plant's life and even after it dies, depends on chemical and physical factors, differences in conditions, and environmental challenges.

First, the edaphic environment is unequivocally the primary reservoir of both beneficial (*Rhizobium* spp., *Bacillus subtilis*, *Azospirillum brasilense*, etc.) and pathogenic (*Agrobacterium tumefaciens*, *Ralstonia solanacearum*, *Pectobacterium carotovorum*, etc.) microorganisms that significantly influence the phytosanitary conditions of crops. Plants take in these microorganisms by simply touching them or actively recognizing them and attracting them with substances released from their roots and leaves (chemotaxis) (Fig. 16.1) (Da Costa et al. 2022).

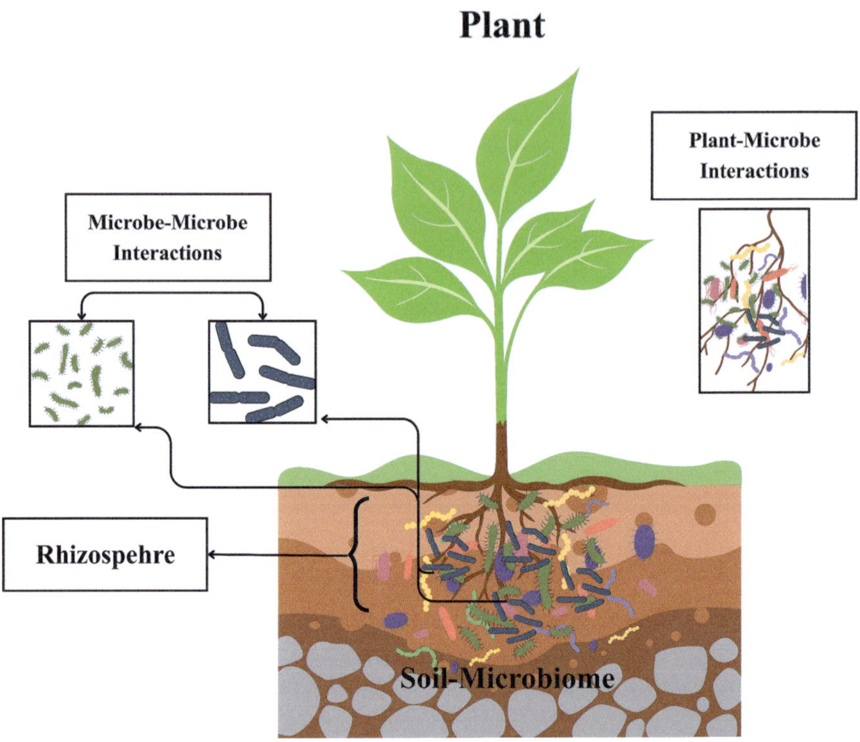

Fig. 16.1 Soil as the primary reservoir of plant-interacting microorganisms (Tharanath et al. 2024)

Second, air functions as a dynamic reservoir and a conduit for microbial transfer among diverse ecological compartments (soil, fauna, flora, anthropogenic surfaces, etc.), yet it is frequently undervalued despite its crucial role in the selection of strains that exhibit tolerance to atmospheric stress and their enduring or transient integration into the plant microbiome, particularly within the aerial regions of the host plant (Phyllosphere) (Compant et al. 2019). The initial colonization and sustained presence of novel microbial taxa (*Pseudomonas* spp., *Escherichia coli*, *Aspergillus* spp., *Candida* spp., *Penicillium* spp., *Cryptococcus* spp., etc.) in plant biomass are contingent upon their adaptive mechanisms and their mutualistic or competitive interactions with the established microbial flora. Airborne microbial pathogens influence the functionality and ecological resilience of holobionts, leading to plant degradation or physiological enhancement through their capacity to produce catalytic enzymes and protective substances, including exopolysaccharides, pigments, and biofilms (Sohrabi et al. 2023). During passive dispersion, air currents conduct abiotic filtering of free bioaerosols or those linked with dust, pollen, water droplets, soil particles, or compost. The dissemination of air pollution in urban and agricultural regions jeopardizes this process by introducing taxa that may harm plant and human health (Sohrabi et al. 2023).

Moreover, the microorganisms associated with plant tissues play a crucial role in the taxonomic composition of the phytomicrobiome, being disseminated to plant organs via dewwater, wastewater, precipitation, runoff, irrigation, mist, groundwater, and spring water. Increased hydration of the plant environment enhances microbial movement and amplifies their impact on the plant microbiome (Singh and Das 2024; Tkacz and Poole 2020).

Furthermore, animals, far from being classified as negative or positive vectors, are intensely involved in the establishment of the plant microbiome through their direct contact with the aerial and underground structures of the plant body or indirectly through fecal droppings and the excretion of biological fluids (saliva, mucus, sweat, urine, etc.) impregnated with microorganisms that positively or negatively affect the stability of the plant microbiome and, consequently, the health status of the plant and its development process (Hashimi et al. 2023; Sánchez-Cañizares et al. 2017).

Additionally, plants can significantly influence the microbial landscape of neighboring plants, as the tangle of their aerial parts generates connectivity corridors conducive to biotic and abiotic exchanges. Meanwhile, the physical proximity and interconnectivity between their root systems, rich in nutrient exudates, promotes a common mycorrhizal network that facilitates the mobilization and spread of a vast beneficial or even pathogenic microbial diversity. Furthermore, crop residues and plant debris act as a real source of living microorganisms that can affect emerging plants, improving or altering the quality of their microbiome (Hashimi et al. 2023; Compant et al. 2019; Sánchez-Cañizares et al. 2017).

Finally, human-derived microorganisms can disrupt the harmony of the plant microbiome, particularly when adapted to the new microenvironment. Humans are often seen as an ecological relationship facilitating inter-vegetative microbial dissemination, resulting in a new diversity and beneficial or detrimental microenvironmental plasticities. The manipulation of agricultural cultures by the toucher is the primary factor promoting the infection of microbes from cutane, buccal, and intestinal flowers. Other factors include irrigation, fertilization, outrage, riffage, land exchange, and other agricultural practices. Under strictly regulated and supervised scientific conditions, researchers can intentionally inject genetically modified, human-derived microorganisms (such as Escherichia coli or other species) into host plant tissues. This technique serves several biotechnological interests, including the transfer of biochemical reactions, the insertion of exogenous genes into plant cell genomes, and the investigation of complex host-microbiome relationships (Hashimi et al. 2023; Berg and Cernava 2022; Compant et al. 2019).

16.3.2 Microbiome of Vertical Origin (Endogenous, Hereditary)

They are exclusively beneficial and symbiotic microorganisms inherited directly from the mother plant by its offspring, ensuring their existence in subsequent generations via the seeds and reproductive tissues that shelter them during fertilization and plant proliferation (Bernabeu et al. 2024).

Microbiome transmission between plant generations can occur as early as the seed stage, serving as a vital biological conduit for transferring endogenous microorganisms to new seedlings during germination, thereby enhancing and regulating their initial growth and fortifying their ecosystem interactions. At the seed level, a rich diversity of microbial organisms (*Pseudomonas*, *Bacillus*, *Neotyphodium*, etc.) derived from various plant compartments (ovary, pollen, ovules, flowers, vascular tissues, etc.) predominantly inhabit embryonic tissues, with a lesser affinity for seed coats and internal tissues (Abdelfattah et al. 2022; Murphy et al. 2023). In addition, plant reproductive structures, including ovules, pollen, and flowers, serve as critical entry points for specific microbial pathogens, enabling their spread within the plant, particularly during fertilization, which facilitates hereditary transmission across generations (Abdelfattah et al. 2022; Murphy et al. 2023).

16.4 Components of the Plant Microbiome

The plant microbiome is a dynamic network of microscopic and interdependent ecological communities that work in synergy to establish neutral, pathogenic, or mutually beneficial cooperations with the host plant (Tkacz and Poole 2020). These microbiotas have a specific, hierarchical, complex, and diverse organization by occupying specific plant niches depending on their species and nutritional needs, the plant variety, its developmental stage, and pedoclimatic conditions. Depending on their location in the host plant biomass, three main actors of the plant microbiome are distinguished (Fig. 16.2) (Xiong et al. 2023; Santoyo 2022; Tkacz and Poole 2020).

16.4.1 Rhizospheric Microbiome

The term rhizosphere is a scientific neologism derived from two Greek roots: rhiza, meaning root, and shaira, meaning sphere. This word denotes the soil region directly around plant roots, which has remarkable microbial diversity. It is defined by a highly dynamic physicochemical gradient arising from root exudations and intensive metabolic exchanges. This essential microecosystem serves as a significant locus of coevolution and vigorous biotic and abiotic interactions among the plant root system, soil-dwelling microbial organisms, and organic and inorganic substances within the soil environment (Dubey and Sharma 2021). In contrast to the adjacent bulk soil, the rhizosphere exhibits a higher density of diverse specific and functional microorganisms drawn by the array of organic compounds (carbohydrates, polyphenols, organic acids, proteins, enzymes, amino acids, etc.) exuded by the host plant's roots, which modifies the pH, nutrient accessibility, and other physicochemical parameters of this localized plant environment. The hosted microbial species are predominantly symbiotic, enhancing plant nutrition by mobilizing and

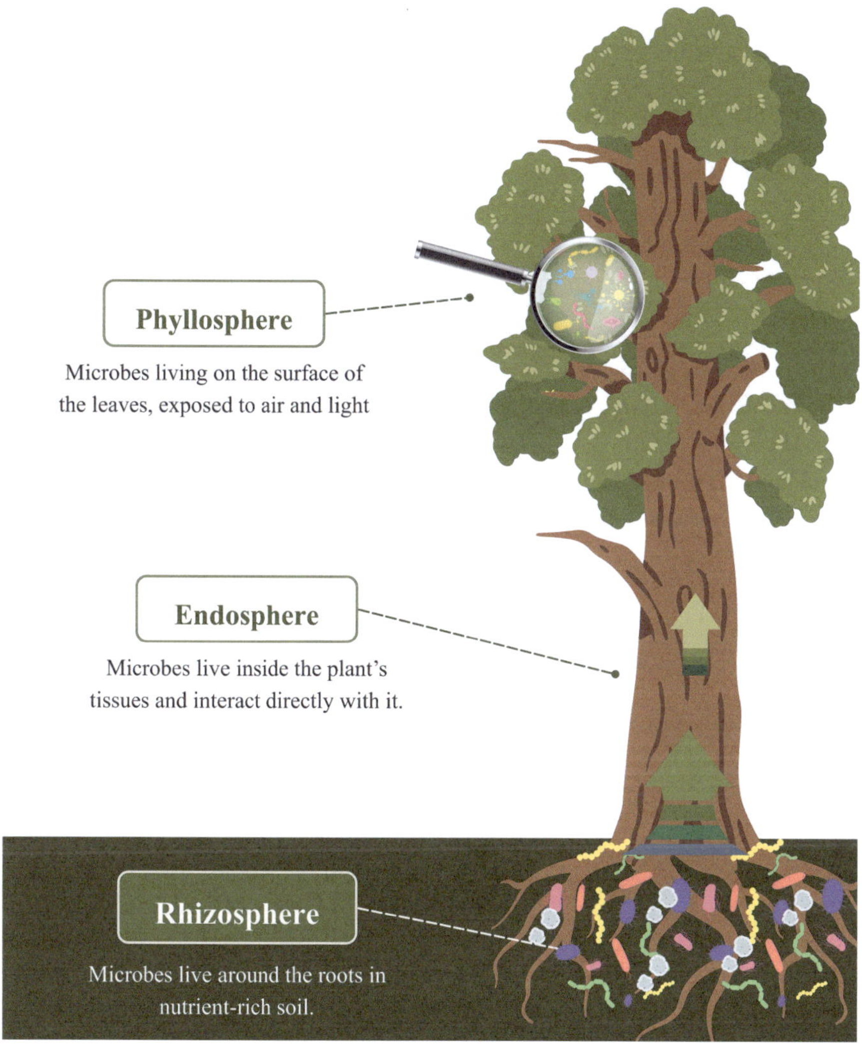

Fig. 16.2 Spatial organization of the plant microbiome (Xiong et al. 2023)

solubilizing organic and inorganic matter and fortifying plant immune and antioxidant defenses against biotic threats and environmental stressors. They function as a biological barrier through competition for space and nutrients, secretion of antibiotics, lytic enzymes, allelopathic compounds, and phytohormonal modulation. The rhizospheric microflora may include neutral or opportunistic pathogenic species (Dubey and Sharma 2021; Xun et al. 2021). The rhizosphere comprises three functional compartments. The rhizoplane, endosphere, and external rhizosphere are critical contacts between the root epidermis and soil particles, facilitating the exchange of water and nutrients. Rhizoplanes represent the direct contact between the root epidermis and soil particles, facilitating water and nutrient exchanges, as well as direct communication among the soil, microbes, and plants. The endosphere is a close intra-root habitat of the endophytic bacteria that integrates with the plant structure, creating nodules without inducing adverse effects on the host plant. It establishes delicate commensal or symbiotic relationships to enhance nutritional dynamics and bioprotection, functioning as a biological barrier against microbial infections, particularly those originating from soil. The exterior rhizosphere encircles the roots, forming a close and intricate microhabitat for many microorganisms. Subsurface emissions, including bioactive and organic chemicals, primarily influence biological and physicochemical features, which are crucial for microbial performance and biogeochemical nutrient cycling in this dynamic interface (Xun et al. 2021, 2024).

16.4.2 Endospheric Microbiome

The term endosphere, originating from the Greek endo, meaning inside, and sphaira, meaning sphere, refers to the internal ecological niche of the microbiota associated with plants, specifically within the stem, leaf, root, and reproductive tissues, facilitating coevolution through non-pathogenic symbiotic interactions (Joshi et al. 2025; Aziz et al. 2021). The diverse topography of the endosphere (xylem, cortex, phloem, mesophyll) and its gradients in pH, oxygen, water, nutrients, and phytohormones significantly affect the taxonomic composition of endophytic microorganisms (chemotaxis), predominantly comprising bacterial phyla such as *Pseudomonas fluorescens, Rhizobium leguminosarum, Micromonospora* spp., *Paenibacillus polymyxa, Flavobacterium* spp., and *Anabaena* spp., as well as filamentous fungi including *Fusarium oxysporum*, *Cladosporium* spp., *Serendipita indica*, *Rhizophagus irregularis*, *Pestalotiopsis* spp., and *Alternaria alternata*. These microorganisms infiltrate host plant tissues either passively through root hairs, germination sites, stomata, lenticels, and natural or anthropogenic wounds or actively by utilizing their degrading enzymes (proteases, cellulases, polygalacturonases, pectinases, etc.) that can disrupt plant cell walls (Guo et al. 2025; Joshi et al. 2025).

After traversing the plant barriers, endophytes subtly establish themselves, mainly at the apoplast level and, in rare instances, at the symplast level, forming biofilms via their effectors (types III/IV) to enhance the plant's immunological protection by circumventing a hypersensitive reaction. Conversely, the endophytic microbiome functions as a biofertilizer that influences the development of the plant phenotype and enhances its photosynthetic efficiency by supplying vital nutrients and phytohormones. It functions as a biocontroller in phytoremediation by combating and preventing infections while also serving as a potent antioxidant by sequestering reactive oxygen species via the production of glutathione and metallothioneins (Goodwin 2022; Compant et al. 2020). Endophytic microorganisms serve as the primary defense for the holobiont against major environmental stresses. They achieve this by modulating osmotic processes (for salinity) and inducing stomatal closure via ABA-like metabolites (for drought), while also protecting against pollution by degrading and neutralizing harmful substances. These natural probiotics can be horizontally transferred between plant generations or via seeds and are presently being deliberately transplanted and domesticated to enhance agricultural sustainability and precision (Guo et al. 2025; Joshi et al. 2025; Goodwin 2022; Compant et al. 2020).

16.4.3 Phyllospheric Microbiome

The phyllosphere, derived from the Greek words phyllo, meaning leaf, and shaira, meaning sphere, refers to the aerial biomass of plants, including all exophytic components, particularly the leaves that are directly exposed to the environment and its continual weather variations (Jiang et al. 2025). The phyllosphere, characterized by its distinct compartments (veins, cuticles, and stomata), represents a critically significant microbial habitat. However, its temporal stability is markedly diminished by phytooxidation, leaf renewal, the interplay between the photochemical phase and the Calvin cycle, precipitation, and temperature fluctuations. These elements catalyze continuous alterations in its microenvironment through cuticular exudations, gas exchanges, and water cycles associated with transpiration, creating a dynamic arena for intricate, multi-faceted interactions in matter and energy between the host plant, microorganisms, and among the microorganisms themselves (Sohrabi et al. 2023; Aziz et al. 2021). The floral microbiota typically consists of a predominant microbial species that stabilizes the trophic pyramid (keystone taxa), auxiliary species that execute specific functions (foliar detoxification, hormonal stimulation of plant growth, mineral provision, and adaptation to environmental stress), and foliar phages essential for the lysogenic and lytic cycles of microorganisms to regulate microbial biomass density and functional resilience (Zhan et al. 2022; Zhu et al. 2021; Aziz et al. 2021). Aerial microorganisms (*Pseudomonas syringae*, *Sphingomonas* spp., *Pantoea agglomerans*, *Epicoccum nigrum*, *Aureobasidium pullulans*, *Sporobolomyces* spp., etc.) inhabiting the interface between the atmosphere and the biosphere exhibit significant ecological resilience due to their adaptable phenotypes, rapid growth and dispersal rates, complex and versatile metabolic systems, and low nutritional requirements, as they derive energy from the catabolism of simple sugars, mannitol, and other aromatic compounds. Furthermore, these microbes possess photoprotective mechanisms, redox sensors, quorum sensing systems, and T3SS and T6SS secretion nanomachines. This sophisticated architecture enables the phyllosphere microbiome to produce hydrophilic exopolysaccharides, lytic enzymes (such as monooxygenase and dioxygenase), and secondary metabolites (including leaf biofilms, photoprotective pigments, and various immune defense effectors) to combat pathogens through mechanisms such as competition, antibiosis, conditioned mutualism, parasitism, or by enhancing induced or acquired plant systemic resistance (Jiang et al. 2025; Sohrabi et al. 2023; Zhan et al. 2022; Zhu et al. 2021).

16.4.4 Carpospheric Microbiome

The term carposphere is derived from Greek, with karpo signifying fruit and "sph-aira" denoting sphere. This delineates a biological interface between the epicarp of the fruit and its associated microflora, collectively constituting a distinct microecological ecosystem that influences the health, organoleptic properties, and economic outcomes of the fruit via a series of significant physicochemical and biological reactions (Aziz et al. 2021). Fruit exudate, abundant in nutrients and signaling molecules, attracts a diverse array of microorganisms (e.g., *Pseudomonas fluorescens*, *Xanthomonas campestris*, *Erwinia carotovora*, *Botrytis cinerea*, *Aspergillus niger*, *Penicillium expansum,* and *Saccharomyces cerevisiae*) that may be either pathogenic or beneficial. These microorganisms can stimulate the expression of genes related to ripening, senescence, and responses to biotic and abiotic stress, serve a sentinel role in fermentation, biopreservation, or biosuppression against biological changes, or exhibit opportunistic behavior that removes dead tissue from fruits (Aziz et al. 2021; Zhang et al. 2021). The taxonomic profile of the fruit microbiome encapsulates the biological imprint of the fruit, documenting its type, ripening stage, geographic origin, agricultural practices, stress factors, postharvest conditions, and microbial assemblage throughout its agri-food traceability and postharvest potential. The fruit skin serves as a vital interface between the fruit and its environment, embodying many metabolic, functional, dynamic, and coevolutionary interactions with its microbial partners (Zhang et al. 2021).

16.4.5 Spermosphere Microbiome

The spermosphere is a scientific neologism derived from two Greek roots: sperma, meaning seed, and sphaira, meaning sphere. It describes a transient, reactive, and declining ecosystem that arises within minutes of seed imbibition. The primary and secondary metabolites present in the seminal exudate, which constitutes the seed's identity code, initiate the initial molecular dialogue unique to each plant genotype. The spermosphere facilitates interaction between the emerging plant and the soil's microbial community (Aziz et al. 2021).

The thickness of this microbiological and biochemical halo, limited to a few millimeters around a sprouting seed, constitutes a genuine molecular battleground for selecting a future microbiome that is both productive and protective of the plant. According to detection, microbial pathogens respond to the chemical signals sent by seeds as either an opportunity or a danger. Beneficial microbes, including plant-growth-promoting Rhizobacteria, congregate, collaborate, colonize, produce protective biofilm structures, and establish symbiotic relationships. Conversely, pathogenic bacteria (*Xanthomonas campestris, Colletotrichum* spp., *Fusarium* spp., etc.) see this as a vulnerability to be exploited, so influencing the health and capacity of dormant organisms to transition from a protective state into an "organic memory" plant. The function of the spermasphere diminishes as the embryonic root system and rhizosphere evolve (transition from microbiota to rhizosphere) (Aziz et al. 2021; Ge and Wei 2021). The convergence of pedomicrobiology and phytomicrobial ecology suggests that the spermasphere is the origin of the future plant holobiont, the initial interface between seeds and soil microbes that initiates coevolution. In this regard, seed coating with plant probiotics has emerged as an effective and promising biocontrol strategy for sustainable plant health preservation (Ge and Wei 2021).

16.5 Function and Mechanisms of Interaction Between the Plant and Its Microbiome

The plant microbiome influences the ecophysiology of the host plant through several mechanisms (Matthews 2024).

16.5.1 Plant Growth Promotion and Modulation of Plant Metabolism

16.5.1.1 Supplying Plants with Mineral and Organic Nutrients

The plant microbiome serves as a dynamic collaborator in soil fertilization. It functions as a quasi-symbiotic system that facilitates nutrient supply to plants through the decomposition of organic matter (lignin, proteins, cellulose, etc.) and the conversion of mineral species into forms readily absorbable by plants (nitrates, selenites, metal ions, etc.) utilizing intra- and extracellular enzymes. The phytomicrobiome is not merely an auxiliary; it is a crucial biochemical engineer that regulates the synthesis, concentration, and transport of nutrients to the plant in return for the hydrocarbons secreted by the plant for microbial nutrition (Table 16.1) (Matthews 2024; Noman et al. 2024).

16.5.1.1.1 Nitrogen

Despite its atmospheric prevalence (78%), plants cannot interact with the inert gaseous form of nitrogen; thus, each nitrogen atom within plant biomass signifies a series of specific, energetically demanding metabolic reactions conducted by diazotrophic microorganisms associated with the plant rhizosphere, which are vital for plant survival (Timofeeva et al. 2023).

Root nodule bacteria and those loosely linked with the rhizosphere, including *Rhizobium*, *Sinorhizobium*, *Bradyrhizobium*, *Azotobacter*, *Cylindrospermum*, and *Beijerinckia*, facilitate the transformation of atmospheric nitrogen into ammonium

Table 16.1 Nutrient–microbiome interactions and mechanisms of action

Nutrient	Microbiome species	Mechanism of action	References
Nitrogen	*Rhizobium*, *Sinorhizobium*, *Bradyrhizobium*, *Azotobacter*, *Cylindrospermum*, *Beijerinckia*	Conversion of atmospheric nitrogen into ammonium (NH^{4+}) via nitrogenase enzymes. Ammonium is assimilated by roots or converted into nitrite and then nitrate (NO^{3-}) by other bacteria.	Matthews (2024), Timofeeva et al. (2023), Singh and Das (2024)
Phosphorus	*Bacillus* spp., *Rhizobium* spp., *Klebsiella* spp., *Trichoderma* spp., *Rhizopus* spp., *Micromonospora* spp.	Release of phosphorus from mineral complexes by producing organic acids (citric acid, gluconic acid, oxalic acid) that lower pH and mobilize phosphorus towards the roots.	Timofeeva et al. (2023), Janati et al. (2021), Rawat et al. (2021)
Organic Phosphorus	*Burkholderia phytofirmans*, *Pseudomonas fluorescens*, *Pseudomonas putida*, *Serratia marcescens*, *Piriformospora indica*	Hydrolysis of organic phosphorus compounds (phytates, phospholipids) into orthophosphates by phosphatases and phytases produced by rhizospheric microbiota.	Singh and Das (2024), Timofeeva et al. (2023)
Iron	*Azospirillum brasilense*, *Bacillus subtilis*, *Rhizobium leguminosarum*, *Pseudomonas fluorescens*, *Burkholderia cepacia*	Acidification of the rhizosphere through secretion of organic acids (malic acid, citric acid, oxalic acid) to dissolve ferric (Fe^{+3}) complexes, reducing them to Fe^{+2}, a bioavailable form for plants.	Timofeeva et al. (2023), Rafique et al. (2022), Hashimi et al. (2023)
Other trace elements	*Shewanella*, *Gobacter*, *Bacillus*, *Pseudomonas* spp.	Acidic dissociation of insoluble metal compounds (zinc, copper) by bacteria, followed by their transport to the roots through fungal filaments, enabling plant absorption.	Hashimi et al. (2023), Timofeeva et al. (2023), Rafique et al. (2022)

(NH^{4+}) via nitrogenase metalloenzymes. The resultant ammonium may be directly assimilated by roots or converted into other nitrogenous compounds, such as nitrite (NO^{2-}), by the action of ammonium monooxygenase and hydroxylamine oxidoreductase enzymes produced by Nitromonas and other nitrifying bacteria. Nitrite may be converted to nitrate (NO^{3-}) by nitrite oxidoreductase produced by Nitrobacter. Nitrate is the most mobile and bioavailable nitrogen compound for plants, which may subsequently convert it to the ammonium form (Singh and Das 2024; Timofeeva et al. 2023).

16.5.1.1.2 Phosphorus

In terrestrial substrate, phosphorus is predominantly stored as poorly soluble mineral complexes, including iron phosphate, aluminum phosphate, and calcium phosphate, which constitute valuable yet inaccessible reserves for plant phosphate nutrition (Timofeeva et al. 2023; Compant et al. 2019). Within rhizospheric microbial communities, we identify phosphorus solubilizers (*Bacillus* spp., *Rhizobium* spp., *Klebsiella* spp., *Trichoderma* spp., *Rhizopus* spp., *Micromonospora* spp., etc.) that enhance plant bioavailability of phosphorus by releasing organic acids such as citric acid, gluconic acid, and oxalic acid. These acids lower the medium's pH by chelating calcium, ferric, and aluminum cations, thereby disrupting their bonds with phosphorus, facilitating its release and mobilization towards root endophytes (*Burkholderia phytofirmans*, *Pseudomonas fluorescens*, *Pseudomonas putida*, *Serratia marcescens*, *Piriformospora indica*, etc.) in exchange for nutrients from root exudates (Janati et al. 2021; Rawat et al. 2021).

Similar to mineral phosphorus, phosphorus within organic matter (such as phytates, phospholipids, and nucleic acids) cannot be directly assimilated by plants without first being liberated as orthophosphate anions (H2PO4- and HPO4-2) through the hydrolytic activity of phosphatases and phytases, primarily generated by the microbiota in the plant rhizosphere (Singh and Das 2024; Timofeeva et al. 2023). The symbiotic association between roots and endomycorrhizal fungi, including *Glomus* and *Rhizophagus,* is crucial for plant phosphorus acquisition, as they form extensive hyphal networks that facilitate the transport of this micronutrient by drawing it from soil regions beyond the host plant's root zone (Janati et al. 2021; Rawat et al. 2021).

16.5.1.1.3 Iron and Other Trace Elements

Specific rhizospheric bacteria and mycorrhizal fungi establish close symbiotic relationships with plant roots to enhance plant health and ecological resilience by providing essential trace elements. These microorganisms function as a potent catalyst for the chemical transformation of minerals that are not readily assimilable by plants into highly bioavailable chemical species, such as iron. The microbial agents elevate the local acidity of the rhizosphere through the secretion of various organic acids, including malic acid, citric acid, and oxalic acid, which destabilize chemically inert ferric complexes (Fe^{+3}) precipitated in the soil, reducing them to divalent iron (Fe^{+2}) capable of permeating plant cell membranes (Timofeeva et al. 2023; Rafique et al. 2022). Various iron-solubilizing bacteria, including *Azospirillum brasilense*, *Bacillus subtilis*, *Rhizobium leguminosarum*, *Pseudomonas fluorescens*, and *Burkholderia cepacia*, can extract biologically inactive ferric iron from the Earth's crust via biochelation with small molecular siderophores, which facilitate diffusion into the plant's edaphic microenvironment. The iron-siderophore molecular conjugation exhibits a high affinity for root cell membrane receptors. Conversely, certain bacteria, such as *Shewanella* and *Gobacter*, play a crucial role in regulating the iron cycle in soil by enhancing the bioavailability of divalent iron through the direct enzymatic reduction of trivalent iron, benefiting plants (Hashimi et al. 2023; Rafique et al. 2022). Rhizopheric microbial organisms are essential in supplying plants with zinc, sulfur, copper, and other trace elements through bacterio–fungal interactions. This process involves two phases: initially, acidifying bacteria facilitate the acidic dissociation of insoluble metallic compounds in the soil, followed by the complexation of these compounds with organic acids, rendering them bioactive and enabling their capture and transport by fungal filaments to the roots of the host plant (Hashimi et al. 2023; Timofeeva et al. 2023; Rafique et al. 2022).

16.5.2 Production of Phytohormones

The plant microbiome, especially the endophyte microbiome, has a complex biochemical repertoire that may biosynthesize endogenous phytohormones or their functional equivalents. These are emitted as secondary metabolites to assist in regulating the physiological dynamics of the host plant (Del Carmen Orozco-Mosqueda et al. 2023).

Auxins, specifically indole-3-acetic acid, are notable phytohormones secreted by various bacterial taxa, including *Pseudomonas* spp., *Rhizobium* spp., *Enterobacter* spp., *Serratia marcescens, Streptomyces* spp., *Frankia* spp., and *Bacillus subtilis*. They facilitate shoot cell elongation and stimulate lateral rhizogenesis through profound sculpting (Oyedele et al. 2024). Zeatin (6-(4-hydroxy-3-methylbut-2--enylamino)purine) is a plant hormone classified within the cytokinin family, mostly generated by *Trichoderma* spp., *Azospirillum brasilense, Pseudomonas fluorescens, Bacillus subtilis,* and *Rhizobium* spp., among others. It functions as a genuine morphogenetic enhancer that fosters the extended vitality of plant organs by stimulating cellular regeneration and meristematic expansion, facilitating bud neoformation, postponing leaf senescence, and bolstering metabolic processes, particularly within the plant's aerial structures (Del Carmen Orozco-Mosqueda et al. 2023; Compant et al. 2020). Gibberellins are signaling molecules belonging to the diterpene family, encompassing approximately 130 phytohormones with diverse chemical structures, primarily synthesized by *Gibberella fujikuroi, Penicillium* spp., *Fusarium fujikuroi, Sphaceloma manihoticola, Acetobacter diazotrophicus, Azospirillum* spp., and *Rhizobium* spp. These plant hormones are specifically involved in surmounting physiological barriers associated with seed dormancy by promoting seed germination, subsequent rapid stem elongation, and the regulation of phenological transitions, including the shift to the floral phase (Oyedele et al. 2024; Hashimi et al. 2023). Certain species within the plant microbiome affect the hormonal function of plants by modulating the concentration of phytohormones in various plant compartments, exemplified by *Pseudomonas putida*, which modifies ethylene biosynthesis in plants by degrading its biological precursor. 1-aminocyclopropane-1-carboxylate via the secretion of the enzyme 1-aminocyclopropane-1-carboxylate deaminase facilitates root advancement in soil, particularly under adverse environmental circumstances, since elevated ethylene levels inhibit rhizospheric development (Oyedele et al. 2024; Del Carmen Orozco-Mosqueda et al. 2023).

16.5.3 Induction of Systemic Resistance

Systemic acquired resistance refers to a comprehensive adaptive immune defense mechanism within the plant's vegetative and reproductive systems, triggered by local pathogenic biotic assaults (such as *Phytophthora infestans*, *Botrytis cinerea, Xanthomonas campestris, Tobacco mosaic virus,* etc.) or symbiotic interactions (including *Bacillus subtilis, Trichoderma* spp., *Azospirillum brasilense, Rhizobium* spp., etc.). It is characterized by a persistent state of vigilance in the plant, representing a kind of plant immunological memory that provides resistance by a rapid response (via analogous signaling pathways) to subsequent biotic assaults (Joshi et al. 2025; Ali et al. 2024). These microorganisms interact with plant roots by releasing specific molecular signatures termed MAMPs (Microbial Associated Molecular Patterns), including flagellin, peptidoglycan, beta-glucans, mannose, and microbial DNA, which are detected by surface receptors known as Pattern Recognition Receptors (PRRs) in plant cells to trigger a signaling cascade. This mechanism resembles a plant immunization that enhances the state of alertness against biotrophic chemicals (Sheoran et al. 2025; Ali et al. 2024). Systemic acquired resistance in plants arises following hormonal activation, notably through salicylic acid released by specific microbial pathogens, which is subsequently metabolized by the plant into methyl salicylate. This compound serves as a signal that disseminates via the phloem to previously uninfected distal regions of the plant, thereby stimulating transcription through the methylation of pathogenesis-related genes and their translation into chitinase, proteinase, and glucanase, all of which are integral to the plant's immune defense (Sheoran et al. 2025; Vincze et al. 2024).

Additional components of the plant microbiome, including the plant-growth-promoting rhizobacteria (PGPR) group, are linked to the enhancement of induced systemic resistance in plants against necrotrophic diseases and insects, depending on non-lesional latent activation of plant biomass. *Pseudomonas fluorescens* and similar bacteria release chemical signals to plant roots, prompting the biosynthesis of specific phytohormones, including ethylene and jasmonic acid, to regulate the expression of PDF1.2 (plant defensin 1.2) immune resistance effector genes by modulating key transcription factors within the AP2/ERF network, such as ERF1 and ORA59. This mechanism is termed transcriptional priming, indicative of epigenetic plasticity that enhances plant biochemical barriers (Joshi et al. 2025; Sheoran et al. 2025; Ali et al. 2024).

16.5.4 Resilience to Abiotic Stress

The plant microbiome is essential for enhancing plant resilience to variations in unfavorable climatic circumstances, such as drought, salt, cold, heat, and nutrient deprivation. It serves as an adaptable functional biological extension that preserves plant life by biochemical collaboration (Fig. 16.3) (Oyedele et al. 2024;

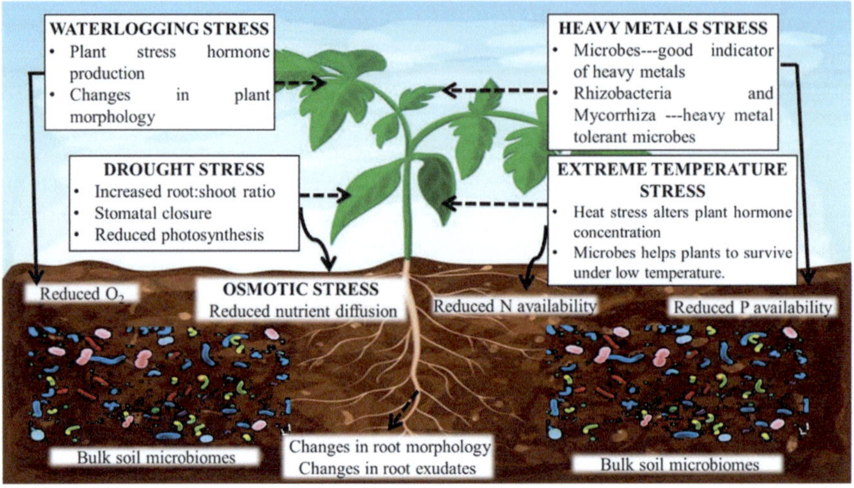

Fig. 16.3 The role of plant microbiome in resisting abiotic stress (Munir et al. 2022)

Srivastav et al. 2024). Throughout its life cycle, the plant holobiont may encounter water and osmotic stress, a potentially lethal condition; however, a significant portion of the resolution relies on the involvement of microbiological partners such as *Bacillus* spp., *Rhizobium* spp., *Azospirillum* spp., *Funneliformis* spp., *Glomus* spp., *Streptomyces* spp., and *Frankia* spp., which produce intricate communication signals with the host plant to stimulate its endogenous synthesis of osmoprotective compounds such as trehalose, glycine-betaine, and proline. At the same time, these microbes regulate the water potential of plant cells by secreting exopolysaccharides that form protective biofilms, retaining moisture around the roots (Zeng et al. 2025; Sultana and Islam 2024). Climatic stressors, such as extreme heat and high salinity, disrupt plant redox balance by weakening antioxidant defenses and increasing the production of reactive oxygen species. In this context, the plant microbiome enhances its antioxidant capacity in response to environmental stresses by modulating the gene expression of certain antioxidant enzymes, including superoxide dismutase, ascorbate peroxidase, and catalase. Certain microbial communities aid in mitigating extreme conditions by producing antioxidant enzymes or by promoting the synthesis of phytohormones like abscisic acid, which reduces water loss by decreasing transpiration while preserving ionic homeostasis by sequestering harmful ions such as chloride and sodium via SOS1 and HKT1 channels and enhancing root water absorption through elongation to alleviate the effects of oxidant/antioxidant imbalance (Ge and Wang 2025; Srivastav et al. 2024; Sultana and Islam 2024).

16.5.5 Biocontrol of Pathogens

Incorporating the plant microbiome into plant health management against biotic stresses is a natural, sustainable, and ecologically benign biocontrol strategy, in contrast to the use of potentially harmful chemical inputs (Vincze et al. 2024).

Certain symbiotic microorganisms within the microbiome employ competition for nutritional resources and ecological habitats as a strategy to obstruct pathogen access to vital resources by suppressing their growth and virulence, particularly in plant niches such as root systems and leaf tissues (Singh and Das 2024). In the context of microbial infections in plants, diverse microorganisms from the plant microbiome, including *Streptomyces roseosporus, Bacillus amyloliquefaciens, Pseudomonas aeruginosa,* and *Burkholderia* spp., excrete an array of bioactive secondary metabolites exhibiting antimicrobial properties via intricate and sophisticated mechanisms, especially during their stationary growth phase when resources are limited. Among these antimicrobial metabolites, we identify antibiotics such as daptomycin and fencing, antimicrobial peptides including pyocin, pyoluteorin, and surfactin, organic acids like acetic acid and lactic acid, polyketides, and terpenoids such as aflatoxins and streptomycin, all aimed at preserving plant health from unseen adversaries (Noman et al. 2024; Zhang et al. 2021). Some beneficial components of the plant microbiome act as direct antipathogens, protecting the host plant. For example, Glomeromycetes produce complex physical barriers, such as vesicles and pseudopodia, that impede pathogens from entering root tissues. Species such as *Streptomyces, Bacillus, Trichoderma,* and *Pseudomonas aeruginosa* utilize complex degradative metabolisms, generating a variety of enzymes that degrade microorganisms' cell walls. Sometimes, the plant microbiome hinders the onset of microbial infections by diverting pathogens without eliminating them. This occurs through modifications in the expression of virulence genes, which enhances pathogen sensitivity to antibiotics and prevents recognition between the pathogen and the host plant. This process is known as quorum sensing inhibition (Sheoran et al. 2025; Zhang et al. 2021).

16.6 Factors Influencing the Balance of Phytomicrobiomes

The phytomicrobiome is a complex system influenced by various factors, including plant traits, soil properties, and climate. It plays a crucial role in maintaining plant health and equilibrium. The impact of plant microbiota can be beneficial, neutral, or harmful, depending on the context and environment (Lyu et al. 2021). A fragile equilibrium exists, where plants have better nutrient absorption, growth, and resistance to stress. Disrupting this equilibrium can reduce their vitality and immunity to illness. Plant species and genotypes also play a role in influencing the microbiome, as they cultivate unique microbial consortia through variations in root exudates and immune signals, creating a non-random, host-specific community with beneficial symbionts (Petrushin et al. 2024; Thijs et al. 2016).

For example, plants can actively recruit microbes with specific functional traits, such as degradative enzymes in polluted soils, from the diverse microbial pool under appropriate conditions provided by the host's exudates (Thijs et al. 2016). The core microbiome consists of hub microorganisms that influence community structure, and the host's genome can indirectly control microbiome balance by modulating the abundance of these hubs. The plant's developmental stage also influences the phytomicrobiome, with the initial microbiome being seeded from the propagule and early life stage, followed by progressive colonization from surrounding soil (Petrushin et al. 2024). The community evolves over time, with seedlings and mature plants typically exhibiting distinct microbial profiles. The host plant's identity and condition are critical determinants of its microbiome composition (Petrushin et al. 2024; Thijs et al. 2016). Soil characteristics and abiotic environmental conditions significantly influence the plant microbiome, as soil serves as a primary reservoir for root-associated microbial communities. The type of soil and its properties determine the thriving microbial communities (Patyal et al. 2025; Bhatt et al. 2020). Beneficial bacteria thrive in nutrient-rich soil with neutral pH, while acidic or nutrient-poor soils favor distinct microbial species, affecting plant microbiome structure. Moisture and external climatic factors like extreme temperatures, salinity, and ultraviolet radiation also impact microbial survival (Petrushin et al. 2024). Under drought or salinity stress, for example, many microbes cannot survive or remain active, but stress-tolerant microorganisms (some of which may help the plant cope by producing osmoprotectants or antioxidants) become relatively more abundant (Singh et al. 2023). Elevated temperatures or cold snaps likewise shift the balance by favoring microbes adapted to those conditions and altering plant root exudation patterns. Indeed, climate change-induced environmental perturbations (e.g., prolonged heat, erratic rainfall, or saltwater intrusion) are causing measurable changes in soil microbial communities and, by extension, plant–microbe interactions (Singh et al. 2023; Maheshwari et al. 2021). In addition, disease outbreaks often correspond with a loss of beneficial microbes or proliferation of opportunistic organisms, whereas healthy plants typically maintain a microbiome with many protective or growth-promoting members. Conversely, plants under attack may alter their exudation ("cry for help") to recruit defensive microbes that help suppress pests and diseases, thereby shifting community composition as a form of adaptive response (Petrushin et al. 2024). In agricultural settings, anthropogenic influences are among the most potent disruptors of microbiome balance. Intensive use of chemical pesticides and fungicides can decimate not only target pests but also non-target beneficial microorganisms in the soil and on plant surfaces, leading to reduced microbial diversity and an imbalance favoring resilient but often less beneficial strains (Bhatt et al. 2020). Likewise, the heavy application of fertilizers alters natural nutrient cycles and soil chemistry, potentially depressing symbiotic microbes (such as nitrogen-fixing bacteria and mycorrhizal fungi) due to high nutrient availability, or encouraging fast-growing microbial decomposers that outcompete more mutualistic species (Maheshwari et al. 2021). Over time, such practices have been shown to create "unbalanced" soils where the microbial community is shifted away from its native equilibrium, sometimes impairing soil health and plant productivity

(Bhatt et al. 2020). Beyond agrochemicals, other human-driven factors like soil contamination (heavy metals, hydrocarbons, and other pollutants) impose strong selective filters on the microbiome. The presence of pollutants in soil acts as a strong selective pressure, allowing only pollutant-tolerant or degradative microbes to persist. This drastically changes the community structure. Nevertheless, plants can still utilize these persistent microbes for phytoremediation, provided the right partners are selected for the task (Thijs et al. 2016; Bhatt et al. 2020). Due to this complex interplay of factors, the plant microbiome is inherently dynamic rather than static. It has been described as a "rather unstable structure" that is constantly influenced by various factors related to the host, microbes, and the environment. These microbial shifts are often in relative abundance (quantitative changes) rather than a complete shift of species, meaning that essential components may remain but their proportions change in response to new conditions (Petrushin et al. 2024). In cases of severe or protracted perturbation, the microbiome may reach a dysbiotic condition, resulting in reduced beneficial activities. This vulnerability to disturbance significantly affects plant health: an unbalanced microbiome, such as one lacking essential plant growth-promoting bacteria, may diminish the plant's capacity to absorb nutrients, withstand stress, and combat disease. In contrast, a well-balanced and functionally diversified microbiome improves plant resilience, effectively protecting the plant from shocks by mechanisms such as producing antioxidants, osmolytes, phytohormones, and other protective compounds (Singh et al. 2023). The plant–microbiome partnership has significantly evolved, making the concept of plant health incomplete without its microbiome. Maintaining or restoring the natural balance of the plant microbiome is critical for plant vitality. Agriculture and biotechnology are working to manage the microbiome through microbiome engineering or probiotic inoculations to support plant growth and stress tolerance in environmental challenges (Patyal et al. 2025; Lyu et al. 2021; Maheshwari et al. 2021).

16.7 Methods for Studying the Plant Microbiome

The plant microbiome, a complex community of plant-associated microorganisms, is a major focus in plant biology and agricultural science. Modern methodologies have improved our understanding of these communities, from their composition to functional interactions. Historically, the rhizosphere has been the primary focus, but recent advances have expanded to include phytosphere, endosphere, and seed-associated microbes (Compant et al. 2019; Orozco-Mosqueda et al. 2018). The integration of metagenomics, amplicon sequencing (16S rRNA and ITS regions), and shotgun metagenomics has allowed for the comprehensive taxonomic and functional profiling of microbial communities (Fitzpatrick et al. 2020; Edwards et al. 2015). Moreover, metatranscriptomics and metaproteomics are increasingly being used to uncover the active functions of these microbial consortia in situ. Recent studies have emphasized the importance of ecological and host-driven factors in shaping microbial community composition (Finkel et al. 2020; Vandenkoornhuyse et al. 2015). Plants actively modulate their associated microbiota through root exudates and immune signaling, thereby influencing microbial colonization and activity (Bulgarelli et al. 2013; Van Der Heijden and Schlaeppi 2015). These interactions are bidirectional and complex, as microbes can in turn influence plant immunity, growth, and stress resilience (Compant et al. 2019; Orozco-Mosqueda et al. 2018).

Microbiome engineering is a strategy to manipulate microbial communities to improve plant health. Techniques like host-mediated selection, microbial inoculants, tissue seeding, and seed or soil inoculation have been explored to enhance beneficial plant–microbe interactions (Orozco-Mosqueda et al. 2018). Genotypic systems and artificial communities allow controlled experiments to analyze functional contributions of taxon or consortium (Fitzpatrick et al. 2020). Bioinformatics pipelines and statistical tools have improved the interpretation of microbiome data, identification of core microbes, and prediction of microbial functions. However, standardization of methodologies and integration of ecological theories are crucial for deepening our understanding (Finkel et al. 2020; Vandenkoornhuyse et al. 2015). Future efforts should focus on high-throughput molecular methods and experimental validation to link microbial community structure to plant phenotypes and environmental outcomes.

16.8 Targeted Modulation of Microbiomes

Modern plant science recognizes the plant microbiome as a key determinant of plant health and productivity. Consequently, there is growing interest in targeted microbiome modulation, deliberately steering microbial community composition and function to promote plant growth and sustainably suppress diseases. The rationale for this approach lies in the concept of the plant holobiont: plants and their associated microbes form integrated ecological units, and beneficial microbes can enhance nutrient uptake, stress tolerance, and immunity (Berg et al. 2021). By manipulating microbiomes, one can potentially boost these positive functions while inhibiting pathogens, offering an ecological alternative to chemical fertilizers and pesticides. Recent studies have even drawn parallels between plant microbiome therapeutics and those in human medicine (Zhang et al. 2019), underlining that restoring a healthy microbiota (from soils to guts) can ameliorate host health when dysbiosis (microbial imbalance) has occurred.

A variety of strategies have emerged for microbiome modulation in agriculture, each targeting the plant microbiota via different mechanisms. Microbial inoculants are widely used, which include plant growth-promoting bacteria (PGPB) and fungi, biostimulants, biofertilizers, and biopesticides introduced to the rhizosphere or endosphere. Such inoculants can directly contribute beneficial functions (e.g., nitrogen fixation, phosphate solubilization, phytohormone production, or antagonism against pests) and can also reshape the indigenous microbiome. For instance, introducing specific *Bacillus* and *Pseudomonas* strains can suppress root pathogens while enriching beneficial taxa in the rhizosphere, thereby creating a disease-suppressive community (Berg et al. 2021; Zhang et al. 2019). In some cases, consortia of multiple microbes (synthetic communities) are designed to provide complementary benefits: one recent approach combined nitrogen-fixing *Azospirillum*, phosphate-solubilizing *Bacillus*, and biocontrol *Trichoderma* to enhance crop growth and protection synergistically (Pascale et al. 2020). The diversity and composition of inoculated communities appear crucial—higher diversity consortia can utilize resources more efficiently and persist longer in soil, though overly complex mixtures may also risk internal competition (Pascale et al. 2020).

Plant secondary metabolites and root exudates can be used to modulate the microbiome, as plants naturally secrete chemicals that promote beneficial microbe growth or deter pathogens. Recent research has identified specific phytochemicals, such as coumarins, glucosinolates, benzoxazinoids, camalexins, and terpenes, that significantly influence microbiome composition. These compounds can act as signals or nutrients to ward off symbionts or as antimicrobials that inhibit susceptible microbes, disrupting microbial balance. For example, phenolic root exudates, like coumarin, have been shown to accumulate under iron deficiency or pathogen attack, selectively stimulating protective *Pseudomonas aeruginosa* strains while inhibiting some soilborne pathogens (Jacoby et al. 2021; Pascale et al. 2020). This phenomenon, known as a plant's chemical "cry for help," illustrates how host-driven metabolic reprogramming can promote a protective microbiome during stress. Modifying root chemistry is now considered a powerful tool for rationally controlling the microbiome to improve crop resilience (Jacoby et al. 2021).

Beyond influencing microbes, the feedback from microbes to plant metabolism is another layer of targeted modulation. Beneficial microbes often induce systemic changes in plant physiology. For instance, PGPB inoculation can rewire the plant metabolome toward stress-alleviating compounds. Ferreira et al. (2024) demonstrated that treating *Salicornia* seeds with certain bacteria triggered the accumulation of sugars, fatty acids, organic acids, phenolics, and flavonoids in the plants, changes that were associated with improved biomass and stress tolerance. Such microbe-induced metabolomic shifts can fortify plant defenses (e.g., by boosting antioxidants or antimicrobial metabolites) and thereby indirectly shape the microbiome by making the habitat more favorable to stress-tolerant mutualists and less hospitable to pathogens. This two-way chemical dialogue between plant and microbiome is central to microbiome-driven disease suppression: a healthy microbiome not only directly antagonizes pathogens but can also prime the plant's immunity (Berg et al. 2021; Pascale et al. 2020). For example, many biocontrol microbes elicit induced systemic resistance in their host plant, heightening the plant's basal defenses against a broad range of diseases.

Deploying microbiome modulation strategies holds promise for sustainable agriculture. In practice, this could mean seed coatings or soil inoculants with tailored microbial consortia, the use of biofertilizers that both nourish the plant and restore soil microbial balance, or crop management that enhances beneficial exudation (such as intercropping with companion plants that release bioactive root exudates). Moreover, the concept of restoring microbiome balance (recovering from dysbiosis) is increasingly relevant: just as fecal microbiota transplants can restore gut health

(Zhang et al. 2019), soil microbiome transplants or successive introductions of beneficial microbes can rehabilitate degraded, pathogen-dominated soils. Indeed, adding specific microbes has been shown to compensate for pathogen-induced community shifts and re-establish a more balanced, disease-suppressive microbiome (Berg et al. 2021). The ultimate goal of these interventions is to create a rhizosphere environment where beneficial microorganisms thrive and support the plant, while opportunistic pathogens are kept in check by microbial competition and antagonism.

Research in the field of microbiome modification is focusing on genetic polymorphism approaches to understand how introduced microbes interact with local communities and host plants. This knowledge is guiding the design of next-generation inoculants, such as genetically engineered strains with enhanced capabilities and functionally robust populations in diverse environments. Tailored strategies are needed, considering plant genotype, soil context, and climate. Scientists aim to manipulate the plant microbiome with precision, akin to "microbial medicine" for crops (Ferreira et al. 2024; Berg et al. 2021). This targeted modification is a significant step towards improving plant health, boosting agricultural productivity, and promoting more resilient and sustainable cropping systems.

16.9 Innovations and Perspectives in Plant Microbiomics

Optimal utilization of agricultural resources and continuous food supply are the two main priorities in modern agriculture (Saifi et al. 2025). The study of plant microbiomes has evolved from descriptive analyses to strategies for improving plant health, productivity, and sustainability. Innovative methods, including synthetic communities, precision inoculation, bioinformatics-driven predictions, and microbiome-informed breeding programs, are being developed (Sasse et al. 2018; Saleem et al. 2019). Synthetic microbial communities (SynComs) are valuable tools for understanding plant–microbe interactions in controlled and reproducible ways. These communities are assembled using native plant microbes and tested for their effects on plant growth, disease resistance, and abiotic stress tolerance. SynComs are designed to restore disease-suppressing microbes to agricultural soils. When designed wisely, they can mimic or enhance natural microbiome functions, enabling personalized biocomposites (Wei et al. 2019; Finkel et al. 2020; Trivedi et al. 2020; Compant et al. 2019). Microbiome engineering involves modifying microbial activities and interactions, not just focusing on their microbial composition. It improves rhizosphere colonization, nutrient uptake, and plant performance through the introduction of functionally synergistic strains. Advances in assembling microbial communities have led to the development of multi-trophic biostimulants that combine bacteria and fungi to enhance resilience. These techniques require advanced computational tools for pattern recognition and predictive modeling in complex microbiome datasets (Gouda et al. 2017; Ferreira et al. 2024). Artificial intelligence (AI) and machine learning are increasingly used to correlate microbiome profiles with plant traits like stress tolerance, growth, and disease resistance. AI has enabled the discovery of microbial biomarkers predictive of crop yield and pathogen suppression. Bioinformatics tools are critical in analyzing and integrating multi-omics datasets, linking microbial community structure with host function. These tools construct functional gene networks in root-associated microbiomes, unravel metabolite exchanges, and model microbiome succession across development stages. New developments in high-throughput culturing and single-cell technologies are expected to further enhance our ability to decode microbial functions and interactions. The host genotype plays a pivotal role in shaping its microbiome, and researchers are intentionally integrating microbiome-aware breeding to select genotypes that enrich beneficial microbes, screening for exudate profiles that support growth-promoting microbes or suppress pathogens (Kimotho and Maina 2024; Mishra et al. 2022; Zhang et al. 2019; Xie et al. 2019; Saleem et al. 2019).

Moreover, QTL mapping and GWAS have identified loci influencing microbiome composition, and CRISPR/Cas9 technologies may enable precision editing of these traits. These advancements lead to a new generation of crops optimized for microbiome compatibility. Root exudates, chemical signals, modulate microbial colonization and community structure. Plants can alter exudation patterns under stress, and microbial inoculation reprograms the host metabolome towards stress-alleviating secondary metabolites (Ferreira et al. 2024; Jacoby et al. 2021; Pascale et al. 2020).

In addition, metabolomics and microbiomics have revealed compounds influencing microbial communities, leading to targeted manipulation through breeding or synthetic biology. Precision microbiome management aims to tailor interventions based on plant species, genotype, environmental context, and developmental stage, mirroring personalized medicine in humans.

Finally, future tools include portable microbiome sensors, microbiome-assisted breeding platforms, and intelligent consortia design. Precision application of microbial inoculants will make microbiome-based agriculture more effective and scalable, but challenges like field stability and regulatory frameworks remain (Pellegrino and Marrassini 2021; Compant et al. 2019).

References

Abdelfattah A, Tack AJ, Lobato C, Wassermann B, Berg G (2022) From seed to seed: the role of microbial inheritance in the assembly of the plant microbiome. Trends Microbiol 31(4):346–355. https://doi.org/10.1016/j.tim.2022.10.009

Afridi MS, Ali S, Salam A, Terra WC, Hafeez A, Sumaira N, Ali B, AlTami MS, Ameen F, Ercisli S, Marc RA, Medeiros FHV, Karunakaran R (2022) Plant microbiome engineering: hopes or hypes. Biology 11(12):1782. https://doi.org/10.3390/biology11121782

Ali S, Tyagi A, Mir ZA (2024) Plant immunity: at the crossroads of pathogen perception and defense response. Plants 13(11):1434. https://doi.org/10.3390/plants13111434

Aziz U, Rehmani MS, Wang L, Luo X, Xian B, Wei S, Wang G, Shu K (2021) Toward a molecular understanding of rhizosphere, phyllosphere, and spermosphere interactions in plant growth

and stress response. Crit Rev Plant Sci 40(6):479–500. https://doi.org/10.1080/07352689.2022.2031728
Berg G, Cernava T (2022) The plant microbiota signature of the Anthropocene as a challenge for microbiome research. Microbiome 10(1). https://doi.org/10.1186/s40168-021-01224-5
Berg G, Kusstatscher P, Abdelfattah A, Cernava T, Smalla K (2021) Microbiome modulation—toward a better understanding of plant microbiome response to microbial inoculants. Front Microbiol 12. https://doi.org/10.3389/fmicb.2021.650610
Bernabeu M, Cabello-Yeves E, Flores E, Samarra A, Summers JK, Marina A, Collado MC (2024) Role of vertical and horizontal microbial transmission of antimicrobial resistance genes in early life: insights from maternal-infant dyads. Curr Opin Microbiol 77:102424. https://doi.org/10.1016/j.mib.2023.102424
Bhatt P, Verma A, Verma S, Anwar MS, Prasher P, Mudila H, Chen S (2020) Understanding phytomicrobiome: a potential reservoir for better crop management. Sustainability 12(13):5446. https://doi.org/10.3390/su12135446
Bulgarelli D, Schlaeppi K, Spaepen S, Van Themaat EVL, Schulze-Lefert P (2013) Structure and functions of the bacterial microbiota of plants. Annu Rev Plant Biol 64(1):807–838. https://doi.org/10.1146/annurev-arplant-050312-120106
Choi K, Khan R, Lee S (2021) Dissection of plant microbiota and plant-microbiome interactions. J Microbiol 59(3):281–291. https://doi.org/10.1007/s12275-021-0619-5
Compant S, Samad A, Faist H, Sessitsch A (2019) A review on the plant microbiome: ecology, functions, and emerging trends in microbial application. J Adv Res 19:29–37. https://doi.org/10.1016/j.jare.2019.03.004
Compant S, Cambon MC, Vacher C, Mitter B, Samad A, Sessitsch A (2020) The plant endosphere world – bacterial life within plants. Environ Microbiol 23(4):1812–1829. https://doi.org/10.1111/1462-2920.15240
Compant S, Cassan F, Kostić T, Johnson L, Brader G, Trognitz F, Sessitsch A (2025) Harnessing the plant microbiome for sustainable crop production. Nat Rev Microbiol. https://doi.org/10.1038/s41579-024-01079-1
Da Costa PB, Benucci GMN, Chou M, Van Wyk J, Chretien M, Bonito G (2022) Soil origin and plant genotype modulate switchgrass aboveground productivity and root microbiome assembly. MBio 13(2). https://doi.org/10.1128/mbio.00079-22
Del Carmen Orozco-Mosqueda MDC, Santoyo G, Glick BR (2023) Recent advances in the bacterial phytohormone modulation of plant growth. Plants 12(3):606. https://doi.org/10.3390/plants12030606
Du Y, Han X, Tsuda K (2024) Microbiome-mediated plant disease resistance: recent advances and future directions. J Gen Plant Pathol. https://doi.org/10.1007/s10327-024-01204-1
Dubey S, Sharma S (2021) Rhizospheric engineering by plant-mediated indirect selection of microbiome for agricultural sustainability. Crit Rev Plant Sci 40(5):379–397. https://doi.org/10.1080/07352689.2021.1959137
Edwards J, Johnson C, Santos-Medellín C, Lurie E, Podishetty NK, Bhatnagar S, Eisen JA, Sundaresan V (2015) Structure, variation, and assembly of the root-associated microbiomes of rice. Proc Natl Acad Sci U S A 112(8). https://doi.org/10.1073/pnas.1414592112
Fadiji AE, Ayangbenro AS, Akanmu AO, Babalola OO (2024) Plant health in the era of metagenomics: current state and future prospects. In: Elsevier eBooks, pp 399–419. https://doi.org/10.1016/b978-0-323-91631-8.00014-7
Ferreira MJ, Veríssimo ACS, Pinto DCGA, Sierra-Garcia IN, Granada CE, Cremades J, Silva H, Cunha Â (2024) Engineering the rhizosphere microbiome with plant growth promoting bacteria for modulation of the plant metabolome. Plants 13(16):2309. https://doi.org/10.3390/plants13162309
Finkel OM, Salas-González I, Castrillo G, Conway JM, Law TF, Teixeira PJPL, Wilson ED, Fitzpatrick CR, Jones CD, Dangl JL (2020) A single bacterial genus maintains root growth in a complex microbiome. Nature 587(7832):103–108. https://doi.org/10.1038/s41586-020-2778-7

Fitzpatrick C, Salas-González I, Conway J, Finkel O, Gilbert S, Russ D, Teixeira P, Dangl J (2020) The plant microbiome: from ecology to reductionism and beyond. Carolina Digital Repository (University of North Carolina at Chapel Hill). https://doi.org/10.17615/6aks-2230

Ge A, Wang E (2025) Exploring the plant microbiome: a pathway to climate-smart crops. Cell 188(6):1469–1485. https://doi.org/10.1016/j.cell.2025.01.035

Ge M, Wei X-L (2021) Research progress in plant spermosphere microbiome. World Forestry Res 34(3):13–18. https://doi.org/10.13348/j.cnki.sjlyyj.2020.0113.y

Goodwin PH (2022) The endosphere microbiome of ginseng. Plants 11(3):415. https://doi.org/10.3390/plants11030415

Gouda S, Kerry RG, Das G, Paramithiotis S, Shin H, Patra JK (2017) Revitalization of plant growth promoting rhizobacteria for sustainable development in agriculture. Microbiol Res 206:131–140. https://doi.org/10.1016/j.micres.2017.08.016

Guo S, Pan R, Zhang Y, Gu Q, Shen Q, Yang J, Huang L, Shen Z, Li R (2025) Plant-microbe interactions influence plant performance via boosting beneficial root-endophytic bacteria. Environ Microbiome 20(1). https://doi.org/10.1186/s40793-025-00680-y

Hashimi MH, Abad Q, Noori GH (2023, February 28) A review of plant microbiome: its functions and components. https://www.ijojournals.com/index.php/ar/article/view/639

Jacoby RP, Koprivova A, Kopriva S (2021) Pinpointing secondary metabolites that shape the composition and function of the plant microbiome. J Exp Bot 72(1):57–69. https://doi.org/10.1093/jxb/eraa424

Janati W, Benmrid B, Elhaissoufi W, Zeroual Y, Nasielski J, Bargaz A (2021) Will phosphate Bio-Solubilization stimulate biological nitrogen fixation in grain legumes? Front Agron 3. https://doi.org/10.3389/fagro.2021.637196

Jiang Y, Luo J, Guo X, Qiao Y, Li Y, Zhang Y, Zhou R, Vaculík M, Li T (2025) Phyllosphere microbiome assists the hyperaccumulating plant in resisting heavy metal stress. J Environ Sci (China) 154:563–574. https://doi.org/10.1016/j.jes.2024.05.032

Joshi D, Kaushik A, Kumar R, Arya A, Santoyo G, Singh VK, Kashyap N, Solanki MK, Kumari M, Bhardwaj N, Kumar A (2025) Improving plant performance through microbiome manipulation: the potential role of current bioengineering approaches. Bacteria 4(1):12. https://doi.org/10.3390/bacteria4010012

Kimotho RN, Maina S (2024) Unraveling plant–microbe interactions: can integrated omics approaches offer concrete answers? J Exp Bot 75(5):1289–1313. https://doi.org/10.1093/jxb/erad448

Lyu D, Zajonc J, Pagé A, Tanney CAS, Shah A, Monjezi N, Msimbira LA, Antar M, Nazari M, Backer R, Smith DL (2021) Plant holobiont theory: the phytomicrobiome plays a central role in evolution and success. Microorganisms 9(4):675. https://doi.org/10.3390/microorganisms9040675

Maheshwari HS, Bharti A, Agnihotri R, Dukare A, Prabina BJ, Gangola S, Sharma MP (2021) Combating the abiotic stress through phytomicrobiome studies. In: Phytomicrobiome interactions and sustainable agriculture. Wiley, Hoboken, pp 45–65. https://doi.org/10.1002/9781119644798.ch4

Matthews JL (2024) Editorial: Holobiont interactions. Frontiers in Ecology and Evolution, 12. https://doi.org/10.3389/fevo.2024.1382169

Mishra AK, Sudalaimuthuasari N, Hazzouri KM, Saeed EE, Shah I, Amiri KMA (2022) Tapping into Plant–Microbiome Interactions through the Lens of Multi-Omics Techniques. Cells 11(20):3254. https://doi.org/10.3390/cells11203254

Munir N, Hanif M, Abideen Z, Sohail M, El-Keblawy A, Radicetti E, Mancinelli R, Haider G (2022) Mechanisms and strategies of plant microbiome interactions to mitigate abiotic stresses. Agronomy 12(9):2069. https://doi.org/10.3390/agronomy12092069

Mukherjee A, Singh BN, Kaur S, Sharma M, De Araújo ASF, De Araujo Pereira AP, Morya R, Puopolo G, Melo VMM, Verma JP (2024) Unearthing the power of microbes as plant microbiome for sustainable agriculture. Microbiol Res 286:127780. https://doi.org/10.1016/j.micres.2024.127780

Murphy KM, Le SM, Wilson AE, Warner DA (2023) The microbiome as a maternal effect: a systematic review on vertical transmission of microbiota. Integr Comp Biol 63(3):597–609. https://doi.org/10.1093/icb/icad031

Noman M, Ahmed T, Wang J, White JC (2024) Micronutrient–microbiome interplay: a critical regulator of soil–plant health. Trends Microbiol 32(4):319–320. https://doi.org/10.1016/j.tim.2024.02.008

Orozco-Mosqueda MDC, Del Carmen Rocha-Granados M, Glick BR, Santoyo G (2018) Microbiome engineering to improve biocontrol and plant growth-promoting mechanisms. Microbiol Res 208:25–31. https://doi.org/10.1016/j.micres.2018.01.005

Oyedele AO, Ezaka E, Taiwo LB (2024) Microbial biosynthesis of the classical phytohormones by plant growth-promoting microorganisms in plants. In: Chauhan PS, Bisht N, Agarwal R (eds) Microbial biostimulants for plant growth and abiotic stress amelioration (Vol. Biostimulants and protective biochemical agents). Elsevier, pp 345–366. https://doi.org/10.1016/B978-0-443-13318-3.00004-5

Pascale A, Proietti S, Pantelides IS, Stringlis IA (2020) Modulation of the root microbiome by plant molecules: the basis for targeted disease suppression and plant growth promotion. Front Plant Sci 10. https://doi.org/10.3389/fpls.2019.01741

Patyal U, Bala R, Kaur M, Faizan M, Alam P (2025) Phyto-microbiome engineering: designing plant-microbe interactions for improved crop performance. The Microbe:100272. https://doi.org/10.1016/j.microb.2025.100272

Pellegrino E, Marrassini V (2021) Plant microbiome for improving productivity and resilience of crops. Acta Sci Agric 5(9):14. https://www.iris.sssup.it/bitstream/11382/540596/1/ASAG-05-1052.pdf

Petrushin IS, Filinova NV, Gutnik DI (2024) Potato microbiome: relationship with environmental factors and approaches for microbiome modulation. Int J Mol Sci 25(2):750. https://doi.org/10.3390/ijms25020750

Rafique E, Mumtaz MZ, Ullah I, Rehman A, Qureshi KA, Kamran M, Rehman MU, Jaremko M, Alenezi MA (2022) Potential of mineral-solubilizing bacteria for physiology and growth promotion of Chenopodium quinoa Willd. Front Plant Sci 13. https://doi.org/10.3389/fpls.2022.1004833

Rawat P, Das S, Shankhdhar D, Shankhdhar S (2021) Phosphate-solubilizing microorganisms: mechanism and their role in phosphate solubilization and uptake. J Soil Sci Plant Nutr 21(1):49–68. https://doi.org/10.1007/s42729-020-00342-7

Saifi R, Kokiçi H, Saifi H, Akça İ, Benabdelkader M, Xhemali B, Çota E, Hadjeb A (2024) Does climate change heighten the risk of *Xylella fastidiosa* infection? In: Abd-Elsalam KA, Abdel-Momen SM (eds) Plant quarantine challenges under climate change anxiety (Chap. 11). Springer, Cham. https://doi.org/10.1007/978-3-031-56011-8_11

Saleem MH, Potgieter J, Arif KM (2019) Plant disease detection and classification by deep learning. Plants 8(11):468. https://doi.org/10.3390/plants8110468

Saifi R, Kulli F, Kokiçi H, Gjinovci G, Çakalli A, Saifi H, Xhemali B (2025) Challenges and opportunities of nanomaterials in sustainable agriculture. In: Rajan R, Ahmad F, Khoshariya A, Seth CS (eds) Sustainable era of nanomaterials (Chap. 3). Springer. https://doi.org/10.1007/978-981-96-4471-1_3

Sánchez-Cañizares C, Jorrín B, Poole PS, Tkacz A (2017) Understanding the holobiont: the interdependence of plants and their microbiome. Curr Opin Microbiol 38:188–196. https://doi.org/10.1016/j.mib.2017.07.001

Santoyo G (2022) How plants recruit their microbiome? New insights into beneficial interactions. J Adv Res 40:45–58. https://doi.org/10.1016/j.jare.2021.11.020

Sasse J, Martinoia E, Northen T (2018) Feed your friends: do plant exudates shape the root microbiome? Trends Plant Sci 23(1):25–41. https://doi.org/10.1016/j.tplants.2017.09.003

Sheoran AR, Lakra N, Saharan BS, Luhach A, Kumar R, Seth CS, Duhan JS (2025) Enhancing plant disease resistance: insights from biocontrol agent strategies. J Plant Growth Regul. https://doi.org/10.1007/s00344-024-11480-y

Singh N, Das PP (2024) Microbial consortium as promising biostimulants for plant health: a future perspective for agriculture. In: Parray JA, Shameem N, Egamberdieva D (eds) Microbiome drivers of ecosystem function. Academic Press, pp 123–143. https://doi.org/10.1016/B978-0-443-19121-3.00019-3

Singh A, Mazahar S, Chapadgaonkar SS, Giri P, Shourie A (2023) Phyto-microbiome to mitigate abiotic stress in crop plants. Front Microbiol 14. https://doi.org/10.3389/fmicb.2023.1210890

Sohrabi R, Paasch BC, Liber JA, He SY (2023) Phyllosphere microbiome. Annu Rev Plant Biol 74(1):539–568. https://doi.org/10.1146/annurev-arplant-102820-032704

Srivastav AK, Das P, Jaiswal P, Parashar V (2024) Plant pathology in the age of climate change: challenges and adaptations. Int J Innov Res Sci Eng Technol 13(10). https://doi.org/10.15680/IJIRSET.2024.1310002

Sultana T, Islam MR (2024) The potential of soil microbiomes in alleviating climate change–associated stresses on crop plants. In: Sa T (ed) Beneficial microbes for sustainable agriculture under stress conditions. Academic Press, pp 81–111. https://doi.org/10.1016/B978-0-443-13193-6.00005-1

Tanger IS, Stefanschitz J, Schwert Y, Roth O (2024) The source of microbial transmission influences niche colonization and microbiome development. Proc R Soc Lond B Biol Sci 291(2016). https://doi.org/10.1098/rspb.2023.2036

Tharanath AC, Upendra RS, Rajendra K (2024) Soil symphony: a comprehensive overview of plant–microbe interactions in agricultural systems. Appl Microbiol 4(4):1549–1567. https://doi.org/10.3390/applmicrobiol4040106

Thijs S, Sillen W, Rineau F, Weyens N, Vangronsveld J (2016) Towards an enhanced understanding of Plant–Microbiome interactions to improve phytoremediation: Engineering the metaorganism. Front Microbiol 7. https://doi.org/10.3389/fmicb.2016.00341

Timofeeva AM, Galyamova MR, Sedykh SE (2023) Plant growth-promoting soil bacteria: nitrogen fixation, phosphate solubilization, siderophore production, and other biological activities. Plants 12(24):4074. https://doi.org/10.3390/plants12244074

Tkacz A, Poole P (2020) The plant microbiome: the dark and dirty secrets of plant growth. Plants People Planet 3(2):124–129. https://doi.org/10.1002/ppp3.10167

Trivedi P, Leach JE, Tringe SG, Sa T, Singh BK (2020) Plant–microbiome interactions: from community assembly to plant health. Nat Rev Microbiol 18(11):607–621. https://doi.org/10.1038/s41579-020-0412-1

Van Der Heijden MGA, Schlaeppi K (2015) Root surface as a frontier for plant microbiome research. Proc Natl Acad Sci U S A 112(8):2299–2300. https://doi.org/10.1073/pnas.1500709112

Vandenkoornhuyse P, Quaiser A, Duhamel M, Van AL, Dufresne A (2015) The importance of the microbiome of the plant holobiont. New Phytol 206(4):1196–1206. https://doi.org/10.1111/nph.13312

Vincze É, Becze A, Laslo É, Mara G (2024) Beneficial soil microbiomes and their potential role in plant growth and soil fertility. Agriculture (Basel) 14(1):152. https://doi.org/10.3390/agriculture14010152

Wang N, Sundin GW, De La Fuente L, Cubero J, Tatineni S, Brewer MT, Zeng Q, Bock CH, Cunniffe NJ, Wang C, Candresse T, Chappell T, Coleman JJ, Munkvold G (2024) Key challenges in plant pathology in the next decade. Phytopathology 114(5):837–842. https://doi.org/10.1094/phyto-04-24-0137-kc

Wei Z, Gu Y, Friman V, Kowalchuk GA, Xu Y, Shen Q, Jousset A (2019) Initial soil microbiome composition and functioning predetermine future plant health. Sci Adv 5(9). https://doi.org/10.1126/sciadv.aaw0759

Xiong Q, Yang J, Ni S (2023) Microbiome-Mediated Protection against Pathogens in Woody Plants. Int J Mol Sci 24(22):16118. https://doi.org/10.3390/ijms242216118

Xie K, Guo L, Bai Y, Liu W, Yan J, Bucher M (2019) Microbiomics and plant health: an interdisciplinary and international workshop on the plant microbiome. Mol Plant 12(1):1–3. https://doi.org/10.1016/j.molp.2018.11.004

Xun W, Shao J, Shen Q, Zhang R (2021) Rhizosphere microbiome: functional compensatory assembly for plant fitness. Comput Struct Biotechnol J 19:5487–5493. https://doi.org/10.1016/j.csbj.2021.09.035

Xun W, Liu Y, Ma A, Yan H, Miao Y, Shao J, Zhang N, Xu Z, Shen Q, Zhang R (2024) Dissection of rhizosphere microbiome and exploiting strategies for sustainable agriculture. New Phytol 242(6):2401–2410. https://doi.org/10.1111/nph.19697

Zeng Q, Hu H, Ge A, Xiong C, Zhai C, Duan G, Han L, Huang S, Zhang L (2025) Plant–microbiome interactions and their impacts on plant adaptation to climate change. J Integr Plant Biol. https://doi.org/10.1111/jipb.13863

Zhan C, Matsumoto H, Liu Y, Wang M (2022) Pathways to engineering the phyllosphere microbiome for sustainable crop production. Nat Food 3(12):997–1004. https://doi.org/10.1038/s43016-022-00636-2

Zhang Z, Tang H, Chen P, Xie H, Tao Y (2019) Demystifying the manipulation of host immunity, metabolism, and extraintestinal tumors by the gut microbiome. Signal Transduct Target Ther 4(1). https://doi.org/10.1038/s41392-019-0074-5

Zhang H, Boateng NAS, Ngea GLN, Shi Y, Lin H, Yang Q, Wang K, Zhang X, Zhao L, Droby S (2021) Unravelling the fruit microbiome: the key for developing effective biological control strategies for postharvest diseases. Compr Rev Food Sci Food Saf 20(5):4906–4930. https://doi.org/10.1111/1541-4337.12783

Zhu Y, Xiong C, Wei Z, Chen Q, Ma B, Zhou S, Tan J, Zhang L, Cui H, Duan G (2021) Impacts of global change on the phyllosphere microbiome. New Phytol 234(6):1977–1986. https://doi.org/10.1111/nph.17928

MIX
Papier aus verantwortungsvollen Quellen
Paper from responsible sources
FSC® C105338

If you have any concerns about our products,
you can contact us on
ProductSafety@springernature.com

In case Publisher is established outside the EU,
the EU authorized representative is:
Springer Nature Customer Service Center GmbH
Europaplatz 3, 69115 Heidelberg, Germany

Printed by Libri Plureos GmbH
in Hamburg, Germany